Introduction to Enumerative and Analytic Combinatorics

This award-winning textbook targets the gap between introductory texts in discrete mathematics and advanced graduate texts in enumerative combinatorics. The author's goal is to make combinatorics more accessible to encourage student interest and to expand the number of students studying this rapidly expanding field.

The book first deals with basic counting principles, compositions and partitions, and generating functions. It then focuses on the structure of permutations, graph enumeration, and extremal combinatorics. Lastly, the text discusses supplemental topics, including error-correcting codes, properties of sequences, and magic squares.

Updates to the **Third Edition** include:

- Quick Check exercises at the end of each section, which are typically easier than the regular exercises at the end of each chapter.
- A new section discussing the Lagrange Inversion Formula and its applications, strengthening the analytic flavor of the book.
- An extended section on multivariate generating functions.

Numerous exercises contain material not discussed in the text allowing instructors to extend the time they spend on a given topic. A chapter on analytic combinatorics and sections on advanced applications of generating functions, demonstrating powerful techniques that do not require the residue theorem or complex integration, and extending coverage of the given topics are highlights of the presentation.

The second edition was recognized as an Outstanding Academic Title of the Year by Choice Magazine, published by the American Library Association.

Miklós Bóna received his PhD in mathematics from the Massachusetts Institute of Technology in 1997. Since 1999, he has taught at the University of Florida, where, in 2010, he was inducted into the Academy of Distinguished Teaching Scholars. Professor Bóna has mentored numerous graduate and undergraduate students. He is the author of five books and more than 100 research articles, mostly focusing on enumerative and analytic combinatorics. His book, *Combinatorics of Permutations*, won a 2006 Outstanding Title Award from *Choice*, the journal of the American Library Association. He is also an editor-in-chief for the *Electronic Journal of Combinatorics*, and for two book series at CRC Press.

Discrete Mathematics and Its Applications
Series editors:
Miklós Bóna, Donald L. Kreher, Douglas B. West

The Mathematics of Chip-Firing
Caroline J. Klivans

Computational Complexity of Counting and Sampling
Istvan Miklos

Volumetric Discrete Geometry
Karoly Bezdek, Zsolt Langi

The Art of Proving Binomial Identities
Michael Z. Spivey

Combinatorics and Number Theory of Counting Sequences
Istvan Mezo

Applied Mathematical Modeling
A Multidisciplinary Approach
Douglas R. Shier, K.T. Wallenius

Analytic Combinatorics
A Multidimensional Approach
Marni Mishna

50 years of Combinatorics, Graph Theory, and Computing
Edited By Fan Chung, Ron Graham, Frederick Hoffman, Ronald C. Mullin, Leslie Hogben, Douglas B. West

Fundamentals of Ramsey Theory
Aaron Robertson

Methods for the Summation of Series
Tian-Xiao He

The Lambert W Function
Its Generalizations and Applications
István Mező

Combinatorics of Permutations, Third Edition
Miklós Bóna

Topics in Graph Theory
Mark Anderson, Jonathan Gross and Jay Yellen

Applied Combinatorics, Third Edition
Fred S. Roberts and Barry Tesman

Discrete Mathematics
An Open Introduction, Fourth Edition
Oscar Levin

https://www.routledge.com/Discrete-Mathematics-and-Its-Applications/book-series/CHDISMTHAPP

Introduction to Enumerative and Analytic Combinatorics

Third Edition

Miklós Bóna

CRC Press
Taylor & Francis Group
Boca Raton London New York

CRC Press is an imprint of the
Taylor & Francis Group, an **informa** business
A CHAPMAN & HALL BOOK

Fourth edition published 2025
by CRC Press
2385 NW Executive Center Drive, Suite 320, Boca Raton FL 33431

and by CRC Press
4 Park Square, Milton Park, Abingdon, Oxon, OX14 4RN

CRC Press is an imprint of Taylor & Francis Group, LLC

First edition published by Cambridge University Press 2011
Second edition published by CRC Press 2016

ISBN: 978-1-032-30270-6 (hbk)
ISBN: 978-1-032-30271-3 (pbk)
ISBN: 978-1-003-30427-2 (ebk)

DOI: 10.1201/9781003304272

Typeset in Latin font
by KnowledgeWorks Global Ltd.

Publisher's note: This book has been prepared from camera-ready copy provided by the authors.
Access the Support Material:

To Linda
To Mikike, Benny, and Vinnie

Contents

Foreword to the first edition

What could be a more basic mathematical activity than counting the number of elements of a finite set? The misleading simplicity that defines the subject of enumerative combinatorics is in fact one of its principal charms. Who would suspect the wealth of ingenuity and of sophisticated techniques that can be brought to bear on such an apparently superficial endeavor? Miklós Bóna has done a masterful job of bringing an overview of all of enumerative combinatorics within reach of undergraduates. The two fundamental themes of bijective proofs and generating functions, together with their intimate connections, recur constantly. A wide selection of topics, including several never appearing before in a textbook, are included that give an idea of the vast range of enumerative combinatorics. In particular, for those with sufficient background in undergraduate linear algebra and abstract algebra, there are many tantalizing hints of the fruitful connection between enumerative combinatorics and algebra that plays a central role in the subject of algebraic combinatorics. In a foreword to another book by Miklós Bóna, I wrote, "This book can be utilized at a variety of levels, from random samplings of the treasures therein to a comprehensive attempt to master all the material and solve all the exercises. In whatever direction the reader's tastes lead, a thorough enjoyment and appreciation of a beautiful area of combinatorics is certain to ensue." Exactly the same sentiment applies to the present book, as the reader will soon discover.

Richard Stanley
Cambridge, Massachusetts
June 2005

Preface to the third edition

There are at least three ways to use this book. If one decides to cover every chapter, and some of the exercises containing new material, then one can teach a two-semester combinatorics course from the book. Instructors looking for a text for a course in enumerative combinatorics can teach such a course by selecting chapters with the strongest focus on counting, such as Chapters 2, 3, 4, 5, 7, and 9. Finally, one can teach a one-semester course with a not-quite-as-strong focus on enumeration by using Chapters 1, 2, 3, and then the desired chapters from the rest of the book.

Our hope is that our book can broaden access to the fascinating topics of enumerative and analytic combinatorics, and will prepare readers for the more advanced, classic books of the field, such as *Enumerative Combinatorics* by Richard Stanley and *Analytic Combinatorics* by Philippe Flajolet and Robert Sedgewick.

There are two main new features in this edition. First, there are three Quick Check exercises at the end of each section, which are typically easier than the regular exercises at the end of each chapter. Second, we added a new section discussing the Lagrange Inversion Formula and its applications, strengthening the analytic flavor of the book. We also extended the section on multivariate generating functions.

Just as before, numerous exercises contain material not discussed in the text, which allows instructors to extend the time they spend on a given topic.

Combinatorics is a rapidly expanding field, and we hope that our book will increase the number of students with an interest in it even further.

Miklós Bóna
Gainesville, FL
November 2024

Acknowledgments

First, I am thankful to my grandfather, the late Dr. János Bóna, who introduced me to counting problems at an early age. I am indebted to the researchers whose work attracted me to enumerative combinatorics. They include Richard Stanley, Herbert Wilf, Bruce Sagan, Catherine Yan, Francesco Brenti, and many others. The second edition of the book has an analytic flavor that would not have been possible if I had not learned analytic combinatorics from, and been influenced by, Philippe Flajolet, Robert Sedgewick, Boris Pittel, Robin Pemantle, and E. Rodney Canfield.

As always, my gratitude is extended to my wife, Linda, and my sons, Miki, Benny, and Vinnie, who tolerated another book project reasonably well.

Frequently used notation

- $A(n,k)$ number of n-permutations with k ascending runs
- $B(n)$ number of all partitions of an n-element set
- $b(n,k)$ number of n-permutations with k inversions
- $c(n,k)$ number of n-permutations with k cycles
- c_n the Catalan number $c_n = \binom{2n}{n}/(n+1)$
- C_n a cycle on n vertices
- $|\mathcal{F}|$ number of edges of the hypergraph $\mathcal{F}$
- $F^{\langle -1 \rangle}(x)$ compositional inverse of the power series $F(x)$
- G_i stabilizer of the element i under the action of the group G
- $G - v$ graph G with its vertex v and all edges adjacent to v removed
- $i(p)$ number of inversions of the permutation p
- i^G orbit of the element i under the action of the group G
- K_n complete graph on n vertices
- $K_{m,n}$ complete bipartite graph with m red and n blue vertices
- χ_G chromatic polynomial of the graph G
- $\chi(G)$ chromatic number of the graph G
- $n!$ $n(n-1)\cdots 1$
- $\binom{n}{k}$ $\frac{(n)_k}{k!}$
- $[n]$ set $\{1, 2, \cdots, n\}$
- $\begin{bmatrix} \mathbf{n} \\ \mathbf{k} \end{bmatrix}$ Gaussian polynomial $\prod_{i=1}^{k} \frac{q^n - q^{i-1}}{q^k - q^{i-1}}$
- $(n)_m$ $n(n-1)\cdots(n-m+1)$
- $p(n)$ number of partitions of the integer n

- $p_k(n)$ number of partitions of the integer n into at most k parts
- $|S|$ number of elements of the set S
- S_n set of all n-permutations
- $S_n(q)$ number of all n-permutations avoiding the pattern q
- $S(n,k)$ number of partitions of the set $[n]$ into k blocks
- $s(n,k)$ $(-1)^{n-k}c(n,k)$

Part I

Methods

1

Basic methods

1.1 When we add and when we subtract

We start our course in enumerative combinatorics by considering a very simple task, namely finding the size of the union of two disjoint sets.

1.1.1 When we add

A group of friends went on a canoe trip. Five of them fell into the water at one point or another during the trip, while seven completed the trip without even getting wet. How many friends went on this canoe trip?

Before the reader laughs at us for starting the book with such a simple question, let us give the answer. Of course, $5 + 7 = 12$ people went on this trip. However, it is important to point out that such a simple answer was only possible because each person on the trip *either* fell into the water *or* stayed dry. There was no middle way, there was no way to belong to both groups, or to neither group. Once you fall into the water, you know it. In other words, each person was included in exactly one of those two groups of people.

In contrast, let us assume that we are not told how many people did or did not fall into the water, but instead are told that five people wore white shirts on this trip and eight people wore brown hats. Then we could not tell how many people went on this trip, as there could be people who belonged to *both groups* (it is possible to wear both a white shirt and a brown hat), and there could be people who belonged to neither.

We can now present the first, and easiest, counting principle of this book. Let $|X|$ denote the number of elements of the finite set X. So, for instance, $|\{2, 3, 5, 7\}| = 4$. Recall that two subsets are called *disjoint* if they have no elements in common.

Theorem 1.1 (Addition principle) *If A and B are two* disjoint *finite sets, then*

$$|A \cup B| = |A| + |B|. \tag{1.1}$$

It may seem somewhat strange that we provide proof for such an extremely simple statement, but we want to set standards.

Proof: Both sides of (1.1) count the elements of the same set, the set $A \cup B$.

DOI: 10.1201/9781003304272-1

The left-hand side does this directly, while the right-hand side counts the elements of A and B separately. In either case, each element is counted exactly once (as A and B are disjoint), so the two sides are indeed equal. ◊

The previous theorem was about *two* disjoint finite sets, but there is nothing magical about the number *two* here. If we had said that each of the participants of the canoe trip had exactly one unfortunate event on the trip, say, five of them fell into the water, three were attacked by hornets, and four got a sunstroke, then we could still conclude that there were $5+3+4=12$ people at this enjoyable excursion. This is an example of the generalized addition principle.

Theorem 1.2 (Generalized addition principle) *Let* $A_1, A_2, \cdots, A_n$ *be finite sets that are pairwise disjoint. Then*

$$|A_1 \cup A_2 \cup \cdots \cup A_n| = |A_1| + |A_2| + \cdots + |A_n|.$$

Proof: Again, both sides count the elements of the same set, the set $A_1 \cup A_2 \cup \cdots \cup A_n$; therefore, they have to be equal. ◊

Needless to say, it is again very important to insist that the sets A_i are pairwise disjoint.

1.1.2 When we subtract

We could have told the story of our canoeing friends of Section 1.1.1 as follows: Twelve friends went on a canoe trip. Five of them fell into the water. How many friends completed the trip without falling into the water? The answer is, of course, $12 - 5 = 7$. This is an example of the *subtraction principle.* In order to make the discussion of this principle easier, we introduce the notion of the *difference of two sets.* If A and B are two sets, then $A - B$ is the set consisting of the elements of A that are not elements of B.

Example 1.3 *Let* $A = \{2,3,5,7\}$*, and let* $B = \{4,5,7\}$*. Then* $A - B = \{2,3\}$.

Note that $A - B$ is defined even when B is not a subset of A. The subtraction principle, however, applies only when B is a subset of A.

Theorem 1.4 (Subtraction principle) *Let A be a finite set, and let $B \subseteq A$. Then* $|A - B| = |A| - |B|$.

Proof: We will first prove the equation

$$|A - B| + |B| = |A|. \tag{1.2}$$

This equation holds true by the addition principle. Indeed, $A - B$ and B are disjoint sets, and their union is A.

In other words, both sides count the elements of A, but the left-hand side first counts those that are not contained in B, then those that are contained in B.

The claim of Theorem 1.4 is now proved by subtracting $|B|$ from both sides of (1.2). ◊

That $B \subseteq A$ is a very important restriction here. The reader is invited to verify this by checking that the subtraction principle does *not* hold for the sets A and B of Example 1.3. For another caveat, let A be the set of all one-digit positive integers that are divisible by 2 and let B be the set of all one-digit positive integers that are divisible by 3. Then $A = \{2, 4, 6, 8\}$, so $|A| = 4$ and $B = \{3, 6, 9\}$, so $|B| = 3$. However, $|A - B| = |\{2, 4, 8\}| = 3$. As $4 - 3 \neq 3$, we see that the subtraction principle does not hold here. The reason for this is that the conditions of the subtraction principle are not fulfilled; that is, B is not a subset of A.

The reader should go back to our proof of the subtraction principle and see why the proof fails if B is not a subset of A.

The use of the subtraction principle is advisable in situations when it is easier to enumerate the elements of B ("bad guys") than the elements of $A-B$ ("good guys").

Example 1.5 *The number of positive integers less than or equal to 1000 that have at least two different digits is* $1000 - 27 = 973$.

Solution: Let A be the set of all positive integers less than or equal to 1000, and let B be the subset of A that consists of all positive integers less than or equal to 1000 that *do not have* two different digits. Then our claim is that $|A-B| = 973$. By the subtraction principle, we know that $|A-B| = |A| - |B|$. Furthermore, we know that $|A| = 1000$. Therefore, we will be done if we can show that $|B| = 27$. What are the elements of B? They are all positive integers of at most three digits in which there are no two distinct digits. That is, in any element of B, only one digit occurs, but that one digit can occur once, twice, or three times. So the elements of B are $1, 2, \cdots, 9$, then $11, 22, \cdots, 99$, and finally, $111, 222, \cdots, 999$. This shows that $|B| = 27$, proving our claim. ◊

Note that using the subtraction principle was advantageous because $|A|$ was very easy to determine and $|B|$ was almost as easy to compute. Therefore, getting $|A| - |B|$ was faster than computing $|A - B|$ directly.

Quick Check

1. How many two-digit positive integers are there that are more than 20, less than 50, and are either prime or squares of a prime?

2. Let us call a positive integer x a *standalone* prime if x is a prime number, but neither $x-2$ nor $x+2$ is a prime. How many standalone primes x are there between 10 and 35?
3. How many positive integers x are there that so that $x \leq 50$ and there is no nonnegative integer k so that $x = p^k$ for some prime number p?

1.2 When we multiply

In this section, we will be considering the number of all possible choices we can make in a situation when we choose objects with more than one relevant parameter.

1.2.1 The product principle

A car dealership sells five different models, and each model is available in seven different colors. If we are only interested in the model and color of a car, how many different choices does this dealership offer to us?

Let us denote the five models by the capital letters A, B, C, D, and E, and let us denote the seven colors by the numbers 1, 2, 3, 4, 5, 6, and 7. Then each possible choice can be totally described by a *pair* consisting of a capital letter and a number. The list of all choices is shown below.

- $A1, A2, A3, A4, A5, A6, A7$,
- $B1, B2, B3, B4, B5, B6, B7$,
- $C1, C2, C3, C4, C5, C6, C7$,
- $D1, D2, D3, D4, D5, D6, D7$, and
- $E1, E2, E3, E4, E5, E6, E7$.

Here each row corresponds to a certain model. As there are five rows, and each of them consists of seven possible choices, the total number of choices is $5 \times 7 = 35$.

This is an example of the following general theorem.

Theorem 1.6 (Product principle) *Let X and Y be two finite sets. Then the number of pairs (x, y) satisfying $x \in X$ and $y \in Y$ is $|X| \times |Y|$.*

Proof: There are $|X|$ choices for the first element x of the pair (x, y); then, regardless of what we choose for x, there are $|Y|$ choices for y. Each choice of x can be paired with each choice of y, so the statement is proved. ◊

Note that the set of all ordered pairs (x, y) so that $x \in X$ and $y \in Y$ is called the *direct product* (or Cartesian product) of X and Y, and is often denoted by $X \times Y$. We call the pairs (x, y) *ordered pairs* because the order of the two elements matters in them. That is, $(x, y) \neq (y, x)$, unless $x = y$.

Example 1.7 *The number of two-digit positive integers is 90.*

Solution: Indeed, a two-digit positive integer is nothing but an ordered pair (x, y), where x is the first digit and y is the second digit. Note that x must come from the set $X = \{1, 2, \cdots, 9\}$, while y must come from the set $Y = \{0, 1, \cdots, 9\}$. Therefore, $|X| = 9$ and $|Y| = 10$, and the statement is proved by Theorem 1.6. ◊

Theorem 1.8 (Generalized product principle) *Let $X_1, X_2, \cdots, X_k$ be finite sets. Then, the number of k-tuples $(x_1, x_2, \cdots, x_k)$ satisfying $x_i \in X_i$ is $|X_1| \times |X_2| \times \cdots \times |X_k|$.*

Informally, we could argue as follows. There are $|X_1|$ choices for x_1; then, regardless of the choice made, there are $|X_2|$ choices for x_2, so by Theorem 1.6, there are $|X_1| \times |X_2|$ choices for the sequence (x_1, x_2). Then there are $|X_3|$ choices for x_3, so again by Theorem 1.6, there are $|X_1| \times |X_2| \times |X_3|$ choices for the sequence (x_1, x_2, x_3). Continuing this argument until we get to x_k proves the theorem.

The line of thinking in this argument is correct, but the last sentence is somewhat less than rigorous. In order to obtain a completely formal proof, we will use the method of mathematical induction. It is very likely that the reader has already seen that method. A brief overview of the method can be found in the appendix.

Proof: (of Theorem 1.8) We prove the statement by induction on k. For $k = 1$, there is nothing to prove, and for $k = 2$, the statement reduces to the product principle.

Now let us assume that we know the statement for $k-1$, and let us prove it for k. A k-tuple $(x_1, x_2, \cdots, x_k)$ satisfying $x_i \in X_i$ can be decomposed into an ordered pair $((x_1, x_2, \cdots, x_{k-1}), x_k)$, where we still have $x_i \in X_i$. The number of such $(k-1)$-tuples $(x_1, x_2, \cdots, x_{k-1})$ is, by our induction hypothesis, $|X_1| \times |X_2| \times \cdots \times |X_{k-1}|$. The number of elements $x_k \in X_k$ is $|X_k|$. Therefore, by the product principle, the number of ordered pairs $((x_1, x_2, \cdots, x_{k-1}), x_k)$ satisfying the conditions is

$$(|X_1| \times |X_2| \times \cdots \times |X_{k-1}|) \times |X_k|,$$

so this is also the number of k-tuples $(x_1, x_2, \cdots, x_k)$ satisfying $x_i \in X_i$. ◊

Example 1.9 *For any positive integer k, the number of k-digit positive integers is $9 \cdot 10^{k-1}$.*

Solution: A k-digit positive integer is just a k-tuple $(x_1, x_2, \cdots, x_k)$, where x_i is the ith digit of our integer. Then x_1 has to come from the set $X_1 = \{1, 2, \cdots, 9\}$, while x_i has to come from the set $X_i = \{0, 1, 2, \cdots, 9\}$ for $2 \leq i \leq k$. Therefore, $|X_1| = 9$ and $|X_i| = 10$ for $2 \leq i \leq k$. The proof is then immediate by Theorem 1.8. ◊

Example 1.10 *How many four-digit positive integers both start and end in even digits?*

Solution: The first digit must come from the 4-element set $\{2, 4, 6, 8\}$, whereas the last digit must come from the 5-element set $\{0, 2, 4, 6, 8\}$. The second and third digits must come from the 10-element set $\{0, 1, \cdots, 9\}$. Therefore, the total number of such positive integers is $4 \cdot 10 \cdot 10 \cdot 5 = 2000$. ◊

An interesting special case of Theorem 1.8 is when all X_i have the same size because all X_i are identical as sets.

If A is a finite alphabet consisting of n letters, then a *k-letter string over A* is a sequence of k letters, each of which is an element of A.

Corollary 1.11 *The number of k-letter strings over an n-element alphabet A is n^k.*

Proof: Apply Theorem 1.8 with $X_1 = X_2 = \cdots = X_k = A$. ◊

1.2.2 Using several counting principles

Life for combinatorialists would be just too simple if every counting problem could be solved using a single principle. Most problems are more complex than that, and one needs to use several counting principles in the right way in order to solve them.

Example 1.12 *A user needs to choose a password for a bank card. The password can use the digits $0, 1, \cdots, 9$ with no restrictions, and it has to consist of at least four and at most seven digits. How many possible passwords are there?*

Solution: Let A_i denote the set of acceptable codes that consist of i digits. Then, by the product principle (or Corollary 1.11), we see that $A_i = 10^i$ for any i satisfying $4 \leq i \leq 7$. So, by the addition principle, we get that the total number of possibilities is

$$A_4 + A_5 + A_6 + A_7 = 10^4 + 10^5 + 10^6 + 10^7 = 11\ 110\ 000.$$

$\diamondsuit$

Example 1.13 *Let us assume that a prospective thief tried to steal the code of the card of Example 1.12. As the rightful user typed in the code, the thief observed that the password consisted of five digits, did not start with zero, and contained the digit 8. If the thief gets hold of the card,* at most *how many attempts will he need to find out the password?*

If we try to compute this number (that, is, the number of five-digit positive integers that contain the digit 8) directly, we risk making our work unduly difficult. For instance, we could compute the number of five-digit integers that *start* with 8, the number of those whose *second digit* is 8, and so on. We would run into difficulties in the next step, however, as the sets of these numbers are *not disjoint.* Indeed, just because the first digit of a number is 8, it could well be that its fourth and fifth digits are also 8. Therefore, if we simply added our partial results, we would count some five-digit integers many times (the number of times they contain the digit 8). For instance, we would count the integer 83885 three times. While we will see in later chapters that this is not an insurmountable difficulty, it does take a significant amount of computation to get around it. It is much easier to solve the problem in a slightly more indirect way.

Solution: (of Example 1.13) Instead of counting the five-digit positive integers that contain the digit 8, we count those that do not. Then, simply apply the subtraction principle by subtracting that number from the number of *all* five-digit positive integers, which is, of course, $9 \cdot 10^4 = 90000$ by Example 1.9.

How many five-digit positive integers do *not* contain the digit 8? These integers can start in eight different digits (everything but 0 and 8); then any of their remaining digits can be one of nine digits (everything but 8). Therefore, by the product principle, their number is $8 \cdot 9^4 = 52488$. Therefore, the number of those five-digit positive integers that do *not* contain the digit 8 is $90000 - 52488 = 37512$. $\diamondsuit$

We would hope that the cash machine will not let anyone take that many guesses.

1.2.3 When repetitions are not allowed

In some situations, we are not allowed to make the same choice more than once. This modifies our counting arguments as we will discuss in this subsection.

1.2.3.1 Permutations

Eight people participate in a long-distance running race. There are no ties. In how many different ways can the competition end?

This question is a little bit more complex than the questions in the two preceding subsections. This is because, while in those subsections we chose elements from sets so that our choices were *independent* of each other, in this section that will no longer be the case. For instance, if runner A wins the race, he cannot finish third, or fourth, or fifth. Therefore, the possibilities for the person who finishes third *depend* on who won the race and who finished second. Fortunately, this will not hurt our enumeration efforts, and we will see why.

Let us start with something less ambitious. How many possibilities are there for the winner of the competition? As there are eight participants, there are eight possibilities. How about the number of possibilities for the *ordered pair* of the winner and the runner-up? Well, there are eight choices for the winner, then, regardless who the winner is, there are seven choices for the runner-up (we can choose any person except the winner). Therefore, by the product principle, there are $8 \cdot 7 = 56$ choices for the winner/runner-up ticket.

We can continue the argument in this manner. No matter who finishes first and who finishes second, there will be six choices for the person finishing third, then five choices for the runner who finishes fourth, and so on. Therefore, by the generalized product principle, the number of total possible outcomes in this competition is

$$8 \cdot 7 \cdot 6 \cdot 5 \cdot 4 \cdot 3 \cdot 2 \cdot 1 = 40320. \tag{1.3}$$

This argument was possible because of the following underlying facts. The *set* of possible choices for the person finishing at position i depended on the choices we made previously. However, the *number* of these choices did not depend on anything but i (and was in fact equal to $9 - i$).

The general form of the number obtained in (1.3) is so important that it has its own name.

Definition 1.14 *Let n be a positive integer. Then the number*

$$n \cdot (n-1) \cdot \dots \cdot 2 \cdot 1$$

is called n-factorial, *and is denoted by* $n!$.

We point out that $0! = 1$, even if that may sound counterintuitive this time. See Exercise 1 for an explanation. Continuing from $n = 1$, the next few values of $n!$ are 1, 2, 6, 24, 120, 720, 5040, and 40320.

The set $\{1, 2, \cdots, n\}$ will be one of our favorite examples in this book because it exemplifies an n-element set, that is, n distinct objects. Therefore, we introduce the shorter notation $[n]$ for this set.

It goes without saying that there was nothing magical about the number eight in the previous example.

Theorem 1.15 *For any positive integers n, the number of ways to arrange all elements of the set $[n]$ in a line is $n!$.*

Proof: There are n ways to select the element that will be at the first place in our line. Then, regardless of this selection, there are $n-1$ ways to select the element that will be listed second, $n-2$ ways to select the element listed third, and so on. Our claim is then proved by the generalized product principle. $\diamondsuit$

We could have again proved our statement by induction, just as we proved the generalized product principle.

The function $f(n) = n!$ grows very rapidly. It is not difficult to see that, if n is large enough, then $n! > a^n$ for any fixed real number a. This is because, while a might be a huge number, it is a fixed number. Therefore, as n grows to $n+1$, the value of a^n gets multiplied by a, while the value of $n!$ gets multiplied by $n+1$, which will eventually be larger than a. Exercise 24 and Supplementary Exercise 11 provide more precise information about the growth rate of the factorial function, while the Notes section contains an even more precise result (1.13), called *Stirling's formula*, without proof.

Note the two simple but important features of the task of arranging all elements of $[n]$ in a line. Namely,

(1) each element occurs in the line, and

(2) each element will occur in the line only once.

In other words, each element will occur in the line exactly once. Arrangements of elements of a set with these properties are so important that they have their own name.

Definition 1.16 *A* permutation *of a finite set S is a list of the elements of S containing each element of S exactly once.*

With this terminology, Theorem 1.15 says that the number of permutations of $[n]$ is $n!$. Permutations are omnipresent in combinatorics, and they are frequently used in other parts of mathematics, such as algebra, group theory, and computer science. We will learn more about them in this book. For now, let us return to basic counting techniques.

1.2.3.2 Partial lists without repetition

Let us return to the eight-person running race. Let us assume that the runners who arrive first, second, or third will receive medals (gold, silver, and bronze), and the rest of the competitors will not receive medals. How many different possibilities are there for the list of medal winners?

We can start our argument as before, that is, by looking at the number of possibilities for the gold medal winner. There are eight choices for this person. Then there are seven choices for the silver medalist, and then six choices for the bronze medalist. So there are $8 \times 7 \times 6 = 336$ possibilities for the list of the medalists. Our task ends here. Indeed, as the remaining runners do not get any medals, their order does not matter.

Generalizing the ideas explained above, we get the following theorem.

Theorem 1.17 *Let n and k be positive integers so that $n \geq k$. Then, the number of ways to make a k-element list from $[n]$ without repeating any elements is*

$$n(n-1)(n-2)\cdots(n-k+1).$$

Proof: There are n choices for the first element of the list, then $n-1$ choices for the second element of the list, and so on; finally, there are $n-k+1$ choices for the last (kth) element of the list. The result then follows by the product principle. ◊

The number $n(n-1)(n-2)\cdots(n-k+1)$ of all k-element lists from $[n]$ without repetition occurs so often in combinatorics that there is a symbol for it, namely,

$$(n)_k = n(n-1)(n-2)\cdots(n-k+1).$$

Note that Theorem 1.15 is a special case of Theorem 1.17, namely, the special case when $n = k$.

Let us discuss a more complicated example, one in which we need to use both addition and multiplication.

Example 1.18 *A student cafeteria offers the following special. For a certain price, we can have our choice of one out of four salads, one out of five main courses, and something for dessert. For dessert, we can either choose one out of five sundaes, or we can choose one of four gourmet coffees and, no matter which gourmet coffee we choose, one out of two cookies. How many different meals can a customer buying this special have?*

Solution: We can argue as follows. The customer has to decide whether he prefers a sundae or a gourmet coffee with a cookie. As he cannot have both, the set of choices containing a sundae is disjoint from the set of choices containing a gourmet coffee and a cookie. Therefore, the total number of choices will be the sum of the sizes of these two sets. Now, let us compute the sizes of these sets separately. If the customer prefers the sundae, he has four choices for the salad, then five choices for the main course, and then five choices for the sundae, yielding a total of $4 \cdot 5 \cdot 5 = 100$ choices. If he prefers the gourmet coffee and the cookie, then he has four choices for the salad, five choices for the main course, four choices for the coffee, and finally two choices for the cookie. This yields a total of $4 \cdot 5 \cdot 4 \cdot 2 = 160$ choices. So the customer has $100 + 160 = 260$ choices.

Alternatively, we could count as follows. The customer has to choose the salad, then the main course. Up to that point, he has $4 \cdot 5 = 20$ choices. Then, he either chooses a sundae, in one of five ways, or a coffee and a cookie, in $4 \cdot 2 = 8$ ways. So he has $5 + 8 = 13$ choices for dessert. Therefore, if dessert is considered the third course, he has $4 \cdot 5 \cdot 13 = 260$ choices, in agreement with what we computed above. ◊

Example 1.19 *A college senior will spend her weekend visiting some graduate schools. Because of geographical constraints, she can either go to the north, where she can visit four schools out of the ten schools in which she is interested, or she can go to the south, where she can visit five schools out of eight schools in which she is interested. How many different itineraries can she set up?*

Note that we are interested in the number of possible itineraries, so the order in which the student visits the schools is important.

Solution: (of Example 1.19) The student can either go to the north, in which case, by Theorem 1.17, she will have $(10)_4$ possibilities, or she can go to the south, in which case she will have $(8)_5$ possibilities. Therefore, by the addition principle, the total number of possibilities is

$$(10)_4 + (8)_5 = 10 \cdot 9 \cdot 8 \cdot 7 + 8 \cdot 7 \cdot 6 \cdot 5 \cdot 4 = 5040 + 6720 = 11760.$$

◊

Quick Check

1. How many three-digit positive integers are there in which all digits are divisible by 3?
2. How many three-digit positive integers are there in which all digits are divisible by 3 and no digit is repeated?
3. What is the number of permutations of the digits 1, 2, 3, 4, 5, and 6 in which each digit in an even position is larger than each digit in an odd position?

1.3 When we divide

We sometimes make choices that look different, but for the purposes of our current counting problem, they are not really different. We have to take this into account so that we do not count the same choice more than once.

1.3.1 The division principle

Let us assume that several families are invited to a children's party. Each family comes with two children. The children then play in a room while the adults take turns supervising them. If a visitor looks in the room where the children are playing, how can the visitor determine the number of families at the party?

The answer to this question is not difficult. The visitor can count the children who are present, then divide that number by two. We included this question, however, as it exemplifies a very often-used counting technique. When we want to count the elements of a certain set S, it is often easier to count elements of another set T so that each element of S corresponds to d elements of T (for some *fixed number* d) while each element of T corresponds to one element of S. In order to describe the relation between the sets T and S more precisely, we make the following definition.

Definition 1.20 *Let S and T be finite sets, and let d be a fixed positive integer. We say that the function $f : T \to S$ is* d-to-one *if for each element $s \in S$ there exist exactly d elements $t \in T$ so that $f(t) = s$.*

Recall that the fact that f is a *function* automatically ensures that $f(t)$ is unique for each $t \in T$. See Figure 1.1 for an illustration.

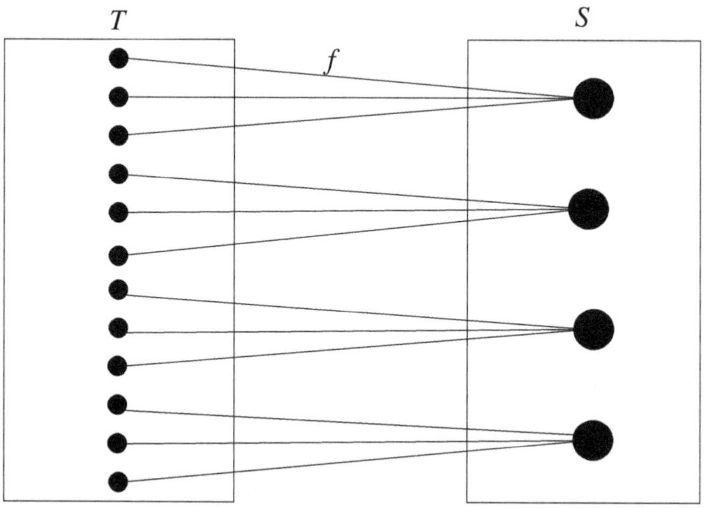

Figure 1.1
Diagram of a three-to-one map.

Theorem 1.21 (Division principle) *Let S and T be finite sets so that a d-to-one function $f : T \to S$ exists. Then*

$$|S| = \frac{|T|}{d}.$$

Proof: This is a direct consequence of Definition 1.20. $\diamondsuit$

In the above example, S was the set of *families* present, but we could not determine $|S|$ directly because we only saw the children and did not know who were siblings. However, T was the set of *children* present. We could easily determine $|T|$, then use our knowledge that each family had two children present (so $d = 2$) and obtain $|S|$ as $|T|/2$.

We will now turn to a classic example that will be useful in Chapter 4. Let us ask n people to sit around a circular table, and consider two seating arrangements identical if each person has the same *left neighbor* in both seatings.

For instance, the two seatings at the top of Figure 1.2 are identical, but the one at the bottom is not, even if each person has the same neighbors in that seating as well. This is because, in that seating, for each person, the former left neighbor becomes the right neighbor. If the food always arrives from one direction, this can be quite some difference.

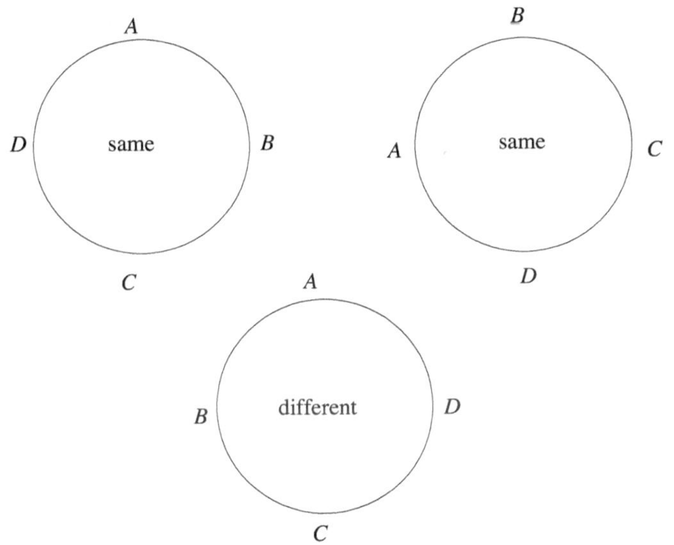

Figure 1.2
Two identical seatings and a different one.

Having made clear when two seating arrangements are considered different, we are ready to discuss our next example.

Example 1.22 *The number of different seating arrangements for n people around a circular table is $(n - 1)!$.*

Solution: (of Example 1.22) If the table were linear, instead of circular, then the number of all seating arrangements would be $n!$. In other words, if T is the set of seating arrangements of n people along a linear table, then $|T| = n!$.

Now let S be the set of seating arrangements around our circular table. We claim that each element of S corresponds to n elements of T. Indeed, take a circular seating $s \in S$, and choose a person p in that seating, in one of n ways. Then turn s into a linear seating by starting the seating with p, then continuing with the left neighbor of p, the left neighbor of that person, then the left neighbor of that person, and so on. As there are n choices for p, each circular seating s can be turned into n different linear seating arrangements.

On the other hand, each linear seating $lins$ corresponds to one circular seating $f(lins)$, because no matter where we "fold" $lins$ into a circle, the left neighbor of each person will not change.

This means that $f : T \rightarrow S$ is an n-to-one function. Therefore, by the division principle,

$$|S| = \frac{|T|}{n} = \frac{n!}{n} = (n-1)!.$$

So this is the number of circular seating arrangements. ◊

1.3.2 Subsets

It is difficult to overstate the importance of the counting problem that we discuss next.

1.3.2.1 The number of k-element subsets of an n-element set

At a certain university, the Department of Mathematics has 55 faculty members. The department is asked to send three of its faculty members to the commencement ceremonies to serve as marshals there. All three people chosen for this honor will perform the same duties. It is up to the department chair to choose the three professors who will serve. How many different possibilities does the chair have?

A superficial *and wrong* argument would go like this: The chair has 55 choices for the first faculty member, then 54 choices for the second faculty member, and, finally, 53 choices for the last one. Therefore, as we explained with Theorem 1.17, the number of all possibilities is $55 \cdot 54 \cdot 53$.

The reader should take a moment here to try to see the problem with this argument. Once you have done that, you can read further. The problem is that this line of thinking counts the same triple of professors many times. Indeed, let A, B, and C be three professors from this department. (The department hires people with short names only.) The above line of thinking considers ABC and BAC as different triples, whereas they are in fact identical. Indeed, we said that all three people chosen will perform the same duties. Therefore, the order in which these people are chosen is irrelevant.

There is no reason to despair, though. While it is true that all triples are counted more than once, we will show that the above line of thinking *can* be corrected. This is because all triples are counted the same number of times.

Indeed, the triple containing professors A, B, and C will be counted $3! = 6$ times since there are six ways these three letters can be listed. This was proved in Theorem 1.15.

Now we resort to the division principle. The number $55 \cdot 54 \cdot 53$, that is, the number of all possible triples in which the order of the people chosen matters, is exactly six times as large as the number of all "triples" in which the order of the people chosen does not matter. Therefore, the latter is equal to $(55 \cdot 54 \cdot 53)/6 = 26235$. Therefore, this is the number of possibilities the department chair has. If she considers each of them for exactly one minute, and works 24 hours a day, she will still need more than 18 days to do this. Nobody says being a chair is easy.

The situation described above is an example of a fundamental problem in combinatorics, that is, *selecting a subset of a set.* The following theorem introduces a basic notation and presents the enumerative answer to this problem.

Theorem 1.23 *Let n be a positive integer, and let $k \leq n$ be a nonnegative integer. Then the number of all k-element subsets of $[n]$ is*

$$\frac{n(n-1)\cdots(n-k+1)}{k!}. \tag{1.4}$$

Proof: As we have seen in Theorem 1.17, the number of ways we can make a k-element list using elements of $[n]$ without repeating any elements is $n(n-1)\cdots(n-k+1) = (n)_k$. Because a k-element subset has $k!$ ways of being listed, each k-element subset will be counted $k!$ times by the number $(n)_k$. Therefore, by the division principle, the number of all k-element subsets of $[n]$ is $\frac{(n)_k}{k!}$. $\diamondsuit$

The number of all k-element subsets of $[n]$ is of quintessential importance in combinatorics. Therefore, we introduce the symbol $\binom{n}{k}$, read "n choose k," for this number. With this terminology, Theorem 1.23 says that $\binom{n}{k} = (n)_k/k!$. The numbers $\binom{n}{k}$ are called *binomial coefficients.*

Note that, by multiplying both the numerator and the denominator of (1.4) by $(n-k)!$, we get the more compact formula

$$\binom{n}{k} = \frac{n!}{k!(n-k)!}. \tag{1.5}$$

1.3.2.2 The binomial theorem for positive integer exponents

Binomial coefficients play a very important role in algebraic computations, because of the *binomial theorem.*

Theorem 1.24 (Binomial theorem) *If n is a positive integer, then*

$$(x+y)^n = \sum_{k=0}^{n} \binom{n}{k} x^k y^{n-k}.$$

Proof: The left-hand side is the product

$$(x+y)(x+y)\cdots(x+y),$$

where the factor $(x+y)$ occurs n times. In order to compute this product, we have to choose one term (that is, x or y) from each of these n factors, multiply the chosen n terms together, then do this in all 2^n possible ways, and then add the obtained 2^n products. It suffices to show that exactly $\binom{n}{k}$ of these products will be equal to $x^k y^{n-k}$. However, this is true, because in order to obtain a product that is equal to $x^k y^{n-k}$, we have to choose an x from exactly k factors. We can do that in $\binom{n}{k}$ ways, and then we must choose a y from all the remaining factors. ◊

So, for the first few positive integer values of n, the polynomials $(x+y)^n$ are as follows:

- $x+y$,
- $(x+y)^2 = \binom{2}{0}x^2 + \binom{2}{1}xy + \binom{2}{2}y^2 = x^2 + 2xy + y^2$,
- $(x+y)^3 = \binom{3}{0}x^3 + \binom{3}{1}x^2y + \binom{3}{2}xy^2 + \binom{3}{3}y^3 = x^3 + 3x^2y + 3xy^2 + y^3$, and
- $(x+y)^4 = \binom{4}{0}x^4 + \binom{4}{1}x^3y + \binom{4}{2}x^2y^2 + \binom{4}{3}xy^3 + \binom{4}{4}y^4 = x^4 + 4x^3y + 6x^2y^2 + 4xy^3 + y^4$.

Life would be, of course, too simple if the division principle used in the proof of Theorem 1.23 were never to be used in conjunction with other principles, such as the addition principle.

Example 1.25 *A city has 110 different bus lines. Passengers are asked to buy tickets before boarding, then to validate them when on the bus by inserting them into a machine. (The tickets have to be inserted in a certain direction; rotations are not allowed.) The machine then punches two or three holes into the ticket within some of the nine numbered squares. Can the city set the machines up so that on each bus line they will punch the tickets differently?*

Solution: We have to determine whether or not the total number of possibilities is at least 110. Since the machines punch either two or three holes, we have to enumerate all two-element subsets of $[9]$ and all three-element subsets of $[9]$. The previous theorem tells us that the total number of these subsets is

$$\binom{9}{2} + \binom{9}{3} = \frac{9 \cdot 8}{2 \cdot 1} + \frac{9 \cdot 8 \cdot 7}{3 \cdot 2 \cdot 1} = 36 + 84 = 120.$$

Therefore, the city can indeed set up the machines in the desired way. ◊

Quick Check

1. What is the number of subsets of [10] that have at least four elements, but not more than six elements?
2. A company has 30 employees, 20 of whom work in production, and ten of whom work in sales. How many ways are there to form a committee of six employees of this company so that four of the committee members work in production and two committee members work in sales?
3. Continuing the previous question, how many ways are there to form a six-member committee if the committee is to have *at least* one member who works in production and *at least* one member who works in sales?

1.4 Applications of basic counting principles

We will now consider more complex situations where the basic counting principles turn out to be useful.

1.4.1 Bijective proofs

The following example will teach us an extremely important proof technique in enumerative combinatorics.

Example 1.26 *In a certain part of our town, the streets form a square grid, and each street is one way to the north or to the east. Let us assume that our car is currently at the southwest corner of this grid, which we will denote by* $O = (0, 0)$.

(a) In how many ways can we drive to the point $X = (6, 4)$*?*

(b) In how many ways can we drive to the point X *if we want to stop at the bakery at* $Y = (4, 2)$*?*

(c) In how many ways can we drive to the point X *if we want to stop at either the ice cream shop at* $U = (3, 2)$ *or at the coffee shop at* $V = (2, 3)$*?*

See Figure 1.3 for an illustration.

Solution:

(a) The car needs to travel ten blocks, namely, six blocks to the east and four blocks to the north. In other words, the driver needs to choose a six-element subset of [10] that will tell him when to drive east.

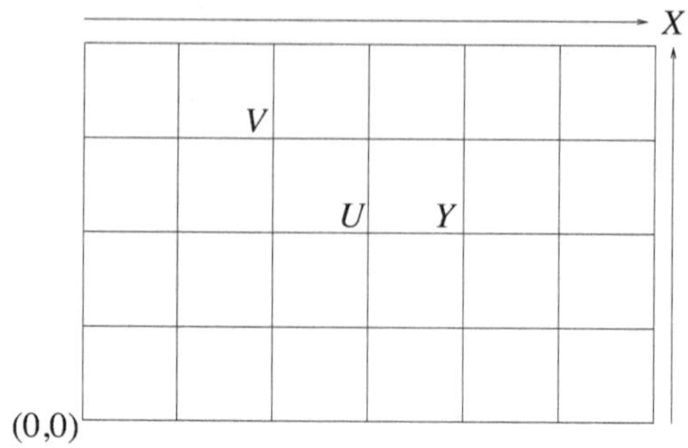

Figure 1.3
The grid of our one-way streets.

If, for instance, he chooses the set $\{2, 3, 5, 7, 8, 9\}$, then his second, third, fifth, seventh, eighth, and ninth streets will go east, and all the rest, that is, his first, fourth, sixth, and tenth streets will go north. Since the number of six-element subsets of $[10]$ is $\binom{10}{6}$, this is the number of ways the car can get to point X.

(b) The car first needs to get to Y. By the argument of part (a), there are $\binom{6}{4} = 15$ ways to do this. Then, it needs to go from Y to X, and, by an analogous argument, there are $\binom{4}{2} = 6$ ways to do it. As any path from $(0, 0)$ to Y can be followed by any path from Y to X, the total number of acceptable paths is $\binom{6}{2}\binom{4}{2} = 15 \cdot 6 = 90$.

(c) Using the argument explained in part (b), there are $\binom{5}{3}\binom{5}{3} = 100$ paths from $(0, 0)$ to X via U, and there are $\binom{5}{2}\binom{5}{4} = 50$ paths from $(0, 0)$ to X via V. Note that no path can go through both U and V. Therefore, the total number of acceptable paths is $\binom{5}{3}\binom{5}{3} + \binom{5}{2}\binom{5}{4} = 150$.

$\Diamond$

Let us analyze the argument of the above example in detail. In part (a), we had to count the possible paths from $(0, 0)$ to $(6, 4)$. We said that this was the same as counting the six-element subsets of $[10]$. Why could we say that? We could say that because there is a *one-to-one* correspondence between the set S of our lattice paths and the set T of six-element subsets of $[10]$. Therefore, we must have $|S| = |T|$. Note that this is the special case of Theorem 1.21 (the division principle) when $d = 1$.

The special case of $d = 1$ of the division principle is so important that it has its own name.

Definition 1.27 *If the map $f : S \to T$ is one-to-one and onto, then we call f a* bijection.

In other words, if $f : S \to T$ is a bijection, then it creates pairs, matching each element of S to a different element of T.

Corollary 1.28 *Let S and T be finite sets. If a bijection $f : S \to T$ exists, then $|S| = |T|$.*

Note that the requirement that S and T be finite can be dropped, and then one can *define* the notion that two infinite sets have the same size if there is a bijection between them. This is a very interesting topic, but it belongs to a textbook on Set Theory; therefore, we will not discuss it here.

See Figure 1.4 for the diagram of a generic bijection.

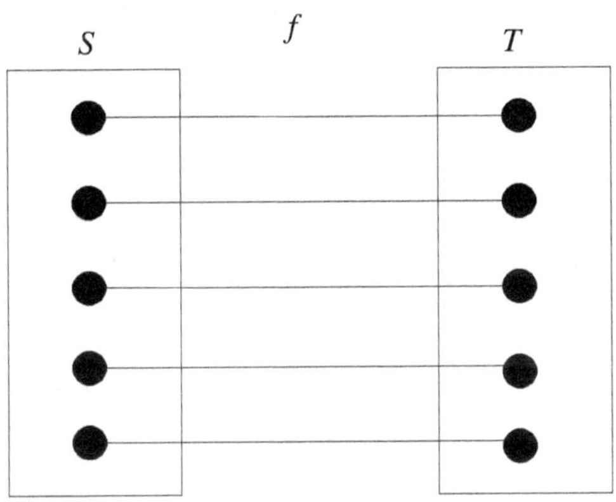

Figure 1.4
The diagram of a generic bijection.

The idea of bijections, or bijective proofs, is used very often in counting arguments. When we want to enumerate elements of a set S, we can instead prove that a bijection $f : S \to T$ exists with some set T whose number of elements we know. Usually, this is done by first defining a function f, then showing that this f is indeed a bijection from S into T. Once that is done, we can conclude that $|S| = |T|$. Sometimes, we do not need the actual number of elements in the sets, just the fact that the two sets have the same number of elements. In that case, the method of bijections can save us the actual counting. The reader will see many examples of this method in the next section, and the rest of the book for that matter. For now, let us see some simple applications of the method.

Proposition 1.29 *For any positive integer n, the number of divisors of n that are larger than $\sqrt{n}$ is equal to the number of divisors of n that are smaller than $\sqrt{n}$.*

As this is our first bijective proof, we will explain it in full detail. In particular, we will show how one can prove that a map is indeed a bijection. The reader should not be discouraged by thinking that this proof is too technical. All we do is show that two sets have the same size by matching their elements, one by one.

Proof: Let S be the set of divisors of n that are larger than $\sqrt{n}$, and let T be the set of divisors of n that are smaller than $\sqrt{n}$. Define $f : S \to T$ by $f(s) = n/s$.

Now comes that heart of the proof, that is, we will show that f is *indeed a bijection from S into T*. First, for all $s \in S$, the equality

$$s \cdot f(s) = n \tag{1.6}$$

holds, so $f(s)$ is indeed a divisor of n. Second, $f(s) < \sqrt{n}$ must hold, otherwise $s \cdot f(s) > \sqrt{n} \cdot \sqrt{n} = n$, contradicting (1.6). Therefore, $f(s)$ is always an element of T, and so f is indeed a function from S into T.

Now we have to show that f is one-to-one, that is, for all $t \in T$, there exists exactly one $s \in S$ so that $f(s) = t$. On one hand, there is at least one such s, namely, $s = n/t$. Indeed, by the definition of f, we have $f(n/t) = \frac{n}{n/t} = t$. On the other hand, this is the only good s. Indeed, if $f(s) = t$, then by (1.6), we must have $s \cdot t = n$, so $s = n/t$. Therefore, f is a bijection, and so S and T have the same number of elements. ◊

Example 1.30 *The integer 1000 has exactly eight divisors that are larger than $\sqrt{1000}$.*

Solution: By Proposition 1.29, it suffices to count the divisors of 1000 that are *smaller* than $\sqrt{1000} = 31.62$. They are 1, 2, 4, 5, 8, 10, 20, and 25, so there are indeed eight of them. ◊

In other words, instead of scanning the interval $[32, 1000]$ for divisors, we only had to scan the much shorter interval $[1, 31]$.

The way in which we showed that the function f was indeed a bijection from S into T in the above example is fairly typical. Let us summarize this method for future reference.

In order to prove that $|S| = |T|$ by the method of bijections, proceed as follows:

1. Define a function f on the set S that has a chance to be a bijection from S into T.
2. Show that for all $s \in S$, the relation $f(s) \in T$ holds.
3. Show that for all $t \in T$, there is exactly one $s \in S$ satisfying $f(s) = t$. This is often done in two smaller steps, namely,

(a) proving that there is at least one s satisfying $f(s) = t$, and
(b) proving there is at most one s satisfying $f(s) = t$.

Example 1.31 *A new house has ten rooms. For each room, the owner can decide whether he wants a ceiling fan for that room, and if yes, whether he wants it to be a cheaper or a more expensive model. How many different possibilities does the owner have for all ten rooms?*

Solution: We claim that the number of possibilities is 3^{10}. We show this by constructing a bijection f from the set S of all possibilities the owner has to the set T of all ten-letter words over the alphabet $\{a, b, c\}$. Then our result will follow from Corollary 1.11.

Let $s \in S$, and define the ith letter of $f(s)$ by

$$f(s)_i = \begin{cases} a \text{ if there is no ceiling fan in room } i, \\ b \text{ if there is a cheap ceiling fan in room } i, \\ c \text{ if there is a more expensive ceiling fan in room } i. \end{cases}$$

Our construction then implies that $f(s) \in T$, so f is indeed a function from S to T. Furthermore, f is indeed a bijection, as any word t in T tells us with no ambiguity what kind of ceiling fan each room needs to have in s if $f(s) = t$. ◊

1.4.1.1 Catalan numbers

For a more involved bijection, let us return to Example 1.26. A drive that is allowed in that example is called a *northeastern lattice path.* Northeastern lattice paths are omnipresent in enumerative combinatorics, since they can represent a plethora of different objects.

Lemma 1.32 *The number of northeastern lattice paths from* $(0, 0)$ *to* (n, n) *that never go above the diagonal* $x = y$ *(the main diagonal) is equal to the number of ways to fill a* $2 \times n$ *grid with the elements of* $[2n]$ *using each element once so that each row and column is increasing (to the right and down).*

For brevity, a $2 \times n$ rectangle whose boxes contain the elements of $[2n]$ so that each element is used once and each row and column is increasing (to the right and down) will be called a *Standard Young Tableau* of shape $2 \times n$.

Example 1.33 *Let* $n = 3$. *Then, there are five northeastern lattice paths from* $(0, 0)$ *to* $(3, 3)$ *that do not go above the main diagonal; they are shown in Figure 1.5. There are five Standard Young Tableaux of shape* 2×3, *which are shown in Figure 1.6.*

Proof: (of Lemma 1.32) It should not come as a surprise that we will construct a bijection from the set S of all northeastern lattice paths from $(0, 0)$ to (n, n)

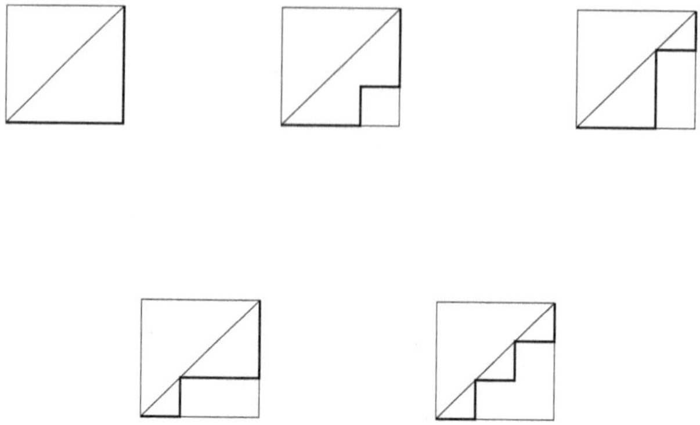

Figure 1.5
The five northeastern lattice paths that do not go above the main diagonal, for $n = 3$.

that do not go above the main diagonal into the set T of all Standard Young Tableaux of shape $2 \times n$.

Our bijection f is defined as follows: Let $s \in S$. Let $e_1, e_2, \cdots, e_n$ denote the positions of the n east steps of f. That is, if the first three east steps of f are in fact the first, third, and fourth steps of f (and so the second, fifth, and sixth steps of f are to the north), then $e_1 = 1$, $e_2 = 3$, and $e_3 = 4$. Similarly, let $n_1, n_2, \cdots, n_n$ denote the north steps of s. That is, keeping our previous example, $n_1 = 2$, $n_2 = 5$, and $n_3 = 6$.

Let $f(s)$ be the array of shape $2 \times n$ whose first row is $e_1, e_2, \cdots, e_n$ and whose second row is $n_1, n_2, \cdots, n_n$. We claim that $f(s) \in T$. The rows of $f(s)$ are increasing, since the ith east step had to happen before the $(i + 1)$st east step, and the same holds for north steps. We claim that the columns are increasing as well. Indeed, otherwise $n_j < e_j$ would hold for some j, meaning

1	2	3
4	5	6

1	2	4
3	5	6

1	2	5
3	4	6

1	3	4
2	5	6

1	3	5
2	4	6

Figure 1.6
The five Standard Young Tableaux of shape $2 \times n$.

that the jth step to the north was completed before the jth step to the east. That is impossible, because that would mean that after the jth north step we are at the point (x, j) for some $x < j$, which would mean that we are *above the main diagonal.*

To see that f is a bijection, let $t \in T$. If there exists an $s \in S$ so that $f(s) = t$, then it must be the lattice path whose east steps correspond to the first row of t and whose north steps correspond to the second row of t. On the other hand, that lattice path s never goes above the main diagonal because the increasing property of the columns assures that the jth north step of s comes after the jth east step of s, for any j. Therefore, f is one-to-one, and the statement is proved. $\diamondsuit$

Example 1.34 *Figure 1.7 shows an example of the bijection of Lemma 1.32.*

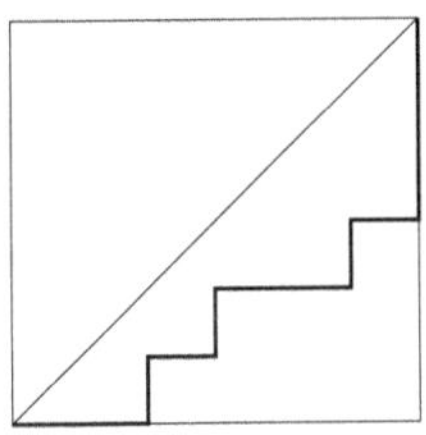

1	2	4	6	7	9
3	5	8	10	11	12

Figure 1.7
Turning a northeastern lattice path that never goes above the main diagonal into a Standard Young Tableau.

Hopefully, you are now asking yourself what the number of such lattice paths and Standard Young Tableaux is. That number c_n turns out to be a very famous number in Combinatorics; and it is called the nth *Catalan number*. These numbers are important because they count over 150 different kinds of objects. These objects come from all branches of Combinatorics, indeed, we will see how Catalan numbers occur in various counting problems (in Chapter 3), in Permutation Enumeration (in Chapter 4), and in Graphical Enumeration (Chapter 5). While finding an explicit formula for c_n is somewhat harder than our problems in this first chapter, we will walk the reader through the process of finding such a formula in the Exercises section.

1.4.2 Properties of binomial coefficients

An interesting application of northeastern lattice paths is proving basic properties of binomial coefficients. We will show a few examples of that here, leaving several others for the Exercises.

Proposition 1.35 *Let n and k be nonnegative integers so that $k \leq n$. Then* $\binom{n}{k} = \binom{n}{n-k}$.

You could ask what the big deal is. We have seen in (1.5) that $\binom{n}{k} = \frac{n!}{k!(n-k)!}$; therefore, replacing k by $n-k$, we get $\binom{n}{n-k} = \frac{n!}{(n-k)!k!}$, proving the previous proposition.

This argument is correct and simple, but not particularly illuminating. It does not show why the result is *really* true. The following proof is meant to provide some deeper insight.

Proof: (of Proposition 1.35) As we have seen in Example 1.26, the number $\binom{n}{k}$ is just the number of all northeastern lattice paths from $O = (0,0)$ to $P = (k, n-k)$. Similarly, the number $\binom{n}{n-k}$ is just the number of all northeastern lattice paths from $O = (0,0)$ to $Q = (n-k, k)$. These two numbers must be the same since reflection through the $x = y$ diagonal turns an OP-path into an OQ-path and vice versa. In other words, reflection through the $x = y$ diagonal is a bijection from the set S of all OP-paths into the set T of all OQ-paths. $\diamondsuit$

See Figure 1.8 for an example.

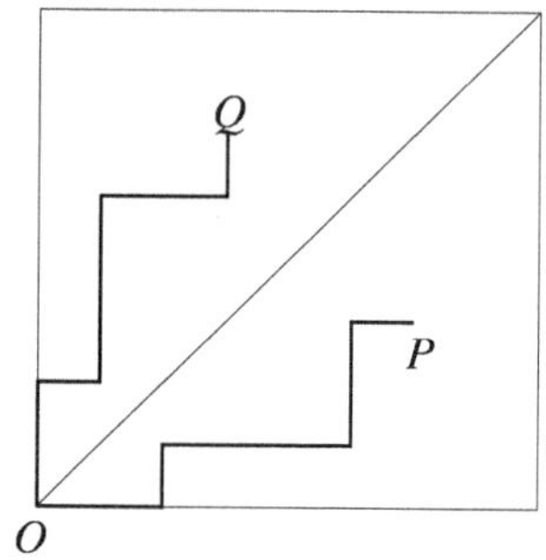

Figure 1.8
Turning an OP-path into an OQ-path and vice versa.

Proposition 1.35 can be proved in a combinatorial way *without* using lattice paths as follows: The left-hand side counts all k-element subsets of $[n]$. The right-hand side counts all $(n-k)$-element subsets of $[n]$. So the proposition will be proved if we can find a bijection $f : S \to T$, where S is the set of all k-element subsets of $[n]$, and T is the set of all $(n-k)$-element subsets of $[n]$. Such a bijection f can be defined as follows: If $A \subseteq [n]$, then the *complement* of A is the set $A^c = [n] - A$. Note that $(A^c)^c = A$ for all $A \in [n]$.

If A is a k-element subset of $[n]$, then we set $f(A) = A^c$. Then f indeed maps from S to T. Furthermore, note that, if $B \in T$, then there is exactly one $A \in S$ satisfying $f(A) = B$, namely, B^c. (If B is the complement of A, then A is the complement of B.) Therefore, f is a bijection, which proves that $|S| = |T|$, and so Proposition 1.35 is proved.

The most frequently used property of the binomial coefficients may be the identity

$$2^n = \sum_{k=0}^{n} \binom{n}{k}. \tag{1.7}$$

At first, this identity may sound surprising. After all, it says that a long sum of fractions whose numerators and denominators both contain factorials has an extremely simple and compact form. Though it is possible to prove this identity computationally, it is a rather tedious procedure. There is, however, a crystal clear combinatorial argument proving (1.7). We suggest finding this argument independently and then checking the solution of Exercise 13.

The following well-known property of binomial coefficients can again be proven computationally, but a combinatorial proof provides deeper understanding and is far more enjoyable to find.

Theorem 1.36 *Let n and k be nonnegative integers so that $k < n$. Then*

$$\binom{n}{k} + \binom{n}{k+1} = \binom{n+1}{k+1}. \tag{1.8}$$

We will prove this statement by a variation of the bijective proof method, which we implicitly applied in our earlier sections. That is, we will show that both the left-hand side and the right-hand side count the elements of the same set. That will prove that the two sides are equal.

Proof: (of Theorem 1.36) The right-hand side is simply the number of all northeastern lattice paths from $O = (0,0)$ to $R = (k+1, n-k)$. Note that each such path arrives at R either via $U = (k, n-k)$, and these paths are counted by the first term of the left-hand side, or via $V = (k+1, n-k-1)$, and these terms are counted by the second term of the left-hand side. Therefore, both sides of our equation count the same objects, and, therefore, they have to be equal. $\diamondsuit$

See Figure 1.9 for an illustration. The proof technique we have just used (showing that two expressions are equal by showing that they count the same objects) is as powerful as it is simple. We will call proofs of this type *combinatorial proofs*, and we will show some additional examples shortly.

In Figure 1.10, we show the number of ways to get from $(0,0)$ to $(k, n-k)$ for small values of n and k. We call the top row the zeroth row, the next row the first row, and so on. In other words, the nth row consists of the binomial coefficients $\binom{n}{k}$, where $0 \leq k \leq n$. The triangle shown in Figure 1.10 is often called the *Pascal triangle.*

Without using lattice paths, we could have said that the right-hand side counts all $(k+1)$-element subsets of $[n+1]$, while the left-hand side counts the same objects, in two steps. The first term of the left-hand side counts

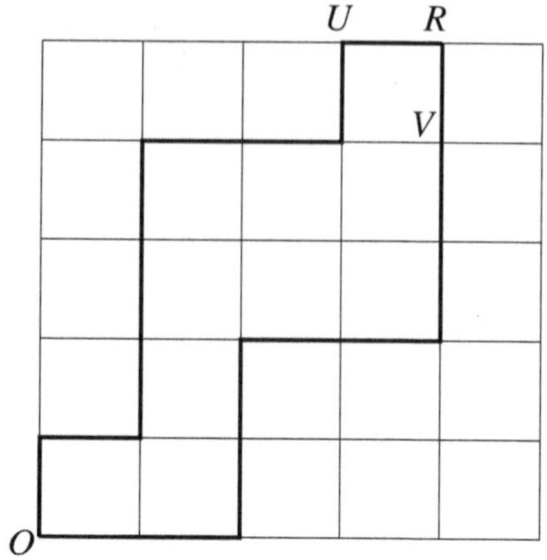

Figure 1.9
Each OR-path is either via U or via V.

$$\begin{array}{ccccccccccc} &&&&&1&&&&& \\ &&&&1&&1&&&& \\ &&&1&&2&&1&&& \\ &&1&&3&&3&&1&& \\ &1&&4&&6&&4&&1& \\ 1&&5&&10&&10&&5&&1 \end{array}$$

Figure 1.10
The values of $\binom{n}{k}$. Note that for fixed n, the values $\binom{n}{0}, \binom{n}{1}, \cdots$ form the nth row of the Pascal triangle.

those subsets that contain the element $n+1$ and are therefore determined by their remaining k elements, which are chosen from [n]. The second term counts those subsets that do not contain $n+1$ and are therefore $(k+1)$-element subsets of $[n]$.

Let us practice our latest techniques with another example.

Theorem 1.37 *For all positive integers n,*

$$\binom{2n}{n} = \sum_{k=0}^{n} \binom{n}{k}^2.$$

Proof: The left-hand side is just the number of northeastern lattice paths from $(0,0)$ to (n,n). These paths all have $2n$ steps. The right-hand side counts the same paths, but in a different way, namely, according to how far east they get in n steps. The number of paths that are on the line $x = k$ in n steps (in other words, the number of paths that contain exactly k east steps among their first n steps) is $\binom{n}{k}$. After the first n steps, each of these paths is at the point $(k, n-k)$. The number of ways to complete such a path is the number of ways to go from $(k, n-k)$ to (n, n), which is $\binom{n}{n-k} = \binom{n}{k}$. Therefore, the number of all northeastern lattice paths from $(0,0)$ to (n,n) that are on line $x = k$ after n steps is $\binom{n}{k}\binom{n}{k} = \binom{n}{k}^2$.

The value of k can be anything from 0 (when the first n steps are all to the north) to n (when the first n steps are all to the east), so to get the total number of paths from $(0,0)$ to (n,n), we have to sum $\binom{n}{k}^2$ over all k from 0 to n, which yields precisely the right-hand side. ◊

We invite the reader to interpret this proof in terms of subsets of the set $[2n]$, instead of lattice paths.

Example 1.38 *Let n be a positive integer. Then*

$$\sum_{k=1}^{n} k\binom{n}{k}^2 = n\binom{2n-1}{n-1}. \tag{1.9}$$

Solution: Let us say a law firm consists of n partners and n associates. We claim then that both sides of (1.9) count the ways of selecting both a committee of n members out of the $2n$ people at the firm and a president for the committee (who must be a member of the committee and a partner).

The right-hand side counts our possibilities as follows. First, choose the president in n ways, then choose the remaining $n-1$ members in $\binom{2n-1}{n-1}$ ways.

On the left-hand side, first choose the k members who are partners in $\binom{n}{k}$ ways, then choose the president in k ways. Finally, choose the $n-k$ members who are associates in $\binom{n}{n-k} = \binom{n}{k}$ ways. As both sides count the same objects, they are equal, and our proof is complete. ◊

We close this section with another example of proving an identity by interpreting both sides as the number of ways to carry out a certain task.

Example 1.39 *For all integers $n \geq 2$, the identity*

$$n \cdot (n-1) \cdot 2^{n-2} = \sum_{k=2}^{n} \binom{n}{k} \cdot k \cdot (k-1) \tag{1.10}$$

holds.

Solution: We claim that both sides count the ways to choose a committee of at least two people from a group of n people and to choose a president and a vice president of the committee who must also be members of the committee.

The left-hand side counts the ways to do so if the president and the vice president are selected first, in one of $n(n-1)$ ways, and the rest of the committee is selected after that, by the selection of one of the 2^{n-2} subsets of the group of the remaining $n-2$ people.

The right-hand side counts the ways to do so if the k-member committee is selected first, in one of $\binom{n}{k}$ ways for a given k, and the president and the vice president are selected after, in one of $k(k-1)$ ways. ◊

In Supplementary Exercise 23 you are asked to deduce (1.10) from the binomial theorem.

1.4.3 Permutations with repetition

Example 1.40 *A car dealership has 30 cars of the same model, 15 of which are white, 10 of which are blue, and 5 of which are red. How many possible ways are there to arrange these cars in a line if cars of the same color are indistinguishable?*

Solution: If there were 30 *different* cars to choose from, then the number of all possibilities would be, by Theorem 1.15, 30!. In order to reduce the task at hand to the simpler task of lining up 30 different cars, let us denote the first white car by the symbol A_1, the second white car by the symbol A_2, and so on, and proceed similarly with the blue cars and the red cars. Then each possible lineup can be identified by a list of 30 symbols in some order, namely, the symbols $A_1, A_2, \cdots, A_{15}$, the symbols $B_1, B_2, \cdots, B_{10}$, and the symbols $C_1, C_2, \cdots C_5$. The number of such lists is, as we said, 30!. The problem is that different lists *do not* necessarily describe different arrangements.

Indeed, if we take a list L, keep all its symbols B_j and C_k fixed, then permute the symbols A_i among themselves, then we get different lists, but they will all correspond to the same arrangement. If you are not sure about this, note that, no matter whether a list starts with A_7A_9 or A_9A_7, the corresponding arrangement will always start with two steps of type A.

The same is true if we permute the symbols B_j among themselves or the symbols C_k among themselves. Briefly, the number of allowed lists is much larger than the number of allowed arrangements. How many times larger? There are 15! ways to permute the symbols A_i among each other, then there are 10! ways to permute the symbols B_j among each other, and finally, there are 5! ways to permute the symbols C_k among each other. There is no other way to rearrange a list without changing an arrangement. So there are $15! \cdot 10! \cdot 5!$ lists corresponding to the same arrangement. Therefore, by the division principle, the number of *different* allowed arrangements is

$$\frac{30!}{15! \cdot 10! \cdot 5!}.$$

◊

The above example is a special case of the following general theorem.

Theorem 1.41 *Assume we want to arrange n objects in a line, the n objects are of k different types, and objects of the same type are indistinguishable. Let a_i be the number of objects of type i. Then the number of different arrangements is*

$$\frac{n!}{a_1!a_2!\cdots a_k!}.$$

Proof: Let us number the objects of type i with integers $1, 2, \cdots, a_i$. We then have n different objects, so the number of ways to arrange them in a line is $n!$. However, if we consider two arrangements identical if they only differ in the numberings of the objects of the same type, then each arrangement will have $a_1!a_2!\cdots a_k!$ identical versions. Therefore, our claim is proved by the division principle. $\diamondsuit$

The following alternative proof of Theorem 1.41 shows the connection between that theorem and binomial coefficients.

Proof: (of Theorem 1.41) An arrangement of our n objects is completely determined by the positions of the a_1 objects of type 1, the positions of the a_2 objects of type 2, and so on. There are $\binom{n}{a_1}$ choices for the positions of the objects of type 1. Then, there are $\binom{n-a_1}{a_2}$ choices for the positions of the objects of type 2, then there are $\binom{n-a_1-a_2}{a_3}$ choices for the positions of the objects of type 3, and so on. At the end, there is $\binom{n-a_1-\cdots-a_{k-1}}{a_k} = \binom{a_k}{a_k} = 1$ choice for the positions of the objects of type k, which is not surprising since these objects must take the positions nobody else took. Therefore, by the generalized product principle, the total number of choices is

$$\binom{n}{a_1} \cdot \binom{n-a_1}{a_2} \cdot \binom{n-a_1-a_2}{a_3} \cdots \binom{n-a_1-\cdots-a_{k-1}}{a_k} =$$

$$\frac{n!}{a_1!(n-a_1)!} \cdot \frac{(n-a_1)!}{a_2!(n-a_1-a_2)!} \cdot \frac{(n-a_1-a_2)!}{a_3!(n-a_1-a_2-a_3)!} \cdots = \frac{n!}{a_1!a_2!\cdots a_k!}.$$

$\diamondsuit$

The expression $\frac{n!}{a_1!a_2!\cdots a_k!}$ is often denoted by the symbol $\binom{n}{a_1, a_2, \cdots, a_k}$ and is called a *multinomial coefficient.* Note that, if $k = 2$, then this multinomial coefficient reduces to the binomial coefficient $\binom{n}{a_1, a_2} = \binom{n}{a_1} = \binom{n}{a_2}$.

Example 1.42 *A quality controller has to visit one factory a day. In the next eight days, she will visit each of the four factories, A, B, C, and D, twice. The controller is free to choose the order in which she visits these factories, but the two visits to factory A cannot be on consecutive days. In how many different orders can the controller proceed?*

Solution: Without the extra restriction on the visits to factory A, the question would be just the number of ways to arrange two copies of A, two copies of B, two copies of C, and two copies of D in a line. By Theorem 1.41, the number of ways to do this is

$$\binom{8}{2,2,2,2} = \frac{8!}{2^4} = 2520.$$

However, some of these arrangements are forbidden by the requirement that the two visits to A not be on consecutive days. In order to count the number of these *bad* arrangements, let us *glue* together the two copies of A. In other words, let us replace the two copies of A by a single symbol A'. Then, the number of ways to list our seven symbols (A', and two copies of each remaining letter) is, by Theorem 1.41,

$$\binom{7}{1,2,2,2} = \frac{7!}{2^3} = 630.$$

Therefore, by the subtraction principle, the number of good arrangements is $2520 - 630 = 1890$. So this is the number of ways in which the controller can plan her visits for the next eight days. ◊

Example 1.43 *Let us assume that in a given year the National Basketball Association has 28 teams, which are split into two conferences of 14 teams each. These conferences are split into divisions. The Atlantic Division consists of five teams. Each of these five teams plays four games against each of the other four teams in the division, three games against each of the remaining teams of the conference, and two games against each team in the other conference, for a total of* $4 \cdot 4 + 3 \cdot 9 + 2 \cdot 14 = 71$ *games. The Orlando Magic plays in the Atlantic Division. In how many different orders can the 71 games of the Magic be scheduled?*

Solution: Let us represent the 27 opponents of the Magic with 27 different letters; then each possible schedule corresponds to a word of length 71. The four letters corresponding to the other teams of the Atlantic Division will occur four times each in this word, the 9 letters corresponding to the remaining teams of the conference will occur three times each, and the remaining 14 letters, corresponding to the 14 teams of the other conference, will occur twice each. Therefore, the total number of possible schedules is

$$\frac{71!}{4!^4 \cdot 3!^9 \cdot 2!^{14}} = \frac{71!}{24^4 \cdot 6^9 \cdot 2^{14}}.$$

◊

Finally, we mention that, just as binomial coefficients had an important algebraic application—the binomial theorem—multinomial coefficients have an analogous application, called the *multinomial theorem.* The reader is invited to check Supplementary Exercise 29 for this theorem, and then the reader is invited to prove that theorem.

Quick Check

1. A frog is sitting on the number line at $x = 0$. At any point of time, the frog can jump one unit to the right, or one unit to the left, but it can never go to the left of $x = 0$. Prove that the number of sequences of $2n$ jumps that end at $x = 0$ is equal to the number of $2 \times n$ Standard Young Tableaux.
2. Prove the identity
$$\sum_{k=0}^{n} \binom{n}{k} 2^k \cdot (-1)^{n-k} = 1.$$
3. What is the number ways to list the entries 1, 1, 1, 2, 2, and 3 so that the first entry of the list is not the same digit as the last entry of the list?

1.5 The pigeonhole principle

The pigeonhole principle is almost as simple to state as the addition principle. However, it turns out to be a very powerful tool that has a plethora of surprisingly strong applications.

Theorem 1.44 (Pigeonhole principle) *Let $A_1, A_2, \cdots, A_k$ be finite sets that are pairwise disjoint. Let us assume that*
$$|A_1 \cup A_2 \cup \cdots \cup A_k| > kr.$$
Then there exists at least one index i so that $|A_i| > r$.

In other words, if the union of a few disjoint sets is "large," then at least one of those sets must also be "quite large." This is something you have probably experienced when scheduling classes for a busy week. If you want to schedule more than 20 hours of classes for a five-day week, then you will have more than four hours of classes on at least one day.

A classic way of thinking about the pigeonhole principle is by boxes and balls (which seems more humane than putting pigeons into holes). If we distribute more than kr balls in k boxes, then at least one box will have more than r balls.

The proof of the pigeonhole principle is an example of a standard proof technique. We prove the theorem by showing that its opposite is impossible. That is, we assume that the opposite of our statement is true (that is, in this case, we assume that there is no index i so that $|A_i| > r$ holds), and we derive a contradiction from this assumption. This procedure is called an *indirect proof* or *proof by contradiction.*

Proof: Let us assume that the statement we want to prove is false. Then $|A_i| \leq r$ holds for each i. Therefore,

$$|A_1 \cup A_2 \cup \cdots \cup A_k| = |A_1| + |A_2| + \cdots + |A_n| \leq kr,$$

which contradicts our original assumption that $|A_1 \cup A_2 \cup \cdots \cup A_k| > kr$. $\diamondsuit$

Example 1.45 *There are at least eight people currently living in New York City who were born in the same hour of the same day of the same year.*

While the population of New York City keeps changing day by day, it is safe to assume that it is always over 7.5 million people.

Solution: We can safely assume that all residents of New York City are no more than 120 years old. Therefore, they were born at most 120 years ago. As each year consists of at most 366 days, the people we are considering are at most $120 \cdot 366 = 43920$ days old. The number of hours in that many days is $k = 24 \cdot 43920 = 1054080$.

Now let A_1 be the set of NYC residents who were born at the first eligible hour, let A_2 be the set of NYC residents who were born at the second eligible hour, and so on, with A_k denoting the set of NYC residents who were born in the last eligible hour (that is, the hour that is ending now). Then we know that

$$|A_1 \cup A_2 \cup \cdots \cup A_k| \geq 7500000, \tag{1.11}$$

because the union of all the A_i is the population of New York City. Let us now apply the pigeonhole principle with $r = 7$. As $k = 1054080$, we see that the left-hand side is larger than $7k$, so at least one of the A_i must contain more than seven people. $\diamondsuit$

A frequently applied special case of the pigeonhole principle is when $r = 1$. In that case, the principle says that if k boxes altogether contain more than k balls, then at least one box has to contain more than one ball. Even this simple special case has interesting applications, as we will see below.

Example 1.46 *Consider the sequence* $1, 3, 7, 15, 31, \ldots$, *in other words, the sequence whose ith element is* $a_i = 2^i - 1$. *Let* q *be any odd integer. Then our sequence contains an element that is divisible by* q.

This is a rather strong statement. We did not say *anything* about q other than it is odd. Therefore, our statement holds true for $q = 17$ just as much as for $q = 2007$, or $q = 3542679$. All these numbers have a multiple that is one less than a power of two.

Solution: Consider the first q elements of our sequence. If one of them is divisible by q, then we are done. If not, then consider their remainders modulo q. That is, let us write

$$a_i = d_i q + r_i,$$

where $0 < r_i < q$, and $d_i = \lfloor a_i/q \rfloor$. As the integers $r_1, r_2, \cdots, r_q$ all come from the open interval $(0, q)$, there are $q - 1$ possibilities for their values. On the other hand, their number is q, so, by the pigeonhole principle, there have to be two of them that are equal. Say these are r_n and r_m, with $n > m$. Then $a_n = d_n q + r_n$ and $a_m = d_m q + r_n$, so

$$a_n - a_m = (d_n - d_m)q$$

or, after rearranging,

$$\begin{aligned} (d_n - d_m)q &= a_n - a_m \\ &= (2^n - 1) - (2^m - 1) \\ &= 2^m(2^{n-m} - 1) \\ &= 2^m a_{n-m}. \end{aligned}$$

As the first expression of our chain of equations is divisible by q, so too must be the last expression. Note that 2^{n-m} is relatively prime to any odd number q, that is, the largest common divisor of 2^{n-m} and q is 1. Therefore, the equality $(d_n - d_m)q = 2^{n-m}a_{n-m}$ implies that a_{n-m} is divisible by q. This completes the solution. ◊

In what follows, we will write $[n]$ for the set $\{1, 2, \cdots, n\}$, that is, the set of the first n positive integers.

Example 1.47 *Let us arbitrarily select $n + 1$ distinct integers from the set $[2n]$. Then there is at least one pair of selected integers whose sum is $2n + 1$, and there is at least one pair of selected integers whose difference is n.*

Solution: Let us split our set into n subsets, namely, the subset $\{1, 2n\}$, the subset $\{2, 2n - 1\}$, and so on, the generic subset being $\{i, 2n + 1 - i\}$, where $1 \leq i \leq n$. As we have selected $n + 1$ integers and have split $[2n]$ into only n two-element subsets, the pigeonhole principle implies that there has to be a two-element subset X so that both elements of X have been selected. The sum of the elements of X is $2n + 1$; therefore, our first claim is proved.

Now let us split $[2n]$ into the n subsets $\{1, n + 1\}$, $\{2, n + 2\}$, and so on, the generic subset being $\{i, n + i\}$, where $1 \leq i \leq n$. Again, by the pigeonhole principle, one of these n subsets, say Y, must consist of two selected integers. However, the difference between the two elements of Y is n, and our second claim is proved. ◊

Based on the previous two examples, the reader might think that the pigeonhole principle can only be applied to problems in which all relevant objects are *integers.* This is far from true, as shown by the following example.

Example 1.48 *Let p be any positive irrational number. Then there exists a positive integer n so that the distance between np and the closest integer is less than* 10^{-10}.

There is nothing magical about 10^{-10} here, we just chose it to represent "tiny number." In other words, the example claims that the multiples of any irrational number can get as close to an integer as we like. This is often expressed by saying that the set of irrational real numbers is *dense* within the set of all real numbers.

Solution: Let us represent the set of all positive real numbers by a *circle*, as shown in Figure 1.11. That is, we think of the circle as having circumference 1, and two real numbers are represented by the same point on the circle if their difference is an integer.

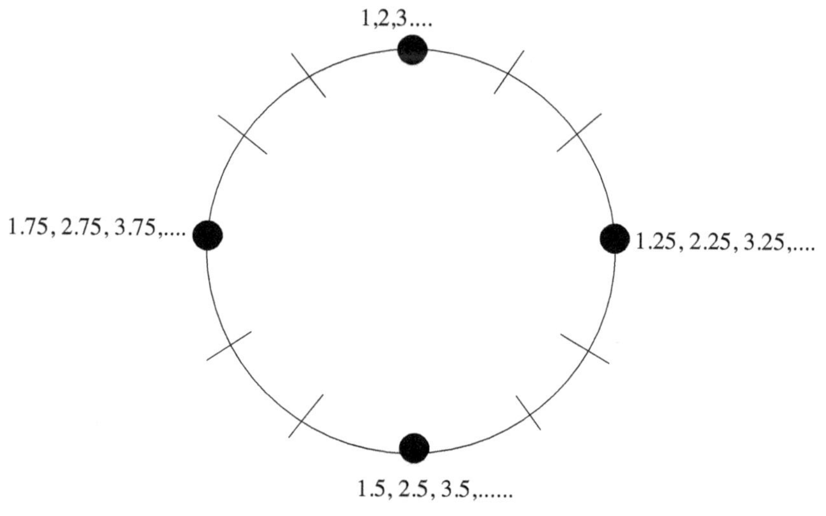

Figure 1.11
A circle representing $\mathbf{R}^+$.

Now let us divide the perimeter of the circle into 10^{10} equal parts. (Figure 1.11 shows a subdivision to 12 parts for better picture quality.) Let us take the first $10^{10} + 1$ elements of the sequence $p, 2p, 3p, \cdots$. Since there are only 10^{10} arcs, by the pigeonhole principle there has to be one arc T that contains at least two elements of this sequence. Let ip and jp be two such elements, with $i < j$. Then this implies that the distance from $(j - i)p$ to the nearest integer is less than the length of the arc T, that is, 10^{-10}. $\diamondsuit$

Quick Check

1. We select 11 distinct elements of the set $[1000]$ at random. Prove that there will be two selected elements whose difference is divisible by ten.
2. We select seven distinct elements of the set $[1000]$ at random. Prove that there will be two selected elements whose sum or difference is divisible by ten.
3. The set S consists of five positive integers, none of which has a prime divisor larger than three. Prove that S contains two elements whose product is a perfect square.

1.6 Notes

While our goal in this chapter was to introduce the reader to the most often used counting techniques, some readers may want to try to test their new enumeration skills on more difficult problems or read related literature.

For more difficult identities involving binomial coefficients, the reader can consult [54]. We have included a few challenging exercises on counting lattice paths (Exercises 25, 26, and Supplementary Exercises 35, 37). In these problems, the reader is asked to count lattice paths that do not go above a certain line with an equation $ky = x$, for some positive integer k. For the more general situation when k can be any positive *real* number, the reader should consult [79] and [80].

Catalan numbers are omnipresent in Enumerative Combinatorics, and will reappear many times in various chapters of this book. Standard Young Tableaux are also a very large topic. A recent survey, written by Ron Adin and Yuval Roichman, can be found in Chapter 14 of [18].

A crucial concept of this chapter was that of factorials, and binomial coefficients. One aspect of these numbers that we did not discuss in the text was just *how large* they are. It is certainly true that $n! < n^n$, and Exercises 24 and 11 will establish some much better bounds. An even more precise result, which we will not prove, is *Stirling's formula.* Let us say that functions $f(n)$ and $g(n)$ are *asymptotically equal* as n goes to infinity if

$$\lim_{n\to\infty} \frac{f(n)}{g(n)} = 1. \tag{1.12}$$

If f and g have this property, we will write $f \sim g$ as $n \to \infty$. With this terminology, Stirling's formula says that, if $n \to \infty$, then

$$n! \sim \left(\frac{n}{e}\right)^n \sqrt{2\pi n}. \tag{1.13}$$

1.7 Chapter review

(A) Counting principles

1. Addition principle. If $A_1, A_2, \cdots, A_n$ are disjoint finite sets, then

$$|A_1 \cup A_2 \cup \cdots \cup A_n| = |A_1| + |A_2| + \cdots + |A_n|.$$

2. Subtraction principle. If $B \subseteq A$, then

$$|A - B| = |A| - |B|.$$

3. Product principle. If $A_1, A_2, \cdots, A_n$ are finite sets, then the number of n-tuples $(a_1, a_2, \cdots, a_n)$ satisfying $a_i \in A_i$ for each i is $|A_1| \times |A_2| \times \cdots \times |A_n|$.
4. Division principle. If T and S are finite sets so that there exists a d-to-one map from $f : T \to S$, then

$$|S| = \frac{|T|}{d}.$$

(B) Enumeration formulae

1. The number of words of length k over an alphabet of n letters is n^k.
2. The number of ways to arrange n distinct objects in a line is $n!$.
3. The number of words of length k consisting of all distinct letters over an alphabet of n letters is $(n)_k$.
4. The number of k-element subsets of an n-element set is $\binom{n}{k}$.
5. The number of ways to arrange n objects, a_i of which are of type i, in a line is $\binom{n}{a_1, a_2, \cdots, a_k}$.

(C) Other important facts

1. Pigeonhole principle. If $A_1, A_2, \cdots, A_k$ are pairwise disjoint finite sets so that

$$|A_1 \cup A_2 \cup \cdots \cup A_k| > kr,$$

then there exists at least one index i so that $|A_i| > r$.
2. Identities on binomial coefficients.
   (a) $\binom{n}{k} = \binom{n}{n-k}$
   (b) $\binom{n}{k} + \binom{n}{k+1} = \binom{n+1}{k+1}$
   (c) $\sum_{k=0}^{n} \binom{n}{k} = 2^n$.

(D) Proof techniques

1. Combinatorial proof: Show that two expressions A and B are equal by showing that they count the number of elements of the same set.
2. Bijective proof: Show that the finite sets S and T have the same number of elements by constructing a bijection $f : S \to T$.
3. Indirect proof: Show that a statement is true by assuming that its contrary is true and deducing a contradiction from that assumption.

1.8 Exercises

1. An airline has two ticket counters. At a given point of time, n people stand in line at the first counter and m people stand in line at the second counter.

   (a) Let us assume that the lines are not necessarily formed on a first come, first serve basis. How many ways are there to rearrange the two lines if nobody is allowed to switch counters?

   (b) Use your answer of part (a) to justify the identity $0! = 1$.

2. Justify the convention $m^0 = 1$ in a way similar to the previous exercise.

3. Find the number of ways to place n rooks on an $n \times n$ chess board so that no two of them attack each other.

4. How many ways are there to place *some* rooks on an $n \times n$ chess board so that no two of them attack each other?

5. A long-distance running race had 15 participants, among them Amy and Bob. How many outcomes are possible if we know that Amy finished ahead of Bob?

6. In one of the Florida lottery games, one has to match six numbers out of 49 numbers to win. How many lottery tickets does one have to buy in order to be sure to have a perfect match?

7. How many four-digit positive integers are there that contain the digit 3 and are divisible by 5?

8. The door of an apartment building can only be opened by a four-digit code. A user forgot the code, but remembers that it uses each of the digits 3, 5, and 9, and no other digits. In the worst case, how many attempts will it take for the user to get into the building?

9. How many ways are there to list the digits 1, 1, 2, 2, 3, 4, 5 so that the digits 3 and 4 are not in consecutive positions?
10. A student works in a bookstore where he is required to work at least four and at most five days a week, at least one of which has to be a weekend day (Saturday or Sunday). How many different weekly work schedules can this student have?
11. A college football coach must choose four new players who will receive scholarships. He can choose among 20 incoming players, half of whom are offensive players, and half of whom are defensive players. The coach can award the scholarships in any way, as long as at least one offensive player and at least one defensive player gets a scholarship. How many possibilities does the coach have?
12. Consider the square grid we discussed in Example 1.26. How many ways are there to drive from $O = (0,0)$ to $A = (6,6)$ if we have to stop at $B = (4,4)$, but also must avoid $C = (3,1)$?
13. Prove formula (1.7).
14. Prove that for all positive integers k and n, with $k \leq n$,

$$\binom{n}{k} = \binom{k-1}{k-1} + \binom{k}{k-1} + \cdots + \binom{n-1}{k-1}.$$

15. +[1] Let n, p, and q be fixed positive integers, so that $p \leq n$, and $q \leq n$. Prove the identity

$$\binom{n}{p}\binom{n}{q} = \sum_{k=0}^{n} \binom{n}{k}\binom{n-k}{p-k}\binom{n-k}{q-k}.$$

16. Let $n = 4k+2$, for some nonnegative integer k. Prove that exactly $1/4$ of all subsets of $[n]$ have a size that is divisible by four.
17. Find a closed formula for the expression

$$\sum_{k=0}^{n} \binom{n}{k} 4^k (-1)^{n-k}.$$

18. A basketball fan looked at the newspaper for a short time and checked the standings of the seven-team division of his favorite team. Later, he tried to remember the standings, but he did not recall every detail. However, he recalled that the Lakers were in the first position, and the Sonics in the fifth position. Furthermore, he remembered that the Kings were ahead of the Trailblazers, who in turn were ahead of the Clippers. How many possibilities does that leave open for the complete standings of this division?

[1] The plus sign indicates an advanced problem.

19. The basketball fan of the previous exercise tried to remember the standings of another seven-team division, but some details escaped him again. All he could remember was that the Rockets and the Grizzlies were in consecutive positions (but he forgot in which order), and the Nuggets were behind the Spurs. How many possibilities does that leave open for the complete standings of this division?

20. Let us revisit Example 1.43 so that now we do take the venues of the games into account. That is, of the four games the Magic play against each team of its division, two have to be played at home, and two have to be played away. Of the two games played against each team of the other conference, one has to be at home, one away. Finally, the Magic can choose five of the remaining opponents whom they will play twice at home, once away, and then they will play the remaining four opponents twice away, and once at home. How many different schedules can the Orlando Magic have?

21. How many ways are there to choose subsets S and T of $[n]$ if there are no conditions whatsoever imposed on these subsets?

22. (a) How many ways are there to choose subsets S and T of $[n]$ so that S contains T?

    (b) How many ways are there to choose *disjoint* subsets R and U of $[n]$?

23. We want to form n pairs from $2n$ tennis players, to play n games. In how many ways can we do this?

24. Let $r > e$. Prove that for all $n \geq 1$, the inequality

$$n! > \left(\frac{n}{r}\right)^n$$

holds. Do not use Stirling's formula. You may want to use the fact learned in calculus that the sequence $a_n = (n/(n+1))^n$ is monotone decreasing and converges to $1/e$.

25. + Find an explicit formula for the Catalan numbers, defined (after Example 1.34) as follows:

    (a) Note that, by the subtraction principle, the number c_n is equal to the number a_n of all northeastern lattice paths from $(0,0)$ to (n,n) minus the number b_n of northeastern lattice paths from $(0,0)$ to (n,n) that go above the main diagonal at some point. Then find an explicit formula for the numbers a_n.

    (b) Find a bijection between the set S of lattice paths enumerated by b_n and the set T of all northeastern lattice paths from $(-1,1)$ to (n,n).

    (c) Find a formula for $|T|$ and apply the subtraction principle.

26. + Let k be a positive integer. Let (a, b) be a point in the first quadrant on or below the line $x = ky$. Prove that the number of northeastern lattice paths from $(0, 0)$ to (a, b) that do not *touch* the line $x = ky$, except for their origin, is
$$\binom{a+b}{a}\frac{a-kb}{a+b}.$$

27. Find a closed formula for the expression
$$\sum_{k=1}^{n}\frac{k}{n^k}\binom{n}{k}.$$

28. + Prove the identity
$$n = \sum_{k=1}^{n}\frac{k}{n^k}\binom{n+1}{k+1}.$$

29. A tennis tournament has 85 participants. Players who lose a game are immediately eliminated; players who win a game keep playing. Still, the organizers have a lot of choices to make. They could give a first-round bye to some players so that after the first round there are $2^6 = 64$ players and no more byes are needed. Or they could give even a second-round bye to the best players, or possibly even a third-round bye to the very best ones. What is the best strategy for the organizers if they want to choose the winner of the tournament using as few games as possible?

30. Sixteen players participated in a round-robin tennis tournament. Each of them won a different number of games. How many games did the player who finished sixth win?

31. Let $A_1, A_2, \cdots, A_k$ be finite sets that are not necessarily pairwise disjoint. Prove that
$$|A_1 \cup A_2 \cup \cdots \cup A_k| \le |A_1| + |A_2| + \cdots + |A_k|.$$

32. Prove that the pigeonhole principle holds, even if we do not assume that the sets $A_1, A_2, \cdots, A_k$ are pairwise disjoint.

33. + All points of the plane that have integer coordinates are colored red, blue, or green. Prove that there will be a rectangle whose vertices are all of the same color.

34. A computer program generated 175 positive integers at random, none of which had a prime divisor larger than 10. Prove that we can always find three numbers among them whose product is the cube of an integer.

1.9 Solutions to exercises

1. There are $n!$ ways for people at the first counter to form a line, and $m!$ ways for people at the second counter to form a line. Therefore, by the product principle, there are altogether $n!m!$ possibilities. Now, if $m = 0$, then the number of total possibilities is $n!$, as people at the first counter still have the same number of ways to form a line. Therefore, $m! = 0! = 1$ is the only choice that leaves the formula $n!m!$ valid.
2. Let us count words consisting of $a + b$ letters, the first a of which comes from an alphabet of n letters, and the last b of which comes from an alphabet of m letters. By the product principle, the number of these words is $n^a m^b$. In particular, if $b = 0$, then we just count a-letter words over an n-element alphabet. We know that the number of these is n^a. Therefore, in this case $n^a = n^a \cdot m^0$, yielding $m^0 = 1$.
3. There has to be one rook in each column. The first rook can be anywhere in its column (n possibilities). The second rook can be anywhere in its column except in the same row where the first rook is, which leaves $n - 1$ possibilities. The third rook can be anywhere in its column, except in the rows taken by the first and second rooks, which leaves $n-2$ possibilities, and so on, leading to $n \cdot (n-1) \cdots 2 \cdot 1 = n!$ possibilities.
4. If we place k rooks, then we first need to choose the k columns in which these rooks will be placed. We can do that in $\binom{n}{k}$ ways. Continuing the line of thought of the solution of the previous exercise, we can then place our k rooks into the chosen columns in $(n)_k$ ways. Therefore, the total number of possibilities is
$$\sum_{k=1}^{n} \binom{n}{k} (n)_k.$$
5. The number of possible outcomes is $15!/2$. Indeed, let S be the set of possible outcomes, and let T be the set of impossible outcomes, that is, those in which Bob finishes ahead of Amy. Then there is a bijection $f : S \to T$, namely, the function that simply switches Amy and Bob.
6. As the order of the numbers played does not matter, this problem simply asks for the number of all six-element subsets of [49]. As we know from Theorem 1.4, this is $\binom{49}{6} = 13983816$.
7. The number of all four-digit integers divisible by five is $9 \cdot 10 \cdot 10 \cdot 2 = 1800$, since such an integer has to end in 0 or 5. Among these, $8 \cdot 9 \cdot 9 \cdot 2 = 1296$ do not contain the digit 3, so $1800 - 1296 = 504$ numbers do contain the digit 3.

8. There will have to be one digit that is used twice, and the remaining two digits are used once. If we used the digit 3 twice, then we have to make $\binom{4}{2,1,1} = 12$ attempts. By symmetry, the same argument applies if the digit 5 or the digit 9 is repeated. Therefore, in the worst case, we will need to make $3 \cdot 12 = 36$ attempts.
9. Without the requirement that 3 and 4 are not in consecutive positions, there would be $\binom{7}{2,2,1,1,1} = 1260$ ways to list the given digits. Let us count those lists in which 3 and 4 are in consecutive positions. To do this, let us *glue* 3 and 4 together to get a superdigit, X. Then there are $\binom{6}{2,2,1,1} = 180$ possible lists. However, each of them corresponds to two original lists because X can be replaced by 34 or 43. Therefore, by the product principle, there are $2 \cdot 180 = 360$ bad lists, and so, by the subtraction principle, there are $1260-360 = 900$ good lists.
10. There are $\binom{7}{4}+\binom{7}{5} = 35+21 = 56$ ways to choose four or five days of a week. Of these, $\binom{5}{4}+\binom{5}{5} = 5+1 = 6$ will not contain any weekend days. Therefore, the number of possible schedules is $56 - 6 = 50$.
11. With no restrictions, the coach would have $\binom{20}{4}$ possibilities. The only bad choices are when all scholarships go to offensive players, or when all scholarships go to defensive players. Therefore, the number of bad choices is $2 \cdot \binom{10}{4}$, and so, by the subtraction principle, the number of good choices is $\binom{20}{4} - 2 \cdot \binom{10}{4}$.
12. Just as in Example 1.26, the number of ways to drive from O to A via B is the number of ways to drive from O to B times the number of ways to drive from B to A. In other words, it is $\binom{8}{4} \cdot \binom{4}{2} = 70 \cdot 6 = 420$. Now we have to subtract the number of bad paths among the paths we just counted, that is, the number of paths that go through the forbidden point C. These bad paths go from O to C to B to A; therefore, their number is $\binom{4}{1} \cdot \binom{4}{3} \cdot \binom{4}{2} = 96$. Consequently, the number of good paths is $420 - 96 = 324$.
13. The left-hand side is the number of all subsets of $[n]$. Indeed, when we select a subset of $[n]$, we make n decisions, and in each of these decisions, we have two choices. That is, for each element of $i\ in[n]$, we decide whether we put i into our subset or not.

    The right-hand side is the same, counted first by the *size* of the subsets. Indeed, $\binom{n}{k}$ is the number of all k-element subsets of $[n]$.

    Alternatively, 2^n is the number of all northeastern lattice paths starting at $(0,0)$ and having n steps (ending on the diagonal $x + y = n$). On the right-hand side, we count these same paths, according to their endpoints $(k, n-k)$.
14. The left-hand side is just the number of ways to go from $(0,0)$ to $(k, n-k)$ by a northeastern lattice path. The right-hand side is the

same, counted by when the *last east step is taken.* As there are altogether k east steps, the last east step has to be preceded by $k-1$ east steps and i north steps, where $0 \leq i \leq n-k$. In other words, i describes how late the last east step is taken (it is the $i+k$th step). The number of ways the i north steps and $k-1$ east steps preceding the last east step could be taken is $\binom{i+k-1}{k-1}$. Summing over all possible values of i, we get the right-hand side.

15. The left-hand side is the number of ways to choose two subsets of $[n]$, one of which is of size p, and the other one of which is of size q. The right-hand side is the same, counted by the size of the *intersection* of these two subsets.

16. We know from Supplementary Exercise 7 that the total number of subsets of $[n]$ that are of *even* size is

$$A = \binom{n}{0} + \binom{n}{2} + \binom{n}{4} + \cdots + \binom{n}{n} = 2^{n-1}.$$

As $n = 4k+2$, we see that, for even m, exactly one of m and $n-m$ is divisible by four, so exactly one of the equal numbers $\binom{n}{m}$ and $\binom{n}{n-m}$ is included in the sum

$$B = \binom{n}{0} + \binom{n}{4} + \binom{n}{8} + \cdots + \binom{n}{n}.$$

Therefore, $A = 2B$, proving that $B = 2^{n-2}$.

17. That expression is equal to 3^n, as can be seen from the binomial theorem, setting $x = 4$ and $y = -1$.

18. As the position of the Lakers and the Sonics are known, we only have to consider the positions of the remaining five teams. Let us replace each of the Kings, the Trailblazers, and the Clippers with a symbol X. These three symbols X and the two remaining teams can be arranged in $\binom{5}{3,1,1} = 5!/3! = 20$ ways. Each such arrangement corresponds to a valid standing, since the first X can be switched back to the Kings, the second X to the Trailblazers, and the third X to the Clippers. So there are ten valid standings.

19. There are 720 solutions. Glue the Rockets and the Grizzlies together. There are $6! = 720$ ways to permute the remaining six symbols. Half of these are wrong, since the Nuggets must be behind the Spurs. On the other hand, the Rockets and the Grizzlies can be unglued in two different ways, multiplying the number of good arrangements by two.

20. Let us first choose the five teams that the Magic will play twice at home and once away. The number of ways to do this is $\binom{9}{5}$.

Once this is done, we can proceed similarly to Example 1.43, except that we now need different symbols to represent home and away games against the same team. Therefore, no symbol occurs more than twice, and the number of symbols that occur twice is $4 \cdot 2 + 9 = 17$. Therefore, the total number of possible schedules is

$$\binom{9}{5} \cdot \frac{71!}{2^{17}}.$$

21. There are 2^n choices for S, there are 2^n choices for T, so there are $2^n \cdot 2^n = 4^n$ choices for (S, T).

22. (a) This number is 3^n. Indeed, each element of $[n]$ can be in T, or in S but not in T, or not in S.

    (b) This number is 3^n again. Indeed, R and U are disjoint if and only if the complement S of R contains U. The result then follows from part (a).

23. Let us ask the tennis players to form a line. This line can be formed in $(2n)!$ ways; let all those possible lines form the set T. Then, let us ask the first person in the line to play with the second, the third person in the line to play with the fourth, and so on. Let S be the set of pairings that are formed this way. We claim that each element of S corresponds to $n!2^n$ different elements of T. Indeed, if we permute the *pairs* of a line in T, the corresponding pairing in S will not change, which explains the $n!$ factor in our claim. Moreover, if we switch the $2i - 1$st and the $2i$th person of a line in T, the corresponding pairing in S will not change, which explains the 2^n factor in our claim. Therefore, by the division principle, there are

$$|S| = \frac{|T|}{d} = \frac{(2n)!}{n! \cdot 2^n}$$

ways to match the tennis players.

24. Let $x_n = \frac{n!}{(n/r)^n}$. Then

$$R_n = \frac{x_{n+1}}{x_n} = r \cdot \left(\frac{n}{n+1}\right)^n = ra_n. \tag{1.14}$$

It follows from the hint given in the exercise that each element of the sequence a_n is larger than $1/e$. Since we know that $e < r$, this means that $R_n > 1$ always, that is, the sequence $n!$ grows faster than the sequence $(n/r)^n$. Since $1! > 1/r$, the result follows.

25. (a) As we have to take $2n$ steps, n of which have to be to the east, $a_n = \binom{2n}{n}$.

    (b) Let p be a path enumerated by b_n. Then p must touch the line $y = x + 1$; let X be the first point where this happens. Our bijection f will take the initial part of p that is between $(0,0)$ and X and reflect it through the line $y = x + 1$. This will turn p into a path $f(p) \in T$. It is easy to see that f is a bijection, since each path $t \in T$ must intersect the $y = x + 1$ line and, therefore, the point X can always be recovered. See Figure 1.12 for an example of our bijection.

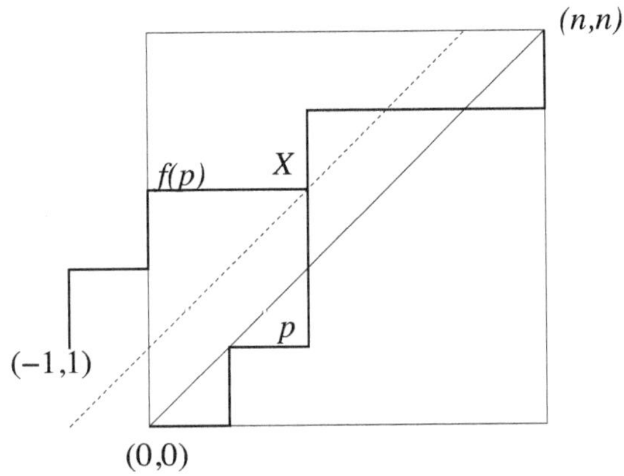

Figure 1.12
The bijection f of Exercise 25.

(c) As we have to take $2n$ steps, $n+1$ of which have to be to the east, $|T| = \binom{2n}{n+1} = \binom{2n}{n-1}$. Therefore, part (b) implies that $|S| = b_n = \binom{2n}{n-1}$, so by part (a),

$$c_n = a_n - b_n = \binom{2n}{n} - \binom{2n}{n-1} = \frac{\binom{2n}{n}}{n+1}. \tag{1.15}$$

26. This result was first proved by Bertrand [9] (who only gave the outline of a proof), and André [5], in 1887. The problem is often called the *ballot problem.*

We prove the statement by induction on $a+b$, the initial case of $a+b=0$ being easy to verify. Note that the statement is also true if $b=0$, or if $kb=a$, that is, if (a,b) is on either boundary line of the allowed domain.

By the addition principle, the number of lattice paths with the described property from $(0,0)$ to (a,b) is certainly equal to the sum of the numbers of such paths from $(0,0)$ to $(a-1,b)$ and from $(0,0)$ to $(a,b-1)$. Therefore, using the induction hypothesis, the number of paths from $(0,0)$ to (a,b) with the described property is $\binom{a-1+b}{a-1}\frac{a-k(b-1)}{a+b-1} + \binom{a+b-1}{a}\frac{a-1-kb}{a+b-1}$. So our statement will be proved if we can prove the identity

$$\binom{a-1+b}{a-1}\frac{a-1-kb}{a+b-1}+\binom{a+b-1}{a}\frac{a-k(b-1)}{a+b-1}=$$
$$\binom{a+b}{a}\frac{a-kb}{a+b}.$$

Dividing both sides by $\binom{a+b}{a}$, then multiplying by $(a+b)(a+b-1)$, produces an identity whose proof requires only routine algebra.

27. Expand $(1+x)^n$ by the binomial theorem, then take derivatives, and multiply both sides by x, to get

$$xn(1+x)^{n-1}=\sum_{k=1}\binom{n}{k}kx^k.$$

Set $x=1/n$ to get the formula

$$\left(1+\frac{1}{n}\right)^{n-1}=\sum_{k=1}^{n}\binom{n}{k}\frac{k}{n^k}. \tag{1.16}$$

28. First, using the formula $\binom{n+1}{k+1}=\binom{n}{k}+\binom{n}{k+1}$, we break up our expression on the right-hand side as

$$\sum_{k=1}^{n}\binom{n+1}{k+1}\frac{k}{n^k}=\sum_{k=1}^{n}\binom{n}{k}\frac{k}{n^k}+\sum_{k=1}^{n}\binom{n}{k+1}\frac{k}{n^k}. \tag{1.17}$$

The result of the previous exercise provides a closed formula for the first term of the right-hand side. The second term can be transformed as follows:

$$\begin{aligned}\sum_{k=1}^{n}\binom{n}{k+1}\frac{k}{n^k} &= n\sum_{k=1}^{n}\binom{n}{k+1}\frac{k}{n^{k+1}}\\ &= n\sum_{k=1}^{n}\binom{n}{k+1}\frac{k+1}{n^{k+1}}-n\sum_{k=1}^{n}\binom{n}{k+1}\frac{1}{n^{k+1}}.\end{aligned}$$

Now note that the sum in the first term of the last line is almost identical to the expression of the previous exercise, and so it is equal to $\left(1+\frac{1}{n}\right)^{n-1}-1$. The sum in the second term of that line equals $\left(1+\frac{1}{n}\right)^{n}-2$ by the binomial theorem, since the terms indexed by 0 and 1 are missing.

This shows that the last term on the right-hand side of (1.16) is equal to

$$n\left[\left(1+\frac{1}{n}\right)^{n-1}-1\right]-n\left[\left(1+\frac{1}{n}\right)^{n}-2\right]=n-\left(1+\frac{1}{n}\right)^{n-1}.$$

The proof is now immediate, since we have computed in the previous exercise that the first term of the left-hand side of (1.16) is precisely $\left(1+\frac{1}{n}\right)^{n-1}$.

29. No matter what the organizers do, it will always take 84 games to determine a winner. Indeed, 84 people need to be eliminated, and that takes 84 games.

30. The possible numbers for victories are $0, 1, 2, \cdots, 15$. This is a list of 16 numbers, so each of them must indeed occur as the number of victories of a player. Therefore, the winner won 15 games, the runner-up won 14 games, the second runner-up won 13 games, and so on, with the person finishing sixth winning 10 games.

31. The left-hand side counts each element of $\cup_{i=1}^{k} A_i$ exactly once. The right-hand side counts these same elements, but if x occurs in t different A_i, then it is counted t times. As $t \geq 1$ for all elements of $\cup_{i=1}^{k} A_i$, the statement follows.

32. The proof is very similar to that of the pigeonhole principle. Simply, (1.11) has to be replaced by the chain of inequalities

$$|A_1 \cup A_2 \cup \cdots \cup A_k| \leq |A_1| + |A_2| + \cdots + |A_n| \leq kr.$$

The previous exercise shows that this chain of inequalities indeed holds. The rest of the proof is unchanged.

33. We claim that it suffices to consider points whose first coordinate is 1, 2, 3, or 4 and whose second coordinate is at least one and at most 82. In other words, we consider points in a rectangular grid that has 82 rows of length 4 each.

As each row has four points, there are $3^4 = 81$ possible colorings for each row. Since there are 82 rows, there will be two rows whose colorings are identical. Finally, at least one color has to be repeated in one (and therefore, the other) of these rows since we can only use three colors to paint four points. Therefore, there will be four points of the same color, and they will form a rectangle.

34. The possible prime divisors of our integers are 2, 3, 5, and 7. So they are all of the form $2^a 3^b 5^c 7^d$, where the exponents are integers. As far as divisibility by 3 is concerned, each exponent can be of the form $3k$, $3k+1$, or $3k+2$. So (a, b, c, d) can be of $3^4 = 81$ different types. Therefore, by the pigeonhole principle, there are three integers among our 175 integers for which the type of (a, b, c, d) is the same. The product of these three integers will be a cube since it will contain each of 2, 3, 5, and 7 raised to an exponent that is divisible by three.

1.10 Supplementary exercises

1. How many four-digit positive integers are there in which at least one digit occurs more than once?
2. A long-distance running race had 15 participants, among them Amy, Bob, and Charlie.
   (a) How many outcomes are possible if we know that Amy finished ahead of Bob and Bob finished ahead of Charlie?
   (b) How many outcomes are possible if we only know that Bob did not finish first or last among the three of Amy, Bob, and Charlie?
3. How many ways are there to choose three subsets A, B, and C of $[n]$ that satisfy $A \subseteq B \subseteq C$?
4. How many ways are there to choose two subsets S and T of $[n]$ that are not disjoint?
5. How many ways are there to choose two subsets A and B of $[n]$ so that $A \subseteq B$, and $|B|$ is even?
6. How many ways are there to choose three subsets A, B, and C of $[n]$ that satisfy $A \subseteq C$ and $B \subseteq C$?
7. Prove, using the method of bijections, that the number of subsets of $[n]$ that consist of an odd number of elements is the same as the number of subsets of $[n]$ that consist of an even number of elements. (Hint: First solve the problem for the case when n is odd.)
8. Generalize the result of Exercise 16 for any powers of 2 instead of 4.
9. Prove that for all positive integers n, the inequality $\binom{2n}{n} < 4^n$ holds.
10. Prove that the result of the previous exercise cannot be improved. That is, prove that, if $a < 4$, then there exists an integer n so that $\binom{2n}{n} > a^n$.
11. Let $r < e$. Prove that there exists an integer N so that, if $N < n$, then
$$n! < \left(\frac{n}{r}\right)^n.$$
Do not use Stirling's formula.
12. A chess competition had six participants and one two-way tie. How many different final rankings are possible?
13. A student has to work five hours in the chemistry lab, six hours in the physics lab, and eight hours in the computer lab. She has to work full one-hour units in the physics and chemistry labs and full two-hour units in the computer lab. In how many different orders can she fulfill these obligations?

14. Give a computational proof of the identity $\binom{n}{k} + \binom{n}{k+1} = \binom{n+1}{k+1}$.

15. Give a combinatorial proof for the identity $\binom{n}{k} = \binom{n}{n-k}$ using the fact that $\binom{n}{k}$ is the number of k-element subsets of $[n]$.

16. Prove (by any method) the identity

$$\sum_{k=0}^{m} (-1)^k \binom{n}{k} = (-1)^m \binom{n-1}{m}.$$

17. (a) Prove the identity

$$\binom{n}{k} 2^{n-k} = \sum_{m=k}^{n} \binom{m}{k} \binom{n}{m}.$$

(b) Prove the identity

$$\sum_{i=k}^{t} \binom{t}{i} \binom{i}{k} = \sum_{i=k}^{t} \binom{t}{k} \binom{t-k}{i-k}.$$

18. Prove that, for all positive integers n, the integer $\binom{2n}{n}$ is *even.*

19. Let n be a power of the prime p. Prove that $\binom{n}{k}$ is divisible by p unless $k = 0$ or $k = n$.

20. Let p be a prime. Find infinitely many values of n so that $\binom{n}{k}$ is not divisible by p for any k.

21. Find a combinatorial argument to prove that, for all positive integers $k \leq n$,

$$\binom{k}{2} + \binom{n-k}{2} + k(n-k) = \binom{n}{2}.$$

22. Prove that, for all positive integers n,

$$\sum_{k=1}^{n} k \binom{n}{k} = n2^{n-1}.$$

23. Deduce identity (1.10) proved in Example 1.39 from the binomial theorem.

24. A small business has eight employees. Every day, the owner designates a team to do a certain task. The owner is careful never to designate a team that is the subset of a former team or a future team, because then some people might feel not really needed in some teams. Prove that, even being so considerate, the owner can choose a different team on each of 70 consecutive days.

25. Solve Exercise 13 using the binomial theorem.

26. We want to paint n chairs using the colors red, blue, and green. In how many ways can we do this if both the number of red chairs and the number of blue chairs have to be odd?

27. Find a closed formula for the sum

$$\sum_{k=0}^{n}(-1)^k\binom{n}{k}.$$

28. Find a closed formula for the sum

$$\sum_{k=0}^{n}\binom{2n+1}{k}.$$

29. Prove the *multinomial theorem*. That is, prove that, for any positive integers n and k,

$$(x_1+x_2+\cdots+x_k)^n = \sum_{a_1+a_2+\cdots+a_k=n}\binom{n}{a_1,a_2,\cdots,a_k}x_1^{a_1}x_2^{a_2}\cdots x_k^{a_k},$$

where the sum on the right-hand side is taken over all possible k-tuples $(a_1,a_2,\cdots,a_k)$ of nonnegative integers satisfying $\sum_{i=1}^{k}a_i = n$.

30. Find a closed formula for the expression

$$\sum_{a_1+a_2+a_3=n}\binom{n}{a_1,a_2,a_3}.$$

31. Prove that

$$\binom{n}{a_1,a_2,a_3,a_4}\le 4^n$$

for any four nonnegative integers a_1, a_2, a_3, a_4 whose sum is n.

32. A traveling salesperson has to visit seven particular major cities. She can visit these cities in any order, except Boston and New York must be consecutive stops (in any order) and Philadelphia cannot be visited before New York. How many different itineraries are possible?

33. A traveling salesperson has to visit each of five cities three times. The only requirement is that he has to finish his tour where he started. In how many ways can he do this?

34. Solve Exercise 23 without using the division principle.

35. A mayoral election in a small town attracted two candidates and $2n$ voters. The result was a tie, which upset candidate A since he never trailed candidate B. How many different voting sequences are possible?

36. + Let $a \geq b \geq 0$. Find the number of northeastern lattice paths from $(0,0)$ to (a,b) that do not go over the line $x = y$. Use the method learned in Exercise 25, not the general formula proved in Exercise 26.

37. + Let k be a positive integer, and let (a,b) be a point with non-negative integer coordinates that is on or below the line given by $ky = x$. Prove (by any method) that the number of northeastern lattice paths from $(0,0)$ to (a,b) that do not go over the line $x = ky$ is
$$\frac{a+1-kb}{a+1} \cdot \binom{a+b}{b}.$$

38. + Let p be a northeastern lattice path from $(0,0)$ to $(k, n-k)$, and let q be a northeastern lattice path from $(-1,1)$ to $(k-1, n-k+1)$. Let us assume furthermore that p and q do not touch each other. Find the number of all such pairs (p,q).

39. Prove that in any company of people, there are two people who know the same number of people in that company. (We assume that if A knows B, then B knows A.)

40. Prove that there are at least four months that have five Sundays in any given year.

41. Five friends run a race every day during the last four months of the year. There are never any ties. Prove that there are two races which end the same way.

42. Seventeen points are chosen within a square whose sides have length 1. Prove that there must be two of them whose distance from each other is less than 0.36.

43. (a) Prove that the infinite sequence of integers $1, 11, 111, 1111, \cdots$ contains an element that is divisible by 2027.

    (b) Prove that the sequence of part (a) contains infinitely many elements that are divisible by 2027.

44. Prove that in a set of 11 positive integers, none of which is divisible by 20, there are always two integers a and b so that at least one of $a+b$ and $a-b$ is divisible by 20.

45. + Let $n \geq k \leq 2n$. Let $A(n,k)$ be the set of northeastern lattice paths from $(1,0)$ to $(k, 2n-k)$. Let $C(n,k)$ be the set of northeastern lattice paths from $(1,0)$ to $(k, 2n-k)$ that touch at main diagonal $x = y$ at some point before their endpoint. (They could only touch

that diagonal at their endpoint in the special case when $k = n$.) Prove that $C(n,k) = A(n,k+1)$.

46. + Prove that the number of northeastern lattice paths that start at $(0,0)$ and end in (n,n) is equal to the number of northeastern lattice paths that start at $(0,0)$, consist of $2n$ steps, and never return to the main diagonal after their initial point.

47. + Prove the identity

$$4^n = \sum_{i=0}^{n} \binom{2i}{i} \binom{2(n-i)}{n-i}.$$

2

Applications of basic methods

2.1 Multisets and compositions

In this section, we are going to prove formulae for the number of ways a positive integer can be written as a sum of nonnegative, or positive, integers. The order of the summands will matter.

2.1.1 Weak compositions

A family of four, Anne, Benny, Charles, and Denise, sat down in their living room to watch a movie. As luck would have it, their home phone rang 10 times during the movie. Furthermore, the answering machine was broken, and the family was expecting an important call, so they answered the phone each time. In order to keep their spirits up, the family agreed that at the end of the movie each person will get a scoop of ice cream for each time that person answered the phone. The 10 scoops of ice cream will certainly be consumed as soon as the movie is over, but how many distributions are possible for those 10 scoops?

Let us start by making two observations. First, the *order* in which the four people pick up the phone does not matter; it is just the number of times each of them answers the phone that matters. This sounds as if we were dealing with a *subset* problem. However, and this is our second observation, no subset of the set $\{A, B, C, D\}$ has 10 elements, so we have to use a different model. We are looking for collections of the letters A, B, C, and D that consist of 10 letters altogether, such as $ABABBBCCDD$.

Such a collection—that is, a collection in which each element has to come from a specific set S, repetitions of elements are allowed, and the order of the elements does not matter—is called a *multiset* over the set S.

In this terminology, our task is to count all 10-element multisets over the four-element set $S = \{A, B, C, D\}$. In order to make this task easier, note that these multisets are totally determined by the *multiplicities* at which each of the four elements occurs in them. That is, if we know that a multiset contains two copies of A, one copy of B, three copies of C, and four copies of D, then we know everything about that multiset. Indeed, there is nothing else to know, since the order of the elements does not matter. This reduces our task

DOI: 10.1201/9781003304272-2

to simply finding the number of ways we can write the integer 10 as a sum of four nonnegative integers (the multiplicities of A, B, C, and D). In other words, we want to find the number of nonnegative integer solutions of the equation $a+b+c+d=10$. Note that the *order of the summands* matters in this language. Indeed, the solution $2+3+1+4=10$ corresponds to a multiset that contains two copies of A, while the solution $1+2+4+3$ corresponds to a multiset that contains one copy of A.

The concept of identifying n-element multisets over a k-element set with ways of obtaining n as a sum of k nonnegative integers leads to an important concept that has its own name.

Definition 2.1 *Let* $a_1, a_2, \cdots, a_k$ *be* nonnegative *integers satisfying*

$$\sum_{i=1}^{k} a_i = n.$$

Then the ordered k-tuple $(a_1, a_2, \cdots, a_k)$ *is called a* weak composition *of n into k parts.*

The discussion above explains that there is a natural bijection from the set of weak compositions of $[n]$ into k parts and n-element multisets over a k-element set. Therefore, we might as well look for the number of weak compositions of n into k parts instead of the number of n-element multisets over k.

These objects are a little bit more difficult to count than k-element subsets of $[n]$, but the formula enumerating them also consists of a binomial coefficient.

Theorem 2.2 *The number of weak compositions of n into k parts is*

$$\binom{n+k-1}{n} = \binom{n+k-1}{k-1}.$$

Proof: Consider n identical balls that have to be distributed into k boxes that are numbered 1 through k. Each such distribution is equivalent to a weak composition of n into k parts, the number of balls in box i being part i.

Now, count the ways of distributing the n balls into the k boxes as follows: Arrange the k boxes in a line so that the numbers on the boxes are increasing order, and boxes i and $i+1$ are separated by a wall (see Figure 2.1).

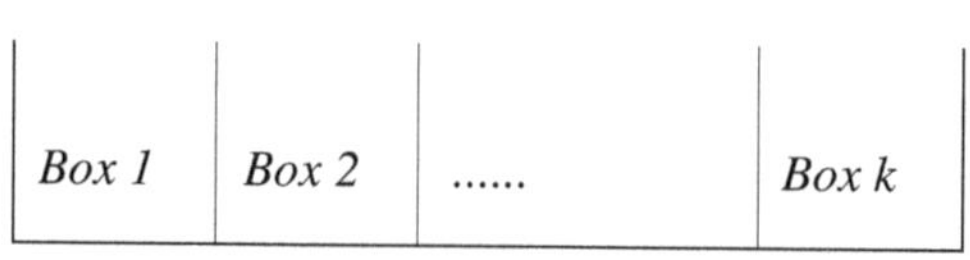

Figure 2.1
Lining up our boxes.

Observe that each distribution corresponds to a unique arrangement of n balls and $k-1$ walls in a line. Indeed, we can think of the walls (except for the left wall of the leftmost box and the right wall of the rightmost box) as objects that were inserted into the line of balls at various points in order to separate some balls from others. Conversely, each arrangement corresponds to a unique distribution of balls in the boxes. Indeed, if the ith and $i+1$st walls are inserted so that there are t balls between them, then there will be t balls in the ith box.

The number of ways to arrange n balls and $k-1$ walls in a line, by Theorem 1.41, is $\binom{n+k-1}{k-1}$. Therefore, this is the number of weak compositions of n into k parts. ♢

Corollary 2.3 *The number of n-element multisets over a k-element set is*

$$\binom{n+k-1}{n} = \binom{n+k-1}{k-1}.$$

Therefore, the four-member family we mentioned at the beginning of this chapter can distribute the 10 scoops of ice cream in $\binom{10+4-1}{4-1} = \binom{13}{3} = 286$ ways.

It is important to note that, in a weak composition, parts *are* allowed to be zero, just as for a multiset S over a set U there could be elements of U that do not occur in S.

We point out that there is an alternative way to prove Theorem 2.2, one that also exemplifies an important technique. The reader should read Exercise 2 and try to complete the proof started in that exercise.

2.1.2 Compositions

At this point, you may be wondering what exactly is *weak* about a weak composition. To stay with our example of a family answering the phone, a weak composition allows for some family members being so "weak" that they *never* pick up the phone. To put it more formally, the *conditions* on the parts of a weak composition are weak in that they allow for the parts to be zero. This is in contrast to our next notion, that of *compositions*.

Definition 2.4 *Let $a_1, a_2, \cdots, a_k$ be* positive *integers satisfying $\sum_{i=1}^{k} a_i = n$. Then the ordered k-tuple $(a_1, a_2, \cdots, a_k)$ is called a* composition *of n into k parts.*

It is not difficult to deduce a formula for the number of compositions of n into k parts from the corresponding formula for weak compositions.

Corollary 2.5 *The number of compositions of n into k parts is $\binom{n-1}{k-1}$.*

Proof: There is a bijection f from the set W of all weak compositions of $n-k$ into k parts into the set C of all compositions of n into k parts. This bijection simply adds one to each part, assuring that each part will become positive. For example, if $a=(3,0,2)$, then $f(a)=(4,1,3)$. The reader is asked to verify that f is a bijection. Therefore, $|W|=|C|$, and our claim follows from Theorem 2.2. Indeed, replacing n by $n-k$ in the result of that corollary yields

$$|C| = |W| = \binom{n-k+k-1}{k-1} = \binom{n-1}{k-1}.$$

◊

What can we say if *any* number of parts is allowed? The reader is invited to use the result of Corollary 2.5 to find a closed formula (that is, one with no summation signs) for the number of *all* compositions of n. We provide two solutions in Exercise 3. The obtained simple formula may encourage the reader to look for a similar formula for the number of all weak compositions of n into any number of parts, but that is a bird of a different feather. Readers should try to explain why that is, then check their answers in the solution of Exercise 4.

Quick Check

1. Find the number of compositions of n in which the first entry is more than 1.
2. Find the number of weak compositions of 20 into four parts that are identical to their reverses.
3. Find the number of compositions of 11 (into any number of parts) that are identical to their reverses.

2.2 Set partitions

Now we are going to consider a more difficult enumeration problem. Accordingly, the counting formulas that we obtain will be more complicated.

2.2.1 Stirling numbers of the second kind

On a day when we felt particularly brave, we invited some friends of our children to play in our house. Altogether, five children were present. Soon it turned out that no room was large enough for all of them, and therefore they split into the three available rooms, using all three of them. In how many different ways could the children do that?

If you try to answer this question, you soon recognize that it is not asked in a precise way. What is missing is the definition of *different*. That is, do we consider the setup where A and B play in one room, C and D in another room, and E in the remaining room *different* from the setup where E plays in the first room, A and B in the second room, and C and D in the last room?

There are no good and bad answers to this question; both interpretations lead to valid and interesting combinatorial problems. However, the two problems are very closely related. Indeed, it is easy to see that if we consider the three rooms different, the answer will be exactly $3! = 6$ times as much as in the case when the rooms are considered identical. This is because we can permute the rooms in that many ways.

Because of this very close connection between the two problems, it suffices to solve the one in which the rooms are considered identical. That is, it is only the playmates that matter. In other words, two playing arrangements are considered different if there is at least one child whose playmates are not the same in the two arrangements. This situation occurs so often in mathematics that it has its own name.

Definition 2.6 *Let n be a positive integer, and let $k \leq n$ be a positive integer. Let $B = \{B_1, B_2, \cdots, B_k\}$, where $B_i \subseteq [n]$ for all $i \in [k]$, the B_i are nonempty and pairwise disjoint, and $\cup_{i=1}^{k} B_i = [n]$. Then we say that B is a* partition *of $[n]$ into k blocks.*

Example 2.7 *There are six partitions of $[4]$ into three blocks, namely,*

- $\{\{1,2\},\{3\},\{4\}\}$,
- $\{\{1,3\},\{2\},\{4\}\}$,
- $\{\{1,4\},\{2\},\{3\}\}$,
- $\{\{2,3\},\{1\},\{4\}\}$,
- $\{\{2,4\},\{1\},\{3\}\}$, *and*
- $\{\{3,4\},\{1\},\{2\}\}$.

Note that the blocks are sets, so the order of elements within each block does not matter. Also note that in Definition 2.6, B itself is a set, so the order of blocks does not matter either. Therefore, it is no surprise that the number of partitions of $[4]$ into three blocks is six. Indeed, all such partitions will have one doubleton and two singleton blocks, so once we know which elements are in the doubleton block, we know the partition. Since there are $\binom{4}{2} = 6$ ways to choose the doubleton block, our claim follows.

Let us practice this type of counting by answering our original question.

Example 2.8 *The number of partitions of $[5]$ into three blocks is 25.*

Solution: There are two ways we can partition $[5]$ into three blocks. The block sizes can be either 3, 1, 1 or 2, 2, 1. In the first case, we need to select the elements that go into the three-element block in $\binom{5}{3} = 10$ ways, and the remaining elements must form singleton blocks.

In the second case, we have five choices for the element that goes into the singleton block. The next step is trickier. We have $\binom{4}{2} = 6$ choices for the elements of *one* of the doubleton blocks. So a superficial line of thought would conclude that, by the product principle, there are $5 \cdot 6 = 30$ possibilities in the second case. This is wrong, however, as each partition got counted twice. Indeed, if first we select 1 for the singleton block, then $\{2, 3\}$ for one of the doubleton blocks, we get the partition $\{\{1\}, \{2, 3\}, \{4, 5\}\}$. However, we would have obtained the same partition if we selected 1 first, then $\{4, 5\}$. This is because the order of the two doubleton blocks is insignificant; it does not matter whether a block is chosen second or last. Therefore, by the division principle, there are $(5 \cdot 6)/2 = 15$ partitions in the second case.

Finally, by the addition principle, the total number of partitions of $[5]$ into three blocks is $10 + 15 = 25$. $\diamond$

So there are 25 ways for the five children to split into three identical rooms, using all three of them. If the three rooms are different, then we can permute the groups of children in six different ways among the rooms, resulting in $6 \cdot 25 = 150$ possibilities.

What can we say in general? That is, what is the number of partitions of $[n]$ into k blocks? We will deduce a formula for that number in a later section of this chapter, when we learn the principle of inclusion and exclusion. For now, we introduce a name for the number in question.

Definition 2.9 *Let n and k be positive integers. Then the number of partitions of $[n]$ into k blocks is denoted by $S(n, k)$ and is called a* Stirling number of the second kind.

It follows from this definition that $S(n, k) = 0$ when $k > n$. We set $S(n, 0) = 0$ if $n > 0$ (since, if there are some elements to partition into blocks, then there will be at least one block) and $S(0, 0) = 1$. Our two previous examples showed that $S(4, 3) = 6$ and $S(5, 3) = 25$. The reader is asked to verify that, for all positive integers n, we have $S(n, 1) = S(n, n) = 1$.

This is all nice and good, you could say, but why Stirling numbers of the *second* kind? What happened to Stirling numbers of the first kind? This is another question we need to postpone a little bit, this time until our chapter on permutations. As a preparation, you should solve Exercise 11 which shows a fascinating property of the Stirling numbers of the second kind.

2.2.2 Recurrence relations for Stirling numbers of the second kind

Although we cannot prove an explicit formula for $S(n, k)$ yet, we can prove some recurrence relations for these numbers. These recurrence relations, and their proofs, will deepen our understanding of partitions, Stirling numbers, and their many occurrences in combinatorics. Let us start with the most frequently used recurrence.

Theorem 2.10 *For all positive integers n and k satisfying $n \geq k$, the equality*

$$S(n,k) = S(n-1,k-1) + kS(n-1,k)$$

holds.

Proof: We show that both sides count the same objects. The left-hand side counts all partitions of $[n]$ into k blocks. The first term of the right-hand side counts all partitions of $[n]$ into k blocks in which the element n forms a block by itself. Indeed, in these partitions, the remaining $n-1$ elements are partitioned into $k-1$ blocks. Finally, the second term of the right-hand side counts all partitions of $[n]$ into k blocks in which the element n does not form a block by itself. Indeed, in these partitions, the remaining $n-1$ elements have to be partitioned into k blocks, then the element n has to be put into one of these k blocks. Therefore, by the addition principle, the two sides enumerate the same set, so they must be identical. $\diamondsuit$

The values of the Stirling numbers of the second kind for the first few nonnegative integers n are shown in Figure 2.2.

Now we are ready to prove that the following problem, which looks unrelated at first sight, also leads to the numbers $S(n,k)$. This problem is a little bit more difficult to state than our earlier problems, but we decided to include it in order to show that Stirling numbers of the second kind indeed occur at unexpected places.

Let $1 \leq k \leq n$, and let k and n be positive integers. Let $h(n,k)$ be the sum of all $\binom{n-1}{k-1}$ products that consist of $n-k$ factors, so that all these factors are elements of $[k]$. Repetition of factors is allowed, but the factors of each product are to be written in increasing order. Note that the definition of $h(n,k)$ means that $h(n,n)$ is a product with no factors, which we will set to be equal to 1.

n=0	1					
n=1	0	1				
n=2	0	1	1			
n=3	0	1	3	1		
n=4	0	1	7	6	1	
n=5	0	1	15	25	10	1

Figure 2.2
Values of $S(n,k)$ for $n \leq 5$.

Theorem 2.11 *Let $h(n,k)$ be defined as above. Then*

$$h(n,k) = S(n,k).$$

As the definition of the numbers $h(n,k)$ is not simple, we will provide two examples before starting the proof of Theorem 2.11.

Example 2.12 *Let $n = 4$ and $k = 2$. Then*

$$h(4,2) = 1 \cdot 1 + 1 \cdot 2 + 2 \cdot 2 = 7 = S(4,2).$$

Example 2.13 *Let $n = 4$ and $k = 3$. Then*

$$h(4,3) = 1 + 2 + 3 = 6 = S(4,3).$$

The result of Theorem 2.11 is fairly surprising. After all, the definition of the numbers $h(n,k)$, while very nice and symmetric, had nothing to do with partitions. Nevertheless, the following lemma shows that the numbers $h(n,k)$ and the numbers $S(n,k)$ share a very important property: They satisfy the same recurrence relation.

Lemma 2.14 *For all positive integers $k \leq n$,*

$$h(n,k) = h(n-1,k-1) + kh(n-1,k).$$

Proof: (of Lemma 2.14) By definition, the left-hand side is the sum of all $(n-k)$-factor products whose factors come from the set $[k]$. These products can be split into two classes: those that contain the factor k, and those that do not. Those that do contain k are the products of k and an $(n-k-1)$-factor product over the set $[k]$. Summing all products of this first class, we get $kh(n-1,k)$. Those products that do not contain k are in fact taken over the set $[k-1]$, so their sum is $h(n-1,k-1)$. ◊

Example 2.15 *Again, let $n = 4$ and $k = 2$. As we saw in Example 2.12, the three products appearing in $h(4,2)$ are $1 \cdot 1$, $2 \cdot 1$, and $2 \cdot 2$. Of these, two contain 2. Their sum is*

$$2 \cdot 1 + 2 \cdot 2 = 2 \cdot (1+2) = 2 \cdot h(3,2).$$

The remaining product, $1 \cdot 1$, does not contain 2, so it is in fact taken over the set $[1]$, and is equal to $h(3,1)$. Indeed, $h(3,1)$ is defined as the sum of all two-factor products over the set $[1]$.

Now we are in a position to prove Theorem 2.11.

Proof: (of Theorem 2.11) We prove the statement by induction on $n+k$. For $n+k=0$, we note that $h(0,0) = S(0,0) = 1$, and the statement is true. Now assume we know that $S(n, k) = h(n, k)$ if $n + k \leq m$, and let $n + k = m + 1$. Then

$$\begin{aligned} S(n,k) &= S(n-1,k-1) + kS(n-1,k) \\ &= h(n-1,k-1) + h(n-1,k) \\ &= h(n,k). \end{aligned}$$

Note that the three equalities above hold because of Theorem 2.10, the induction hypothesis, then Lemma 2.14. This completes the proof. $\diamondsuit$

We would like to point out that we used a particular version of the method of mathematical induction here. We proved that two arrays of numbers were equal in two steps, namely,

1. first we showed that they satisfied the same initial conditions, and
2. then we showed that they satisfied the same recurrence relation.

We will use this method several times to solve upcoming problems.

There is another interpretation of the Stirling numbers of the second kind, and that interpretation does not need Theorem 2.10. The reader is invited to solve Exercise 11 for this interesting interpretation.

Taking a look at Figure 2.2, we see that Theorem 2.10 provides us with a *triangular* recurrence for the numbers $S(n, k)$. That is, the three Stirling numbers involved in that recurrence relation form a triangle in Figure 2.2.

Let us now prove a *vertical* recurrence, that is, one that uses elements of the same NE-SW diagonal of the triangle shown in Figure 2.2 to compute an entry of the following diagonal.

Theorem 2.16 *For all positive integers n and k satisfying $n \geq k$,*

$$S(n+1,k) = \sum_{i=0}^{n} \binom{n}{i} S(n-i,k-1).$$

You could ask what is so great about computing $S(n + 1, k)$ from $n + 1$ different values of the form $S(n-i, k-1)$ when Theorem 2.10 could accomplish the same using just *two* values. The answer is that, in Theorem 2.16, all the values we need are of the form $S(n-i, k-1)$, that is, their second argument is always $k-1$. So if we have a program that can compute the values of $S(m, k-1)$ for some fixed k and any m, then Theorem 2.16 can turn those values into values of $S(m, k)$ for any m. Theorem 2.10 could not be used for the same purpose because in that theorem the two values we needed were $S(n-1, k-1)$ and $S(n-1, k)$. (It is nevertheless possible to use that recurrence to write a computer program to compute $S(n, k)$; one would just need to write a different program.)

Proof: (of Theorem 2.16) We prove that both sides count partitions of $[n+1]$

into k blocks. For the left-hand side, this follows from the definition of Stirling numbers of the second kind. On the right-hand side, the term $\binom{n}{i}S(n-i,k-1)$ counts the partitions of $[n+1]$ into k blocks in which the element $n+1$ is in a block of size $i+1$. Indeed, there are $\binom{n}{i}$ ways to choose the i elements that will share a block with $n+1$; then there are $S(n-i,k-1)$ ways to partition the remaining elements into $k-1$ blocks. The proof then follows by the addition principle. ◊

The reader should take a minute and explain why the above proof works even in the special cases when on the right-hand side one or both arguments of $S(n-i,k-1)$ are equal to 0. This will provide additional justification for our decisions to set $S(n,0)=0$ for positive n and to set $S(0,0)=1$.

2.2.3 When the number of blocks is not fixed

What can we say about the number of *all* partitions of the set $[n]$? We promised that we would soon prove a formula for the numbers $S(n,k)$. Once that is done, we will be able to compute the number of all partitions of n by taking the sum $\sum_{k=1}^{n} S(n,k)$. Until then, we will treat the numbers of all partitions of $[n]$ as we treated the numbers of partitions of $[n]$ into k blocks, that is, we will give them a name and prove a recurrence relation for them.

Definition 2.17 *The number of all partitions of* $[n]$ *is denoted by* $B(n)$ *and is called a* Bell number.

The first few Bell numbers are shown in Figure 2.3.

n	$B(n)$
1	1
2	2
3	5
4	15
5	52
6	203
7	877
8	4140

Figure 2.3
The values of $B(n)$ for $n \leq 8$.

Just as for Stirling numbers of the second kind, there are several explicit formulae for the Bell numbers. They are beyond our reach at this point. We can, however, easily prove the following recurrence relation.

Theorem 2.18 *Set $B(0) = 1$. Then, for all positive integers n,*

$$B(n+1) = \sum_{k=0}^{n} B(k)\binom{n}{k}.$$

The reader should try to prove this result on his own, following the line of thinking we used in the proof of Theorem 2.16. We provide a proof in the solution to Exercise 15.

Quick Check

1. Find the number of partitions of $[n]$ into two blocks in which 1 and 2 are in different blocks.
2. Prove that for all positive integers $k \leq n$, the inequality $S(n,k) \leq k^n$ holds.
3. Prove that for all positive integers $n \geq 3$, the inequality $B(n) < n!$ holds.

2.3 Partitions of integers

In this section, we return to the study of the number of ways to write a positive integer as the sum of positive integers. However, the parts will have to be in nonincreasing order.

2.3.1 Nonincreasing finite sequences of positive integers

Last time Ms. Smith moved, she had to unload a truck, which involved carrying 20 identical medium-sized boxes into her new house. She wanted to be done quickly and tried to carry several boxes at the same time. However, as time passed, she became more and more tired, so she was happy if she could carry the same number of boxes as on the previous trip from the truck to the house. She could not even think of increasing the number of boxes carried at any given point in time. In how many different ways could she carry the 20 boxes into her new house?

This problem may remind the reader of compositions. Indeed, as the boxes are all identical, we are only concerned with their number. That is, we would like to decompose 20 into a sum of positive integers. However, in contrast to compositions, the *order* of the parts has to be nonincreasing. So $12+8$ is acceptable, but $3+4+1+10$ is not. This shows that we are dealing with a different (and we will see, much more difficult) problem than in the previous section. The following definition starts setting the framework for handling this new kind of enumeration problem.

Definition 2.19 *If a finite sequence* $(a_1, a_2, \ldots, a_k)$ *of positive integers satisfies* $a_1 \geq a_2 \cdots \geq a_k$ *and* $a_1 + a_2 + \cdots + a_k = n$, *then we call that sequence a* partition *of the integer* n.

Note that, somewhat regrettably, the word *partition* is used both for partitions of the integer n and for partitions of the set $[n]$, which are, of course, very different notions. Some other languages solved this problem by using different words for these two notions. When reading in English, the reader should be careful not to confuse these two concepts. We try to help that effort by stressing that n is an integer, but $[n]$ is a set.

The number of partitions of the integer n is denoted by $p(n)$.

Example 2.20 *The positive integer* 4 *has five partitions, namely,* (4), $(3,1)$, $(2,2)$, $(2,1,1)$, *and* $(1,1,1,1)$. *Therefore,* $p(4) = 5$.

See Figure 2.4 for the first few values of $p(n)$.

n	$p(n)$
1	1
2	2
3	3
4	5
5	7
6	11
7	15
8	22

Figure 2.4
The values of $p(n)$ for $n \leq 8$.

At this point, the reader probably cannot wait for us to prove an elegant formula for $p(n)$. While an exact formula for $p(n)$ actually exists, it is by no means simple, to say the least. It involves an infinite sum, complex numbers, the function sinh, and a lot more. The formula was obtained, in various forms, by Hardy, Ramanujan, and Rademacher. The interested reader should consult Chapter 5 of the book by George Andrews, *Theory of Partitions* [6], or the book by George Andrews and Kimmo Eriksson, *Integer Partitions* [7], for details.

While a detailed discussion of this formula for $p(n)$ is way beyond the scope of this book, let us mention at least one reason for which this enumeration problem is more difficult than the previous ones. In the previously discussed problems, the enumeration formulae that we proved were either *exponential* (or even larger) in the relevant variable n, such as 2^{n-1}, or $n!$, or k^n, or *polynomial*, such as $(n)_k$, or $\binom{n}{k}$. However, we will explain in Exercises 19 and 20 that $p(n)$ grows faster than any polynomial function $q(n)$. It would be slightly more cumbersome, but still not too difficult, to prove that, on the other hand, $p(n)$ grows slower than any exponential function a^n, for $a > 1$. (One rough upper bound for $p(n)$ is provided by the number of all compositions of n.) Therefore, $p(n)$ is neither exponential nor polynomial, implying that it must be a more "exotic," less well-known function.

The following theorem, which we will not prove, tells us more about the growth rate of $p(n)$. Recall the notation $\sim$, which was introduced immediately following formula (1.12).

Theorem 2.21 *As $n \rightarrow \infty$, the function $p(n)$ satisfies*

$$p(n) \sim \frac{1}{4\sqrt{3}} \exp\left(\pi\sqrt{\frac{2n}{3}}\right). \tag{2.1}$$

At this point the reader might say that this formula does contain an exponential expression, whereas we promised that $p(n)$ will be smaller than all exponential functions of n. Note that the right-hand side of the above formula is an exponential function of $\sqrt{n}$, and not n. No matter how small a constant a is, the function e^{an} will always grow faster than the right-hand side of the above formula and, therefore, faster than $p(n)$.

Chapter 5 of [6] explains how to deduce this asymptotic formula from the mentioned complicated exact formula for $p(n)$.

At this point, the reader could say that maybe it was too ambitious to start with the quest for a formula of $p(n)$, the number of all partitions of n, without first attacking the problem of counting the partitions of n into a *given* number of parts. This is essentially equivalent to the problem of counting partitions of n into *at most* k parts. Let the number of such partitions be $p_k(n)$. Then Exercises 19 and 20 show that, for any fixed k, the function $p_k(n)$ is a function that is very close to a polynomial. The reader is invited to solve those exercises in order to make that statement more precise. In the rest of this section, we will concentrate on a fascinating proof technique applicable to integer partitions, that of *Ferrers shapes*.

2.3.2 Ferrers shapes and their applications

Ferrers shape representation is a simple but powerful tool for visualizing partitions of integers. Ferrers shapes (or Ferrers diagrams, or Ferrers boards) are named in honor of the American mathematician Norman MacLeod Ferrers, who was one of the first people to work with these objects. The Ferrers shape of the partition $(a_1, a_2, \cdots, a_k)$ is a diagram consisting of squares that form part of a rectangular grid. The first row of the Ferrers shape consists of a_1 squares (or boxes), the second row consists of a_2 squares, and so on, ending with the last, kth row, that consists of a_k boxes.

Example 2.22 *The Ferrers shape of the partition* $(5, 3, 2)$ *is shown in Figure 2.5.*

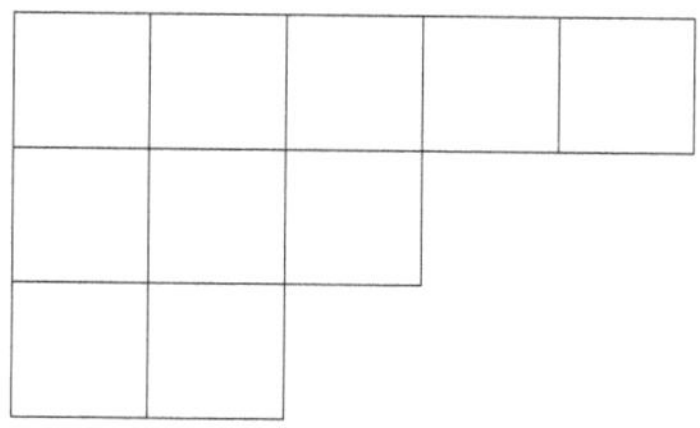

Figure 2.5
The Ferrers shape of $(5, 3, 2)$.

One great advantage of the Ferrers shape representation of partitions is that it defines an elegant involution on the set of all partitions that would be far more cumbersome to define otherwise. Recall that an involution is a map whose square is the identity map. The involution in question maps each Ferrers shape into its *conjugate*, that is, its reflected image through the main (NW-SE) diagonal. With a slight abuse of terminology, we will also talk about the conjugate of a *partition*, not just its Ferrers shape. See Figure 2.6 for an example.

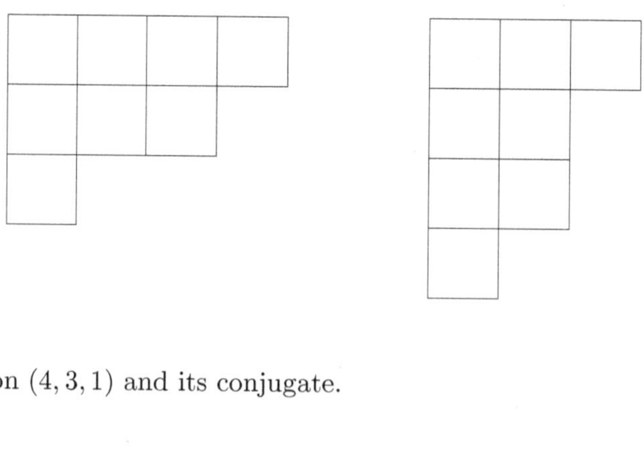

Figure 2.6
The partition $(4, 3, 1)$ and its conjugate.

Example 2.23 *As we can see in Figure 2.6, the conjugate of the partition* $(4,3,1)$ *is the partition* $(3,2,2,1)$.

Let us use the notion of conjugate partitions to prove some identities that would be more difficult to prove otherwise.

Proposition 2.24 *For all positive integers* $k \leq n$*, the number of partitions of* n *that have at least* k *parts is equal to the number of partitions of* n *in which the largest part is at least* k.

For instance, if $n = 5$ and $k = 3$, then the number of partitions of n that have at least k parts is 4 (for the partitions $(3,1,1)$, $(2,2,1)$, $(2,1,1,1)$, and $(1,1,1,1,1)$), and the number of partitions of n in which the largest part is at least k is also 4 (for the partitions (5), $(4,1)$, $(3,2)$, and $(3,1,1)$).

Proof: A partition p of n that has at least k parts corresponds to a Ferrers shape on n boxes that has at least k rows. A partition p' of n whose largest part is at least k has at least k columns (for its first row is of length at least k). So, taking conjugates establishes a bijection between these two classes of partitions, showing that they are equinumerous. $\diamondsuit$

Proposition 2.25 *For every positive integer* n*, the number of partitions of* n *in which the first two parts are equal is equal to the number of partitions of* n *in which each part is at least 2.*

Proof: The first two parts of a partition of n are equal if the first two *rows* of its Ferrers shape are equal. Each part of a partition of n is at least 2 if the first two *columns* of its Ferrers shape are equal. Therefore, the same trick applies; taking conjugates establishes a bijection between these two classes of partitions, showing that they are equinumerous. $\diamondsuit$

Taking conjugates, however useful, is not the only trick that helps us prove partition identities. The following lemma will help the reader solve one of the most interesting problems of this chapter. The lemma might sound a little bit technical at first, but it is worth the effort. It will lead us to an interesting (and more than 200-year-old) theorem by Euler, his famous *pentagonal number theorem.*

Lemma 2.26 *Let* $m > k \geq 1$*. Let* S *be the set of partitions of* n *into* m *parts, the smallest of which is equal to* k*, and let* T *be the set of partitions of* n *into* $m-1$ *parts, in which the* k*th part is larger than the* $(k+1)$*st part and the smallest part is at least* k.

Then $|S| = |T|$.

Note that, if $k = m-1$, then in the definition of T we will say that the $(k+1)$st part of our partition is 0 (as in that particular case, partitions in T have only k parts).

Example 2.27 *Let* $n = 10$*,* $m = 4$*, and* $k = 2$*. Then the set* S *consists of the partitions* $(4, 2, 2, 2)$ *and* $(3, 3, 2, 2)$*, while* T *consists of the partitions* $(5, 3, 2)$ *and* $(4, 4, 2)$*.*

Proof: (of Lemma 2.26) We define a map $f : S \to T$ as follows: Let $s \in S$. Remove the last row of the Ferrers shape of s, and distribute the obtained k boxes equally to the ends of the first k rows of the remaining shape. The obtained Ferrers shape will still have n boxes, and its kth row will be strictly longer than its $(k + 1)$st row, since the former got a new box and the latter did not. Its last row has at least k boxes since it could not be shorter than the removed row. Let this new shape be the Ferrers shape of $f(s)$.

We claim that f is indeed a bijection from S to T. We have seen in the previous paragraph that $f(s) \in T$ for all $s \in S$. What is left to do is to show that f has an inverse. That is, if $t \in T$, then there exists exactly one $s \in S$ so that $f(s) = t$. In other words, we can *recover* the original partition from its image. This is not terribly difficult. Take any $t \in T$. Remove a box from the end of each of its first k rows. We can do this as $t \in T$. Then form a new row from these k boxes and add it to the bottom of the shape of t. If we call the partition corresponding to this new shape s, then $f(s) = t$. Furthermore, for any other shape s', the shapes $f(s)$ and $f(s')$ cannot be identical. Indeed, the reader is invited to observe that if row i of s and s' are different, then so are row i of $f(s)$ and $f(s')$. So f is indeed a bijection, proving our claim. $\Diamond$

Figure 2.7 illustrates our map f in the case discussed in Example 2.27, that is, when $n = 10$, $m = 4$, and $k = 2$.

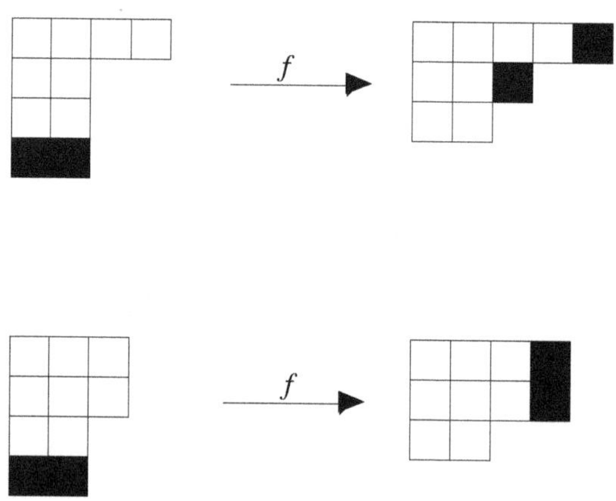

Figure 2.7
How f reuses the boxes of the last row.

2.3.3 Excursion: Euler's pentagonal number theorem

In this section, we cover our most involved theorem so far. This is the first theorem in this book that is complicated enough to have been named after its author, who was probably the most prolific mathematician of all time. The proof of the theorem will be somewhat more complicated than that of our previous results, but it is well worth the effort. We will see a great mathematician at work.

To start, let us compare the number $p_{d,odd}(n)$ of partitions of n into *an odd number of parts, all distinct,* and the number $p_{d,even}(n)$ of partitions of n into *an even number of parts, all distinct.* For $n = 3$ and $n = 4$, both of these numbers will be equal to 1. Then, for $n = 5$, we find that $p_{d,odd}(5) = 1$, but $p_{d,even} = 2$ (for the partitions $(4, 1)$ and $(3, 2)$). For $n = 6$, we find $p_{d,odd}(6) = 2$ (for the partitions (6) and $(3, 2, 1)$). Interestingly, we also find $p_{d,even}(6) = 2$ (for the partitions $(4, 2)$, and $(5, 1)$). For $n = 7$, we find $p_{d,odd}(7) = 2$ (for the partitions (7) and $(4, 2, 1)$), but $p_{d,even}(7) = 3$ (because of $(6, 1)$, $(5, 2)$, and $(4, 3)$). From all these data, we might suspect either that $p_{d,odd}(n)$ and $p_{d,even}(n)$ are equal or that the latter is 1 larger. This is not quite true, as we can see by looking at the trivial cases of $n = 1$ and $n = 2$. In those cases, $p_{d,odd} = 1$ and $p_{d,even} = 0$, so the *former* is 1 larger. This happens again in the case of $n = 12$, when $p_{d,odd} = 8$ (for the partitions (12), $(9, 2, 1)$, $(8, 3, 1)$, $(7, 4, 1)$, $(6, 5, 1)$, $(7, 3, 2)$, $(6, 4, 2)$, and $(5, 4, 3)$), while $p_{d,even} = 7$ (for the partitions $(11, 1)$, $(10, 2)$, $(9, 3)$, $(8, 4)$, $(7, 5)$, $(6, 3, 2, 1)$, and $(5, 4, 2, 1)$).

If we write a computer program and compute the numbers $p_{d,odd}(n)$ and $p_{d,even}(n)$ for the first 34 positive integers n, we see that the two numbers will be equal *most of the time.* In fact, we find that

$$p_{d,even}(n) - p_{d,odd}(n) = \begin{cases} -1 \text{ if } n = 1, 2, 12, 15, \\ 1 \text{ if } n = 5, 7, 22, 26, \\ 0 \text{ otherwise.} \end{cases}$$

These data are explained by the famous theorem of Euler that we are about to state and prove. The numbers for which $p_{d,even}(n) - p_{d,odd}(n) \neq 0$ are the *pentagonal numbers*, that is, positive integers n that satisfy the equality $n = j(3j \pm 1)/2$ for some positive integer j. We will explain this terminology soon. Now we are ready to state Euler's theorem.

Theorem 2.28 (Pentagonal number theorem) *Let n be a positive integer. Then*

$$p_{d,even}(n) - p_{d,odd}(n) = \begin{cases} 0 \textit{ if } n \textit{ is not pentagonal,} \\ \\ (-1)^j \textit{ if } n = \frac{1}{2}(3j^2 \pm j). \end{cases}$$

As the pentagonal number theorem is our most serious theorem so far, we will give a very detailed argument. The details will make our discussion longer, but that should not discourage the reader. While some notions needed for this proof are a touch more complex than those we saw before, we will only need them for simple operations.

One difficulty we face when trying to prove the pentagonal number theorem is that, because equality does not hold for all positive integers n, we cannot hope to find a bijection between the two relevant sets of partitions that works for all n.

Nevertheless, let S (resp. T) be the set of partitions enumerated by $p_{d,even}(n)$ (resp. $p_{d,odd}(n)$). For the case when n is not pentagonal, we will construct a bijection $g : S \to T$. We will then show that, if n is pentagonal, then g is *almost* a bijection between S and T, that is, it is possible to remove one single element of S or T so that g is a bijection between the remaining sets.

In order to construct such a map g, we recall the bijection f of the proof of Lemma 2.26. That bijection changed the number of parts of a given partition by one. That is excellent for our purposes since it changes the *parity* of the number of parts of our partition. However, the task at hand is more difficult since we do not know how large the last part of s is.

The main idea of the bijection g is this: When we can, we want to remove the last part of $s \in S$ and distribute its boxes to the first few rows of s, adding one box to the end of them. We can only do this if s has a sufficient number of rows. So when s has fewer rows than needed, we will do the opposite. We will remove one box from the end of the first few rows and form a new last row with the removed boxes. However, we have to be careful that our removal of boxes does not create equal rows.

The previous two paragraphs described the main idea behind the bijection to be constructed. Now we have to pay attention to the details. We want our map g to be *bijective*, that is, given $g(s)$, we must be able to find out what s is. In order to ensure this, we will be very careful when making the ideas of the previous paragraph formal.

Proof: (of Theorem 2.28) Let $s \in S$, and let $last(s)$ denote the last part of s. Furthermore, let $s = (s_1, s_2, \cdots, s_t)$. We know that the sequence $s_1, s_2, s_3, \cdots$ is a strictly decreasing sequence, but let us see whether it is a decreasing sequence made up by *consecutive* integers. That is, let us see how far we can go in this sequence, starting up front, so that each number is exactly one less than the previous one. Let $stair(s)$ be the number of terms of the sequence satisfying this requirement, since consecutive integers in a Ferrers shape look like a staircase. So, for the partition $s = (5, 3, 1)$, for instance, this yields $stair(s) = 1$, since the second term is not just one less than the first. Similarly, for $s = (4, 3, 1)$, this yields $stair(s) = 2$, and for $s = (6, 5, 4, 3, 1)$, this yields $stair(s) = 4$.

Now we are ready to define g, which we will do by distinguishing three different cases.

1a. First assume that $last(s) < stair(s)$. Remove the last part of s and distribute its boxes among the first $stair(s)$ rows of s, adding one to each of them. Call the obtained partition $g(s)$. Then $g(s)$ has an odd number of parts (one less part than s did). As $g(s)$ has all distinct parts, we conclude that $g(s) \in T$. See Figure 2.8 for an illustration.

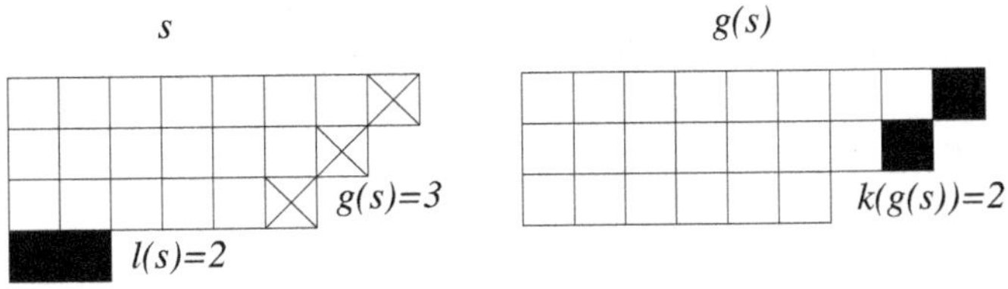

Figure 2.8
If $s = (8, 7, 6, 2)$, then $last(s) = 2$, and $stair(s) = 3$, so $g(s) = (9, 8, 6)$.

The following proposition is a simple consequence of the definition of $stair(s)$. We encourage the reader to prove it on his own. The proposition will be important when we prove that g is a bijection. We provide a proof in the solution of Exercise 21.

Proposition 2.29 *If $last(s) < stair(s)$, then*

$$stair(g(s)) = last(s), \text{ and also, } stair(g(s)) < last(g(s)).$$

1b. Now let us assume that $last(s) = stair(s) = a$. If s has more than a parts, then proceed just as in the previous case. However, if s has exactly a parts, then all parts of s form a decreasing sequence of consecutive integers, so we must have $s = (2a - 1, 2a - 2, \cdots, a)$. In this case, the above definition of g does not work since there are not enough rows to distribute the boxes of the last row. So for this one exceptional case we will not define g. Note that, in this case, $n = (2a - 1) + (2a - 2) + \cdots + a = a(3a - 1)/2$, a pentagonal number, and one in which the *minus* sign is used. Note also that, because s had exactly a parts, a has to be an *even* number here. See Figure 2.9 for an illustration (and also a visual explanation of the term *pentagonal number*). We point out that the Ferrers shape

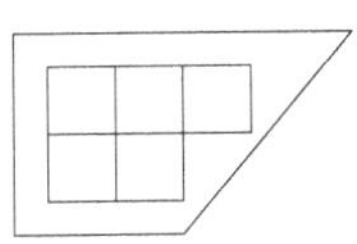

Figure 2.9
The definition of g does not work for $s = (3, 2)$.

of a partition of the type $n = (2a-1)+(2a-2)+\cdots+a$ or $n = 2b+(2b-1)+\cdots+(b+1)$ resembles a pentagon.

If you followed our advice and proved Proposition 2.29 on your own, then the following proposition will be a breeze for you.

Proposition 2.30 *If $last(s) = stair(s) = a$, and s has more than a parts, then $stair(g(s)) = last(s)$, and also, $stair(g(s)) < last(g(s))$.*

2 Finally, let us assume that $last(s) > stair(s)$. As we said, in this case we want to remove the last boxes of the first $stair(s)$ rows and create a new last row of length $stair(s)$ using the removed boxes. Call the obtained partition $g(s)$. See Figure 2.10 for an example.

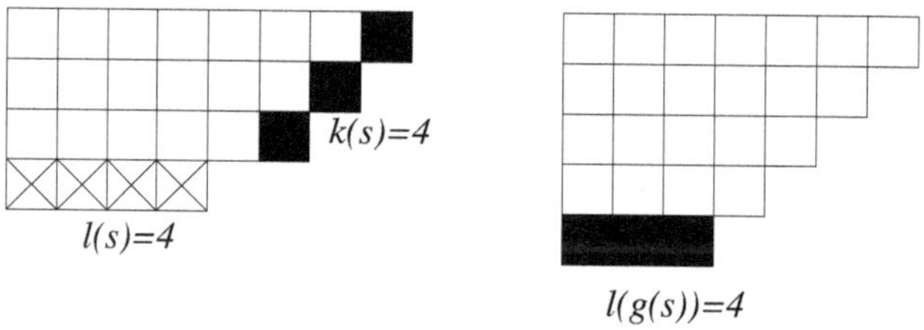

Figure 2.10
If $s = (8,7,6,4)$, then $last(s) = 4$, and $stair(s) = 3$, so $g(s) = (7,6,5,4,3)$.

Proposition 2.31 *If $last(s) > stair(s)$, then*

$$last(g(s)) = stair(s) \text{ and, also, } last(g(s)) \leq stair(g(s)).$$

Again, you should try to explain why by yourself, then check the solution of Exercise 22.

Just as in cases 1a. and 1b., the partition $g(s)$ has an odd number of parts (one more than s), and its parts are all distinct, except possibly when the new last row is equal to the row preceding it. So $g(s) \in T$, except in the case mentioned above. In that case, we must have had $b = last(s) = stair(s)+1$ (this is how the last two rows of $g(s)$ can be equal), so again, all parts of s form a decreasing sequence of consecutive integers, implying that $s = (2b, 2b-1, \cdots, b+1)$. Therefore, in this one case, $g(s) \notin T$. Note that in this case, $n = (2b)+(2b-1)+\cdots+(b+1) = b(3b+1)/2$, a pentagonal number, and one in which the *plus* sign is used. Also note that b is an even number, as s has b parts. See Figure 2.11 for an illustration.

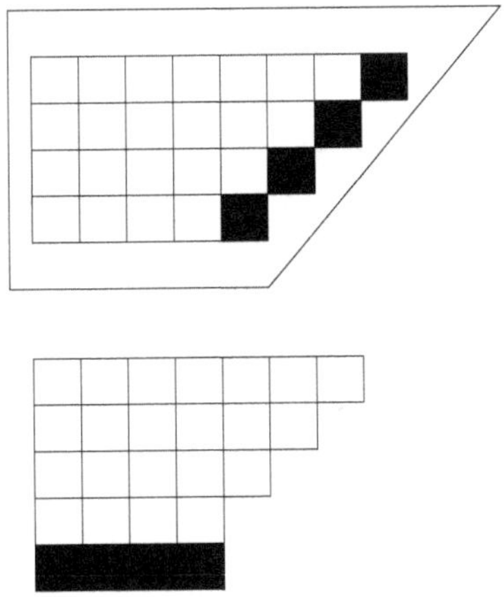

Figure 2.11
If $s = (8, 7, 6, 5)$, then $g(s) = (7, 6, 5, 4, 4)$, which is not in T.

The table shown in Figure 2.12 contains the important properties of the map g. This concise table will come in handy in the next step of our proof, when we show that g is a bijection for most integers n.

We are more than halfway through this proof! Note how nicely and carefully the map g is defined. That is, g does not cause any loss of information about the original partition s, as we will see below. The fact that g works only *almost* always forces us to pay attention to more details than usual, but it also makes the construction even more remarkable. Even in the exceptional cases when g does not work, we can describe what happens.

Let us return to the proof of Theorem 2.28. We must show that g is almost always a bijection from S to T and that, when it is not, it only misses that one element.

We have seen that, if n is not pentagonal, then g is indeed a map from S to T. We claim that, in that case, g is a bijection. Indeed, let $t \in T$. We want to show that there exists exactly one $s \in S$ so that $g(s) = t$. To that end, compare $last(t)$ and $stair(t)$. This comparison can have two different results:

(A) If $last(t) > stair(t)$, then Propositions 2.29, 2.30, and 2.31 imply that, if there is $s \in S$ satisfying $g(s) = t$, then that s must belong to case 1a. or 1b. Indeed, those are the only cases in which the image of s satisfies $last(g(s)) > stair(g(s))$. The unique preimage s of t is then obtained by removing the last entries of the first $stair(t)$ rows of t and using them for a new last row of $s = g^{-1}(t)$. This partition works, that is, it satisfies $g(s) = t$, and since t must be obtained by the rule of cases 1a. and 1b., this is the only partition that will work.

Case 1.a	Case 1.b	Case 2
last(s)< stair(s)	last(s)=stair(s)	last(s)>stair(s)
stair(t)<last(t)	stair(t)<last(t)	stair(t)$\geq$last(t)

Figure 2.12
Comparing the values of *last* and *stair* before and after the application of g. We set $g(s) = t$ for brevity.

The only time this procedure will not yield a partition in S is when the created new last row has c boxes and so does the row preceding it. Similarly to case 2, this can only happen if t has c rows and their lengths are all consecutive integers, that is, when the equality $t = (2c, 2c-1, \cdots, c+1)$ holds for an *odd* integer c. In that case, $n = c(3c+1))/2$, with c odd, and, as we have just seen, one element $t \in T$ has no preimage under f.

(B) If $last(t) \leq stair(t)$, then Propositions 2.29, 2.30, and 2.31 imply that if there is $s \in S$ satisfying $g(s) = t$, then that s must belong to case 2. Indeed, that is the only case in which the image of s satisfies $last(g(s)) \leq stair(g(s))$. The unique preimage s of t is then obtained by removing the last row of t and adding one box to the first $last(t)$ rows of t to get the partition $s = g^{-1}(t)$. Again, this partition works, that is, it satisfies $g(s) = t$, and because t must be obtained by the rule of case 2 here, this is the only s that will work.

The only time this procedure cannot be carried out is when there are simply not enough rows to absorb the boxes of the last row of t. In that case, just as in case 1, $last(t) = stair(t) = d$ must hold, and t must have d rows for some odd integer d. Then $t = (2d-1, 2d-2, \cdots, d)$, so $n = d(3d-1)/2$, with d odd. This shows that there is one element $t \in T$ that has no preimage under f, just as in case (A).

This shows that if n is not pentagonal, then g is a bijection, and so $|S| = |T|$.

As you probably realize, we are not quite done yet. We have to consider the exceptional case when n is pentagonal. First, we claim that in that case there is only one integer j so that n satisfies $n = j(3j \pm 1)/2$. This is fairly simple, and we will show the proof in Exercise 23. Let us now continue with the proof of our theorem.

If n is pentagonal, then the above arguments showing the bijectivity of g will still work, with two exceptions. That is, if $n = a(3a \pm 1)/2$, and a is even, then, as we saw in cases 1b. and 2, there is one particular partition in $s \in S$ that is not matched with a partition in T, so $|S| = |T| + 1$. On the other hand, if $n = a(3a \pm 1)/2$, and a is odd, then there is one particular partition in $t \in T$ that is not matched with a partition in S, so $|T| = |S| + 1$. We have mentioned that, for any given n, only one of the exceptional situations can occur. Therefore, the proof is complete. ◊

We will return to the pentagonal number theorem in Exercise 16 of Chapter 3, where we will show how it can be used to deduce a recurrence relation for the number of all partitions of n.

Quick Check

1. Prove that the number of partitions of n in which the integer 1 is not a part is equal to $p(n) - p(n-1)$.
2. We select eight partitions of the set $[5]$. Is it true that there will be two among them so that those two set partitions agree in their number of blocks of size m, for each positive integer m?
3. We select twelve compositions of 6 at random. Is it true that there will be two among them so that if we rearrange the parts of each of them in nonincreasing order, then we get two identical integer partitions?

2.4 The inclusion–exclusion principle

In this section, we are going to investigate situations in which direct methods would lead to an *overcount*, that is, certain objects would be counted more than once, leading to an incorrect final answer.

2.4.1 Two intersecting sets

Let us revisit our friends on the canoe trip whose travails we discussed in Chapter 1. Undaunted by the problems last summer, they are going on a trip again. We hear that this excursion also turns out to be eventful. To be more precise, we hear that five of them fell in the water at one point or another, while nine of them saw their breakfast stolen by raccoons. None of the friends on the trip managed to escape without either of these two experiences. How many friends went on the canoe trip this year?

Before the reader jumps to the conclusion that we want to bore her with the addition principle again, let us point out that this question is *different* from those in Section 1.1. The difference lies in the fact that this time we do *not* know that the two sets of unfortunate friends are disjoint. Indeed, unless we are told otherwise, we cannot assume that falling in the water will protect somebody's breakfast from raccoons.

Continuing this line of thinking, we easily see that we cannot even solve this problem without getting more information. Indeed, let A be the set of people who fell in the water, and let B be the set of those whose breakfast was impounded by raccoons. It could happen that $A \subset B$, which would imply that nine friends went on the trip this year, or it could happen that A and B are disjoint, which, by the addition principle, would imply that 14 friends did, or it could be that the number of people who had both type of problems was not zero, but less than five, which would result in a final answer larger than nine, but less than 14.

As the last paragraph suggests, the missing piece of information is indeed the number of people who belong to both A and B, that is, the number of those friends who belong to $A \cap B$. Let this number be k. Then we could answer the question as follows: There were five people who fell into the water, and there were $9 - k$ whose breakfast was stolen, but who did not fall in the water, therefore, by the addition principle, the number of all people on the trip is $5 + (9 - k) = 14 - k$.

Generalizing the ideas in the above argument, we get the following useful lemma.

Lemma 2.32 *Let A and B be finite sets. Then we have*

$$|A \cup B| = |A| + |B| - |A \cap B|. \tag{2.2}$$

Proof: We give a slightly different proof from the one we sketched above, since we will want to generalize our proof later. The left-hand side of (2.2) counts the elements of $A \cup B$. The first two members of the right-hand side do the same, except that elements that belong to both A and B are counted *twice*, once in $|A|$ and once in $|B|$. However, the third term corrects this anomaly, by subtracting precisely the number of elements that belong to both A and B. Therefore, the right-hand side also counts the elements of $A \cup B$, and each of them once, so the two sides of (2.2) count the same objects and are therefore equal. ◊

Example 2.33 *The number of positive integers less than or equal to* 300 *that are divisible by at least one of* 2 *and* 3 *is* 200.

Solution: Let A be the set of those eligible integers that are divisible by 2, and let B be the set of those eligible integers that are divisible by 3. Then $|A| = 150$ and $|B| = 100$. To compute $|A \cap B|$, note that n is divisible by both 2 and 3 if and only if n is divisible by 6. This shows that $|A \cap B| = 50$. So, by Lemma 2.32,

$$|A \cup B| = 150 + 100 - 50 = 200.$$

◊

Two intersecting sets, and their sizes, can be visualized with the help of a Venn diagram, shown in Figure 2.13.

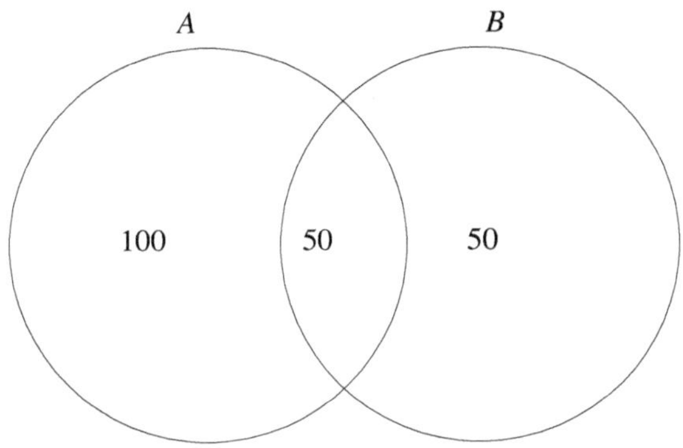

Figure 2.13
The two intersecting sets of Example 2.33.

The previous example showed how to use Lemma 2.32 to compute the number of elements that belong to *at least one* of two sets. Combining that with the subtraction principle, we can find the number of ways to compute the number of elements that do not belong to either one of two sets. We suggest that the reader solve Exercise 13 before reading further.

Example 2.34 *We have invited 30 guests to a wedding. We would like to split them into three groups of sizes of 10, in preparation of seating assignments. However, there are certain guests who do not like each other, namely, U does not like V, and X does not like Y. Therefore, these people cannot be put in the same group. In how many ways can we proceed?*

Solution: Let us count the *bad* partitions first, that is, those partitions of $[30]$ in which each block is of size 10, but either 1 and 2 are in the same block, or 3 and 4 are in the same block, or both of those unfortunate events occur.

Let A be the set of eligible partitions of $[30]$ in which the first bad thing happens, that is, 1 and 2 are in the same block. Removing 1 and 2 from such a partition, we get a partition of $[28]$ into two blocks of size 10, and one block of size 8. In fact, we have just created a bijective mapping from A into the set of partitions of [28] of the mentioned type. Using the method of Exercise 13, we get that this number is $|A| = \frac{28!}{10!10!8!2}$.

Similarly, if B is the set of eligible partitions of [30] in which the second bad thing happens, then, by the very same argument, we get that $|B| = \frac{28!}{10!10!8!2}$.

In order to apply Lemma 2.32, we need to compute $|A \cap B|$. This is a little bit more complex. Let us remove the elements 1, 2, 3, and 4 from any partition belonging to $A \cap B$. This results in a partition of [26], but the type of that partition is not uniquely determined. Indeed, if 1, 2, 3, and 4 are all in the same block, then we get a partition of block sizes 10, 10, and 6. There are $\frac{26!}{10!10!6!2}$ partitions of this type. As the map consisting of removing 1, 2, 3, and 4 from a partition of [30] is bijective, this means that this was the number of partitions containing 1, 2, 3, and 4 all in the same block.

However, it could also happen that 1 and 2 were in one block, and 3 and 4 were together in another block. Removing these four elements, we get a partition with block sizes 10, 8, and 8, and, by the method of Exercise 13, there are $\frac{26!}{10!8!8!2}$ partitions of this type. Crucially, the map consisting of omitting 1, 2, 3, and 4 is *not* a bijection this time; it is a 2-to-1 map. Indeed, there are two ways to reassign the pairs $\{1,2\}$ and $\{3,4\}$ to our partitions of [26]. Therefore, originally there were twice this many partitions of [30], that is, $\frac{26!}{10!8!8!}$ partitions.

Summarizing the results computed in the last two paragraphs, we get that

$$|A \cap B| = \frac{26!}{10!10!6!2} + \frac{26!}{10!8!8!}.$$

This completes the first part of our plan, namely, by Lemma 2.32, we get that the number of bad partitions—that is, partitions in which at least one bad event occurs—is

$$\begin{aligned} |A \cup B| &= |A| + |B| - |A \cap B| \\ &= \frac{28!}{10!10!8!} - \frac{26!}{10!10!6!2} - \frac{26!}{10!8!8!}. \end{aligned}$$

In order to get the number of partitions in which nothing bad happens, we simply need to use the subtraction principle, that is, we need to subtract $|A \cup B|$ from the number of all eligible partitions. That number is, again by the method used in Exercise 13, $\frac{30!}{10!10!10!\cdot 3!}$, so the number of good partitions is

$$\frac{30!}{10!10!10! \cdot 3!} - \frac{28!}{10!10!8!} + \frac{26!}{10!10!6!2} + \frac{26!}{10!8!8!}.$$

◊

2.4.2 Three intersecting sets

This subsection is meant as an intermediate step between the simple case of two intersecting sets and the general case of any number of intersecting sets, and it is meant to provide an opportunity for further practice. Once the reader feels comfortable with the techniques of the principle of inclusion–exclusion learned so far, she can skip to the next subsection.

Let us assume that this time we receive even more complex intelligence about our canoeing friends. We are told that five of them fell in the water, nine had their breakfast impounded by raccoons, and three had their tents blown away by a storm. Furthermore, we know that one of them had the misfortune to go through the first two of these trials, two of them went through the last two, and one person went through the first and the third. There was nobody who got away without being hit by at least one of these calamities. How many friends went on the canoe trip this year?

This is certainly a lot of data, but is it sufficient data to answer the question? The following lemma shows that it is not quite enough; in fact, one more vital number is needed.

Lemma 2.35 *Let A, B, and C be finite sets. Then*

$$|A \cup B \cup C| = |A| + |B| + |C| - |A \cap B| - |A \cap C| - |B \cap C| + |A \cap B \cap C|. \quad (2.3)$$

Proof: The left-hand side counts the elements that belong to $A \cup B \cup C$. The first three terms of the right-hand side count these same elements, but with multiplicities. That is, some elements are counted more than once. However, we will show that, on the whole, the right-hand side counts each element of $A \cup B \cup C$ exactly once.

1. If $x \in A \cup B \cup C$ is contained in just one of our three sets, say A, then the only term of the right-hand side that counts x is $|A|$, and our claim is proved.
2. If $x \in A \cup B \cup C$ is contained in exactly two of our three sets, say A and B, then x is counted twice, by $|A|$ and $|B|$, and then that overcount is corrected when x is counted once, with a negative sign, by $|A \cap B|$. So our claim holds true again.
3. If $x \in A \cup B \cup C$ is contained in all of our sets, then x is first counted three times, by $|A|$, $|B|$, and $|C|$, and then x is counted three times with a negative sign, by $|A \cap B|$, $|A \cap C|$, and $|B \cap C|$, and finally x is counted again once, with a positive sign, by the term $|A \cap B \cap C|$. So x is counted $3 - 3 + 1 = 1$ times, proving our claim.

Because both sides of (2.3) are equal to the number of elements that belong to at least one of A, B, and C, they must be equal, and our lemma is proved. ◊

Example 2.36 *A group of 10 tourists is planning to see the interesting sites of a town in a foreign country. In order to minimize waiting times, they consider splitting into subgroups. However, three of them, X, Y, and Z, do not* *speak the language of the locals. In how many ways can the tourists split into subgroups if none of X, Y, and Z is willing to form a subgroup by himself?*

Note that X, Y, and Z do not refuse to be in a group in which nobody speaks the language of the locals as long as there are at least two people in that group.

Solution: Mathematically speaking, the problem asks for the number of partitions of $[10]$ in which none of 1, 2, and 3 forms a singleton block. We could answer this question by computing the number of bad partitions of $[10]$, that is, those in which at least one of 1, 2, or 3 forms a singleton block; then we could subtract that number from the number of all partitions of $[10]$, that is, $B(10)$.

For $i \in [3]$, let A_i be the set of partitions of $[10]$ in which i forms a singleton block. We are then interested in the size of the union $A_1 \cup A_2 \cup A_3$. How will we compute that size? The reader will not be surprised by learning that we will use Lemma 2.35 to do that.

Indeed, by Lemma 2.35, we get

$$\begin{aligned} |A_1 \cup A_2 \cup A_3| &= |A_1| + |A_2| + |A_3| - |A_1 \cap A_2| - |A_1 \cap A_3| && (2.4) \\ &- |A_2 \cap A_3| + |A_1 \cap A_2 \cap A_3|. && (2.5) \end{aligned}$$

Before you think that we will make you go through seven different pieces of computation to get the seven terms of the right-hand side, let us reassure you that we will not. In fact, we will get away with as few as three different counting arguments, thanks to the symmetries that we will explain shortly.

First, A_1 is the set of partitions of $[10]$ in which 1 is in a singleton block. As there are no more restrictions, all $B(9)$ partitions are allowed on the remaining elements, showing that $|A_1| = B(9)$.

Now note similarly that $|A_2| = B(9)$ and $|A_3| = B(9)$. Indeed, all we need to do is replace the element 1 by the element 2 (or 3) in the argument of the previous paragraph.

Second, we claim that $|A_1 \cap A_2| = B(8)$. Indeed, $A_1 \cap A_2$ is the set of partitions of $[10]$ in which both 1 and 2 are in singleton blocks. There are no restrictions on the partition of the remaining eight entries, so our claim holds true. Replacing 2 by 3 in this argument, we see that $|A_1 \cap A_3| = B(8)$, and, after replacing 1 by 2, we see that $|A_2 \cap A_3| = B(8)$.

Finally, by an analogous argument, we get $|A_1 \cap A_2 \cap A_3| = B(7)$. Substituting our results into (2.4), we get

$$|A_1 \cup A_2 \cup A_3| = 3B(9) - 3B(8) + B(7).$$

Therefore, by the subtraction principle, the number of acceptable partitions is $B(10) - 3B(9) + 3B(8) - B(7)$. $\diamond$

Venn diagrams can be used to visualize three intersecting sets as well. We can fill the diagram from the inside out. That is, we know that there are $B(7)$ partitions that belong to all three sets A_i. Therefore, as there are $B(8)$ partitions belonging to $A_1 \cap A_2$, there must be $B(8) - B(7)$ that belong to $(A_1 \cap A_2) - A_3$. Then a similar argument yields the number of partitions belonging to $A_1 - (A_2 \cup A_3)$. See Figure 2.14.

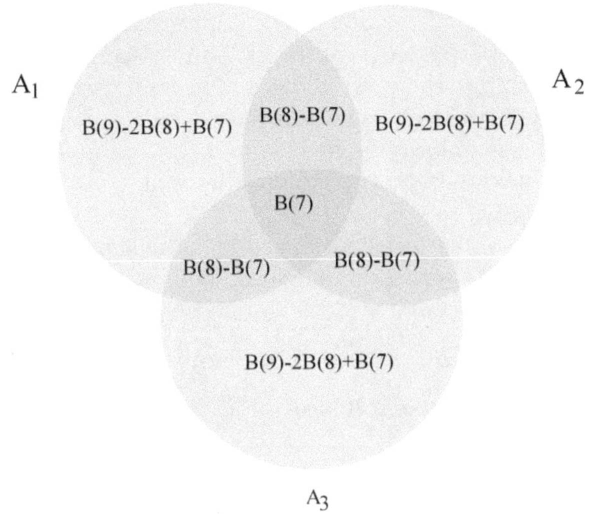

Figure 2.14
The three sets of Example 2.36.

See Exercise 28 for a continuation of this example.

If the reader noticed that the coefficients in our final answer look very similar to those in $(x - 1)^3$, then the reader has made a relevant observation. In our upcoming study of exponential generating functions, we will explain what causes this similarity.

Since working with Lemma 2.35 can be cumbersome, the reader should always check whether there is a more direct way to go before starting such a long computation. This more direct way, when it exists, often goes through the subtraction principle.

Example 2.37 *Ten people arrived at a dentist's office at different times, and they are called in a random order. A person is considered* unlucky *if that person was the* i*th to arrive but is not among the first* i *people to be called.*

Three members of a family, Albert, Bella, and Claudia, arrived third, fifth, and eighth. In how many of the 10! *possible orders will at least one of them be unlucky?*

Solution: Instead of resorting to Lemma 2.35, let us find out the number of orders in which none of the three family members is unlucky, then use the subtraction principle.

If Albert is not unlucky (in which case we will call him lucky), then he has to be among the first three people to be called. Once Albert is lucky, no matter at which point of time he was called, there are four possible times for Bella to be called. Indeed, she has to be called among the first five, but Albert took up one of those possibilities. By a similar argument, if Albert and Bella are both lucky, then Claudia could be called at six (eight minus two) different times. The remaining seven people then can be called in 7! various ways. Therefore, by the product principle, there are $3 \cdot 4 \cdot 6 \cdot 7! = 9!$ possible orders in which none of the three family members is unlucky. So, by the subtraction principle, at least one of them is unlucky in $10! - 9! = 9 \cdot 9!$ orders. In other words, all three of them are lucky just ten percent of the time. ◊

The reader will be able to better appreciate the shortness of this proof if she tries to arrive at this answer by applying Lemma 2.35.

2.4.3 Any number of intersecting sets

Theorem 2.38 (Inclusion–exclusion principle) *Let $A_1, A_2, \cdots, A_n$ be finite sets. Then*

$$|A_1 \cup A_2 \cdots \cup A_n| = \sum_{j=1}^{n} (-1)^{j-1} \sum_{i_1, i_2, \cdots, i_j} |A_{i_1} \cap A_{i_2} \cap \cdots \cap A_{i_j}|,$$

where $(i_1, i_2, \cdots, i_j)$ ranges all j-element subsets of $[n]$.

Proof: We are going to prove the following two claims:

(a) We show that if x is contained in the union of all the A_i, then the right-hand side counts x exactly once, no matter how complicated the right-hand side looks. To that end, assume that x is contained in exactly k of the sets A_i, with $k > 0$. That means that x is certainly not included in any j-fold intersections of the A_i for $j > k$. If, on the other hand, $j \leq k$, then x is contained in exactly $\binom{k}{j}$ different j-fold intersections of the A_i. If we take the signs into account, this means that the right-hand side counts x exactly

$$m = \sum_{j=1}^{k} (-1)^{j-1} \binom{k}{j}$$

times. So our claim will be proved if we show that $m = 1$. Subtracting 1 from both sides of the last equation, then multiplying by (-1), and applying the binomial theorem backward, we get that

$$1 - m = \sum_{j=0}^{k} (-1)^j \binom{k}{j} = (1-1)^k = 0, \tag{2.6}$$

since k is positive. Therefore, $m = 1$, and our claim is proved. If this last step is less than convincing, try to expand $(1-1)^k$ by the binomial theorem, and you will see that you get the expression that appears in (2.6).

(b) Now we show that, if x is not contained in any of the A_i, then the right-hand side counts x zero times. Indeed, repeating the above argument with $k = 0$, the far right-hand side of (2.6) becomes $(1-1)^0 = 1$, implying that $m = 0$, and our claim is proved.

Therefore, the left-hand side and the right-hand side count the same objects, so they are equal. ◇

See Exercise 31 for an alternative proof for part (b).

2.4.3.1 An explicit formula for the numbers $S(n, k)$

As an example of an application of the inclusion–exclusion principle, assume that we have n different toys and that we want to distribute them among k children, where $k \leq n$. However, we do not want to cause severe trauma to any children; therefore, we insist that each child get at least one toy. In how many ways can we do this?

As in many applications of Theorem 2.38, here we want to count distributions in which *nothing goes wrong*, but we will instead first compute the number of distributions in which *something* goes wrong. What *could* go wrong? It could happen that the first child does not get any toy at all (let the set of distributions with that property be A_1), or it could happen that the second child gets no toy (let the set of those distributions be A_2), and so on. In other words, the set of distributions in which *something* goes wrong is $A_1 \cup A_2 \cup \cdots \cup A_k$. If we can compute the size of this set, then we can simply resort to the subtraction principle, since we know that the number of all possible distributions is k^n.

In order to be able to compute $|\cup_{i=1}^k A_i|$ by Theorem 2.38, we need to compute the sizes of all j-fold intersections of the A_i. Let us start with the one-fold intersections, that is, the sets A_i themselves. First, we know that all A_i have the same size since it is just as easy to shortchange one child as it is another (in mathematics, anyway). The number of distributions in which child i gets nothing is $(k-1)^n$, since each toy can go to any of the other $k-1$ children. As all A_i have the same size, this leads to

$$\sum_{i=1}^{k} |A_i| = k|A_1| = k(k-1)^n.$$

Next, let us look at the two-fold intersections. Again, the size of $A_{i_1} \cap A_{i_2}$ does not depend on $\{i_1, i_2\}$ as long as, of course, $i_1 \neq i_2$. If the first and second children do not get any toy, then each toy can go to $k-2$ different children.

This leads to

$$\sum_{i_1,i_2} |A_{i_1} \cap A_{i_2}| = \binom{k}{2} |A_1 \cap A_2| = \binom{k}{2} (k-2)^n,$$

where, it goes without saying, the summation is taken over all two-element subsets $\{i_1, i_2\} \subseteq [k]$.

The above argument can be extended to j-fold intersections. We get that

$$\sum_{i_1,i_2,\cdots,i_j} |A_{i_1} \cap A_{i_2} \cap \cdots \cap A_{i_j}| = \binom{k}{j} |A_1 \cap A_2 \cap \cdots \cap A_j| = \binom{k}{j} (k-j)^n.$$

Again, the summation is over all j-element subsets of $[k]$.

Now we have all ingredients to use Theorem 2.38. Indeed, all we needed were the sizes of various intersections of the A_i, and we computed all of them. Substituting the computed values, we get that

$$|A_1 \cup A_2 \cup \cdots \cup A_k| = \sum_{j=1}^{k} (-1)^{j-1} \binom{k}{j} (k-j)^n.$$

Again, this is the number of distributions in which something goes wrong, that is, there is at least one child who does not receive any toys. Therefore, subtracting this number from the number k^n of all distributions, we obtain that the number of ways to distribute n different toys among k children so that each child gets at least one toy is

$$\sum_{j=0}^{k} (-1)^{j} \binom{k}{j} (k-j)^n. \tag{2.7}$$

Note that each distribution is in fact equivalent to a function from $[n]$ to $[k]$, since it sends each toy to a child. However, these functions have the special property that *each element of $[k]$ has at least one preimage.* This is a very important class of functions in mathematics and therefore has its own name.

Definition 2.39 *A function $f : X \to Y$ is called a* surjection *if, for all $y \in Y$, there exists at least one $x \in X$ so that $f(x) = y$.*

With this terminology, (2.7) yields the following theorem.

Theorem 2.40 *The number of surjections from $[n]$ to $[k]$ is equal to*

$$\sum_{j=0}^{k} (-1)^{j} \binom{k}{j} (k-j)^n.$$

We are finally in a position to fulfill our promise of proving an exact formula for the Stirling numbers of the second kind.

Corollary 2.41 *For all positive integers $k \leq n$,*

$$S(n,k) = \frac{1}{k!} \sum_{j=0}^{k} (-1)^j \binom{k}{j} (k-j)^n.$$

Proof: Corollary 2.41 claims that $S(n,k)$ is equal to $1/k!$ times the number of surjections from $[n]$ onto $[k]$. What is the connection between such a surjection f and a partition counted by $S(n,k)$? The surjection f also defines a partition of $[n]$ into k blocks, namely, the elements of $[n]$ whose image under f is 1 form a block, the elements of $[n]$ whose image under f is 2 form a block, and so on. Clearly, we will obtain each partition of $[n]$ into k blocks from exactly $k!$ surjections $f : [n] \rightarrow [k]$, since the order of the blocks does not matter in a partition. Our claim is then proved by the division principle. ◊

This is an interesting formula, even if it contains a summation sign. Note that, if we did not know that $\frac{1}{k!} \sum_{j=0}^{k} (-1)^j \binom{k}{j} (k-j)^n$ was a Stirling number, it would not even be clear why this number should be an *integer*.

2.4.3.2 An application involving linear orders

Let us practice applications of the general form of the inclusion–exclusion principle on another example.

Example 2.42 *In a small town, n married couples attend a town hall meeting. Each of these $2n$ people wants to speak exactly once. In how many ways can we schedule the $2n$ participants to speak if no married couple can take two consecutive slots?*

Solution: Let A_i be the set of schedules in which married couple i takes consecutive slots. Then $\cup_{i=1}^{n} A_i$ is the set of all *forbidden* schedules, that is, schedules where a couple takes consecutive slots. If we can find the size of this set, we can also find the size of the set of schedules that are *not forbidden* by a simple application of the subtraction principle.

We claim that $|A_i| = 2(2n-1)!$. Indeed, we can make sure that couple i gets consecutive slots by "glueing" them together and then schedule the obtained $(2n-1)$-element set of speakers. Finally, there are two ways in which couple i can give their two speeches, so our claim follows by the product principle. (This type of argument was discussed in Example 1.42, and the reader may return to that example for details.)

By an analogous argument, we get that, for any j couples $i_1, i_2, \cdots, i_j$,

$$|A_{i_1} \cap A_{i_2} \cap \cdots \cap A_{i_j}| = 2^j (2n-j)!.$$

Note that $|A_{i_1} \cap A_{i_2} \cap \cdots \cap A_{i_j}|$ is again independent of the set $(i_1, i_2, \cdots, i_j)$, just as in the previous example; it is only the *size* of that set that matters. Therefore, the inclusion–exclusion principle yields that

$$\begin{aligned} |A_1 \cup A_2 \cdots \cup A_n| &= \sum_{j=1}^{n} (-1)^{j-1} \sum_{i_1, i_2, \cdots, i_j} |A_{i_1} \cap A_{i_2} \cap \cdots \cap A_{i_j}| \\ &= \sum_{j=1}^{n} (-1)^{j-1} \binom{n}{j} 2^j (2n-j)!. \end{aligned}$$

This is the number of forbidden schedules. Consequently, by the subtraction principle, the number of good schedules is

$$(2n)! - \sum_{j=1}^{n} (-1)^{j-1} \binom{n}{j} 2^j (2n-j)! = \sum_{j=0}^{n} (-1)^{j} \binom{n}{j} 2^j (2n-j)!.$$

$\diamond$

2.4.3.3 Euler's ϕ function

For any positive integer n, let $\phi(n)$ denote the number of positive integers $k \leq n$ so that the greatest common divisor of k and n is 1. That is, $\phi(n)$ is the number of positive integers that are relatively prime to n and are not larger than n. If the reader took abstract algebra, the reader probably remembers that $\phi(n)$ is the number of elements of order n in the cyclic group Z_n. The reader also may remember from number theory that, for fixed positive integers n and k, the equation $xn + yk = r$ has an *integer solution* (x, y) for all integers r if and only if the greatest common divisor of n and k is 1. This motivates us to find a formula for $\phi(n)$. We mention that $\phi(n)$ is called *Euler's* ϕ function, or *totient function*.

A little thought shows that $\phi(1) = 1$, $\phi(2) = 1$, $\phi(3) = 2$, $\phi(4) = 2$, $\phi(5) = 4$, and $\phi(6) = 2$. We soon realize that the equality $\phi(p) = p - 1$ holds for all primes p, since all positive integers less than p are relatively prime to p. The reason this equality is so easy to prove is that primes only have *one* prime divisor, and other integers only have to be checked for divisibility by that one prime divisor. Based on this observation, the next class of integers we address is those that have only *two* prime divisors.

Proposition 2.43 *Let $n = pq$, where p and q are distinct primes. Then $\phi(n) = (p-1)(q-1)$.*

Proof: There are p integers in the set $[pq]$ that are divisible by q (namely, $q, 2q, \cdots, pq$), and there are q integers in $[pq]$ that are divisible by p. There is one integer, pq, in $[pq]$ that is divisible by both p and q. Therefore, by the inclusion–exclusion principle, there are $p+q-1$ elements of $[pq]$ that are divisible by at least one of p and q. So by the subtraction principle, there are

$$pq-(p+q-1)=(p-1)(q-1)$$

elements of $[pq]$ that are not divisible by either p or q and, therefore, are relatively prime to pq. $\diamondsuit$

There are two ways to continue from this point. One is to consider integers n with more than two prime divisors (with no prime divisors repeated).

Proposition 2.44 *Let $n = p_1p_2\cdots p_t$, where the p_i are pairwise distinct primes. Prove that*

$$\phi(n)=\prod_{i=1}^{t}(p_i-1).$$

The reader will be asked to prove this result in Exercise 37.

The other way to generalize Proposition 2.43 is by noting that it says that $\phi(n)=\phi(p)\phi(q)$. It turns out that this is true for a wider class of integers than primes.

Lemma 2.45 *Let a and b be two positive integers whose greatest common divisor is 1, and let $n = ab$. Then $\phi(n) = \phi(a)\phi(b)$.*

Proof: The reader is invited to verify that k is relatively prime to n if and only if k is relatively prime to both a and b.

For easier understanding, let us write the elements of $[n]$ into an $a\times b$ matrix N, the first row of which is $1, 2, \cdots, b$, the second row of which is $b+1, b+2, \cdots, 2b$, and so on.

First, we claim that each row of N contains exactly $\phi(b)$ entries that are relatively prime to b. For the first row, this is true by the definition of $\phi(b)$. For the jth row, this is true since $jb+i$ is relatively prime to b if and only if i is.

Note that we have proved that the $\phi(b)$ entries of each row that are relatively prime to b are always in the same positions. That is, the entries of N that are relatively prime to b form an $a\times\phi(b)$ submatrix N_b of N.

Now we claim that each column of N_b contains exactly $\phi(a)$ entries that are relatively prime to a. Let $jb+i$ and $kb+i$ be two entries of the same column of N_b, with $j>k$. Then we claim that their difference is not divisible by a. This is true because their difference is $(j-k)b$, where b is relatively prime to a, and $0<j-k<a$. Therefore, the a entries of any column of N_b have all different remainders when divided by a, so their set of remainders modulo a agrees with the set of remainders of the set $\{1,2,\cdots,a\}$. However, we know that $\phi(a)$ elements of the latter are relatively prime to a, which proves our claim. $\diamondsuit$

Example 2.46 *Let $n = 12$, $a = 3$, and $b = 4$. Then $\phi(12) = \phi(3)\phi(4) = 2 \cdot 2 = 4$. Indeed,*

$$N = \begin{pmatrix} 1 & 2 & 3 & 4 \\ 5 & 6 & 7 & 8 \\ 9 & 10 & 11 & 12 \end{pmatrix} \qquad \text{and } N_b = \begin{pmatrix} 1 & 3 \\ 5 & 7 \\ 9 & 11 \end{pmatrix}.$$

Each column of N_b contains one entry that is divisible by 3 and two entries that are relatively prime to 3.

As we know that each positive integer can be uniquely decomposed into a product of prime powers, we will be able to compute $\phi(n)$ for any n as soon as we can compute $\phi(p^d)$ for any prime p.

Proposition 2.47 *For any prime p, and any positive integer d,*

$$\phi(p^d) = (p-1)p^{d-1}.$$

Proof: An element of $[p^d]$ is relatively prime to p^d if and only if it is not divisible by p. As exactly $1/p$ of all the elements of $[p^d]$ are divisible by p, the claim is proved. ◊

Lemma 2.45 and Proposition 2.47 imply the following theorem.

Theorem 2.48 *Let $n = p_1^{d_1} p_2^{d_2} \cdots p_t^{d_t}$, where the p_i are distinct primes. Then*

$$\phi(n) = \prod_{i=1}^{t} p_i^{d_i - 1}(p_i - 1).$$

For some generalizations of the inclusion–exclusion principle, note that Theorem 2.38 counts the elements that belong to *at least one* A_i. See Exercises 33–35 for refinements of this, that is, for formulae on the number of elements that belong to *exactly* k sets. These exercises gradually become harder, so we suggest the reader start with the easiest one. Then, Supplementary Exercises 44–46 ask for formulae for the number of elements belonging to *at least* a certain number of sets, generalizing Theorem 2.38.

2.4.3.4 Derangements

Example 2.49 *A country has n universities. A national exchange program takes one student from each university and sends each of them to a university different from their original university, making sure that each university receives exactly one exchange student. In how many different ways is this possible?*

Solution: This is the famous problem of *fixed point-free permutations*, or *derangements*. We will learn more about these objects in Chapter 4. Indeed, denoting the students by the elements of the set $[n]$, we are looking for permutations $p = p_1p_2\cdots p_n$ in which there is no i so that $p_i = i$. (That would mean that a student is sent back to his original university, which is not the most exciting exchange program.)

Let A_i denote the set of permutations of $[n]$ in which $p_i = i$. Then we can use the inclusion–exclusion principle to compute the size of the union of the sets A_i, that is, the number of *bad* permutations of $[n]$. Let $S \subseteq [n]$ so that $|S| = j$. Then we get

$$|\cap_{i\in S} A_i| = (n-j)!,$$

since we have no freedom to choose the entries p_i if $i \in S$, but we can choose any of the possible $(n-j)!$ permutations on the remaining entries. By Theorem 2.38, this implies

$$\begin{aligned} |\cup_{i=1}^n A_i| &= \sum_{j=1}^n (-1)^{j-1}\binom{n}{j}(n-j)! \\ &= \sum_{j=1}^n (-1)^{j-1}\frac{n!}{j!}. \end{aligned}$$

By the subtraction principle, the number of good permutations is $n!$ minus the number of bad permutations, that is,

$$D(n) = \sum_{j=0}^n (-1)^j \frac{n!}{j!} = n! \cdot \sum_{j=0}^n (-1)^j \frac{1}{j!}.$$

$\Diamond$s

The reader is invited to spend a minute on proving that as

$$\lim_{n\to\infty} \frac{D(n)}{n!} = \frac{1}{e} \approx 0.368.$$

That is, a random permutation has a rather large chance to be a derangement.

2.4.3.5 Excursion: Another proof of the inclusion–exclusion principle

We are going to present a somewhat less well-known proof of the inclusion–exclusion principle that was implicit in a paper of Garsia and Milne [36] and then published in an explicit form by Zeilberger [86]. The proof itself is shorter than the proof we have previously presented, and it does not use the binomial theorem, but it is conceptually less straightforward.

Let $A_1, A_2, \cdots, A_n$ and A be finite sets, so that $\cup_{i=1}^n A_i \subseteq A$. For $I = \{i_1, i_2, \cdots, i_k\}$, set the notation $A_I = A_{i_1} \cap A_{i_2} \cap \cdots \cap A_{i_k}$. We would like to prove that the number N of elements $a \in A$ that are *not* contained in any A_i is

$$N = |A - \cup_{i=1}^n A_i| = \sum_I (-1)^{|I|}|A_I|, \tag{2.8}$$

where I ranges over all nonempty subsets of $[n]$. The reader is invited to spend a moment on verifying that this is indeed equivalent to Theorem 2.38.

The first crucial idea is this: Instead of simply counting elements of A, we will consider *pairs* (a, J), where $a \in A$, and J is a set of indices $j \in [n]$ so that if $j \in J$, then $a \in A_j$. For example, if a is contained in A_2 and A_3 (and in no other A_i), then the allowed pairs containing a are $(a, \{2\})$, $(a, \{3\})$, $(a, \{2, 3\})$, and $(a, \{\emptyset\})$. Note in particular that the pair (a, J) is allowed if and only if $a \in A_J$.

Let us call a pair (a, J) odd (respectively, even), if $|J|$ is odd (respectively, even). Then the right-hand side of (2.8) is equal to the number $EVEN$ of even pairs minus the number ODD of odd pairs. Indeed, $|A_I|$ is equal to the number of allowed pairs (a, A_I), and it is counted with a positive sign if and only if $|I|$ is even. Therefore, we have

$$EVEN - ODD = \sum_{\substack{I\subseteq[n] \\ |I|\geq 1}} (-1)^{|I|}|A_I|. \tag{2.9}$$

Let $\mathcal{A}$ denote the set of all allowed pairs (a, J), and let i_a be the *smallest* index i so that $a \in A_i$. In our running example, that means that $a_i = 2$.

Define the map $f : \mathcal{A} \rightarrow \mathcal{A}$ by setting

$$f(a, J) = \begin{cases} (a, J \cup i_a) \text{ if } i_a \notin J, \\ (a, J - i_a) \text{ if } i_a \in J. \end{cases}$$

In our running example, $f(a, \{2\}) = (a, \emptyset)$, and $f(a, \{3\}) = (a, \{2, 3\})$.

Note that f is always defined except when a is not contained in any A_i. In that case, i_a is not defined, and therefore f is not defined. Note that in this case, we must have $J = \emptyset$, since $a \in A_i$ never holds. So the number of pairs (a, J) for which f is not defined is simply the number of pairs $(a, \emptyset)$ so that $a \notin A_i$ for any i, that is, just the number N of elements $a \in A$ that do not belong to any A_i.

The crucial property of f is that it turns an even pair into an odd pair, and vice versa. Indeed, f always changes the size of J by one. Furthermore, it is easy to verify that $f(f(a, J)) = (a, J)$, that is, f is an *involution.* That means that f matches odd pairs to even pairs for which f is defined, and therefore that there are as many odd pairs as even pairs for which f is defined. Consequently, $EVEN - ODD$ is equal to the number of even pairs (a, J) for which f is not defined, but we have seen in the previous paragraph that this number was equal to N. Therefore, by transitivity, $N = EVEN - ODD$. Comparing this with (2.9), our statement is proved.

Quick Check

1. Let $A_1, A_2, \cdots, A_n$ be finite sets. Prove that

$$\left| \cup_{i=1}^{n} A_i \right| \leq \sum_{i=1}^{n} |A_i|.$$

A combinatorial argument is preferred.

2. Let $A_1, A_2, \cdots, A_n$ be finite sets. Prove that

$$\left| \cup_{i=1}^{n} A_i \right| \geq \sum_{1 \leq i \leq j \leq n} \left| A_i \cap A_j \right|.$$

A combinatorial argument is preferred.

3. What is the number of partitions of $[n]$ into three blocks in which each block is of size at least 2?

2.5 The twelvefold way

In this section, we will review how the counting techniques learned in the first two chapters imply the solutions of 12 counting problems that are easy to formulate in terms that are similar to each other. In each of these problems, we are placing n balls into k boxes. What makes these problems all different is whether the balls are all the same or all distinguishable (for instance, labeled by different numbers), whether the boxes are all the same or all distinguishable, and the conditions for the placement of the balls, which will sometimes require that at most (or at least) one ball be placed in each box.

Gian-Carlo Rota named this collection of problems the *twelvefold way.* In what follows, we will go through these 12 problems one by one. At this point, the reader has all the necessary tools to solve these problems, so the reader should feel free to try to come up with a solution before reading our argument.

In the first three problems, we will assume that our n balls are all *identical,* and that our n boxes are all *identical.*

Problem 2.50

(a) How many ways are there to distribute n identical balls in k identical boxes?

(b) How many ways are there to distribute n identical balls in k identical boxes if each box has to get at least one ball?

(c) How many ways are there to distribute n identical balls in k identical boxes if each box has to get at most one ball?

Solution:

(a) As both the boxes and the balls are identical, all that matters is how many balls get into the individual boxes. In other words, we are dealing with *partitions of the integer n* here. Since the boxes are identical, we might as well order them in nonincreasing order of the balls contained in them. Therefore, the number of all possibilities is $p_k(n)$, since there will be at most k boxes actually containing balls.

(b) Because distributions with less than k nonempty boxes are not allowed, we get $p_k(n) - p_{k-1}(n)$ possibilities. This number is 0 if $n < k$.

(c) This number is 0 if $n > k$ and 1 if $n \leq k$, since each box gets 0 or 1 ball(s). As the boxes are identical, it does not matter which boxes get 0 and which boxes get 1.

◊

In the next three problems, the boxes will still be *identical*, but the balls will be *distinguishable.*

Problem 2.51

(a) How many ways are there to distribute n distinguishable balls in k identical boxes?

(b) How many ways are there to distribute n distinguishable balls in k identical boxes if each box has to get at least one ball?

(c) How many ways are there to distribute n distinguishable balls in k identical boxes if each box has to get at most one ball?

Solution:

(a) Since the balls are all different, we can label them by the elements of $[n]$. Then the balls are put in identical boxes, so we are dealing with *partitions of the set* $[n]$ here. If there are no other restrictions, then we are free to use any number of boxes not exceeding k, so the answer is $\sum_{i=1}^{k} S(n,i)$ by the addition principle.

(b) In this case, we have to use exactly k boxes, so the answer is $S(n,k)$.

(c) If $n > k$, then the answer is 0. Otherwise, the answer is 1, since each nonempty box will have one ball, and it does not matter which box has which ball as the boxes are all the same. Note the similarity with part (c) of the preceding problem.

◊

In the next three problems, we reverse the conditions: The boxes will be *distinguishable*, but the balls will be *identical.*

Problem 2.52

(a) How many ways are there to distribute n identical balls in k distinguishable boxes?

(b) How many ways are there to distribute n identical balls in k distinguishable boxes if each box has to get at least one ball?

(c) How many ways are there to distribute n identical balls in k distinguishable boxes if each box has to get at most one ball?

Solution:

(a) Since the balls are identical, all that matters is how many of them get into each box. However, the boxes are distinguishable, so their order matters. In other words, we are dealing with *weak compositions* of n into k parts. Therefore, the answer is $\binom{n+k-1}{k-1}$.

(b) As no box can be empty, we are dealing with *compositions* of n into k parts, so the answer is $\binom{n-1}{k-1}$.

(c) If $n > k$, then the answer is 0. Otherwise, we simply need to choose the n boxes that will contain one ball each, which we can do in $\binom{k}{n}$ ways.

◊

In the last three problems, both the boxes and the balls will be *distinguishable.*

Problem 2.53

(a) How many ways are there to distribute n distinguishable balls in k distinguishable boxes?

(b) How many ways are there to distribute n distinguishable balls in k distinguishable boxes if each box has to get at least one ball?

(c) How many ways are there to distribute n distinguishable balls in k distinguishable boxes if each box has to get at most one ball?

Solution:

(a) As any ball can go into any box, the number of possibilities is k^n.

(b) This is similar to partitioning $[n]$ into k blocks, but now the order of the blocks matters, so the answer is $S(n,k)k!$.

(c) Again, the answer is 0 if $n > k$. Otherwise, there are k choices for the first ball, then $k-1$ choices for the second ball, and so on, so the answer is $(k)_n$.

◊

If the reader worked through all 12 problems, then she will have a few useful questions to ask when trying to solve an enumeration problem. Are the objects we are counting distinct? Does their order matter? Is repetition allowed? Answering these questions will usually provide the first step of the solution.

Quick Check

1. What is the number of ways to distribute n indistinguishable balls into k distinguishable boxes if each box is to receive at least two balls?
2. Prove that for all positive integers $k \leq n$, the inequality $k!S(n,k) \leq k^n$ holds.
3. What is the number of ways to distribute n indistinguishable balls into k indistinguishable boxes if each box is to receive at least two balls?

2.6 Notes

We do not want to give the false impression that Ferrers shapes are the only way to prove partition identities. For instance, the reader is invited to verify that the number of partitions of n into odd parts is equal to the number of partitions of n into distinct parts. While this remarkable fact does have a bijective proof, that proof is not easy. On the other hand, with the more computational technique of *generating functions*, which we will cover soon, the proof requires almost no new ideas. An even stronger example is that of partitions in which each successive rank (see Exercise 24) is positive. Using generating functions, Euler proved that the number of such partitions of n is $p(n) - p(n-1)$.

Furthermore, we do not want to give the false impression that all possible distribution problems that are naturally defined are handled in this early chapter.

Let us assume, for instance, that we have 60 toys for three children—20 identical cars, 20 identical blocks, and 20 identical balls—and we want to distribute them among the three children so that each child gets exactly 20 toys. How many ways are there to do this? Or even more strongly, assume that the children return the toys at the end of the day, and come back to play the following day, and the day after that. We require, on one hand, that on each day the previously mentioned conditions hold, and, on the other hand, that in the course of the three days each child altogether has 20 cars, 20 blocks, and 20 balls. In how many ways can we do that? These problems are substantially more difficult, and we will treat them in our chapter on Magic Squares and Cubes.

Finally, we mention that, if F is a Ferrers shape on n boxes, and we bijectively label the boxes of F by the elements of $[n]$ so that the labels increase going down and to the right, we get a *Standard Young Tableau.* (We have seen a very special case of this after Lemma 1.32.) These objects have an extremely rich combinatorial structure and are the subject of several books on their own, such as *Young Tableaux* [34], by William Fulton and *The Symmetric Group* [65] by Bruce Sagan.

2.7 Chapter review

(A) Explicit enumeration formulae

1. Number of n-element multisets over $[k]$ is $\binom{n+k-1}{n} = \binom{n+k-1}{k-1}$.
2. Number of weak compositions n into k parts is $\binom{n+k-1}{n} = \binom{n+k-1}{k-1}$.
3. Number of compositions of n into k parts is $\binom{n-1}{k-1}$.

(B) Recurrence relations related to set partitions

1. Stirling number of the second kind, triangular recurrence:
$$S(n,k) = S(n-1,k-1) + kS(n-1,k).$$
2. Stirling number of the second kind, vertical recurrence:
$$S(n+1,k) = \sum_{i=0}^{n} \binom{n}{i} S(n-i,k-1).$$
3. Bell numbers:
$$B(n+1) = \sum_{k=0}^{n} B(k)\binom{n}{k}.$$

(C) Integer partitions

1.
$$p(n) \sim \frac{1}{4\sqrt{3}} \exp\left(\pi\sqrt{\frac{2n}{3}}\right).$$

2.8 Exercises

1. A function $f : S \to T$ is called an *injection* if it maps different elements into different elements, that is, if $x \neq y$ implies $f(x) \neq$

$f(y)$. A function $f : S \rightarrow T$ is called a *surjection* if, for all $t \in T$, there is an $s \in S$ so that $f(s) = T$.

(a) If a function f is both an injection and a surjection, then it has a name that we have already defined. What is that name?

(b) How many injections $f : [n] \rightarrow [k]$ are there?

(c) How many surjections $f : [n] \rightarrow [k]$ are there?

2. Prove that the number of n-element multisets over $[k]$ is $\binom{n+k-1}{n}$ using the following approach. For each n-element multiset A over k, write the entries a_i of A in nondecreasing order so that they satisfy the chain of inequalities

$$1 \leq a_1 \leq a_2 \leq \cdots \leq a_n \leq k.$$

Now observe that n-element subsets $B = \{b_1, b_2, \cdots, b_n\}$ of $[n+k-1]$ satisfy the chain of inequalities

$$1 \leq b_1 < b_2 < \cdots < b_n \leq n + k - 1.$$

Complete the proof by finding a bijection between the sets of n-tuples that satisfy the two chains of inequalities above.

3. Prove a closed formula (no summation signs) for the number of compositions of n into *any* number of parts by

(a) using the formula for the number of compositions of n into a given number of parts, or

(b) by induction on n.

4. Is there a closed formula for the number of all weak compositions of n into any number of parts?

5. We want to divide 12 children into four playgroups. However, there are two pairs of siblings among the children, and we do not want to put siblings in different groups. How many possibilities do we have?

6. What is the number of compositions of 50 into four odd parts?

7. We want to distribute 20 thousand dollars in bonus payments among six workers. Three of the workers have contracts that stipulate that they get at least two thousand dollars each, while each of the remaining workers must get at least one thousand dollars each. How many possibilities do we have if the amount of any bonus will have to be a multiple of 1000?

8. What is the number of weak compositions of 10 into four parts, all of which are smaller than nine?

9. What is the number of *monotone* functions from $[n]$ into $[n]$? We call a function monotone if $x < y$ implies $f(x) \leq f(y)$.

10. What is the number of compositions of 24 into any number of parts so that each part is divisible by three?

11. (a) Prove that, for all positive integers x,

$$x^n = \sum_{k=0}^{n} S(n,k)(x)_k. \tag{2.10}$$

    (b) Prove that the result of part (a) holds for all *real numbers* x, not just for positive integers.

12. (Basic knowledge of linear algebra required.) Interpret the result of the previous exercise in terms of bases of the vector space of all polynomials with real coefficients.

13. I want to partition the set $[10]$ into three blocks, so that two blocks are of size 3, and one block is of size 4. In how many different ways can I do this?

14. Five people participate in a long-jump competition. There are no four-way or five-way ties, but there can be two-way or three-way ties. How many possible rankings are there?

15. Prove Theorem 2.18.

16. Prove that the number of partitions of n into at most k parts is equal to the number of partitions of n into parts that are not larger than k.

17. What is the number of partitions of 14 into four distinct parts?

18. A partition that is equal to its own conjugate is called *self-conjugate*. Which positive integers have no self-conjugate partitions?

19. Recall that $p_k(n)$ is the number of partitions of n into at most k parts.

    (a) Prove that, for any fixed k, the function $p_k(kn)$ is a polynomial function of n. What is the degree of this polynomial?

    (b) Prove that, for any fixed k, and for any fixed $r < k$, the function $p_k(kn+r)$ is a polynomial function of n.

20. Use the results of the previous exercise to show that $p(n)$ grows faster than any polynomial function $q(n)$. That is, show that for any polynomial function $q(n)$, there exists a positive integer N_q so that

$$p(n) > q(n) \text{ for all } n > N_q. \tag{2.11}$$

21. Prove Proposition 2.29.

22. Prove Proposition 2.31.

23. If n is a pentagonal number, and $n = a(3a+1)/2$ or $n = a(3a-1)/2$ for a positive integer a, then we will say that a makes n pentagonal. Prove that, for each pentagonal number n, there is only one positive integer a that makes n pentagonal.

24. The *Durfee square* of an integer partition p is the largest square that fits in the northwest corner of the Ferrers shape of p. See Figure 2.15 for an illustration.

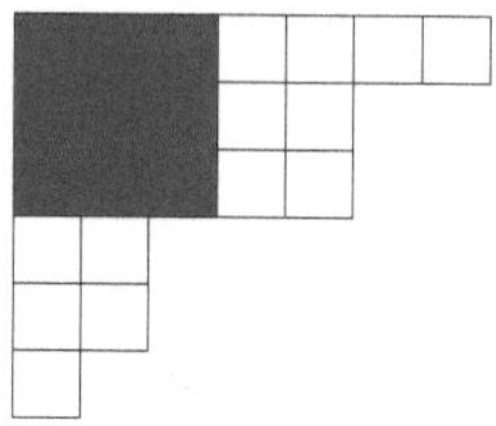

Figure 2.15
The Durfee square of this partition is of size 3×3.

Let the Durfee square of p have side length k. Then we define the *successive ranks* $r_1, r_2, \cdots, r_k$ of p as follows: The successive rank r_i is the difference of the ith row and the ith column of the Ferrers shape of p. So sometimes the successive ranks could be negative. For instance, the successive ranks are 1, 0, 2 for the partition shown in Figure 2.15.

Prove that the number of partitions of n into distinct parts is equal to the number of partitions of $2n$ in which each successive rank is equal to 1.

25. In a company of people, each person writes down the number of other people present that the given person knows. (We assume that if A knows B, then B knows A.)

   (a) Prove that the sum of all the numbers written down is even.

   (b) Let us arrange the numbers written down in nonincreasing order, obtaining a partition p of $2m$ for some integer m. Can the first successive rank r_1 of p be nonnegative?

   (c) Can it happen that $r_1 + r_2 \geq -1$? (Assume that r_2 is defined, that is, the Durfee square of p is of side length at least 2.)

26. Prove that if n is odd, then a partition of n whose third part is 2 cannot be self-conjugate.

27. What is the number of ways to write the digits 1, 1, 2, 2, 3, 4, and 5 in a line so that identical digits are not in consecutive positions?

28. Let us return to Example 2.36. How many ways are there to split the tourists into subgroups if each subgroup is to contain at least one person who speaks the language of the locals?

29. How many positive integers not larger than 1000 are relatively prime to both 7 and 8?

30. The employees of a big company (that means a company with at least three employees) altogether speak four languages. For any three employees, there is at least one language that all of them speak. There is no language that all employees speak. Prove that each employee speaks at least three languages. Construct an example where this happens.

31. Find an alternative argument, not involving computation, for part (b) of the proof of Theorem 2.38.

32. Let $D(n)$ be the number of derangements as defined in Example 2.49. Find a closed formula for the sum

$$\sum_{i=0}^{n} \binom{n}{i} D(i).$$

You can set $D(0) = 1$.

33. Let $A_1, A_2, \cdots, A_n$ be finite sets. Introduce the notation

$$D_k = \sum_{(i_1, i_2, \cdots, i_k)} |A_1 \cap A_2 \cap \cdots \cap A_{i_k}|,$$

where $(i_1, i_2, \cdots, i_k)$ ranges all k-element subsets of $[n]$. In other words, D_k is the sum of the sizes of all k-fold intersections of the sets A_i. Prove that the number of elements that belong to exactly $n-2$ of the A_i is

$$B_{n-2} = D_{n-2} - (n-1)D_{n-1} + \binom{n}{2} D_n.$$

34. Keeping the notations of the previous exercise, prove that the number of elements that belong to exactly $n-3$ of the A_i is

$$B_{n-3} = D_{n-3} - (n-2)D_{n-2} + \binom{n-1}{2} D_{n-1} - \binom{n}{3} D_n.$$

35. State and prove a formula for the number B_k of elements that belong to exactly k sets A_i for some $k \in [n]$.

36. Show an example for an infinite family F of sets so that each infinite subfamily of F has an empty intersection and every finite subfamily of F has an infinite intersection. Try to find at least two solutions.

37. Prove Proposition 2.44. Do not use Theorem 2.48. Use the general version of the inclusion–exclusion principle.

38. Six chess players participate in a tournament. How many final rankings are possible if two-way ties are allowed but ties of more than two players are not?

39. Prove that the average number of blocks taken over all partitions of $[n]$ is
$$\frac{B(n+1)}{B(n)} - 1.$$

40. + Let $f_{n,k}$ be the number of all surjections from $[n]$ to k. Prove that, for any fixed positive integers n, we have
$$\sum_{k=1}^{n}(-1)^k f_{n,k} = (-1)^n.$$

2.9 Solutions to exercises

1. (a) Such a function f is called a "bijection." Indeed, the fact that f is surjective assures that each $t \in T$ has at least one preimage under f, and the fact that f is injective ensures that each $t \in T$ has at most one preimage under f.

   (b) We can consider the elements of $[n]$ as distinguishable balls and the elements of $[k]$ as distinguishable boxes. Then, by part (c) of Problem 2.52, the answer is $(k)_n$.

   (c) Using the same model as in part (b), the answer is $S(n,k)k!$ by part (b) of Problem 2.52.

2. Set $b_i = a_i + i - 1$. The reader is invited to fill in the details.

3. (a) We claim that this number is 2^{n-1}. Indeed, use the result of Exercise 13 of Chapter 1, replacing n by $n-1$.

   (b) For $n = 1$, the statement is true. Now assume it is true for $n-1$, and prove it for n. We are going to construct a 2-to-1 map f from the set $C(n)$ of all compositions of n into the set $C(n-1)$ of all compositions of $n-1$. The map f is defined as follows: Let $c = (c_1, c_2, \cdots, c_k) \in C(n)$. Set
$$f(c) = \begin{cases} (c_1 - 1, c_2, \cdots, c_k) \text{ if } c_1 > 1, \\ \\ (c_2, \cdots, c_k) \text{ if } c_1 = 1. \end{cases}$$

It is easy to see that f is indeed 2-to-1. Indeed, each d element of $C(n-1)$ has two preimages under f; one can be obtained by increasing the first element of d by 1, and the other can be obtained by prepending d by a 1. The reader is asked to verify that the 2^{n-1} compositions of n we obtain this way are indeed all different.

4. There are infinitely many weak compositions of n if the number of parts is not specified. Indeed, we are free to add as many zeros to the end of any weak composition of n as we would like.

5. As the pairs of siblings cannot be separated, let us consider each pair as one "superchild." Then we have 10 children, and we can divide them into four playgroups in any way we want, so we have $S(10, 4)$ possibilities.

6. We will consider the equivalent problem of distributing 50 identical balls into four distinguishable boxes so that each box gets an odd number of balls. First we put one ball in each box. Then we have 46 balls left, which we need to distribute into four boxes so that each of the boxes gets an even number of balls. Note that it is not required that each box gets a ball. This is the same as the number of ways to distribute 23 identical balls into four distinguishable boxes, so the number we are looking for is $\binom{23+4-1}{4-1} = \binom{26}{3}$.

7. There is a bijection f from the set S of acceptable distributions into the set T of all weak compositions of 11 into six parts. Indeed, if $a = (a_1, a_2, \cdots, a_6)$ denotes an acceptable distribution, that is, the first three a_i are at least two, and the last three a_i is at least one, then we set
$$f(a) = (a_1 - 2, a_2 - 2, a_3 - 2, a_4 - 1, a_5 - 1, a_6 - 1).$$
Since the sum of the a_i is 20, the sum of the arguments of $f(a)$ is 11. Therefore, the number of all possibilities is $\binom{11+6-1}{6-1} = \binom{16}{5}$.

8. We apply the subtraction principle here. The number of *all* weak compositions of 10 into four parts is $\binom{13}{3} = 286$. Let us count the *bad* ones among these, that is, those that contain a part 9 or larger. Clearly, there are four weak compositions that contain 10. There are 12 that contain 9. Indeed, 9 can be at four places within the composition, and then the remaining 1 can be at three places, and the product principle implies our claim. Therefore, by the addition principle, there are $4 + 12 = 16$ bad weak compositions of 10, so there are $286 - 16 = 270$ good ones.

9. This is the same as distributing n identical balls (the elements of the domain of our functions) into n distinguishable boxes (the elements of the range of our functions). Once the balls are distributed, we number them 1 through n, but because of the requirement that our function be monotone, we must label them so that the balls in the

first nonempty box get the smallest numbers, the balls in the second nonempty box get the next smallest numbers, and so on. Therefore, the number of monotone functions from $[n]$ into $[n]$ is $\binom{2n-1}{n-1}$.

10. There is a bijection between the set of these compositions of 24 and the set of all compositions of 8. The bijection simply divides each part by three. Therefore, the number we are looking for is simply the number of compositions of 8, which is $2^7 = 128$.

11. (a) We show that both sides count all words of length n that can be built over an x-element alphabet. For the left-hand side, this is a direct consequence of the product principle, or, more precisely, Corollary 1.11. The right-hand side is a little bit more complicated. Let us count those words in which exactly k different letters appear. There are $S(n,k)$ ways to split up the n positions of the word among the k kinds of letters; then there are $(x)_k$ ways to choose the actual letters for these positions. Note that the order of the letters matters; this is why we have not only $\binom{x}{k}$ but $(x)_k$ choices. So, by the product principle, there are $S(n,k)(x)_k$ such words, and then the claim is proved by summing for all k.

    (b) The result of part (a) shows that the two polynomials x^n and $\sum_{k=0}^{n} S(n,k)x^k$ of degree n agree for infinitely many values of x, that is, for all positive integers. That means that their difference is zero infinitely many times. However, their difference is also a polynomial of degree at most n, so the only way it can have more than n roots is when it is the zero polynomial.

12. The set of all polynomials with real coefficients forms a vector space over the field of real numbers. One basis of this field is $\{1, x, x^2, \cdots\}$, and another one is $\{1, (x)_1, (x)_2, \cdots\}$. The result of the previous exercise shows that the *transition matrix* between these two bases is the infinite matrix S formed from the Stirling numbers of the second kind. That is, $S_{n,k} = S(n,k)$ for all nonnegative integers k and n. The first few entries of S are shown below.

$$S = \begin{pmatrix} 1 & 0 & 0 & 0 & 0 & \cdots \\ 0 & 1 & 0 & 0 & 0 & \cdots \\ 0 & 1 & 1 & 0 & 0 & \cdots \\ 0 & 1 & 3 & 1 & 0 & \cdots \\ 0 & 1 & 7 & 6 & 1 & \cdots \\ \cdots & \cdots & \cdots & \cdots & \cdots & \cdots \end{pmatrix}.$$

In other words, if we set $\mathbf{x} = \begin{pmatrix} 1 \\ x \\ x^2 \\ \cdots \end{pmatrix}$ and $\mathbf{x}' = \begin{pmatrix} 1 \\ (x)_1 \\ (x)_2 \\ \cdots \end{pmatrix}$, then we will have $S\mathbf{x}' = \mathbf{x}$.

13. Let us write the elements of [10] in a line, in one of 10! ways, then let us insert a bar after the third and the sixth element. This will certainly give us a partition of the kind we need. However, each partition will be obtained many times. That is, we will get the same partition if we permute the first three elements among each other in 3! ways, the second three elements among each other in 3! ways, and the last four elements among each other in 4! ways. Finally, we can switch the block of the first three elements with the block of the second three elements without changing the partition. Therefore, by the product principle, each partition is obtained in $3! \cdot 3! \cdot 4! \cdot 2$ ways. So, by the division principle, the number of partitions of [10] of the desired kind is
$$\frac{10!}{3! \cdot 3! \cdot 4! \cdot 2}.$$

14. If there are no ties at all, then there are $5! = 120$ possible outcomes.

    If there is one two-way tie and no other ties, then there are $\binom{5}{2} = 10$ choices for the two people in the tie, then there are 4! possible rankings, leading to 240 possibilities by the product principle. If there are two two-way ties, then there are $\binom{5}{2}\binom{3}{2}/2 = 15$ choices for the two pairs in the ties, so there are six possible rankings, leading to 90 different possibilities.

    If there is one three-way tie and no other ties, then we have $\binom{5}{3} = 10$ choices for the triple in the tie, so we have six possible rankings, leading to 60 possibilities. Finally, if there is a three-way tie and a two-way tie, then we have $\binom{5}{3} = 10$ choices for the triple in the tie, so we have two possible rankings, leading to 20 possibilities.

    Therefore, the total number of possible outcomes is
$$120 + 240 + 90 + 60 + 20 = 530.$$

15. Say the block containing the element $n+1$ has $n-k$ additional elements. Then there are $\binom{n}{n-k} = \binom{n}{k}$ ways to choose these elements, so there are $B(k)$ ways to partition the rest of $[n+1]$. The result then follows by the addition principle and summation over all k.

16. We claim that there is a simple bijection between these two sets of partitions, more precisely, between their Ferrers shapes. Indeed, partitions of n into at most k parts correspond to Ferrers shapes on n boxes that have at most k rows. Partitions of n into parts not larger than k correspond to Ferrers shapes on n boxes that have at most k columns. So taking conjugates is a bijection between these two sets.

17. Let (a_1, a_2, a_3, a_4) be such a partition. Then $(a_1-3, a_2-2, a_3-1, a_4)$ is a partition of 8. Vice versa, if (b_1, b_2, b_3, b_4) is a partition of 8,

then $(b_1+3, b_2+2, b_3+1, b_4)$ is a partition of 14 into four distinct parts. Therefore, the number we are looking for is the number of partitions of 8 into four parts. There are five such partitions, namely, $(5,1,1,1)$, $(4,2,1,1)$, $(3,3,1,1)$, $(3,2,2,1)$, and $(2,2,2,2)$.

18. The only such integer is 2. Indeed, if $n = 2k+1$, then the partition consisting of a row and a column of length $k+1$ is self-conjugate. If $n = 2k+4$, then the partition consisting of a row of length $k+2$, a column of length $k+2$, and a second row of length 2 is self-conjugate. Both constructions work for $k \geq 0$, leaving $n = 2$ as the only exception. The reader is encouraged to find an alternate proof based on the result of Supplementary Exercise 19.

19. (a) We claim that the degree of this polynomial is $k-1$, and we are going to prove this, and the fact that $p_k(kn)$ is a polynomial, by induction on k. For $k = 1$, the equality $p_k(kn) = 1$ holds, and the statement is true. Otherwise, we claim that

$$p_k(kn) = p_{k-1}(kn) + p_k((n-1)k). \tag{2.12}$$

Indeed, a partition enumerated by the left-hand side either has fewer than k parts, and then it is enumerated by the first term of the right-hand side, or has exactly k parts, and then subtracting 1 from each of its parts we get a partition enumerated by the second term of the right-hand side. After rearranging, we get

$$p_k(kn) - p_k((n-1)k) = p_{k-1}(kn). \tag{2.13}$$

In other words, the difference of two consecutive values of $p_k(kn)$ is, by the induction hypothesis, a polynomial of degree $k-2$. This implies that $p_k(kn)$ is a polynomial of degree $k-1$. The reader is encouraged to prove this proposition by induction on k.

(b) Analogous to part (a), only (2.12) needs to be replaced by

$$p_k(kn+r) = p_{k-1}(kn+r) + p_k((n-1)k+r). \tag{2.14}$$

20. Let us assume that the opposite is true; that is, let us assume that there is a polynomial $q(n)$ for which (2.11) does not hold. Let q have degree $k-2$. Note that $p(n) > p_k(n)$ for any $n > k$. On the other hand, the previous exercise shows that $p_k(n)$ can be described by k polynomials, each of degree $k-1$, one for each residue class modulo k. Therefore, even $p_k(n)$ grows faster than $q(n)$, implying that so too does $p(n)$.

Note that a function like $p_k(n)$, that is, a function defined by

$$f(n) = \begin{cases} f_0(n) \text{ if } n = kr, \\ f_1(n) \text{ if } n = kr + 1, \\ \cdots \\ f_{k-1}(n) \text{ if } n = kr + (k-1) \end{cases}$$

where the f_i are polynomials, is called a *quasi-polynomial.* Some people in group theory tend to call them *polynomials on residue classes* or *porcs.*

21. Let $s = (s_1, s_2, \cdots, s_u)$. We added 1 to the first $least(s)$ parts of a. Before that, these parts were consecutive integers; therefore, they will be consecutive after this addition. On the other hand, because of $least(s) < stair(s)$, we originally had $stair(s) = least(s) + 1$, so we had $s_{least(s)+1} = s_{least(s)} - 1$. We did not increase $s_{least(s)+1}$, but we did increase $s_{least(s)}$, so we broke the line of parts that are consecutive integers at the $least(s)$th part. This shows that $k(g(s)) = least(s)$.

    To prove the second statement, note that the parts of s are all distinct. In particular, the next-to-last part s_{u-1} of s is larger than $least(s) = k(g(s))$. The statement then follows as either $l(g(s)) = s_{u-1}$ or $l(g(s)) = s_{u-1} + 1$.

22. The first statement is trivial because we defined the last row of $g(s)$ by taking one box from each of the first $stair(s)$ rows. To see the second one, note that decreasing each element of a decreasing sequence of consecutive integers by 1 will result in another decreasing sequence of consecutive integers.

23. First, it is impossible to have $a(3a+1) = b(3b+1)$ or $a(3a-1) = b(3b-1)$ because both functions $f(m) = m(3m+1)$ and $g(m) = m(3m-1)$ are strictly monotone on the positive integers. Second, we cannot have $a(3a+1) = b(3b-1)$ either, since that equation is equivalent to $b - a = \frac{1}{3}$.

24. The number of partitions of n into distinct parts is the same as the number of partitions of $2n$ into distinct even parts (just multiply each part by 2). Let S be the set of the latter, and let T be the set of partitions of $2n$ in which each successive rank is 1. We define a bijection $f : S \rightarrow T$ as follows: Let $s \in S$, and let s_1 be the first part of s_1. Then s_1 is even, say $s_1 = 2h$. Start building a Ferrers shape by taking a first row of length $h+1$ and a first column of length h. This will take precisely $2h = s_1$ boxes because of the overlap in the corner. Then continue on this way. That is, if the second part of s is $s_2 = 2i$, then add a second row of length $i+1$ and a second column of length i (not counting the boxes already in the shape) to

the shape we build, and so on. When we use up all rows of s, we will get a Ferrers shape on $2n$ boxes in which each successive rank is 1. Let $f(s)$ be the partition that belongs to that Ferrers shape. See Figure 2.16 for an example. We leave the easy task of proving that f is a bijection to the reader.

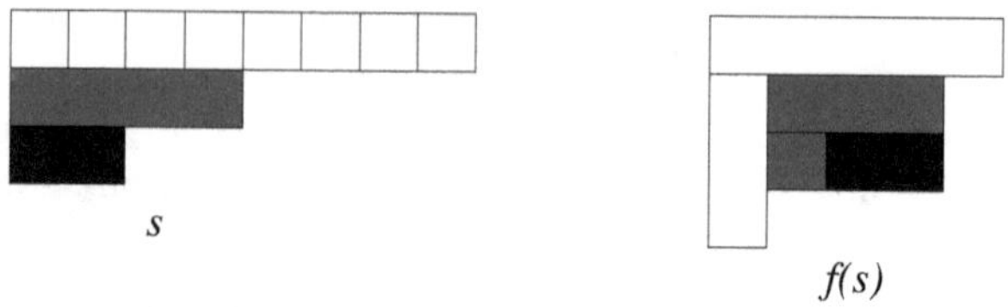

Figure 2.16
Parts become stripes.

25. (a) Let us say that during the party every pair of people who know each other shakes hands once. If the total number of handshakes is m, then the sum we are interested in is $2m$ since each handshake takes two people. In other words, when we sum the numbers written down, each handshake gets counted twice, once for each person who participated in that handshake.
    (b) No. Let A be the person who knows most people in the company, and let the company have n people. Then the length of the first row of p is simply the number of other people A knows (so less than n), while the length of the first column is simply n. Therefore, $r_1 \leq -1$.
    (c) No. Let B be the person who knows most people in the company except for A. Then the length of the second row is the number $|B|$ of people that B knows. Let $|A| + |B| = n + t$. That means there are at least t people who know both A and B. Therefore, the length of the second column is at least $t + 2$ (since A and B know at least two people as well). Therefore, the sum of the first two rows is $n + t$, while the sum of the first two columns is at least $n + t + 2$, so $r_1 + r_2 \leq -2$.

    Note that even more is true. The famous Erdős–Gallai criterion says that the partitions that can play the role of p are precisely those in which

$$r_1 + r_2 + \cdots + r_i \leq -i,$$

    for all $i \in [k]$, where k is the side length of the Durfee square of p.

26. Because the third part of p is 2, the Durfee square of p is of size 2. Since taking conjugates leaves the 2×2 Durfee square fixed, it follows that the number of boxes in a self-conjugate Ferrers shape with this Durfee square is even.

27. We can instead count the permutations of these digits in which two identical digits get into consecutive positions. Let us first count those, the elements of the set A, in which the two 1s are in consecutive positions. Replacing the two 1s by a symbol S, we see that $|A| = 6!/2$. Similarly, if B is the set of permutations of our digits in which the two 2s are in consecutive positions, then $|B| = 6!/2$ by the same argument. Finally, $|A \cap B| = 5!$, since we can replace the two 1s by an S, and the two 2s by a T. So Lemma 2.32 shows that $|A \cup B| = 6! - 5!$. Therefore, by the subtraction principle, the number of permutations of our multiset in which identical digits are not in consecutive positions is
$$\frac{7!}{2 \cdot 2} - (6! - 5!).$$

28. Clearly, all partitions of [10] that were bad for the purposes of Example 2.36 are bad here, too. However, there are additional bad partitions, namely, those in which $\{1,2\}$ is a block, or $\{1,3\}$ is a block, or $\{2,3\}$ is a block, or $\{1,2,3\}$ is a block. Fortunately, these sets of partitions are pairwise disjoint, so, by the addition principle, their total number is $3B(8) + B(7)$. Therefore, the total number of bad partitions is this number plus the number of bad partitions from Example 2.36, that is, $3B(9) + 2B(7)$. So the number of good partitions is $B(10) - 3B(9) - 2B(7)$.

29. Let A_1 be the set of integers in [1000] that are not relatively prime to 7; let A_2 be the set of those that are not relatively prime to 8, or, equivalently, are even. Then $|A_2| = 500$. And since A_1 consists of positive integers divisible by 7, we have $|A_1| = 1000/7 = 142$. Finally, $A_1 \cap A_2$ is the set of positive integers in [1000] that are divisible by 14, so $|A_1 \cap A_2| = 1000/14 = 71$. Therefore, by Lemma 2.32, we get
$$|A_1 \cup A_2| = 142 + 500 - 71 = 571.$$

30. We claim that each employee has to speak at least three languages. Let us assume that the converse is true, that is, Albert speaks only English and Spanish. There is someone at the company who does not speak English, say Bella. Then Bella must speak Spanish, or Albert, Bella, and any third person would violate the conditions. However, there is someone at the company who does not speak Spanish, say Christine, and then Albert, Bella, and Christine have no language in common.

    For an example for such a company, let the company have any number of employees, and divide them into five nonempty groups, A, B, C, and D. Let us say that A is the group of people not speaking English, B is the group of people not speaking Spanish,

C is the group of people not speaking German, D is the group of people not speaking French, and E is the group of people speaking all four. Because everyone speaks at least three languages, nobody belongs to two groups. Then, no matter how we choose three people there will be at least one group among the first four from which we did not choose anyone, so all three chosen people will speak the language not spoken in that group.

31. If x does not belong to any A_i, then none of the terms of the right-hand side counts it, so nor does their sum.

32. We have

$$\begin{aligned} \sum_{i=0} \binom{n}{i} D(i) &= \binom{n}{n-i} D(i) \\ &= n!. \end{aligned}$$

Indeed, there are $n!$ permutations of length n. There are $\binom{n}{n-i}D(i)$ of them in which there are exactly $n-i$ fixed points and in which the remaining i entries form a derangement.

33. We are going to show that the right-hand side also counts the elements that belong to exactly $n-2$ of the A_i. If x is contained in exactly $n-2$ of the A_i, then x is counted once by the first term of the right-hand side and zero times by the two other terms. If x is contained by exactly $n-1$ of the A_i, then x is counted $\binom{n-1}{n-2} = n-1$ times by the first term, and then subtracted $n-1$ times by the second term. The third term counts x zero times, so, altogether, x is counted once. Finally, if x belongs to all of the A_i, then the first term counts x exactly $\binom{n}{n-2} = \binom{n}{2}$ times, the second term subtracts x exactly $(n-1)\binom{n}{n-1} = (n-1)n$ times, and the third term counts x exactly $\binom{n}{2}$ times. However, we have

$$\binom{n}{2} - (n-1)n + \binom{n}{2} = 0,$$

completing the proof of our claim that the right-hand side counts each element x with the desired property exactly once.

34. The proof is very similar to that of the previous exercise. One checks the same way that each element x that is contained in exactly $n-3$, or $n-2$, or $n-1$ of the A_i is counted exactly zero times on the right-hand side. For elements that belong to all A_i, this follows from the identity

$$\binom{n}{3} - (n-2)\binom{n}{2} + \binom{n-1}{2} n - \binom{n}{3} = 0.$$

35. We claim that

$$B_k = \sum_{i=k}^{n} (-1)^{i-k} \binom{i}{k} D_i.$$

Let x be an element that is contained in exactly t sets A_i. Then there are three possibilities.

(a) If $t < k$, then none of the D_i on the right-hand side counts x, so x is counted by the right-hand side zero times.

(b) If $t = k$, then the only term of the right-hand side that counts x is the one indexed by $i = k$, and that term is equal to $\binom{k}{k} D_k = D_k$, so x is counted exactly once by the right-hand side.

(c) If $t > k$, then, for any i satisfying $k \leq i \leq t$ there are $\binom{i}{k}$ different i-fold intersections of the A_i counted by D_i. Therefore, x is counted

$$\sum_{i=k}^{t} \binom{t}{i} \binom{i}{k} (-1)^{i-k}$$

times by the right-hand side. Using Supplementary Exercise 16b of Chapter 1, we see that the above expression is equal to

$$\begin{aligned}\sum_{i=k}^{t} \binom{t}{k} \binom{t-k}{i-k} (-1)^{i-k} &= \binom{t}{k} \left(\sum_{i=k}^{t} \binom{t-k}{i-k} (-1)^{i-k} \right) \\ &= \binom{t}{k} (1-1)^{t-k} = 0,\end{aligned}$$

where in the last step we used the binomial theorem and the fact that $t - k$ is positive.

36. Let F_i be the set of all positive integers divisible by i. Then no matter how we select an infinite number of F_i, their intersection is always empty since no positive integer has infinitely many divisors. On the other hand, if we only select a finite number of F_i, say $i_1, i_2, \cdots, i_n$, their intersection is infinite since it contains all positive integers divisible by $i_1 i_2 \cdots i_n$.

Alternatively, let $F_i = [i, \infty)$. It is straightforward to verify that both requirements are satisfied.

37. Let A_i be the set of elements of $[n]$ that are divisible by p_i. Then $|A_i| = n/p_i$, and, in general,

$$|A_{i_1} \cap A_{i_2} \cap \cdots \cap A_{i_r}| = \prod_{\substack{s \\ s \notin \{i_1, i_2, \cdots, i_r\}}} p_s.$$

Applying the principle of inclusion–exclusion, the result now follows, since, by standard algebraic computations,

$$\prod_{i=1}^{t} (p_i - 1) = \sum_{j=0}^{t} (-1)^j \sum_{K} \prod_{k \in K} p_k,$$

where K ranges over all k-element subsets of $[t]$.

38. If there are no ties, then there are $6! = 720$ possible outcomes. If there is one two-way tie, then there are $\binom{6}{2} = 15$ ways to choose the two players in that tie, then $5! = 120$ ways to complete the ranking, for a total of $15 \cdot 120 = 1800$ possibilities. If there are two two-way ties, then there are $\binom{6}{2} \cdot \binom{4}{2}/2 = 45$ ways to choose those pairs, then $4! = 24$ ways to complete the ranking, for a total of $45 \cdot 24 = 1080$ possibilities. Finally, if there are three two-way ties, then there are $\binom{6}{2} \cdot \binom{4}{2} \cdot \binom{2}{2}/6 = 15$ ways to select those three pairs, then six ways to rank them, for a total of $15 \cdot 6 = 90$ possible outcomes. So the total number of possible rankings is $720 + 1800 + 1080 + 90 = 3690$.

39. The total number of blocks in all partitions of $[n]$ is $\sum_{k=1}^{n} kS(n,k)$. Now use Theorem 2.10.

40. Set $x = -1$ in (2.10).

2.10 Supplementary exercises

1. A student has to take 18 hours of classes per week. He must have at least two hours of classes on each of the five weekdays. Furthermore, classes on Tuesday come in two-hour blocks, so the number of hours of classes he takes on Tuesday must be even. How many ways are there for this student to distribute his weekly workload?

2. What is the number of compositions of 14 into four distinct parts?

3. What is the number of compositions of 50 into three parts of which none is divisible by three?

4. Let n and k be odd positive integers. Find a formula for the number of all compositions of n into k odd parts.

5. A patient is recovering in a hospital after having serious surgery, and her physician wants to increase her daily exercise gradually. On the first day, the patient is allowed one walk to the cafeteria, on the second day, two walks, and so on. The patient stays in the hospital for n days, during which she walks to the cafeteria only three times. In how many different ways is this possible?

6. A writing competition has 12 thousand dollars available for prizes. The jury wants to honor contestants A and B with an equal amount and give C and D some lesser prizes (not necessarily equal). How many possibilities are there if every prize has to be a multiple of 1000?

7. Find explicit formulae for $S(n,2)$ and $S(n,3)$.

8. Find explicit formulae for $S(n, n-1)$ and $S(n, n-2)$.
9. How many ways are there to partition $[12]$ into two blocks of size 3 and three blocks of size 2?
10. We want to divide 12 children into any number of playgroups. However, in order to avoid some predictable disagreements, we want Adam and Brenda to be in different playgroups. How many ways are there to do this? (Playgroups with one child in them are allowed.)
11. We want to divide 10 children into playgroups so that no group has more than 7 children in it. In how many ways can we do this?
12. A small-business owner wants to reward some of his five employees with extra days off. He wants to give away a total of 10 paid holidays to his five workers. No worker is to receive more than five of these holidays. How many choices does the owner have?
13. Prove that, for positive integers $k \leq n$, we have
$$S(n+1,k) = \sum_{i=0}^{n} \binom{n}{i} S(i, k-1).$$
14. Prove that, for positive integers $k \leq n$, we have
$$S(n,k) = \sum_{m=1}^{n} k^{n-m} S(m-1, k-1).$$
15. The number $f(n) = B(n) - \sum_{k=1}^{n} B(n-k)\binom{n}{k}$ counts certain partitions of $[n]$. Which partitions?
16. Let S be a partition of $[n]$, and let us write the blocks of S in non-increasing order of their sizes. If the block sizes are $a_1, a_2, \cdots, a_k$, then let us say that S has *type* $a = (a_1, a_2, \cdots, a_k)$. For instance, partition $\{\{1,3\},\{2,4,\},\{5\}\}$ has type $(2,2,1)$. Find a formula for the number of all partitions of $[n]$ having type $a = (a_1, a_2, \cdots, a_k)$.
17. How many different types are possible for a partition of $[n]$? (The type of a partition of $[n]$ is defined in the previous exercise.)
18. Find all self-conjugate partitions of 16 and 19.
19. Prove that the number of partitions of n into distinct odd parts is equal to the number of self-conjugate partitions of n.
20. What is the number of partitions of $2n$ into exactly n parts?
21. Prove that the number of partitions of n in which the first k parts are distinct (and possibly zero) is equal to $p\left(n - \binom{k}{2}\right)$.
22. Prove that, for all positive integers n, we have $p(n) < p(n+1)$.

23. Prove that, for all positive integers $n \geq 2$, we have

$$2p(n) \leq p(n-1) + p(n+1).$$

When does equality hold?

24. Find exact formulae for $p_{n-2}(n)$ and $p_{n-3}(n)$.

25. Let $p(n,k)$ be the number of partitions of n into exactly k parts. Find three distinct numbers k, l, and m so that, if $n \geq \max(k,l,m)$, then

$$p(n,k) = p(n,l) = p(n,m).$$

26. Prove, with a direct combinatorial argument, that for all positive integers n, we have

$$q(1) + q(2) + \cdots + q(n) \leq 2^n,$$

where $q(n)$ denotes the number of partitions of n into distinct parts.

27. Let $c > 0$ be any real number. Prove that $p_{cn}(n) \sim p(n)$. In other words, for large n, almost all partitions of n have at most cn parts. Note that c may be very small, such as 10^{-6}.

28. Find an explicit formula for $p_2(n)$.

29. Find an explicit formula for $p_3(n)$.

30. Let $p'_k(n)$ denote the number of partitions on n into *exactly* k parts. Prove the identity

$$p'_{k+1}(n) = p'_k(n-1) + p'_{k+1}(n-k-1).$$

31. A library is closed on Sunday, closed on the first of every month, and is otherwise open every day. If this year is not a leap year, and January 1 is a Sunday, then how many days will the library be open this year?

32. Prove that the number of partitions of n in which the size of the average part is at most two is equal to the number of partitions of n in which the largest part is at least as large as the sum of all the other parts. (A precise description of the second category is that we consider partitions $(a_1, a_2, \cdots, a_k)$ in which $a_1 \geq \sum_{i=2}^{k} a_i$.)

33. Find the number of positive integers with at most three digits that are not divisible by five and are not divisible by six.

34. A group of 10 people includes a child and the two parents of the child. How many ways are there to partition this set of people so that the child is in the same block as at least one of his parents?

35. Find a formula for the number of all partitions of $[n]$ that have no singleton blocks.

36. Let n and k be fixed positive integers. Find a formula for the number of partitions of $[n]$ into k blocks, none of which are singletons.

37. How many ways are there to arrange the elements of $[n]$ in a line so that the first element is smaller than the last element, but larger than the next-to-last element?

38. Find a formula for the number of positive integers not larger than 300 that are divisible by at least one of 2, 3, and 5.

39. A group of n people went to a theater, and each of them left his coat and his suitcase in the coatroom. When the group wanted to leave, the coatroom attendant handed each person a coat and a suitcase in a random fashion.

    (a) In how many ways could the attendant return the coats and suitcases if nobody got his own coat back and nobody got his own suitcase back?

    (b) In how many ways could the attendant return the coats and suitcases if some people might have gotten *one* of their belongings back, but nobody got *two* of his belongings back?

In Supplementary Exercises 40–43, the number of boxes is not specified in advance. Therefore, we require that each box get at least one ball, since otherwise we could keep adding empty boxes, and the answer to each question would be "infinite."

Your answers should not contain a summation sign, except for Supplementary Exercise 42.

40. Find the number of ways to distribute n identical balls in identical boxes.

41. Find the number of ways to distribute n identical balls in distinguishable boxes.

42. Find the number of ways to distribute n distinguishable balls into distinguishable boxes.

43. Find the number of ways to distribute n distinguishable balls into identical boxes.

In Supplementary Exercises 44–46, your answer should contain only one summation sign.

44. Let $A_1, A_2, \cdots, A_n$ be finite sets. Find a formula for the number of elements that belong to at least $n-2$ of the A_i.

45. Let $A_1, A_2, \cdots, A_n$ be finite sets. Find a formula for the number of elements that belong to at least $n-3$ of the A_i.

46. Let $A_1, A_2, \cdots, A_n$ be finite sets, and let $k \leq n$. Find a formula for the number of elements that belong to at least k of the A_i.

47. Let F_n be the number of partitions of $[n]$ that do not contain a singleton block. Find a simple relation between the numbers F_n and the Bell numbers.

48. Let b_n be the number of ways to place any number of non-attacking rooks on the boxes of an $(n-1, n-2, \cdots, 1)$-Ferrers shape. Describe the numbers b_n.

3

Generating functions

In this chapter, we will learn our most advanced counting technique so far. We will learn how to encode *all elements* of a possibly *infinite* sequence by *one* single function, the generating function of the sequence. Often, we will first obtain the generating function of a sequence and then decode it, that is, we will then compute the elements of the sequence from the generating function. This idea is a powerful example of using continuous objects in discrete mathematics.

Let us start with a very short review of the type of functions we will use. Rest assured that the book will return to discussing combinatorics very soon.

3.1 Power series

The reader is undoubtedly a good student, and, therefore, the reader surely remembers *power series* from calculus. These are much like polynomials, that is, sums of various powers of the variable x multiplied by constants. But, unlike polynomials, power series can be infinitely long. So, for instance, $\sum_{k=0}^{n} x^k$ is a polynomial, but $\sum_{k=0}^{\infty} x^k$ is a power series.

3.1.1 Generalized binomial coefficients

One frequently encountered example of power series is the binomial theorem with real exponents. In order to review that theorem, we first have to extend the notion of the binomial coefficient $\binom{a}{k}$ to all *real* numbers a (the number k still has to be a nonnegative integer). After all, we cannot define $\binom{-2.5}{k}$ to be the number of all k-element subsets of the set $[-2.5]$, because the set $[-2.5]$ does not exist.

While the combinatorial definition of $\binom{a}{k}$ does not survive if a is not a nonnegative integer, the analytic definition will.

Definition 3.1 *Let a be any real number, and let k be a nonnegative integer. Then we set*

$$\binom{a}{k} = \frac{a(a-1)\cdots(a-k+1)}{k!}.$$

DOI: 10.1201/9781003304272-3

For example,

$$\binom{\frac{1}{2}}{3} = \frac{\frac{1}{2} \cdot \frac{-1}{2} \cdot \frac{-3}{2}}{3!} = \frac{3}{48} = \frac{1}{16}.$$

We are now ready to recall the binomial theorem for real exponents from calculus.

Theorem 3.2 *Let a be a real number, and let $|x| < 1$. Then*

$$(1+x)^a = \sum_{n\geq 0} \binom{a}{n} x^n = 1 + \binom{a}{1} x + \binom{a}{2} x^2 + \cdots . \tag{3.1}$$

Proof: Recall that the *Taylor series expansion* of an analytic function $f(x)$ about $x = 0$ is given by

$$f(x) = \sum_{n\geq 0} f^{(n)}(0) \frac{x^n}{n!}.$$

Let us apply this fact with $f(x) = (1+x)^a$. Computing the first few values of $f^n(0)$, we see that $f^{(0)}(0) = f(0) = 1^a = 1$, $f^{(1)}(0) = a$, $f^{(2)}(0) = a(a-1)$, $f^{(3)}(0) = a(a-1)(a-2)$, and so on. From this, one can see the trend $f^{(n)} = (a)_n$, which is then straightforward to prove by induction, using only the rules of differentiation. Since $(a)_n/n! = \binom{a}{n}$, this proves the theorem. ◊

If a happens to be a nonnegative integer, then $\binom{a}{k} = 0$ for $k > a$, so the right-hand side is a finite sum of terms of the type $c_i x^i$; in other words, it is a polynomial. If a is not a positive integer, however, then the right-hand side is an *infinite sum* of the form $\sum_{i\geq 0} c_i x^i$ and is called a *power series.*

In calculus, you must have spent many fun-filled hours trying to find out for which values of x certain power series converged. Some power series converged for all values of x, such as $\sum_{n\geq 0} x^n/n!$. Some others only converged for one value of x, and that value was zero in many examples, such as for $\sum_{n\geq 0} n!x^n$. You learned various tests, such as the ratio test and the root test, that help to determine the values of x for which a given power series is convergent.

When a power series converges for all $x \in [a, b]$, then there is a chance that we can write that power series in a closed form.

For example, the reader may remember that the power series $F(x) = \sum_{n\geq 0} x^n$ converges for $x \in (-1, 1)$ and, in that interval,

$$F(x) = \frac{1}{1-x}. \tag{3.2}$$

The reader is invited to verify that this is in fact a special case of the binomial theorem, with $-x$ replacing x, and $a = -1$.

Power series will be our main tool in this chapter. We will like them so much that we will not even mind if they are not convergent at any interval; we will still work with them.

3.1.2 Formal power series

Now we will explain a rather subtle difference between power series and the objects we will use in this chapter, called *formal power series.* At this point, the reader can take two courses of action. For a faster, but mathematically less rigorous coverage, the reader can simply accept that, most of the time, we will not be interested in the convergence intervals of our power series, skip the rest of this section (with some abuse of language "we will pretend that they are always convergent"), and continue with Section 3.2. For a somewhat slower, but mathematically more justified coverage, the reader should continue with this section. We prefer the latter. If the reader chooses the former and later encounters some problems understanding the material, then the reader should return to this section and read the skipped parts.

A *formal power series* is an expression

$$F(x) = f_0 + f_1 x + f_2 x^2 + \cdots .$$

If we had not added the word "formal," then purists could protest that, in certain cases, for instance, if $f_n = n!$, this definition would lead to the sum $F(x) = \sum_{n\geq 0} n!x^n$, which is not finite (convergent) for any $x \neq 0$, so $F(x)$ would be undefined for any $x \neq 0$.

We, however, do not care. As you will see in this chapter, the use of formal power series will typically not be connected to substituting values of x into them as much as it will be connected to observing their coefficients. These coefficients f_n always exist.

We can add and multiply formal power series as if they were polynomials. That is, if $A(x) = \sum_{n\geq 0} a_n x^n$ and $B(x) = \sum_{n\geq 0} b_n x^n$ are two formal power series, then

$$(A + B)(x) = \sum_{n\geq 0} (a_n + b_n)x^n,$$

and also

$$(A \cdot B)(x) = \sum_{n\geq 0} \left(\sum_{i=0}^{n} a_i b_{n-i} \right) x^n. \tag{3.3}$$

This rule agrees with the rule from elementary algebra that states that the product of two sums is equal to the sum of all two-factor products whose first and second factors are terms of the first and second sum, respectively. Because of the central role and importance of this rule in this chapter, we ask the reader to take a moment and digest it. Figure 3.1 is meant to help the reader see what the coefficient of x^n is in $A \cdot B$.

If $A(x)$ is a formal power series, then $B(x) = 1/(A(x))$ is the expression that satisfies $A(x)B(x) = 1$. This expression $B(x)$ will sometimes be a formal power series, such as when $A(x) = \sum_{n\geq 0} x^n = \frac{1}{1-x}$, leading to $B(x) = (1-x)$, which is a polynomial (and so a formal power series). There will be times when $B(x)$ will not be a formal power series, such as when $A(x) = x$. In that case, we would have $B(x) = \frac{1}{x}$, which is not a formal power series because a

$$a_0 + a_1x + \ldots\ldots\ldots\ldots + a_nx^n$$

$$b_nx^n + b_{n-1}x^{n-1} + \ldots\ldots\ldots\ldots + b_0$$

Figure 3.1
The product of the two terms in column i is $a_ib_{n-i}x^n$, so that the coefficient of x^n in $A \cdot B$ is $\sum_{i=0}^{n} a_ib_{n-i}$.

formal power series cannot contain a term with a negative exponent. (Series that can contain terms with negative exponents are called *Laurent series.*) In Supplementary Exercise 1, you will be asked to characterize formal power series $A(x)$ for which $1/A(x)$ is also a formal power series.

If $A(x)$ is a formal power series, and p is a positive integer, then $A(x)^{1/p}$ is the expression $D(x)$ satisfying $(D(x))^p = A(x)$. Again, $D(x)$ may or may not be a formal power series.

We will *define* the derivative of a formal power series by setting

$$f'(x) = \left(\sum_{n\geq 0} f_nx^n\right)' = \sum_{n\geq 1} nf_nx^{n-1}.$$

Note that we had to *define* this as f is a *formal* power series, so it could well be that $f(x)$ is not convergent in any interval, and then it is of course not differentiable in the traditional sense. Similarly, we can define

$$\int f(x) = \sum_{n\geq 0} \frac{a_n}{n+1}x^{n+1} + C.$$

The reader is invited to verify that both differentiation and integration (with $C = 0$) of formal power series is a linear operation, that is, $(f+g)' = f' + g'$, and we also have $(cf)' = cf'$ for any real number c, and the same holds for integration.

Note that differentiation and integration are defined so that they *coincide* with the traditional notion of differentiation and integration in the case when our power series are differentiable or integrable in the traditional sense.

In order to practice operations on formal power series, let us prove, in two different ways, that

$$\sum_{n\geq 0}(n+1)x^n = 1+2x+3x^2+\cdots = \frac{1}{(1-x)^2}. \tag{3.4}$$

Indeed, on one hand, we know from (3.2) that $\sum_{n\geq 0} x^n = 1/(1-x)$. Differentiating both sides, we get the statement to be proved.

On the other hand, the right-hand side is the product of $A(x) = 1/(1-x) = \sum_{n\geq 0} x^n$ and $B(x) = 1/(1-x) = \sum_{n\geq 0} x^n$. That is, we have

$$\frac{1}{(1-x)^2} = \left(\sum_{n\geq 0} x^n\right)\cdot\left(\sum_{n\geq 0} x^n\right).$$

Using formula (3.3), with $a_n = b_n = 1$ for all n, we see that the coefficient of x^n in $A(x)B(x)$ is $\sum_{i=0}^{n} a_i b_{n-i} = \sum_{i=0}^{n} 1 = n+1$, again proving our statement.

If the reader is wondering whether there is a third way to prove (3.4), then our answer is in the affirmative; a third proof can be deduced from the binomial theorem (with real exponents).

Quick Check

1. Find the formal power series form of $1/(1+x)$.
2. Find the formal power series form of $\ln(1+x)$.
3. Find the formal power series form of $\ln(1/(1-x))$.

3.2 Warming up: Solving recurrence relations

We start our study of the applications of generating functions by using them to solve recurrence relations. While many recurrence relations can be solved without generating functions, this will be a good warm-up for the more sophisticated applications of our techniques.

3.2.1 Ordinary generating functions

Five people wearing red hats start spreading a rumor that, if you do not put on a red hat, you will become sick. Every hour, each of them tells the rumor to two other people, who immediately all believe it, put on red hats, and enlist to spread the rumor further, with the same speed. No person will be told the rumor twice; this is what the red hats are for. At the same time, one person each hour will discontinue to believe in the rumor, will take his red hat off, and will stop spreading the rumor. Otherwise, nobody else will stop spreading the rumor. How many people will wear red hats after n hours?

When hearing this question, it is natural to try and see what happens within the first few hours. Let a_n denote the number of people wearing red hats at the end of hour n. Then we know that $a_0 = 5$, and that

$$a_n = 3a_{n-1} - 1 \tag{3.5}$$

if $n \geq 1$. This is because each person brings in two followers *in addition to himself*, and because one person takes his red hat off.

Formula (3.5) is enough to compute the values of $a_1, a_2, \cdots$. Indeed, $a_1 = 3a_0 - 1 = 14$, and $a_2 = 3a_1 - 1 = 41$. We could certainly continue this way, but that would become boring and tedious rather soon. Furthermore, it would not be very efficient either. Indeed, assume we only want to know what a_{200} is. Then, using the above step-by-step method, we would first have to compute $a_1, a_2, \cdots, a_{199}$, and it would be only *then* that we could compute a_{200}. Our goal is to find a formula that provides the value of a_n without asking for the value of a_{n-1} or other elements of the sequence $\{a_i\}_{i\geq 0}$. (For the rest of this book, we use the notation $\{a_i\}_{i\geq 0}$ for the sequence $a_0, a_1, \cdots$.)

The crucial idea is that we can encode all elements of the sequence $\{a_i\}_{i\geq 0}$ by one single power series. This leads to the technique of generating functions, which is the central topic of this chapter and an extremely useful method of combinatorial enumeration.

Definition 3.3 *Let $f_0, f_1, \cdots$ be a sequence of real numbers. Then the formal power series*

$$F(x) = \sum_{n\geq 0} f_n x^n$$

is called the ordinary generating function *of the sequence $\{f_i\}_{i\geq 0}$.*

The following example just repeats a fact we already know and have recently mentioned, this time in the language of generating functions.

Example 3.4 *Let $f_n = 1$ for all $n \geq 0$. The ordinary generating function of this sequence is*

$$F(x) = \sum_{n\geq 0} f_n x^n = \sum_{n=0} x^n = \frac{1}{1-x}.$$

We urge the reader to take the time and justify the summation above. (Hint: Multiply both sides of $\sum_{n=0} x^n = \frac{1}{1-x}$ by $1-x$ and see what cancels on the left-hand side.)

Example 3.5 *Let $f_n = n$. Then*

$$\begin{aligned} F(x) &= \sum_{n=0} f_n x^n = x \sum_{n\geq 1} n x^{n-1} = x\left(\frac{1}{1-x}\right)' \\ &= \frac{x}{(1-x)^2}. \end{aligned}$$

In the above two examples, we were given a sequence by an explicit formula, and we then computed the generating function using that formula. Often, as in our running example with red hats and rumors, we have to do the converse, that is, find an explicit formula for a sequence if its generating function is known. Before we attack the problem of spreading rumors, let us see an example for this reverse computation.

Example 3.6 *Find an explicit formula for f_n if*

$$F(x) = \sum_{n\geq 0} f_n x^n = \frac{x}{1-3x}.$$

Solution: The crucial observation is that f_n is nothing other than the coefficient of x^n in $F(x)$. So we have to find a formula for that coefficient or, equivalently, for the coefficient of x^{n-1} in $\frac{1}{1-3x}$. However, replacing x with $3x$ in Example 3.4, we have

$$\frac{1}{1-3x} = \sum_{n\geq 0}(3x)^n = \sum_{n\geq 0} 3^n \cdot x^n.$$

Therefore, the coefficient we are looking for is $f_n = 3^{n-1}$. $\diamond$

Our techniques in this chapter will very often ask for the coefficient of x^n in a certain formal power series $A(x)$. Therefore, we introduce the notation $[x^n]A(x)$ for that coefficient. So the result of the previous example can be expressed with this notation by writing $[x^n]F(x) = 3^{n-1}$.

In our running example, we are looking for an explicit formula for the numbers a_n. To that end, we first need to compute their ordinary generating function $A(x) = \sum_{n\geq 0} a_n x^n$. In order to be able to do that, we need an identity that is satisfied by $A(x)$. We can get such an identity as follows: Take the defining recurrence relation of our sequence, that is, the equation $a_n = 3a_{n-1} - 1$, and multiply both sides by x^n. Then sum over all values of n for which the defining equation holds; in our case, that means for $n \geq 1$. We then get

$$\sum_{n\geq 1} a_n x^n = 3\sum_{n\geq 1} a_{n-1}x^n - \sum_{n\geq 1} x^n.$$

Now note that the left-hand side is almost exactly $A(x)$; just the term a_0 is missing. That is, the left-hand side is $A(x) - a_0 = A(x) - 5$. Similarly, note that the first term of the right-hand side is nothing but $3xA(x)$, and that the second term of the right-hand side is just $x/(1-x)$. (Consider Example 3.4 again.) So the last identity implies

$$A(x) - 5 = 3xA(x) - \frac{x}{1-x},$$

$$A(x)(1-3x) = 5 - \frac{x}{1-x},$$

or, solving for $A(x)$,

$$A(x) = \frac{5}{1-3x} - \frac{x}{(1-x)(1-3x)}. \tag{3.6}$$

Just as in Example 3.6, we can now find an explicit formula for a_n by finding an explicit formula for $[x^n]A(x)$. According to (3.6), this coefficient is equal to the coefficient of x^n in $\frac{5}{1-3x}$ minus the coefficient of x^n in $\frac{x}{(1-x)(1-3x)}$, so we need to compute these two coefficients. The former is very easy to compute since

$$\frac{5}{1-3x} = 5 \cdot \frac{1}{1-3x} = 5 \cdot \sum_{n\geq 0} 3^n x^n,$$

so the coefficient of x^n in this power series is $5 \cdot 3^n$.

It is a little bit more difficult to compute the coefficient of x^n in the second term of the right-hand side of (3.6), that is, in $\frac{x}{(1-x)(1-3x)}$.

There are several ways to get around this problem. Perhaps the fastest is to decompose $\frac{x}{(1-x)(1-3x)}$ into the sum of partial fractions. That is, we can look for real numbers P and Q so that

$$\frac{x}{(1-x)(1-3x)} = \frac{P}{1-x} + \frac{Q}{1-3x}.$$

After multiplying both sides by $(1-x)(1-3x)$, we get

$$x = P + Q - x(3P + Q).$$

So we must have $P + Q = 0$, and $3P + Q = -1$, leading to $P = -1/2$ and $Q = 1/2$. Therefore, we have

$$\begin{aligned} \frac{x}{(1-x)(1-3x)} &= \frac{1}{2} \cdot \frac{1}{1-3x} - \frac{1}{2} \cdot \frac{1}{1-x} \\ &= \frac{1}{2} \sum_{n\geq 0} 3^n x^n - \frac{1}{2} \sum_{n\geq 0} x^n = \sum_{n\geq 0} \frac{3^n - 1}{2} x^n. \end{aligned}$$

The coefficient of x^n in this power series is $(3^n - 1)/2$, so that is the coefficient of x^n in $\frac{x}{(1-x)(1-3x)}$.

Therefore,

$$[x^n]A(x) = a_n = 5 \cdot 3^n - \frac{3^n - 1}{2}. \tag{3.7}$$

So this is the number of people wearing red hats after n hours.

Note that we could have used the fact that

$$[x^n]\frac{x}{(1-x)(1-3x)} = [x^{n-1}]\frac{1}{(1-x)(1-3x)},$$

and then we could have decomposed the latter into a sum of partial fractions.

We have therefore succeeded in finding an explicit formula for a_n. The reader is invited to verify that this formula is indeed correct, first by computing the first few values of a_n by this formula and finding that we obtain 5, 14, 41, as we should, and then by a more general argument that shows that our formula is indeed correct for all n. See Exercise 2.

Why did the idea of generating functions help? The defining equation (3.5) had the drawback of containing two unknowns, a_n and a_{n-1}. The generating function $A(x)$, however, comprised *all* values of a_n, so when we translated (3.5) into an equation containing $A(x)$ (namely, (3.6)), then the only unknown in that equation was $A(x)$. Therefore, we could get an explicit expression for $A(x)$ from that equation. Once we knew $A(x)$, the determination of a_n was a matter of computation.

Let us practice this method by considering another example.

Example 3.7 *Five people wearing red hats start spreading a rumor. In the first hour, each of them tells the rumor to one person who did not wear a red hat before. These new converts will put on red hats and enlist to spread the rumor further. Then the same trend continues, following the rule that each person will tell the rumor to one other person in his first hour after enlisting, and to nine other people in each subsequent hour. Nobody will ever take their red hat off. How many people will wear red hats after n hours?*

Let b_n be the number of people wearing a red hat after n hours. Then $b_0 = 5$, and $b_1 = 10$. Just as before, we start by finding a recursive rule satisfied by the numbers b_n. What do we know about b_n? First, in the nth hour, every person counted by b_{n-1} tells the story to one other person, resulting in $2b_{n-1}$ people wearing red hats. Second, in the nth hour, those people who are also counted by b_{n-2}, that is, those who are not fresh converts, tell the story to eight more people each (besides those already accounted for), resulting in $8b_{n-2}$ additional people wearing red hats. This leads to the identity

$$b_n = 2b_{n-1} + 8b_{n-2}, \tag{3.8}$$

for all $n \geq 2$.

The next step is to define the ordinary generating function $B(x) = \sum_{n\geq 0} b_n x^n$ and to translate (3.8) into an equation containing $B(x)$. To that end, let us multiply both sides of (3.8) by x^n and sum over all n for which

(3.8) holds, that is, for $n \geq 2$. We get

$$\begin{aligned}\sum_{n\geq 2} b_n x^n &= 2\sum_{n\geq 2} b_{n-1}x^n + 8\sum_{n\geq 2} b_{n-2}x^n \\ &= 2x\sum_{n\geq 2} b_{n-1}x^{n-1} + 8x^2\sum_{n\geq 2} b_{n-2}x^{n-2}.\end{aligned}$$

Now is the time to look for expressions close to $B(x)$ on both sides of the last equation. We find that the last equation is equivalent to

$$B(x) - 15x - 5 = 2x(B(x) - 5) + 8x^2 B(x),$$

$$B(x)(1 - 2x - 8x^2) = 5,$$

$$B(x) = \frac{5}{1 - 2x - 8x^2}.$$

In order to find the power series form of $B(x)$, we use the technique of partial fractions as learned in calculus and sketched in our previous example. If you feel confident that you remember that technique, you can skip the following paragraph.

We are in the fortunate situation that the denominator factors nicely, namely, $1 - 2x - 8x^2 = (1 + 2x)(1 - 4x)$. Therefore, we have

$$\frac{5}{1 - 2x - 8x^2} = \frac{C}{1 + 2x} + \frac{D}{1 - 4x},$$

$$5 = C(1 - 4x) + D(1 + 2x),$$

$$5 = x(2D - 4C) + C + D.$$

The only way that the (constant) polynomial 5 can be equal to the polynomial $x(2D - 4C) + C + D$ is by the coefficient of x in the latter being 0 and the constant term in the latter being 5. This leads to $D = 10/3$ and $C = 5/3$. Therefore, we get

$$B(x) = \frac{5}{1 - 2x - 8x^2} = \frac{5}{3} \cdot \frac{1}{1 + 2x} + \frac{10}{3} \cdot \frac{1}{1 - 4x}. \tag{3.9}$$

Using (3.4), this leads to

$$\begin{aligned}B(x) &= \frac{5}{3}\sum_{n\geq 0}(-2x)^n + \frac{10}{3}\sum_{n\geq 0}(4x)^n \\ &= \frac{5}{3}\sum_{n\geq 0}(-1)^n \cdot 2^n x^n + \frac{10}{3}\sum_{n\geq 0} 4^n x^n \\ &= \sum_{n\geq 0}\left(\frac{5}{3}(-1)^n \cdot 2^n + \frac{10}{3}4^n\right)x^n.\end{aligned}$$

Finally, we obtain $b_n = [x^n]B(x)$, that is,

$$b_n = \frac{5}{3}(-1)^n \cdot 2^n + \frac{10}{3}4^n. \tag{3.10}$$

One can check again that this formula is indeed correct for all $n \geq 0$. The reader is asked to do so in Supplementary Exercise 2.

The last two recurrence relations, that is, (3.5) and (3.8), were reasonably easy to solve because, when we multiplied them by x^n and added over all values of n for which these equations held, we got expressions in which it was quite easy to recognize the generating function of the relevant sequence. It was either just multiplied by a constant or multiplied by a power of x. Sometimes, we have to do a little more to find our generating functions in the last step. The following example illustrates this.

Example 3.8 *Let $a_0 = 1$, and let us assume that*

$$\binom{n+2}{2} = \sum_{i=0}^{n} a_i a_{n-i} \tag{3.11}$$

for all integers $n \geq 1$. Find an explicit formula for a_n.

This recurrence relation is somewhat unusual in that a_n depends on all previous values of a_i, not just a few. We can compute directly that $a_1 = 3/2$ and $a_2 = 15/8$.

Solution: Note that (3.11) holds even for $n = 0$. Now multiply both sides of (3.11) by x^n, and add over all $n \geq 0$, that is, all values of n for which the equality holds. We get

$$\sum_{n\geq 0} \binom{n+2}{2} x^n = \sum_{n\geq 0}\sum_{i=0}^{n} a_i a_{n-i} x^n.$$

This is the moment to look for the generating function $A(x) = \sum_{n\geq 0} a_n x^n$ on both sides. However, we have to be a little more resourceful than before. Observe that the left-hand side is precisely $(1-x)^{-3}$. (Details are given in Exercise 1.) As far as the right-hand side is concerned, observe that this is precisely $A(x)^2$. Indeed, apply formula (3.3) with $B(x) = A(x)$. Therefore, the last displayed equation simply says that

$$(1-x)^{-3} = A(x)^2,$$
$$(1-x)^{-3/2} = A(x).$$

In order to compute the coefficients of $A(x)$, we will need to compute the binomial coefficient $\binom{-3/2}{n}$. We get that

$$\binom{-3/2}{n} = \frac{\frac{-3}{2} \cdot \frac{-5}{2} \cdot \cdots \cdot \frac{-2n-1}{2}}{n!} = (-1)^n \frac{1 \cdot 3 \cdots (2n+1)}{2^n \cdot n!}.$$

Note that the numerator of the fraction on the right-hand side is just the product of all *odd* integers from 1 to $2n+1$. This product is similar to a factorial, but we take *every other* integer from 1 to $2n+1$ instead of every integer. Therefore, we call this product $(2n+1)$-*semifactorial* and denote it by $(2n+1)!!$.

Using the binomial theorem and our new notation, we obtain that

$$\begin{aligned} A(x) &= (1-x)^{-3/2} = \sum_{n\geq 0} \binom{-3/2}{n} (-x)^n \\ &= \sum_{n\geq 0} \frac{(2n+1)!!}{2^n \cdot n!} x^n. \end{aligned}$$

So $a_n = \frac{(2n+1)!!}{2^n \cdot n!}$ for all $n \geq 0$. ◊

In particular, the explicit formula yields $a_0 = 1$, $a_1 = 3/2$, and $a_2 = 15/8$, agreeing with our computations before the solution.

3.2.2 Exponential generating functions

The following example is meant to show the limits of the technique of ordinary generating functions.

Example 3.9 *Let us assume that our red hat–wearing friends switch into a higher gear in their recruiting efforts. The five of them start spreading the rumor that, unless you wear red hats, you will fall ill, and they do this in an ever more efficient manner. That is, for all positive integers n, in the nth hour of recruiting, each person already wearing a red hat will tell the rumor to $n-1$ other people, who will immediately put on red hats, and, in addition to this, $2n$ people will simply join the red hat movement without being recruited. Just as before, nobody will hear the rumor twice, as recruiters will not waste time talking to people already wearing red hats. How many people will wear red hats after n hours?*

As ordinary generating functions served us so well in similar situations, let us try to apply them again. Let a_n denote the number of people wearing red hats after n hours. Then the above discussion leads to the recurrence relation

$$a_n = na_{n-1} + 2n \tag{3.12}$$

for $n \geq 1$ and with $a_0 = 5$. The reader is invited to verify that $a_1 = 7$, $a_2 = 18$, and $a_3 = 60$. Let us multiply both sides of (3.12) by x^n, then sum over all $n \geq 1$, to get

$$\sum_{n\geq 1} a_n x^n = x \sum_{n\geq 1} na_{n-1}x^{n-1} + 2x \sum_{n\geq 1} nx^{n-1}.$$

Note that the left-hand side is almost $A(x)$ and the last term of the right-hand side is easy to compute, so we get

$$A(x) - 5 = x^2 \sum_{n\geq 2} na_{n-1}x^{n-2} + 5x + \frac{2x}{(1-x)^2}. \tag{3.13}$$

At this point, we cannot proceed in the usual way, since the right-hand side cannot be written as a sum of some multiples of $A(x)$ and some other known functions. One possibility would be to try to use the identity $A'(x) = \sum_{n\geq 2}(n-1)a_{n-1}x^{n-2}$, but that would turn our equation for $A(x)$ into a *differential equation*, and we would like to avoid that if at all possible, because differential equations are sometimes much more difficult to solve than functional equations in which the only unknown is $A(x)$.

Why did our reliable technique fail us this time? We claim that its failure was no surprise. Recurrence relation (3.12) shows that $a_n > n!$ for all positive n. Therefore, by the comparison test, $A(x) = \sum_{n\geq 0} a_n x^n$ cannot be convergent for any $x \neq 0$. This means that we cannot find a closed form for the generating function $A(x) = \sum_{n\geq 0} a_n x^n$. The following definition introduces a new kind of generating function that will take care of this problem.

Definition 3.10 *Let $f_0, f_1, \cdots$ be a sequence of real numbers. Then the formal power series*

$$F(x) = \sum_{n\geq 0} f_n \frac{x^n}{n!}$$

is called the exponential generating function *of the sequence $\{f_n\}_{n\geq 0}$.*

Note that the coefficients are divided by $n!$ compared to what they are divided by ordinary generating functions. This allows exponential generating functions to be convergent in some cases when the ordinary generating functions are not.

Just as for ordinary generating functions, it is useful to look at a really simple example first.

Example 3.11 *Let $b_n = 1$ for all nonnegative integers n. Then the exponential generating function of the sequence $\{b_n\}_{n\geq 0}$ is*

$$B(x) = \sum_{n\geq 0} b_n \frac{x^n}{n!} = \sum_{n\geq 0} \frac{x^n}{n!} = e^x.$$

Now we have the necessary tools to find the explicit solution to the question of Example 3.9.

Solution: (of Example 3.9) Let $D(x) = \sum_{n\geq 0} a_n \frac{x^n}{n!}$ be the exponential generating function of the sequence $\{a_n\}$. Let us multiply both sides of (3.12) by $x^n/n!$ and sum over all $n \geq 1$. We get

$$\sum_{n\geq 1} a_n \frac{x^n}{n!} = \sum_{n\geq 1} na_{n-1}\frac{x^n}{n!} + 2 \cdot \sum_{n\geq 1} n \frac{x^n}{n!}.$$

We would like to recognize $D(x)$ on both sides. On the left-hand side, this is easy since the only term missing is a_0. On the right-hand side, note that $\frac{n}{n!} = \frac{1}{(n-1)!}$ to get

$$\begin{aligned} D(x) - 5 &= x \sum_{n \geq 1} a_{n-1} \frac{x^{n-1}}{(n-1)!} + 2x \sum_{n \geq 1} \frac{x^{n-1}}{(n-1)!} \\ &= xD(x) + 2xe^x. \end{aligned}$$

Therefore,

$$D(x) = \frac{5}{1-x} + \frac{2x}{1-x} e^x. \tag{3.14}$$

Recall that $D(x) = \sum_{n \geq 0} a_n \frac{x^n}{n!}$; therefore, a_n is the coefficient of $\frac{x^n}{n!}$ in $D(x)$. To find this coefficient, we simply have to find the coefficient of $\frac{x^n}{n!}$ in each term of the right-hand side of (3.14) and add them. On one hand,

$$\frac{5}{1-x} = 5 \sum_{n \geq 0} x^n = \sum_{n \geq 0} \frac{5 \cdot n!}{n!} x^n, \tag{3.15}$$

so the coefficient of $\frac{x^n}{n!}$ in $\frac{5}{1-x}$ is $5 \cdot n!$. On the other hand,

$$\begin{aligned} \frac{2x}{1-x} e^x &= \left(2x \sum_{n \geq 0} x^n \right) \cdot \left(\sum_{n \geq 0} \frac{x^n}{n!} \right) \\ &= 2 \left(\sum_{n \geq 1} x^n \right) \cdot \left(\sum_{n \geq 0} \frac{x^n}{n!} \right). \end{aligned}$$

Considering the last expression, we see that the summand x^i from the first sum needs to be multiplied by the summand $\frac{x^{n-i}}{(n-i)!}$ of the second sum. This shows that

$$\begin{aligned} \frac{2x}{1-x} e^x &= \sum_{n \geq 1} \left(\sum_{i=1}^{n} \frac{1}{(n-i)!} \right) x^n \\ &= \sum_{n \geq 1} \left(\sum_{i=1}^{n} \frac{n!}{(n-i)!} \right) \frac{x^n}{n!}. \\ &= \sum_{n \geq 1} \left(\sum_{i=0}^{n-1} \frac{n!}{i!} \right) \frac{x^n}{n!}. \end{aligned}$$

Comparing this to (3.14) and (3.15), we see that the coefficient of $\frac{x^n}{n!}$ in $D(x)$ is

$$a_n = 5n! + 2 \sum_{i=0}^{n-1} \frac{n!}{i!}.$$

◊

Quick Check

1. Let $a_0 = 1$ and let $a_n = 5a_{n-1} - 1$ if $n \geq 1$. Find an explicit formula for a_n.
2. Let $b_1 = 0$, and let $b_{n+1} = (n+1)b_n + n$ if $n \geq 1$. Find an explicit formula for b_n.
3. Let $a_0 = 0$, let $a_1 = 1$, and let $a_n = 4a_{n-1} - 3a_{n-2}$ if $n \geq 2$. Find an explicit formula for a_n.

3.3 Products of generating functions

In this section, we treat problems where generating functions are even more essential than they were in solving recurrence relations in the previous section.

3.3.1 Ordinary generating functions

Example 3.12 *A utility crew is painting n poles next to a road, starting at one end of the road and advancing one by one. For each pole, they randomly choose a color out of their three possibilities (red, blue, and green). At a certain time, they notice that their shift will end soon, and they will not be able to paint more poles that day. In order to celebrate the end of the workday, they choose one of the remaining poles at random and paint a smiling face on it, in red. How many different looks can the poles of this road have after the happy crew has left?*

The difference between the situation of this example and the situations studied in the previous sections is that the workers here proceed according to *two* different rules, one before realizing time was up, and one after that. Furthermore, we do not know when this change occurs, that is, after how many painted poles the crew will choose an unpainted pole for the smiling face. Let us assume that this happens after i poles were painted. That means the first i poles are painted in one of 3^i ways; then the crew chooses a remaining pole, in one of $n-i$ ways. So by the product principle the street can have $3^i(n-i)$ different looks. See Figure 3.2.

Let a_n denote the number of ways the poles can be after the crew left. Using the addition principle for $i \in [0, n]$, the argument of the previous paragraph shows that

$$a_n = \sum_{i=0}^{n} 3^i(n-i). \tag{3.16}$$

With the natural initial condition $a_0 = 0$ (if there are no poles, there is no pole for the smiling face, and the crew will not come anyway), this leads to the first values $a_1 = 1$ and $a_2 = 5$.

i painted poles	one chosen pole
3^i	$n-i$

Figure 3.2
The product principle at work.

We would now like to find an explicit formula for a_n. To that end, note that the right-hand side of (3.16) looks very much like the coefficient of x^n in a *product of two generating functions.* In order to make this observation more precise, set $A(x) = \sum_{n\geq 0} a_n x^n$, and also, $B(x) = \sum_{n\geq 0} 3^n x^n = \frac{1}{1-3x}$, and $D(x) = \sum_{n\geq 0} nx^n = \frac{x}{(1-x)^2}$. As we mentioned, the crucial observation is that

$$A(x) = B(x)D(x). \tag{3.17}$$

Indeed, the coefficient of x^n on the right-hand side is $\sum_{i=0}^{n} 3^i(n-i)$, which is in turn equal to a_n by (3.16).

Therefore, we can obtain a_n as $[x^n]A(x) = [x^n]B(x)D(x)$. Fortunately, we know both $B(x)$ and $D(x)$ explicitly. So (3.17) yields

$$\begin{aligned} A(x) &= B(x)D(x) = \frac{1}{1-3x}\cdot\frac{x}{(1-x)^2} \\ &= \frac{x}{(1-3x)(1-x)^2}. \end{aligned}$$

In order to find the power series form of $A(x)$, we decompose the last expression as a sum of partial fractions as we learned in calculus. That is, we are looking for real numbers P, Q, and R so that

$$A(x) = \frac{x}{(1-3x)(1-x)^2} = \frac{P}{1-3x} + \frac{Q}{1-x} + \frac{R}{(1-x)^2}.$$

A routine computation yields that $P = 3/4$, $Q = -1/4$, and $R = -1/2$. Thus we have

$$\begin{aligned} A(x) &= \frac{3}{4}\frac{1}{1-3x} - \frac{1}{4}\frac{1}{1-x} - \frac{1}{2}\frac{1}{(1-x)^2} \\ &= \frac{3}{4}\sum_{n\geq 0} 3^n x^n - \frac{1}{4}\sum_{n\geq 0} x^n - \frac{1}{2}\sum_{n\geq 0}(n+1)x^n \\ &= \sum_{n\geq 0} \frac{3^{n+1} - 1 - 2(n+1)}{4} x^n. \end{aligned}$$

Therefore, we get that $a_n = \frac{3^{n+1}-1-2(n+1)}{4}$. The reader is invited to verify that this formula is indeed correct.

The most important feature of the previous example was that the crew *started doing something*, then, at one point, it *did something else*. This idea of cutting the interval $[1, n]$ into two parts and then considering distinct structures on each part occurs in many problems. Fortunately, the *product* of the ordinary generating functions counting the two structures is a very powerful tool in these situations.

Let us consider a similar problem where the same line of thinking helps.

Example 3.13 *A student makes a study plan for the final exam period. She splits the period into two parts. In the first part, she devotes each day either to physics or to abstract algebra, except for one day, when she will only study the chosen subject of the day in the morning, and relax in the afternoon. She devotes all days of the second part to combinatorics, except for two days, when she takes breaks. These two days do not have to be consecutive days. If the exam period consists of n days, in how many different ways can the student plan her studies?*

Solution: Let us assume again that the first part of studies will consist of k days and the second part will consist of $n-k$ days. That means that the student has $k2^k$ possibilities for the first part and $\binom{n-k}{2}$ possibilities for the second part. So, by the product principle, the student has $k2^k\binom{n-k}{2}$ possibilities for this fixed k. Summing over all allowed k (which means $1 \leq k \leq n-2$, otherwise there are not enough days for breaks), we see that the total number of possibilities she has is $\sum_{k=1}^{n-2} k2^k\binom{n-k}{2}$. Following the line of thinking of the previous example, consider the generating functions of the numbers of possibilities in each part. That is, consider the generating functions

$$B(x) = \sum_{n\geq 1} n2^n x^n = \frac{2x}{(1-2x)^2},$$

and

$$D(x) = \sum_{n\geq 2} \binom{n}{2} x^n = \frac{x^2}{(1-x)^3}.$$

Let $A(x) = B(x)D(x)$. Similarly to the previous example, the number of all possibilities for the student, that is, the number $\sum_{k=1}^{n-2} k2^k\binom{n-k}{2}$, is precisely the coefficient of $[x^n]A(x)$. So all we have to do is to find that coefficient. Using the previous two equations, we get

$$\begin{aligned} A(x) &= B(x)D(x) = \frac{2x}{(1-2x)^2} \cdot \frac{x^2}{(1-x)^3} \\ &= \frac{2x^3}{(1-2x)^2(1-x)^3}. \end{aligned}$$

Converting this generating function into partial fraction form is a tedious task.

However, there are various software packages that can do it for us. Either way, we get that

$$\begin{aligned} A(x) &= \frac{2}{(1-x)^3} + \frac{2}{(1-x)^2} + \frac{6}{1-x} + \frac{2}{(1-2x)^2} - \frac{12}{1-2x} \\ &= \sum_{n\geq 0}(n+2)(n+1)x^n + \sum_{n\geq 0} 2(n+1)x^n + \sum_{n\geq 0} 6x^n \\ &+ \sum_{n\geq 0}(n+1)2^{n+1}x^n - \sum_{n\geq 0} 12\cdot 2^n x^n \\ &= \sum_{n\geq 0}(2^{n+1}(n-5) + (n+1)(n+4) + 6)x^n. \end{aligned}$$

Therefore, the number of possibilities the student has is $a_n = 2^{n+1}(n-5) + (n+1)(n+4) + 6$. The reader is invited to verify that this is indeed the correct formula, in particular, $a_1 = a_2 = 0$ and $a_3 = 2$. ◊

Our result illustrates the power of the method very well. After all, the formula $a_n = 2^{n+1}(n-5) + (n+1)(n+4) + 6$ would have been quite difficult to guess, and it would have been quite difficult to prove the formula combinatorially. There seems to be no easy explanation for the summands $2^{n+1}(n-5)$, or $(n+1)(n+4)$, or 6 in the formula. However, our generating function method worked fairly mechanically, and we did not need any bright ideas.

We have seen two examples of the same counting principle, and now we are going to formulate the general theorem applicable to similar situations.

Theorem 3.14 (Product formula) *Let f_n be the number of ways one can carry out a certain task on the set $[n]$. Let g_n be the number of ways one can carry out another task on $[n]$. Let $F(x)$ and $G(x)$ be the ordinary generating functions of the sequences $\{f_n\}_{n\geq 0}$ and $\{g_n\}_{n\geq 0}$.*

Let h_n be the number of ways to split the set $[n]$ into the intervals $\{1, 2, \cdots, i\}$ and $\{i+1, i+2, \cdots, n\}$, and then carry out the first task on the first interval and the second task on the second interval. Let $H(x)$ be the ordinary generating function of the sequence $\{h_n\}_{n\geq 0}$. Then

$$H(x) = F(x)G(x). \tag{3.18}$$

Proof: The definition of the numbers h_n yields the recurrence relation

$$h_n = \sum_{i=0}^{n} f_i g_{n-i}.$$

Indeed, for any fixed i there are f_i ways to carry out the first task, g_{n-i} ways to carry out the second task, and so there are $f_i g_{n-i}$ ways to carry out both of them. Summing over all i, the claim follows by the addition principle.

Now note that h_n is the coefficient of x^n in $H(x)$, and that $\sum_{i=0}^{n} f_i g_{n-i}$ is the coefficient of x^n in $F(x)G(x)$. As these coefficients are equal for all n, the power series $H(x)$ and $F(x)G(x)$ are also equal. ◊

Note that Example 3.12 was a straightforward application of the product formula. Indeed, the first task, to be carried out on the first i poles, was to color each of them red, blue, or green, leading to the generating function $F(x) = \sum_{n\geq 0} 3^n x^n = \frac{1}{1-3x}$. The second task, to be carried out on the last $n-i$ poles, was to choose one pole, leading to the generating function $G(x) = \sum_{n\geq 0} nx^n = \frac{x}{(1-x)^2}$. As we now know without needing further explanation, the generating function of the combined task was then simply $F(x)G(x) = \frac{x}{(1-x)^2(1-3x)}$.

Let us consider one more simple application of the product formula before attacking more difficult problems.

Example 3.15 *A section in a combinatorics textbook contains n exercises. The book is used by a dutiful professor who wants the students to practice the basic methods discussed in the section, but also wants to provide a challenge for those who are interested in more difficult questions. Therefore, she will assign the first $2i$ problems for homework (for some integer $i \geq 2$) and also one of the remaining exercises as a bonus problem. In how many different ways can she proceed?*

Solution: In the language of the product formula, the professor splits the set $[n]$ into two intervals, then she assigns all problems in the first interval as homework (which she can do in one way if the size of that interval is even and at least four, and in zero ways otherwise), and chooses one problem from the second interval (which she can do in j ways if that is the size of that interval).

Therefore, again with the notations of the product formula,

$$F(x) = x^4 + x^6 + \cdots = \sum_{i\geq 2} x^{2i} = x^4 \sum_{i\geq 0} (x^2)^i = \frac{x^4}{1-x^2}$$

and

$$G(x) = \sum_{j\geq 0} jx^j = \frac{x}{(1-x)^2}.$$

By the product formula, this yields

$$H(x) = F(x)G(x) = \frac{x^5}{(1-x^2)(1-x)^2}.$$

While there is still work to be done, we can see that we are on the right track since $H(x)$ will not have terms with exponents less than five; indeed, the professor could not make her choices if the section had less than five exercises.

In order to find $[x^n]H(x)$, it suffices to find $[x^{n-5}]\frac{1}{(1-x^2)(1-x)^2}$. This is a routine, if tedious computation, which can also be done by computer. (We will talk about that a little bit more in the next example.) Either way, we get

$$\frac{H(x)}{x^5} = \frac{1}{2(1-x)^3} + \frac{1}{4(1-x)^2} + \frac{1}{8(1-x)} + \frac{1}{8(1+x)}.$$

Using the identity proved in Exercise 1, this yields

$$\begin{aligned}\frac{H(x)}{x^5} &= \frac{1}{2}\sum_{n\geq 0}\binom{n+2}{2}x^n + \frac{1}{4}\sum_{n\geq 0}(n+1)x^n + \\ &+ \frac{1}{8}\left(\sum_{n\geq 0}x^n + \sum_{n\geq 0}(-1)^n x^n\right) \\ &= \sum_{n\geq 0}\frac{4\binom{n+2}{2}+2(n+1)+1+(-1)^n}{8}x^n.\end{aligned}$$

So the number h_n of possibilities that the professor has is the coefficient of x^{n-5} in the latest expression, that is, $h_n = \frac{4\binom{n-3}{2}+2(n-4)+1+(-1)^{n-5}}{8}$. ◊

Let us now practice the use of the product formula on a classic problem that is more difficult. Once we see the numerical solution of the problem, it will remind us of something we have seen before.

Example 3.16 *Two soccer teams played an exciting game last weekend. The game ended in a draw, with each team scoring n goals. Throughout the game, it often seemed likely that the home team would win, since they often had the lead, while the visitors never did. In how many different orders could the $2n$ goals be scored?*

Solution: So that we can understand the problem better, let us compute the answer h_n to the above question for small values of n. Denote by A a goal scored by the home team, and denote by B a goal scored by the visiting team. If $n = 1$, then the only possible order of goals is AB, since the home team never trailed. If $n = 2$, then there are two possibilities, namely, $AABB$ and $ABAB$. If $n = 3$, then we have five possibilities, namely, $AAABBB$, $AABABB$, $AABBAB$, $ABAABB$, and $ABABAB$. So $h_1 = 1$, $h_2 = 2$, and $h_3 = 5$. Also note that $h_0 = 1$ since there is one way for both teams not to score (by not scoring). Let $H(x) = \sum_{n\geq 0} h_n x^n$.

What makes this problem more difficult than the previous problems or some of the standard counting problems of Chapter 1 is the condition that the home team *never trailed.* Without that condition, there would be $\binom{2n}{n}$ possibilities, since we would only need to count words of length $2n$ consisting of n copies of A and n copies of B. Now we also have the condition that no initial segment of our words can contain more Bs than As.

We would like to use the product formula to compute $H(x)$. The difficulty lies in finding out how to decompose $H(x)$ into the product of two generating functions. What are the two tasks mentioned in the product formula?

Let us assume the game did not end in a scoreless tie (0–0). Let us call the moment when the visitors first tied the score the *critical moment* of the game. Since the game did end in a draw, we know for sure that there was a critical moment. Let us say that, at the critical moment, the score was a tie at i goals each.

We will now break the game up into two parts, the part before the critical moment and the part after the critical moment.

First, we claim that there are h_{n-i} ways the game could be completed *after* the critical moment. Indeed, simply cancel all goals scored before the critical moment, and then the remainder of the game has to end in a tie of $n-i$ goals for each team, with the home team never trailing. Therefore, the number of ways the game could be completed after the critical moment is enumerated by the generating function

$$G(x) = H(x).$$

It is a little bit more interesting to figure out how many ways the goals could be scored *before* the critical moment. Note that it would be incorrect to say that the number of ways this could happen is h_i since the visiting team never tied before the critical moment, and h_i does not take that into account.

Instead, observe the following: The home team must have scored the first goal, and the visiting team must have scored the last goal before the critical moment. Let us call the time period *between* these two goals, and excluding them, the *middle period* of the game. So the first goal of the game came before the middle period, and the critical moment came after the middle period. (Soccer games do not really have middle periods; they have two halves, but never mind.) Then the middle period resulted in a tie at $i-1$ goals for each team, so that the home team never trailed *when counting goals scored in this period.* Indeed, to say that the home team never trailed when counting goals scored in the middle period is the same as saying that considering the whole game the visitors did not tie the game before the critical moment. This is because the home team has the lead when the middle period starts. Therefore, the number of ways to reach the critical moment with an $i-i$ tie is the same as the number of middle periods resulting in an $(i-1)-(i-1)$ tie, that is, h_{i-1}.

So there are h_{i-1} ways to go from the start of the game to the critical moment if the critical moment comes at the score of $i-i$. Therefore, the generating function for the number of possibilities for the part of the game before the critical moment is

$$F(x) = \sum_{i \geq 1} h_{i-1}x^i = xH(x).$$

We are now ready to use the product formula. Do not forget that our decomposition only works if $n > 0$ (if there are no goals at all, there is no critical moment). The generating function for the sequence $h_1, h_2, \cdots$ is $H(x)-1$, since $h_0 = 1$. Therefore, the product formula yields

$$H(x) - 1 = xH(x)^2. \tag{3.19}$$

This is a quadratic equation for $H(x)$. Solving it, we get

$$H(x) = \frac{1-\sqrt{1-4x}}{2x}. \tag{3.20}$$

We hope that the alert reader is upset that we did not explain how $\pm\sqrt{1-4x}$ in the solution of the quadratic equation (3.19) became simply $-\sqrt{1-4x}$. Here is our explanation: The numbers h_n are, of course, uniquely determined by n; therefore, the generating function $H(x)$ is also unique. So only one of the power series $\frac{1-\sqrt{1-4x}}{2x}$ and $\frac{1+\sqrt{1-4x}}{2x}$ can be equal to $H(x)$. In order to see which one, we need to look no further than at $H(x)$. From the definition of $H(x)$, we see that $H(0) = h_0 = 1$. On the other hand, the function $\frac{1+\sqrt{1-4x}}{2x}$ is not even defined in $x = 0$ since its denominator is 0 while its numerator is 2. The function $\frac{1-\sqrt{1-4x}}{2x}$ can, however, be extended to $x = 0$, since at that point both the numerator and the denominator of the fraction are equal to 0. One easily checks by l'Hôpital's rule or otherwise that $\lim_{x=0} \frac{1-\sqrt{1-4x}}{2x} = 1$, so the choice of the negative sign is justified. (In fact, once we know that the positive sign is not the right one, we know that the negative sign is the right one since $H(x)$ does exist, and it does have to be one of $\frac{1-\sqrt{1-4x}}{2x}$ and $\frac{1+\sqrt{1-4x}}{2x}$.)

In order to find the numbers h_n, we need to find $[x^n]H(x)$. To that end, we first compute $\sqrt{1-4x}$ by the binomial theorem. We get

$$(1-4x)^{1/2} = \sum_{n\geq 0} \binom{1/2}{n} (-4x)^n. \tag{3.21}$$

Now we need to compute $\binom{1/2}{n}$. By the definition of binomial coefficients, we get

$$\binom{1/2}{n} = \frac{\frac{1}{2}\cdot\frac{-1}{2}\cdot\frac{-3}{2}\cdots\frac{-2n+3}{2}}{n!} = (-1)^{n-1}\frac{(2n-3)!!}{2^n\cdot n!}.$$

Comparing this with (3.21) yields

$$\sqrt{1-4x} = -\sum_{n\geq 0} \frac{2^n\cdot(2n-3)!!}{n!} x^n. \tag{3.22}$$

$$= 1 - 2\sum_{n\geq 0} \frac{\binom{2n-2}{n-1}}{n} x^n. \tag{3.23}$$

Finally, we can substitute the obtained expression into (3.20) to get the power series form of $H(x)$. This yields

$$H(x) = \frac{1 - \sqrt{1-4x}}{2x} = \sum_{n \geq 0} \frac{\binom{2n}{n}}{n+1} x^n. \tag{3.24}$$

Therefore, $H_n = \frac{\binom{2n}{n}}{n+1}$. Note that this shows that the numbers h_n are in fact equal to the Catalan numbers c_n, which we defined after Example 1.34. ◊

Remarks:

1. Note that (3.19) is equivalent to the recurrence relation

$$h_n = \sum_{i=0}^{n-1} h_i h_{n-1-i} \tag{3.25}$$

   for $n \geq 1$ and $h_0 = 1$. In fact, an alternative way to prove the result of Example 3.16 is to prove (3.25) first, then multiply both sides by x^n, sum for $n \geq 1$, then recognize that the left-hand side is $H(x) - 1$ and the right-hand side is $xH(x) \cdot H(x)$.
2. We have mentioned that the scoring sequences counted in this example are equinumerous to northeastern lattice paths from $(0, 0)$ to (n, n) that never go above the diagonal $x = y$. In fact, there is a simple bijection between these two sets, with letters A corresponding to an east step, and letters B corresponding to a north step. Note that the *critical moment* corresponds to the first time a northeastern lattice path touches the diagonal $x = y$.

In order to extend the range of the product formula, note that there is nothing magical about the number *two* when we split $[n]$ into two intervals. If we split $[n]$ into three consecutive intervals so that they are disjoint and their union is $[n]$, and then we carry out various tasks on each interval, we can argue similarly. That is, if $A(x)$, $B(x)$, and $C(x)$ are the generating functions enumerating the number of ways we can carry out the three tasks, then, by the product formula, $A(x)B(x)$ is the generating function for the number of ways to carry out the *first two* tasks. Then, applying the product formula again, for $A(x)B(x)$ and $C(x)$ we get that $A(x)B(x)C(x)$ is the generating function for carrying out all three tasks. We can proceed similarly if we have more than three tasks. This leads to a new array of applications, which are introduced by the following example.

Example 3.17 *Find the number h_n of ways one can pay n dollars using only 1-dollar, 2-dollar, and 5-dollar bills. The order in which we use the bills does not matter, only the number of bills of each denomination does.*

Solution: Let a_n (resp. b_n, c_n) be the number of ways to pay n dollars using only 1-dollar bills (resp. only 2-dollar bills, only 5-dollar bills). Let $A(x)$ (resp. $B(x)$, $C(x)$) be the ordinary generating functions of the corresponding sequences. Then $a_n = 1$ for all $n \geq 0$, while $b_n = 1$ if n is divisible by two and $b_n = 0$ otherwise, and $c_n = 1$ if n is divisible by five, and $c_n = 0$ otherwise. This leads to the generating functions $A(x) = \sum_{n\geq 0} x^n = \frac{1}{1-x}$, $B(x) = \sum_{n\geq 0} x^{2n} = \sum_{n\geq 0}(x^2)^n = \frac{1}{1-x^2}$, and $C(x) = \sum_{n\geq 0} x^{5n} = \sum_{n\geq 0}(x^5)^n = \frac{1}{1-x^5}$.

Therefore, by the product formula, the generating function $H(x) = \sum_{n\geq 0} h_n x^n$ of the numbers h_n is

$$H(x) = A(x)B(x)C(x) = \frac{1}{(1-x)(1-x^2)(1-x^5)}. \tag{3.26}$$

◊

Several comments are in order. First, you could ask what is the use of this result since we did not get a formula for the numbers h_n, and getting one by finding the partial fraction decomposition of $\frac{1}{(1-x)(1-x^2)(1-x^5)}$ by hand would be quite some work. As we mentioned before, standard mathematics software, such as Mathematica or Maple, can do this tedious work for us very easily. For instance, in Maple we can type

```
convert(1/((1-x)*(1-x^2)*(1-x^{5})), parfrac);
```

and the computer will return

$$\frac{1}{4(x-1)^2} - \frac{13}{40(x-1)} + \frac{x^3+2x^2+x+1}{5(x^4+x^3+x^2+x+1)} + \frac{1}{8(x+1)} - \frac{1}{10(x-1)^3}.$$

Now it is relatively painless to find the power series form of h_n, especially if we notice that we can multiply both the numerator and the denominator of the middle term by $x-1$, turning the denominator into $5(x^5-1)$.

If we are simply interested in the first 15 values of the sequence, we can get that by simply typing

```
series(1/((1-x)*(1-x^2)*(1-x^{5})), x=0, 16);
```

in Maple and hitting Return. Maple will then return

$$1 + x + 2x^2 + 2x^3 + 3x^4 + 4x^5 + 5x^6 + 6x^7 + 7x^8 + 8x^9 + 10x^{10}$$

$$+11x^{11} + 13x^{12} + 14x^{13} + 16x^{14} + 18x^{15} + O(x^{16}),$$

and we find $h(n)$ as the coefficient of x^n in the above expression for $n \leq 15$.

Second, and this is the more far-reaching of our comments, note that we only found the generating function for the numbers of *partitions* of the non-negative integer n into parts equal to one, two, or five. It goes without saying that it is easy to generalize our result for the cases when other parts are allowed. Because of the broad applications of this method, it is worth spending some time exploring the connection between $H(x)$ and partitions of n into parts of size one, two, and five.

We have seen that

$$H(x) = (1 + x + x^2 + \cdots)(1 + x^2 + x^4 + \cdots)(1 + x^5 + x^{10} + \cdots).$$

After carrying out all multiplications, a typical summand on the right-hand side will be of the form $x^i x^{2j} x^{5k} = x^n$. That means that $i + 2j + 5k = n$, or, in other words, that n has a partition into $k + j + i$ parts, k of which are 5, j of which are 2, and i of which are 1. Each partition of n into parts of allowed sizes will correspond to one summand equal to x^n, showing again that the coefficient of x^n in $H(x)$ is the number of partitions of n into parts of size one, two, and five.

You should verify your understanding of the above example by trying to find the summands on the right-hand side that correspond to partitions of 5 into allowed parts.

There are plenty of fascinating theorems on partitions of integers that can be proved by generating functions. The reader is invited to solve Supplementary Exercises 9, 10, 11, and 12 to warm up. Then the reader is invited to solve Exercises 12, 13, and 14 to obtain a very interesting identity involving partitions. Exercises 27, 28, and 29 generalize that identity.

Some of these exercises will involve *infinite* products of infinite sums. This may be unusual for the reader. Therefore, we will discuss the concept here. We will do this using an example, as opposed to in full generality, in order to avoid excessive notation.

Let $p_{even}(n)$ be the number of partitions of n into even parts only, and let $p_{even}(0) = 1$. We claim that then the equality

$$\sum_{n \geq 0} p_{even}(n) x^n = (1 + x^2 + x^4 + \cdots)(1 + x^4 + x^8 + \cdots) \cdots \quad (3.27)$$

$$= \prod_{i \geq 1} \frac{1}{1 - x^{2i}} \quad (3.28)$$

holds.

First of all, how do we know that the right-hand side, which is an infinite product of infinite sums, is well-defined? Let us turn that question around. What could possibly go wrong? When would the right-hand side not make sense? The right-hand side is a product of sums. We would start computing this sum as if it were a finite sum, that is, by computing the (finite) term-by-term products and then adding them. As we compute finite term-by-term products, such as $x^2 \cdot x^{10} \cdot x^{14}$, each of these products is defined. The problem could come afterward, when we add these products. If infinitely many of these products were equal to x^k for some k, then we could not add them. (We will *not* identify infinite sums with their limit here, even if that limit exists.)

Indeed, what would be the coefficient of x^k in the sum? Therefore, the final product would not be defined. For example, the infinite product

$$(1+x)(1+x+x^2)(1+x+x^2+x^3)\cdots = \prod_{n\geq 2} \frac{x^n-1}{x-1}$$

is not defined because infinitely many of the term-by-term products are equal to x.

Let us return to formula (3.27). The reader should explain why that formula holds, which will include an explanation of the fact that the right-hand side is defined. There is no significant difference between the argument we used to prove (3.26) in Example 3.17 and this equation, except that there are infinitely many terms on the right-hand side of the latter.

3.3.2 Exponential generating functions

When we manipulate exponential generating functions, we often need to multiply two power series of the kind $\sum_{n\geq 0} a_n \frac{x^n}{n!}$. Fortunately, such products are not too difficult to handle. This is the content of the next proposition, whose full importance will be explained shortly.

Proposition 3.18 *Let* $F(x) = \sum_{n\geq 0} f_n \frac{x^n}{n!}$ *and let* $G(x) = \sum_{n\geq 0} g_n \frac{x^n}{n!}$. *Then the equality*

$$(F\cdot G)(x) = \sum_{n\geq 0} \left(\sum_{k=0}^{n} \binom{n}{k} f_k g_{n-k} \right) x^n$$

holds.

Proof: Let us multiply F by G as power series. As we saw in (3.3), the coefficient of x^n in this product is equal to

$$\sum_{k=0}^{n} \frac{f_k}{k!} \cdot \frac{g_{n-k}}{(n-k)!} = \frac{1}{n!} \sum_{k=0}^{n} \frac{n!}{k!(n-k)!} f_k \cdot g_{n-k} = \frac{1}{n!} \sum_{k=0}^{n} \binom{n}{k} f_k \cdot g_{n-k},$$

which is equivalent to our claim. ◊

Proposition 3.18 shows that the difference between the product of two exponential generating functions and two ordinary generating functions is somehow related to binomial coefficients. We will soon understand the cause of this phenomenon better.

Example 3.19 *A small liberal arts college has n students and two writing competitions. Each student must select exactly one competition to enter. Those who participate in the first contest must write a short story that is related to* *one of three given topics. Those participating in the other contest must write an essay about one of four given topics.*

At the end of the contest, participants are ranked in both contests, and the topic of their work is noted next to their name. How many final rankings are possible? (The final rankings consist of the rankings of students in both competitions, with the topic of their work noted.)

The reader probably notices that this problem does bear some similarity to the problems we solved with the product formula for ordinary generating functions, but does the reader see that it is also significantly different from those problems?

The similarity is that, again, we have a set of n students, which we split into two parts, the first contest and the second contest, and then we carry out a task on each part. On the first part, the task is to establish a ranking and mark the topic (1–3) chosen by each person next to the name of that person. On the second part, the task is to establish a ranking, then mark the chosen topic 1–4 next to the name of each person. The question is, just as before, the number of ways to carry out the combination of both tasks.

However, there is a very important difference *in the way we split* $[n]$ into two parts. In the problems of the previous subsection, we had to split $[n]$ into two *intervals*, whereas here we simply have to split $[n]$ into two *subsets*. While there are $n+1$ ways to split $[n]$ into two (possibly empty) intervals, there are 2^n ways to split $[n]$ into an ordered pair of (possibly empty) sets. If we want the sizes of the parts to be i and $n-i$, then there is only one way to split in the first scenario, but $\binom{n}{i}$ ways in the second scenario. This extra factor $\binom{n}{i}$ and Proposition 3.18 suggest that we should look at the *product of the exponential generating functions* of the numbers of ways to carry out the two tasks.

Solution: Let us assume that i students decide to go for the first contest. There are $\binom{n}{i}$ ways for this to happen. Then there are 3^i ways for the chosen students to pick their topic, and, finally, there are $i!$ ways to rank these students. So there are $a_i = 3^i i!$ ways for this contest to end. The remaining $n-i$ students can choose their topics in 4^{n-i} ways, and they can be ranked in $(n-i)!$ ways, meaning that the second contest can end in $b_{n-i} = 4^{n-i}(n-i)!$ different ways. Looking at the composite final results for this fixed i, it follows that there are $\binom{n}{i} \cdot 3^i i! \cdot 4^{n-i}(n-i)!$ possible outcomes. Summing over all i, we get that the total number of outcomes is

$$d_n = \sum_{i=0}^{n} \binom{n}{i} \cdot 3^i i! \cdot 4^{n-i}(n-i)!. \tag{3.29}$$

In order to find a formula for d_n, set

$$A(x) = \sum_{n \geq 0} 3^n n! \frac{x^n}{n!} = \frac{1}{1-3x}$$

and

$$B(x) = \sum_{n\geq 0} 4^n n! \frac{x^n}{n!} = \frac{1}{1-4x}.$$

That is, $A(x)$ and $B(x)$ are the exponential generating functions of the two individual tasks. Set $D(x) = \sum_{n\geq 0} d_n \frac{x^n}{n!}$.

The crucial observation, and the main reason for which exponential generating functions are useful in similar situations, is that $A(x)B(x) = D(x)$. To see this, either recall Proposition 3.18 or note that the coefficient of $x^n/n!$ in $A(x)B(x)$ is

$$n! \sum_{i=0}^{n} \frac{a_i}{i!} \cdot \frac{b_{n-i}}{(n-i)!} = \binom{n}{i} a_i b_{n-i},$$

in agreement with (3.29).

Therefore, we get an explicit formula for $D(x)$, namely,

$$D(x) = A(x)B(x) = \frac{1}{1-3x} \cdot \frac{1}{1-4x}.$$

Converting this last expression to partial fractions, we obtain

$$\begin{aligned} D(x) &= \frac{4}{1-4x} - \frac{3}{1-3x} = 4\sum_{n\geq 0} 4^n x^n - 3\sum_{n\geq 0} 3^n x^n \\ &= \sum_{n\geq 0} (4^{n+1} - 3^{n+1}) x^n = \sum_{n\geq 0} n!(4^{n+1} - 3^{n+1}) \frac{x^n}{n!}. \end{aligned}$$

Therefore, we get $d_n = n!(4^{n+1} - 3^{n+1})$ as the coefficient of $x^n/n!$ in $D(x)$. ◊

We will often use the notation $[x^n/n!]A(x)$ to denote the coefficient of $x^n/n!$ in $A(x)$. In particular, this means that $[x^n/n!]A(x) = n!([x^n]A(x))$.

The ideas used in this example work in a more general setup as well. This is the content of the following theorem.

Theorem 3.20 (Product formula) *Denote by f_n the number of ways to carry out a task on $[n]$, and denote by g_n the number of ways to carry out another task on $[n]$. Let $F(x)$ and $G(x)$ be the exponential generating functions of the sequences $\{f_n\}$ and $\{g_n\}$.*

Let h_n be the number of ways to choose a subset S of $[n]$, carry out the first task on S, and then carry out the second task on the set $[n]-S$. Let $H(x)$ be the exponential generating function of the sequence $\{h_n\}$.

Then

$$H(x) = F(x)G(x).$$

Proof: If S is to be of size i, then there are $\binom{n}{i} f_i g_{n-i}$ ways to carry out the combined task. Summing over all allowed i, we get

$$h_n = \sum_{i=0}^{n} \binom{n}{i} f_i g_{n-i}.$$

Our claim now follows, since the left-hand side of the last equation is $[x^n/n!]H(x)$, while the right-hand side is $[x^n/n!](F(x)G(x))$, as we saw in Proposition 3.18. ◊

Figure 3.3 is meant to illustrate the difference between the situations when the product formula is applied with ordinary generating functions and the situations when the product formula is applied with exponential generating functions.

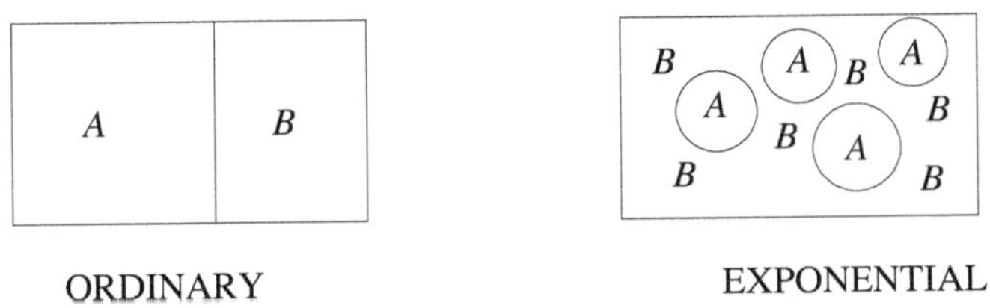

Figure 3.3
When the tasks are carried out on intervals, we use ordinary generating functions. When they are carried out on subsets, we use exponential generating functions.

Let us practice the product formula with exponential generating functions on some further examples.

Example 3.21 *A football coach has n players on his roster. For today's practice, he selects an even number of players for scrimmage. Then he assigns each player who is not selected for scrimmage to one of two other practicing groups. In how many ways can the coach make his choices?*

Solution: Once the set S of players who will participate in scrimmage is selected, there is no additional task to be carried out on S. However, $|S|$ must be even. So, in the language of the product formula, we have $f_i = 1$ if i is even, and $f_i = 0$ otherwise. In the set $[n] - S$, each player is assigned to one of two groups, leading to $g_j = 2^j$ for all $j \geq 0$. Therefore,

$$F(x) = 1 + \frac{x^2}{2} + \cdots = \sum_{n \geq 0} \frac{x^{2n}}{(2n)!} = \frac{e^x + e^{-x}}{2},$$

and

$$G(x) = 1 + 2x + 4\frac{x^2}{2} + \cdots = \sum_{n \geq 0} \frac{2^n x^n}{n!} = e^{2x}.$$

Therefore, by the product formula, we get

$$\begin{aligned} H(x) &= F(x)G(x) = \frac{e^x + e^{-x}}{2} \cdot e^{2x} \\ &= \frac{e^{3x} + e^x}{2} = \frac{1}{2}\sum_{n\geq 0} \frac{3^n x^n}{n!} + \frac{1}{2}\sum_{n\geq 0} \frac{x^n}{n!}. \end{aligned}$$

Thus, the number of ways the coach can proceed is $h_n = (3^n + 1)/2$. ◊

Just as in the case of ordinary generating functions, we can repeatedly apply the product formula to count the ways a multiple task can be carried out when that task has a fixed number $k > 2$ parts.

Example 3.22 *A group of n senators wants to form three committees—Appropriations, Business, and Conference—so that each of the senators is a member of exactly one committee. In addition, the Appropriations committee needs to have an even number (possibly zero) of members, the Business committee needs to have an odd number of members, and the Conference committee needs to have a president. How many different selections are possible?*

Solution: For a set S of senators, there is one way for S to form the Appropriations committee if $|S|$ is even, and zero ways otherwise. This leads to

$$A(x) = \sum_{n\geq 0} \frac{x^{2n}}{(2n)!} = \frac{e^x + e^{-x}}{2},$$

where $A(x)$ is the exponential generating function for the number of ways the Appropriation committee can have i members.

By similar arguments for the other two committees, we get

$$B(x) = \sum_{n\geq 0} \frac{x^{2n+1}}{(2n+1)!} = \frac{e^x - e^{-x}}{2}$$

and

$$C(x) = \sum_{n\geq 0} n\frac{x^n}{n!} = x\sum_{n\geq 1} \frac{x^{n-1}}{(n-1)!} = xe^x.$$

Therefore, if h_n is the number of ways to select the three committees and the president of the Conference committee, then the product formula yields

$$\begin{aligned} H(x) &= A(x)B(x)C(x) = \frac{e^x + e^{-x}}{2} \cdot \frac{e^x - e^{-x}}{2} \cdot xe^x \\ &= \frac{e^{2x} - e^{-2x}}{4} \cdot xe^x = x \cdot \frac{e^{3x} - e^{-x}}{4} \\ &= \frac{1}{4}\sum_{n\geq 0} \frac{3^n x^{n+1}}{n!} - \frac{1}{4}\sum_{n\geq 0} (-1)^n \frac{x^{n+1}}{n!} \\ &= \frac{1}{4}\sum_{n\geq 1} (3^{n-1} - (-1)^{n-1}) \frac{x^n}{(n-1)!}. \end{aligned}$$

Therefore, the number of ways to make all necessary selections is $h_n = n(3^{n-1} - (-1)^{n-1})/4$, for all $n \geq 1$. $\diamondsuit$

Quick Check

1. How many words of length n are there over the alphabet $\{A, B, C\}$ that consist of an initial part of letters A only, but contain no letters A after that? The initial part does not have to be proper, that is, it is allowed to be of length 0 or n.
2. How many words of length n are there over the alphabet $\{A, B, C\}$ that consist of an initial part and ending part of letters A only, and a middle part that does not contain the letter A? We allow each of the three parts to be empty.
3. How many ways are there to select an odd-sized subset of $[n]$, and then color each of the unselected elements of $[n]$ red, blue, or green?

3.4 Compositions of generating functions

In this section, we extend the reach of our generating function techniques even further.

3.4.1 Ordinary generating functions

We start by exploring a situation when we still subdivide the set $[n]$ into intervals, but in a more complex way.

Example 3.23 *The dissertation of a doctoral student in combinatorics has to consist of exactly n pages. The dissertation can have* as many chapters as the student likes, *but each chapter must contain at least one page of text and at least one page of illustrations. Furthermore, the number of text pages as well as the number of illustration pages in any chapter must be an integer.*

Find an explicit formula for the number b_n of ways in which the student can arrange the pages of his dissertation.

What is difficult in this problem is that we do not know the number of chapters the dissertation will have. If we knew that there would be two chapters, or three chapters, we could easily apply the product formula (for ordinary generating functions) to get the answer.

Indeed, let a_n be the number of ways *one* chapter can be arranged. We claim that $a_n = 2^n - 2$ for $n \geq 2$, and $a_n = 0$ for $n < 2$. This is because each page can be a text page or an illustration page, but the student is not allowed to have text pages only or illustration pages only. Define $A(x) = \sum_{n \geq 0} a_n x^n$, the ordinary generating function of the sequence $\{a_n\}$. Then

$$A(x) = \sum_{n \geq 1} (2^n - 2)x^n = \sum_{n \geq 1} 2^n x^n - 2 \sum_{n \geq 1} x^n \tag{3.30}$$

$$= \frac{2x}{1-2x} - \frac{2x}{1-x} = \frac{2x^2}{(1-2x)(1-x)}. \tag{3.31}$$

If the thesis were to consist of two chapters, then the product formula would imply that the possible arrangements the student has are enumerated by the generating function $A^2(x)$.

Therefore, if the thesis were to contain *one or two* chapters, then the possibilities would be enumerated by the generating function $A(x) + A^2(x)$. We can certainly continue this way; that is, if the thesis is to consist of k chapters, then the number of possibilities is given by the formal power series $A^k(x)$, and if there are at most k chapters, then it is given by the power series $\sum_{n=1}^{k} A^n(x)$. If we agree that there is one way to write a thesis with no chapters, then we can say that $A^0(x) = 1$, and then $\sum_{n=0}^{k} A^n(x)$ is the generating function for the number of possibilities of a thesis of at most k chapters.

This is all very good, you might say, but where does this lead us? After all, we do not know how many chapters the thesis will have, and the above method seems to depend on that information. Fortunately, that is not really true. In order to understand why, note that the infinite sum

$$\sum_{n=0}^{\infty} A^n(x) = 1 + A(x) + A^2(x) + A^3(x) + \cdots \tag{3.32}$$

is actually a *well-defined* power series. That is, even if we add infinitely many power series, in their sum the coefficient of x^m is *finite* for all m. (This discussion of infinite sums of power series will probably remind the reader of the discussion of infinite products of power series, which can be found following (3.27).) Indeed, the term of the lowest exponent in $A^n(x)$ is x^{2n}. Therefore, for any m, there is only a finite number of power series of the form $A^n(x)$ that do have an x^m-term, namely, those in which $2n \leq m$. So the coefficient of x^m in our infinite sum $\sum_{n=0}^{\infty} A^n(x)$ is just a sum of *a finite number* of coefficients.

Therefore, the infinite sum $\sum_{n=0}^{\infty} A^n(x)$ does make sense, and, by our previous argument, it is the generating function for the number of ways an n-page long thesis can be arranged with zero chapters, or one chapter, or two chapters, or *any number of chapters*, since the summands do not stop coming. So at the very least, we did find the generating function of the numbers we were looking for. The power series $B(x) = \sum_{n=0}^{\infty} A^n(x)$ is precisely the ordinary generating function $\sum_{n=0}^{\infty} b_n x^n$, where b_n is the number of ways an n-page dissertation can be arranged into any number of chapters, each of which contains at least one text page and at least one page of illustrations.

Again, you might ask, what good is that? What can we do with an infinite sum of infinite sums (power series)? Fortunately, the answer to this question is not nearly as bad as one might expect. The key observation is that

$$B(x) = \sum_{n=0}^{\infty} A^n(x) = \frac{1}{1-A(x)}.$$

However, we do have an exact formula for $A(x)$, since we computed it in (3.30). Using that formula, we get

$$B(x) = \frac{1}{1-A(x)} = \frac{1}{1-\frac{2x^2}{(1-2x)(1-x)}}.$$

In other words, now we do have an explicit formula for the generating function of the numbers b_n! Finding the numbers b_n is now just a matter of computation. Using partial fractions, we find that

$$\begin{aligned} B(x) &= \frac{1}{1-\frac{2x^2}{(1-2x)(1-x)}} = \frac{(1-x)(1-2x)}{1-3x} \\ &= \frac{2}{9}\cdot\frac{1}{1-3x} - \frac{6x-7}{9} \\ &= \frac{2}{9}\cdot\left(\sum_{n\geq 0} 3^n x^n\right) - \frac{2}{3}x + \frac{7}{9} \\ &= 1 + 2\sum_{n\geq 2} 3^{n-2}x^n. \end{aligned}$$

That is, if the dissertation will have n pages, with $n \geq 2$, then there are $b_n = 2\cdot 3^{n-2}$ ways to arrange its pages into chapters.

This is a very nice and compact formula for b_n, and we challenge the reader in Exercise 20 to find a proof for it that does not use generating functions.

The existence of a proof that does not use generating functions for a formula does not decrease the value of a proof for that same formula that does use generating functions. It is often the case, just as in Exercise 20, that a combinatorial proof simply *verifies* the truthfulness of a formula, while a generating function proof *deduces* it. In other words, for a combinatorial proof, we may need to know the formula in advance, while for a generating function proof, we do not.

The novelty of the above computation was that we substituted the *power series* $A(x)$ for x in the power series $1/(1-x)$. That is, we *composed* generating functions. Let us formalize this concept.

Definition 3.24 *Let $F(x) = \sum_{n=0}^{\infty} f_n x^n$ be a formal power series, and let $A(x)$ be a formal power series having constant term 0. Then the* composition *of these power series is the power series*

$$F(A(x)) = \sum_{n=0}^{\infty} f_n A^n(x),$$

where we set $A^0(x) = 1$.

Note that this is a meaningful definition since the coefficient of x^n in $F(A(x))$ is finite for all n. Indeed, the smallest exponent in $A^m(x)$ is at least m, so, if $m > n$, then $A^m(x)$ will not have an x^n-term. So $[x^n]F(A(x))$ is the sum of a finite number of finite summands.

Now that we have formalized the technique of composing formal power series, we may generalize the central idea behind the previous example.

Theorem 3.25 *Let a_k be the number of ways to carry out a certain task on a k-element set, with $a_0 = 0$, and let $A(x) = \sum_{k\geq 0} a_k x^k$.*

Let b_n be the number of ways to split the interval $[n]$ into any number of disjoint nonempty subintervals, then carry out the task enumerated by the numbers a_i on each of the subintervals. Set $b_0 = 1$, and $B(x) = \sum_{n\geq 0} b_n x^n$.

Then

$$B(x) = \frac{1}{1 - A(x)}.$$

Proof: By the product formula, the number of ways to split $[n]$ into k nonempty intervals and then carry out the task counted by $A(x)$ on each of them is given by the generating function $A^k(x)$.

There is a subtle point here. The product formula does *not* require that the intervals into which we split $[n]$ be nonempty. However, by requiring this here, we are not cheating since we have $a_0 = 0$ anyway. That is, the formal power series $\sum_{k\geq 0} a_k x^k$ and $\sum_{k\geq 1} a_k x^k$ are identical. On the other hand, for the present problem we must require that the intervals be nonempty, otherwise there would be infinitely many ways to split $[n]$ into disjoint intervals by just adding empty intervals. This is one reason why the requirement that $a_0 = 0$ is essential.

Summing over all k, we get that, if there is no restriction on the number of possibilities, the number of ways to split $[n]$ into disjoint nonempty subintervals and then carry out the original task on each of them is counted by the power series

$$\sum_{k\geq 0} A^k(x) = \frac{1}{1 - A(x)} = B(x).$$

Here, the left-hand side is a well-defined sum as $a_0 = 0$. This shows the other reason for the requirement that $a_0 = 0$. $\diamondsuit$

Example 3.26 *This fall, the United States Congress will have n working days, split into an unspecified number of sessions. Within each session, one day will be designated for a plenary session, and each of the remaining days (if there are any remaining days) will be designated for either committee work or subcommittee work.*

Find an explicit formula for the number b_n of ways in which Congress can schedule its season.

Solution: This is a situation in which Theorem 3.25 applies, the scheduling of each session being the first task. Let a_n denote the number of ways to schedule one session. Then $a_n = n \cdot 2^{n-1}$, and, therefore,

$$A(x) = \sum_{n\geq 0} a_n x^n = \sum_{n\geq 0} n \cdot 2^{n-1} x^n = \frac{x}{(1-2x)^2}.$$

The last step can be obtained from the identity $\sum_{n\geq 0} nx^{n-1} = 1/(1-x)^2$, which we proved after stating formula (3.4).

Therefore, if b_n is the number of ways to carry out the composite task (splitting the fall into sessions, then specifying the type of work to be done on each working day) and $B(x) = \sum_{n\geq 0} b_n x^n$, then, by Theorem 3.25,

$$\begin{aligned} B(x) &= \frac{1}{1-A(x)} = \frac{1}{1-\frac{x}{(1-2x)^2}} \\ &= \frac{(1-2x)^2}{(1-2x)^2 - x} = 1 + \frac{x}{(1-2x)^2 - x} \\ &= 1 + \frac{x}{4x^2 - 5x + 1} = 1 + \frac{1}{3} \cdot \frac{1}{1-4x} - \frac{1}{3} \cdot \frac{1}{1-x} \\ &= 1 + \frac{1}{3} \cdot \sum_{n\geq 0} 4^n x^n - \frac{1}{3} \cdot \sum_{n\geq 0} x^n = 1 + \sum_{n\geq 0} \frac{4^n - 1}{3} x^n. \end{aligned}$$

Consequently, the number of ways to schedule the fall legislative session is $b_n = (4^n - 1)/3$ for $n \geq 1$. $\Diamond$

Now that the reader knows the formula for b_n, we challenge the reader to find an inductive proof for it without the use of generating functions in Supplementary Exercise 19.

Example 3.27 *Let f_n be the number of compositions of n into parts that are 1 or 2. Set $f_0 = 1$. Find the closed form of the generating function $F(x) = \sum_{n\geq 0} f_n x^n$.*

Solution: We cut $[n]$ into nonempty intervals, and then we "identify" each obtained interval as an interval of length one or length two. We can do that in one way if the given interval is *actually* of length one or two, and in zero ways otherwise. That shows that $A(x) = x + x^2$, and therefore, by Theorem 3.25,

$$F(x) = \frac{1}{1-x-x^2} = 1 + x + 2x^2 + 3x^3 + 5x^4 + \cdots .$$

$\diamondsuit$

The numbers f_n are called the *Fibonacci numbers.* The reader is asked to prove that they satisfy the recurrence relation $f_n = f_{n-1} + f_{n-2}$ for $n \geq 2$.

Theorem 3.25 can be generalized in the following way.

Theorem 3.28 *(Compositional formula for ordinary generating functions) Let a_n be the number of ways to carry out a task on an n-element set, and set $a_0 = 0$. Let b_n be the number of ways to carry out another task on an n-element set, with $b_0 = 1$. Finally, let c_n be the number of ways to split $[n]$ into nonempty intervals, then carry out the first task on each interval, and then carry out the second task* on the set of these intervals. *Then, if $A(x)$, $B(x)$, and $C(x)$ denote the ordinary generating functions of the three sequences, then the equality*

$$C(x) = B(A(x)) \tag{3.33}$$

holds.

Proof: If we split $[n]$ into k nonempty intervals, then, by the product formula, the number of ways to carry out the first task on each interval is given by the generating function $A(x)^k$. Then there are b_k ways to carry out the second task on the set of the given k intervals, leading to the generating function $b_k A(x)^k$. Summing over all k, we get that

$$C(x) = \sum_{k \geq 0} b_k A(x)^k = B(A(x)).$$

$\diamondsuit$

Example 3.29 *A soccer coach has her n players stand in a line. Then she breaks the line at a few places, to form nonempty units, and chooses a leader in each unit. Finally, she chooses one of the units for a specific task. Find the generating function for the number c_n of ways she can proceed.*

Solution: Clearly, there are m ways to choose a leader in a unit of m players, so $a_m = m$ for $m > 0$. Similarly, there are k ways to choose a unit out of k units, so $b_k = k$ for $k \geq 1$, with $b_0 = 1$. Therefore, we have $A(x) = \sum_{m \geq 1} m x^m = \frac{x}{(1-x)^2}$, and $B(x) = 1 + \sum_{k \geq 1} k x^k = 1 + \frac{x}{(1-x)^2}$. So Theorem

3.28 yields

$$C(x) = B(A(x) = 1 + \frac{\frac{x}{(1-x)^2}}{\left(1 - \frac{x}{(1-x)^2}\right)^2}$$
$$= \frac{1 - 5x + 9x^2 - 5x^3 + x^4}{(1 - 3x + x^2)^2}.$$

◊

3.4.2 Exponential generating functions

Exponential generating functions are even more susceptible to applications involving compositions than ordinary generating functions. We start with the analog of Theorem 3.25 in the exponential case.

3.4.2.1 The exponential formula

Recall that, in Theorem 3.25, we split an interval into an unspecified number of nonempty subintervals, then carried out the same task on each of those intervals, and then computed the number of ways all this could be done. Just as with the product formulae, we can replace ordinary generating functions by exponential generating functions if we replace subintervals by subsets. The following theorem makes this statement more precise.

Theorem 3.30 (Exponential formula) *Let a_n be the number of ways to carry out a certain task on an n-element set, and set $a_0 = 0$. For $n \geq 1$, let h_n be the number of ways to partition $[n]$ into an arbitrary number of nonempty blocks, and then to carry out the first task on each of these blocks. Set $h_0 = 1$. If $A(x) = \sum_{n \geq 0} a_n \frac{x^n}{n!}$ and $H(x) = \sum_{n \geq 0} h_n \frac{x^n}{n!}$, then*

$$H(x) = e^{A(x)}.$$

Remark:
There is another way to state this theorem that is slightly more technical, but also somewhat more general. Let **P**, respectively, **N**, and **R** denote the set of positive integers, respectively nonnegative integers, and real numbers. Let $a : \mathbf{P} \to \mathbf{R}$ be a function, and let $h : \mathbf{N} \to \mathbf{P}$ be another function, defined by $h(0) = 1$ and

$$h(n) = \sum_{\pi} a(|\pi_1|)a(|\pi_2|) \cdots a(|\pi_m|),$$

where the sum is taken over all partitions $\pi = (\pi_1, \pi_2, \cdots, \pi_m)$ of $[n]$, and where $|\pi_i|$ denotes the size of the block π_i. Then we have $H(x) = e^{A(x)}$.

The reader should take a moment and explain why this second version implies the first one and why the second version is slightly more general than the first one. (Hint: What are the possible values of a in this version, and what are the possible values of the corresponding function in Theorem 3.30?)

We will prove the first version of the exponential formula, but the same argument could be used to prove the second version as well.

Proof: Let $H_k(x)$ be the generating function for the number of ways the composite task can be done if we partition n into k blocks. Then, by the product formula, we have

$$H_k(x) = \frac{A(x)^k}{k!}.$$

Indeed, in the product formula the *order* of the blocks mattered, whereas here it does not, and this explains the division by $k!$. Furthermore, the product formula allowed for empty blocks; this theorem does not. That is fine, however, since we set $a_0 = 0$. Therefore, we can say that we allow empty blocks, but if a block is empty, then there are zero ways to carry out the composite task.

Summing over all $k \geq 1$, and recalling that we set $b_0 = 1$, we get

$$\begin{aligned} H(x) &= 1 + H_1(x) + H_2(x) + \cdots = \sum_{k=0}^{\infty} \frac{(A(x))^k}{k!} \\ &= e^{A(x)}. \end{aligned}$$

◊

Perhaps the most classic illustration of the structures connected to the exponential formula is the following example.

Example 3.31 *The number of ways to partition a set of n people and then have each block of people sit around a circular table is $n!$. (Two seating arrangements are considered identical if each person has the same left neighbors in both of them.)*

Solution: The first task, that is, the task to be carried out within each block, is to have people sit around a circular table. We saw in Example 1.22 that this can be done in $(k-1)!$ different ways if we have k people for a given table. Therefore, we have $a_k = (k-1)!$ for $k \geq 1$, leading to

$$\begin{aligned} A(x) &= \sum_{n\geq 1} (n-1)! \frac{x^n}{n!} = \sum_{n\geq 1} \frac{x^n}{n} \\ &= \ln\left(\frac{1}{1-x}\right). \end{aligned}$$

Let h_n denote the number of ways the composite task can be carried out, and let $H(x) = \sum_{n\geq 0} h_n \frac{x^n}{n!}$ be the exponential generating function of the numbers $H(x)$. Then, by the exponential formula,

$$\begin{aligned} H(x) &= e^{A(x)} = e^{\ln\left(\frac{1}{1-x}\right)} \\ &= \frac{1}{1-x} \\ &= \sum_{n\geq 0} x^n. \end{aligned}$$

So the coefficient h_n of $x^n/n!$ in $H(x)$ is indeed $n!$ as claimed. ♢

This is a surprising result (if you have not taken abstract algebra before, that is). We do not even know how many tables will be used, so we might think the answer would contain a summation sign indexed by the number of tables with guests. However, the answer has a very simple form, $n!$, suggesting that there should be a bijection between the set of all possible seating arrangements and all permutations of $[n]$. Such a bijection indeed exists, and it will be discussed in great detail in the following chapter.

The expression $e^{A(x)}$ in the exponential formula often leads to generating functions whose coefficients are not easy to express by an explicit formula. The main exception to this is when it is applied to circular seating arrangements with various restrictions, as in the example above. As we mentioned, these arrangements in fact correspond to permutations, and we will study their enumeration in great detail in the following chapter. Until then, let us consider some examples in which we can at least easily find the generating function $H(x)$.

In Example 3.31, we identified two seating arrangements where each person had the same left neighbor in both of them. This can be useful when food is passed around the table in a given direction and you are wondering whether the big eaters will help themselves before you can. However, if there is no food to pass around and having entertaining conversations is the main goal, we may want to identify two seating arrangements where each person has the same set of neighbors in both of them. This situation is discussed in the next example.

Example 3.32 *Let h_n be the number of ways to have n people sit around an unspecified number of circular tables, where all tables are used. Two seating arrangements are considered identical if each person has the same set of neighbors in both of them. Find an explicit formula for the exponential generating function $H(x) = \sum_{n\geq 0} h_n \frac{x^n}{n!}$.*

Solution: Let a_n be the number of seating arrangements for n people around one table, with $a_0 = 0$. Then $a_1 = a_2 = 1$, and, by Example 1.22 and the division principle, $a_n = (n-1)!/2$ for $n \geq 3$. Therefore, we get

$$\begin{aligned} A(x) &= \sum_{n\geq 0} a_n \frac{x^n}{n!} = x + \frac{x^2}{2} + \sum_{n\geq 3} \frac{x^n}{2\cdot n} \\ &= \frac{x}{2} + \frac{x^2}{4} + \frac{1}{2}\cdot \ln\left(\frac{1}{1-x}\right). \end{aligned}$$

So the exponential formula shows that

$$\begin{aligned} H(x) &= e^{A(x)} = e^{\frac{x}{2}+\frac{x^2}{4}+\frac{1}{2}\cdot\ln\left(\frac{1}{1-x}\right)} \\ &= \frac{e^{\frac{x}{2}+\frac{x^2}{4}}}{\sqrt{1-x}}. \end{aligned}$$

◇

It is interesting to see that such a minor modification in the formulation of the problem leads to such a substantial change in the result. Indeed, the result is far more complicated than that of Example 3.31. This is because the minor change (essentially, a division by two) happened *before* the application of the exponential formula. In other words, it is the exponent that got divided by 2, which led to the final result to be raised to the (1/2)th power.

The following is a very general application of the exponential formula, and one that we will revisit in the next chapter.

Theorem 3.33 *Let $S = \{s_1, s_2, \cdots\}$ be a set of positive integers. Let $h_S(n)$ be the number of partitions of the set $[n]$ into blocks so that each block size is an element of S. Let $H_S(x)$ be the exponential generating function of the sequence $\{h_S(n)\}$. Then*

$$H_S(x) = e^{\sum_{i\geq 1} \frac{x^{s_i}}{s_i!}}.$$

Proof: If we note that $\sum_{i\geq 1} \frac{x^{s_i}}{s_i!}$ is precisely the exponential generating function for the number of choices we have on one block, then it is easy to see that this is a direct consequence of the exponential formula. Indeed, if the size m of a block is not in S, then, in the language of the exponential formula, $a_m = 0$. Likewise, if the size of the block is in S, then $a_m = 1$. Note that this means that the contribution of a partition to $h_S(n)$ is one if and only if all blocks of the partition are elements of S. ◇

The set S of Theorem 3.33 can be infinite. The following example shows an instance of that.

Example 3.34 *Find the exponential generating function for the numbers of partitions of $[n]$ into blocks of even size.*

Solution: We will use Theorem 3.33, with $S = \{2, 4, 6, \cdots\}$. This yields

$$\sum_{i \geq 1} \frac{x^{s_i}}{s_i!} = \sum_{i \geq 1} \frac{x^{2i}}{(2i)!} = \frac{e^x + e^{-x}}{2} - 1$$
$$= \cosh x - 1.$$

Therefore, Theorem 3.33 implies that

$$H_S(x) = e^{\cosh x - 1}.$$

$\diamondsuit$

Example 3.35 *A teacher has n students in his classroom. He splits the set of students into an arbitrary number of nonempty groups, then names a leader in each of the groups. In how many ways can the teacher proceed?*

Solution: Within a group of k students, there are $a_k = k$ ways to choose a leader. Therefore,

$$A(x) = \sum_{k=1}^{\infty} k \frac{x^k}{k!} = x \sum_{k=1}^{\infty} \frac{x^{k-1}}{(k-1)!} = xe^x.$$

So the exponential formula yields that, if the teacher can proceed in h_n ways, then

$$H(x) = \sum_{n \geq 0} h_n \frac{x^n}{n!} = e^{xe^x}.$$

Writing

```
series(exp(x*exp(x)), (x=0,10));
```

in Maple and hitting return, we get the first 10 terms of the generating function $H(x)$.

These terms are

$$1 + x + \frac{3}{2}x^2 + \frac{5}{3}x^3 + \frac{41}{24}x^4 + \frac{49}{30}x^5 + \frac{1057}{720}x^6 + \frac{3161}{2520}x^7 + \frac{41393}{40320}x^8 + \frac{5243}{6480}x^9.$$

$\diamondsuit$

3.4.2.2 The compositional formula

The following theorem is the most general one in this subsection.

Theorem 3.36 (Compositional formula) *Let a_n be the number of ways to carry out a certain task on an n-element set, and set $a_0 = 0$. Let b_n be the number of ways to carry out a second task on an n-element set. Finally, let h_n be the number of ways to partition $[n]$ into an unspecified number of nonempty blocks, carry out the first task on each of the blocks, then carry out the second task on* the set of the blocks. *Let $A(x)$, $B(x)$, and $H(x)$ denote the exponential generating functions of the sequences a_n, b_n, and h_n. Then*

$$H(x) = B(A(x)).$$

Proof: (Similar to the proof of the exponential formula.) Let us assume first that we split $[n]$ into k blocks, and then let $H_k(x)$ be the generating function for the number of ways to carry out the composite task. We have seen in the proof of the exponential formula that without the second task the numbers of possibilities would be enumerated by the generating function $\frac{A(x)^k}{k!}$. That is, this power series counts the number of ways to partition $[n]$ into k parts, then carry out the first task on each part. We now have b_k ways to carry out the second task. This means that the number of ways to carry out the composite task is enumerated by the power series

$$H_k(x) = b_k \frac{A(x)^k}{k!}.$$

Summing over all k, we get that

$$\begin{aligned} H(x) &= 1 + H_1(x) + H_2(x) + \cdots = \sum_{k=0}^{\infty} b_k \frac{(A(x))^k}{k!} \\ &= B(A(x)). \end{aligned}$$

◊

Note that Theorem 3.30 is a special case of Theorem 3.36, namely, the special case when $b_n = 1$ for all n.

To illustrate this theorem, let us start with a surprising application.

Example 3.37 *The number of ways to split a set of n people into an unspecified number of nonempty groups, have each group sit down around a circular table, and then serve exactly one choice of red wine or white wine to each table is $(n + 1)!$.*

Solution: The first task, just as in Example 3.31, is to find a seating arrangement for each table. We have seen in Example 1.22 that this can be done in $a_k = (k - 1)!$ ways if there are $k > 0$ people at a given table, leading to the generating function $A(x) = \ln\left(\frac{1}{1-x}\right)$. (See Example 3.31 for details.)

The second task is to decide which tables will get red wine. If there are k nonempty tables, then this decision can be made in $b_k = 2^k$ different ways, leading to the generating function

$$B(x) = \sum_{k\geq 0} 2^k \frac{x^k}{k!} = e^{2x}.$$

Let h_n be the number of ways to carry out the composite task, and set $H(x) = \sum_{n\geq 0} h_n \frac{x^n}{n!}$. Then, by the compositional formula,

$$\begin{aligned} H(x) &= B(A(x)) \\ &= e^{2\ln\left(\frac{1}{1-x}\right)} \\ &= \frac{1}{(1-x)^2} \\ &= \sum_{n\geq 0}(n+1)x^n. \end{aligned}$$

The coefficient of $x^n/n!$ in the last expression is $(n+1)n! = (n+1)!$, proving our claim. ◊

The result is surprising not only because of its compactness, but also because it does not contain any powers of 2, which we might have expected since each table got one of two different wines.

Example 3.38 *A professor has n students who must take an oral exam. She lets the students take the exam on any number of consecutive days as long as there is at least one student for each day. On each exam day, the students take the exam one by one. How many different exam schedules are possible?*

Solution: If k students take the exam on a given day, then they can do it in $a_k = k!$ different ways for all $k \geq 1$, and zero ways otherwise. This leads to

$$A(x) = \sum_{k\geq 1} k! \frac{x^k}{k!} = \frac{x}{1-x}.$$

Once the set of students is partitioned into m blocks so that students taking the exam on the same day are in the same block, then there are $m!$ ways to order the set of these blocks. This leads to

$$B(x) = 1 + \sum_{m=1}^{\infty} m! \frac{x^m}{m!} = \frac{1}{1-x}.$$

Therefore, by the compositional formula, we get

$$\begin{aligned} H(x) &= B(A(x)) = \frac{1}{1-\frac{x}{1-x}} = \frac{1-x}{1-2x} \\ &= \frac{1}{1-2x} - \frac{x}{1-2x} = \sum_{n\geq 0} 2^n x^n - \sum_{n\geq 1} 2^{n-1} x^n. \end{aligned}$$

Therefore, $h_0 = 1$, and $h_n = n!(2^n - 2^{n-1}) = n!2^{n-1}$ for $n \geq 1$. ◊

At this point, the reader might say, Big deal, I could have done that without generating functions. This is true (see Supplementary Exercise 24), but note the following: Suppose the professor is not willing to come in and hold exams just for the sake of one student, but she requires instead that at least t students come to take the exam each day. In that case, a purely combinatorial proof is more difficult, while a generating function proof is immediately obtained by changing $A(x)$ to $x^t/(1-x)$ in the above argument.

Just as in the case of the exponential formula, it happens here too that we get a generating function for $H(x)$ whose coefficients are not easy to express by an explicit formula.

Example 3.39 *Let us revisit the teacher of Example 3.35. Now he splits the set of his n students into nonempty groups, names a leader in each group, then assigns the order in which the groups will give their presentations. Find the exponential generating function for the number of ways he can proceed.*

Solution: There are k ways to name a leader in a group of k students, and, as we have computed in Example 3.35, the corresponding generating function is $A(x) = \sum_{k\geq 1} k\frac{x^k}{k!} = xe^x$.

The extra feature in this example, as opposed to Example 3.35, is that here the teacher carries out a task on the *set of groups of students* as well, by linearly ordering them. If there are m groups, then this can be done in $b_m = m!$ different ways, leading to the generating function

$$B(x) = \sum_{m\geq 0} b_m \frac{x^m}{m!} = \sum_{m\geq 0} x^m = \frac{1}{1-x}.$$

Therefore, by the compositional formula, we get a formula for the number of ways to carry out the composite task. That is, if h_n is the number of ways to proceed, and $H(x)$ is the exponential generating function of the numbers h_n, then

$$H(x) = B(A(x)) = \frac{1}{1-xe^x}.$$

◊

Quick Check

1. Let $h(n)$ be the number of ways to split the interval $[1, n]$ into nonempty segments, then color each integer from 1 through n red or blue so that in each segment, both colors are represented. Find ordinary generating function of the numbers $h(n)$.

2. Let $h(n)$ be the number of all partitions of the set $[n]$ in which each block is of size two. Find the exponential generating function of the sequence of the numbers $h(n)$.
3. Let $h(n)$ be the number of ways to select a partition of $[n]$, arrange the elements of each block circularly, and finally arrange these circles in a line. Find the exponential generating function of the sequence of the numbers $h(n)$.

3.5 A different type of generating functions

While ordinary and exponential generating functions are the generating functions that are used most often in practice, they are not the only ones that are useful in enumeration. The interested reader can find various other generating functions in [84]. We will just show one more type of generating function, *doubly exponential generating functions*, which we will use in one of the later chapters.

Example 3.40 *Let* $a_0 = 1$*, and let* $a_n = n^2 a_{n-1} + n!$ *if* $n \geq 1$*. Find an explicit formula for* a_n*.*

A little experimentation shows that the two kinds of generating functions we learned so far will not help in this case, as this sequence grows too fast even for the technique of exponential generating functions. Therefore, we define

$$A(x) = \sum_{n\geq 0} a_n \frac{x^n}{n!^2},$$

and we call $A(x)$ the *doubly exponential generating function* of the sequence $\{a_n\}_{n\geq 0}$.

The rest is not difficult. Let us multiply both sides of the recurrence relation of Example 3.40 by $x^n/n!^2$, and sum over all $n \geq 1$. We get

$$\sum_{n\geq 1} a_n \frac{x^n}{n!^2} = \sum_{n\geq 1} a_{n-1} \frac{x^n}{(n-1)!^2} + \sum_{n\geq 1} \frac{x^n}{n!},$$

or, recognizing $A(x)$ on both sides,

$$A(x) - 1 = xA(x) + e^x - 1,$$

$$A(x) = \frac{e^x}{1-x}.$$

It is now simple to express $A(x)$ explicitly. Indeed, the last equation yields

$$\begin{aligned} A(x) &= \left(\sum_{i\geq 0} \frac{x^i}{i!}\right) \cdot \left(\sum_{k\geq 0} x^k\right) \\ &= \sum_{n\geq 0} \left(\sum_{i=0}^{n} \frac{1}{i!}\right) x^n. \end{aligned}$$

Therefore, the coefficient of $x^n/n!^2$ in $A(x)$ is

$$a_n = n!^2 \sum_{i=0}^{n} \frac{1}{i!}.$$

Quick Check

1. Let $a_0 = 1$, and let $a_n = n^2 a_{n-1} + (-1)^n n!$ if $n \geq 1$. Find an explicit formula for the numbers a_n.
2. Let $a_0 = 2$, and let $a_n = n^2 a_{n-1} + n!$ if $n \geq 1$. Find an explicit formula for the numbers a_n.
3. Let $a_0 = 1$, and let $a_n = 2n^2 a_{n-1} + n!$ if $n \geq 1$. Find an explicit formula for the numbers a_n.

3.6 Notes

The interested reader should consult Herb Wilf's book *generatingfunctionology* [84] for a very readable book on generating functions. A more advanced treatment of the techniques explained in this chapter can be found in Richard Stanley's book *Enumerative Combinatorics,* Volume 2 [72]. Readers with a particular interest in identities involving integer partitions should take a look at *Theory of Partitions* [6], by George Andrews.

In all the theorems discussed in this chapter, the coefficients of the generating functions that we studied were defined as the number of ways to carry out a particular task. That meant that these coefficients were all *nonnegative integers.* However, if we take a closer look at the proofs of our theorems, we realize that, in the algebraic sense, the proofs did not use that fact; they would have worked just as well if the coefficients were complex numbers, or even variables. This suggests that the theorems covered in this chapter should have more general versions, in which the relevant generating functions are defined in a manner that keeps the proofs valid, but does not imply that the coefficients of these generating functions are integers. We leave the thorough treatment of this topic to more advanced books, such as [72], but we will take a few excursions into this topic in subsequent chapters.

In this chapter, we used generating functions to deduce *exact* enumeration formulae. We will keep using them for that purpose in the next three chapters, when we focus on permutations, graphs, and extremal combinatorics. Then, in Chapter 7, generating functions will again become the center of our attention as we use them to prove *asymptotic* enumeration results, in situations when an exact formula is not possible, or not necessary.

In subsequent chapters, we will learn further uses of generating functions, such as applications of their *derivatives* and generating functions in *several variables*.

3.7 Chapter review

(A) Formal power series

1. If $F(x) = \sum_n f_n x^n$ and $G(x) = \sum g_n x^n$, then

$$(F \cdot G)(x) = \sum_{n \geq 0} \left(\sum_{k=0}^{n} f_k g_{n-k} \right) x^n.$$

2. If $F(x) = \sum_n f_n \frac{x^n}{n!}$ and $G(x) = \sum g_n \frac{x^n}{n!}$, then

$$(F \cdot G)(x) = \sum_{n \geq 0} \left(\sum_{k=0}^{n} \binom{n}{k} f_k g_{n-k} \right) \frac{x^n}{n!}.$$

(B) Generating functions of the sequence $a_0, a_1, \cdots$

1. Ordinary generating function: $A(x) = \sum_n a_n x^n$.
2. Exponential generating function: $A(x) = \sum_n a_n \frac{x^n}{n!}$.

(C) Product formulae

1. Ordinary generating functions: If $A(x)$ counts the ways to carry out a certain task, $B(x)$ counts the ways to carry out another task, then $A(x)B(x)$ counts the ways to split an interval into two disjoint subintervals, then carry out the first task on the first subinterval, and the second task on the second subinterval.
2. Exponential generating functions: If $A(x)$ counts the ways to carry out a certain task, $B(x)$ counts the ways to carry out another task, then $A(x)B(x)$ counts the ways to split a set into two disjoint subsets, then carry out the first task on the first subset, and the second task on the second subset.

(D) Compositional formulae

1. Ordinary generating functions:
   (a) If $A(x)$ counts the ways to carry out a certain task on nonempty sets, and $B(x)$ counts the ways to split $[n]$ into nonempty and disjoint subintervals, and then carry out the task counted by $A(x)$ on each of these subintervals, then
   $$B(x) = \frac{1}{1 - A(x)}.$$
   (b) If $A(x)$ counts the ways to carry out a certain task on nonempty sets, $B(x)$ counts the ways to carry out another task on nonempty sets, and $C(x)$ counts the ways to split $[n]$ into nonempty subintervals, then to carry out the first task on each interval and the second task at the set of the intervals, then
   $$C(x) = B(A(x)).$$
2. Exponential generating functions:
   (a) Exponential formula. If $A(x)$ counts the ways to carry out a certain task on nonempty sets, and $B(x)$ counts the ways to partition $[n]$ into nonempty and disjoint blocks, and then carry out the task counted by $A(x)$ on each of these blocks, then
   $$B(x) = \exp A(x).$$
   (b) Compositional formula. If $A(x)$ counts the ways to carry out a certain task on nonempty sets, and $B(x)$ counts the ways to carry out a second task on nonempty sets, and $H(x)$ counts the ways to partition $[n]$ into nonempty blocks, then carry out the task counted by $A(x)$ on each of these blocks, and then carry out the task counted by $B(x)$ on the set of blocks, then
   $$H(x) = B(A(x)).$$

3.8 Exercises

1. Prove that, for all positive integers k,
$$(1-x)^{-k} = \sum_{n\geq 0} \binom{n+k-1}{k-1} x^n.$$

2. How can we quickly verify that formula (3.7) is indeed correct for all n?
3. Find an explicit formula for a_n if $a_0 = 2$ and if, for all integers $n \geq 1$, we have $a_n = 4a_{n-1} - 3$.
4. Find an explicit formula for a_n if $a_0 = 0$, and $a_1 = 1$, and if, for all integers $n \geq 2$, we have $a_n = 4a_{n-1} - 4a_{n-2}$.
5. Find an explicit formula for a_n if $a_0 = 0$, and $a_1 = 1$, and if, for all integers $n \geq 2$, we have $a_n = 4a_{n-1} - 5a_{n-2}$.
6. (Basic knowledge of linear algebra required.) Let V be the set of all sequences of complex numbers $\{a_n\}$ that satisfy the recurrence relation
$$a_n = pa_{n-1} + qa_{n-2} \tag{3.34}$$
for all $n \geq 2$, where p and q are fixed real numbers.
   (a) Prove that V is a vector space over the field of real numbers.
   (b) What is the dimension of V?
7. Let V be defined as in the previous exercise. Find all elements $\{a_n\}$ of V whose terms are of the form $a_n = a^n$ for all nonnegative integers n.
8. + Let V be defined as in Exercise 6. Find a basis for V, including explicit formulae for the elements of the basis you found.
9. Solve the recurrence relation $a_0 = 1$ and $a_n = 3\sum_{i=0}^{n-1} a_i$ for all $n \geq 1$.
10. Let $F_1, F_2, \cdots$ be an infinite sequence of formal power series satisfying $F_i(0) = 1$ for all i. Prove that, if there exists a positive integer n so that infinitely many elements of the sequence contain an x^n-term (with a nonzero coefficient), then the infinite product $\prod_{i\geq 1} F_i$ is not defined.
11. Is the converse of the statement of the previous exercise true?
12. Let $p_{odd}(n)$ be the number of partitions of the integer n into odd parts, with $p_{odd}(0) = 1$. Find the ordinary generating function of the numbers $p_{odd}(n)$.
13. Let $p_d(n)$ be the number of partitions of the integer n into distinct parts, with $p_d(0) = 1$. Find the ordinary generating function of the numbers $p_d(n)$.
14. + Compare the results of the previous two exercises. What do they tell us about the connection between $p_{odd}(n)$ and $p_d(n)$?
15. + Express the infinite product
$$\prod_{i\geq 1}(1 - x^i)$$
as an infinite sum, using only one summation sign.

16. Use the result of the previous exercise to prove a recurrence relation for the numbers $p(n)$.

17. Find an explicit formula for a_n if $a_n = na_{n-1} + (n+1)!$ for $n \geq 1$, and $a_0 = 0$.

18. Find an explicit formula for a_n if $a_n = na_{n-1} + (-1)^n$ and $a_0 = 1$.

19. Find a combinatorial proof for the result of Example 3.22. Do not use induction.

20. Let b_n be defined as in Example 3.23. Prove that

$$b_n = 2(b_{n-1} + b_{n-2} + \cdots + b_0).$$

Then deduce from this that $b_n = 2 \cdot 3^{n-2}$ for $n \geq 2$. Do not use generating functions.

21. Let $OS(n,3)$ be the number of *ordered partitions* of $[n]$ into three nonempty blocks. That is, $OS(n,3) = 3!S(n,3)$ since now we can arrange the blocks of any partition counted by $S(n,3)$ in six different ways.

    (a) Find an explicit formula for the exponential generating function

$$OS_3(x) = \sum_{n\geq 0} OS(n,3)\frac{x^n}{n!}.$$

    (b) Deduce a formula for the numbers $OS(n,3)$ and the numbers $S(n,3)$.

In the following three exercises, the reader may want to translate the problem into the language of lattice paths.

22. + Two grandmasters played a series of chess games using the following scoring system: If a game has a winner, the winner gets one point and the loser gets no points. If a game is a tie, both players get one point. Their series ended in a tie of $n - n$. At the end, a journalist noticed that tie games only occurred when the current aggregate score was a tie. Let d_n be the number of possible ways this could happen, and set $d_0 = 1$.

    Find the ordinary generating function $D(x) = \sum_{n\geq 0} d_n x^n$.

23. + Two grandmasters play a series of n chess games using the traditional scoring system. That is, a win is one point, a tie is a half point for both players, and a loss is zero points. Let M_n be the number of ways the series can end up in an aggregate tie (that is, a tie of $\frac{n}{2} - \frac{n}{2}$) if grandmaster A never trails. Find the ordinary generating function of the numbers M_n.

24. \+ Let us modify the scoring system of the previous exercise so that, in case of a tie, both players get one point. The rest of the conditions are unchanged. Let r_n be the number of ways the series can now end in an aggregate tie of $n-n$, and set $r_n = 1$. Find the ordinary generating function of the numbers r_n.

25. Let $f(n)$ be the number of ways we can tile a $1 \times n$ rectangle using red and blue tiles of size 1×1, and black, white, and yellow tiles of size 1×2. Find the ordinary generating function for the numbers $f(n)$, then find an explicit formula for these numbers.

26. Let $r \geq 2$ be a fixed positive integer, and let $h_r(n)$ be the number of compositions of n into parts that are of the form $rk+1$. Find the closed form for the generating function $H_r(x) = \sum_{n \geq 0} h_r(n)x^n$. Explain the result in the special case of $r = 2$.

27. Let $d > 2$ be a fixed positive integer. Find the ordinary generating function for the number of partitions of the integer n in which each part occurs less than d times.

28. Let $d \geq 2$ be a fixed positive integer. Find the ordinary generating function for the number of partitions of the integer n in which no part is divisible by d.

29. Compare the results of the previous two exercises. What is your conclusion?

30. We deduced formula (3.19) for the generating function of the Catalan numbers using the Product formula. Deduce (3.19) using the Compositional formula.

3.9 Solutions to exercises

1. Expanding the left-hand side by the binomial theorem, we see that the coefficient of x^n is $(-1)^n\binom{-k}{n}$. If we can show that this is equal to $\binom{n+k-1}{k-1}$, we will be done. We have

$$\begin{aligned}
(-1)^n\binom{-k}{n} &= (-1)^n \cdot \frac{(-k)(-k-1)\cdots(-k-n+1)}{n!} \\
&= (-1)^{2n} \cdot \frac{k(k+1)\cdots(n+k-1)}{n!} = \binom{n+k-1}{n} \\
&= \binom{n+k-1}{k-1}.
\end{aligned}$$

2. For $n = 0$, the formula is correct as it gives $a_0 = 5$. Now assume the formula is correct for n. Recall that the sequence $\{a_n\}$ is defined by

$a_{n+1} = 3a_n - 1$ for $n \geq 0$. Therefore, the fact that (3.7) is correct for n implies

$$a_{n+1} = 3\left(5 \cdot 3^n - \frac{3^n - 1}{2}\right) - 1 = 5 \cdot 3^{n+1} - \frac{3^{n+1} - 1}{2},$$

so (3.7) is correct for $n+1$ as well. So, by induction, (3.7) is correct for all nonnegative integers n.

3. Let $A(x)$ be the ordinary generating function of the sequence. Multiply both sides of the recursion by x^n, then sum over $n \geq 1$ to get

$$A(x) - 2 = 4xA(x) - \frac{3x}{1-x}.$$

This yields

$$\begin{aligned} A(x) &= \frac{2}{1-4x} - \frac{3x}{(1-x)(1-4x)} = \frac{2}{1-4x} - \frac{1}{1-4x} + \frac{1}{1-x} \\ &= \frac{1}{1-4x} + \frac{1}{1-x} \\ &= \sum_{n \geq 0} 4^n x^n + \sum_{n \geq 0} x^n. \end{aligned}$$

Therefore, the coefficient of x^n in $A(x)$ is $a_n = 4^n + 1$.

4. Start as in Example 3.7. If $A(x)$ is the ordinary generating function of the sequence, then the usual steps will lead us to the functional equation

$$A(x) - x = 4xA(x) - 4x^2 A(x). \tag{3.35}$$

Therefore, we get

$$\begin{aligned} A(x) &= \frac{x}{1 - 4x + 4x^2} = \frac{x}{(1-2x)^2} \\ &= x \sum_{n \geq 0} \binom{-2}{n} (-2x)^n = x \sum_{n \geq 0} \binom{n+1}{n} (-1)^n (-2)^n x^n \\ &= x \sum_{n \geq 0} (n+1) \cdot 2^n x^n \sum_{n \geq 0} (n+1) \cdot 2^n x^{n+1} = \sum_{n \geq 1} n \cdot 2^{n-1} x^n. \end{aligned}$$

Therefore, we have $a_n = n \cdot 2^{n-1}$.

5. Proceeding as in both Example 3.7 and the previous exercise, we get the equation

$$A(x) - x = 4xA(x) - 5x^2 A(x),$$

which looks similar to (3.35), but is actually quite different from it, as we will see. Expressing $A(x)$ from the previous equation, we get

$$A(x) = \frac{x}{1 - 4x + 5x^2}.$$

In this case, it is somewhat harder to find the partial fraction decomposition of the right-hand side, since the denominator has *complex roots*, namely, $0.4 \pm 0.2i$. Set $\alpha = 0.4 + 0.2i$ and $\beta = 0.4 - 0.2i$. Then, after simplifying by 5, we are looking for numbers C and D so that

$$\frac{C}{x-\alpha} + \frac{D}{x-\beta} = \frac{0.2x}{0.2(1-4x+5x^2)},$$

$$C(x-\beta) + D(x-\alpha) = 0.2x.$$

This leads to $C + D = 0.2$ and $C\beta + D\alpha = 0$. Solving this system, we get $C = 0.1 - 0.2i$ and $D = 0.1 + 0.2i$. Therefore, we have

$$\begin{aligned} A(x) &= \frac{0.1-0.2i}{x-\alpha} + \frac{0.1+0.2i}{x-\beta} = -\frac{0.1-0.2i}{\alpha(1-\frac{x}{\alpha})} - \frac{0.1+0.2i}{\beta(1-\frac{x}{\beta})} \\ &= -\frac{0.1-0.2i}{\alpha}\sum_{n\geq 0}\frac{x^n}{\alpha^n} - \frac{0.1+0.2i}{\beta}\sum_{n\geq 0}\frac{x^n}{\beta^n} \\ &= \frac{i}{2}\sum_{n\geq 0}\frac{x^n}{\alpha^n} - \frac{i}{2}\sum_{n\geq 0}\frac{x^n}{\beta^n}. \end{aligned}$$

We now find a_n as the coefficient of x^n in a_n, that is,

$$a_n = \frac{i}{2}\left(\frac{1}{\alpha^n} - \frac{1}{\beta^n}\right).$$

There are several ways to bring this equation to a nicer form. To start, note that $1/\alpha = 2 + i$ and $1/\beta = 2 - i$, so we have

$$a_n = \frac{i}{2}\left((2+i)^n - (2-i)^n\right).$$

If we expand $(2+i)^n$ and $(2-i)^n$ by the binomial theorem, and then we compute their difference, we see that the terms in which the exponents of i are even will cancel. Therefore, we get

$$\begin{aligned} a_n &= \frac{i}{2}\left(\sum_{\substack{j=1 \\ j \text{ odd}}}^{n} 2\binom{n}{j}2^{n-j}i^j\right) \\ &= \sum_{\substack{j=1 \\ j \text{ odd}}}^{n} \binom{n}{j}2^{n-j}i^{j-1} \\ &= n\cdot 2^{n-1} - \binom{n}{3}2^{n-3} + \binom{n}{5}2^{n-5} - \cdots. \end{aligned}$$

6. (a) These sequences form a vector space, since, if $\{a_n\}$ and $\{b_n\}$ both satisfy the recurrence relation (3.34), then so does their sum, that is, the sequence $\{a_n + b_n\}$. Similarly, if $\{a_n\}$ satisfies (3.34), then so does its constant multiple, that is, the sequence $\{ca_n\}$ for any $c \in \mathbf{R}$.

(b) The dimension of this vector space is two. Indeed, let $\{a_n\}$ be the sequence in V for which $a_0 = 1$ and $a_1 = 0$, and let $\{b_n\}$ be the sequence in V for which $b_0 = 0$ and $b_1 = 1$. These terms uniquely determine these sequences. Then $\{a_n\}$ and $\{b_n\}$ are linearly independent. Furthermore, any other sequence in V can be obtained as their linear combination. Indeed, if $\{c_n\}$ is a sequence in V with initial terms c_0 and c_1, then we have $\{c_n\} = c_0\{a_n\} + c_1\{b_n\}$.

7. Let us assume that $a_n = a^n$. Then (3.34) turns into

$$a^n = pa^{n-1} + qa^{n-2}.$$

Excluding the uninteresting possibility of $a = 0$, this is equivalent to

$$a^2 - pa - q = 0. \tag{3.36}$$

That is, $a_n = a^n$ will be a solution for (3.34) if and only if a is a solution of (3.36).

8. We have seen in the solution of the previous exercise that we must look for the solutions of (3.36). If it has two distinct solutions (real or complex), say a_1 and a_2, then the sequences $\{a_1^n\}$ and $\{a_2^n\}$ are two linearly independent elements of V. We have seen in the solution of Exercise 6 that $\dim V = 2$, so it follows that $\{a_1^n\}$ and $\{a_2^n\}$ form a basis of V.

If (3.36) has repeated real root a, then we know that $\{a^n\}$ is a solution. To find a second linearly independent solution, note that a being a repeated root implies that

$$a^2 - pa - q = (a - p/2)^2 = 0,$$

so $a = p/2$ and $q = -p^2/4$. In this case, a routine computation shows that $\{na^n\}$ is a solution, and this solution is linearly independent of the solution $\{a^n\}$.

While the reader may be asking how one is supposed to guess this, the reader has probably taken a course in differential equations, where the very same idea is used when the characteristic equation of a second-order equation with constant coefficients has repeated real roots.

9. Let $A(x)$ be the generating function of the sequence we are looking for. Multiply both sides of the defining equation by x^n, and sum over all $n \geq 1$. We get

$$A(x) - 1 = 3\sum_{n\geq 1}\left(\sum_{i=0}^{n-1} a_i\right)x^n,$$

$$A(x) - 1 = 3\frac{xA(x)}{1-x}.$$

In the last step, we used the fact that $\sum_{i=0}^{n-1}$ is the coefficient of x^{n-1} in $A(x)/(1-x)$. Solving for $A(x)$,

$$\begin{aligned} A(x) &= \frac{1-x}{1-4x} = \frac{1}{1-4x} - \frac{x}{1-4x} \\ &= \sum_{n\geq 0} 4^n x^n - \sum_{n\geq 0} 4^{n-1}x^n = \sum_{n\geq 0} 3\cdot 4^{n-1}x^n. \end{aligned}$$

Therefore, $a_n = 3\cdot 4^{n-1}$ for $n \geq 1$.

Alternatively, we could have written up the defining recurrence relation for a_{n+1}, and then subtracted the recurrence relation for a_n from it, to get the formula $a_{n+1} - a_n = 3a_n$, or $a_{n+1} = 4a_n$ for $n \geq 1$, with $a_0 = 1$ and $a_1 = 3$, and then solved that recurrence relation.

10. What would $[x^n]\prod_{i\geq 1} F_i(x)$ be?
11. Yes. If there is no such n, then for every m only a finite number of the power series contain an x^n-term for all $n \leq m$, so $[x^m]\prod_{i\geq 1} F_i(x)$ is finite.
12. Using the same technique as in Example 3.17, we get

$$\sum_{n\geq 0} p_{odd}(n)x^n = \prod_{\substack{i\geq 1\\ i\ odd}} \frac{1}{1-x^i}. \tag{3.37}$$

13. Following the discussion after Example 3.17, we see that we have

$$\sum_{n\geq 0} p_d(n)x^n = (1+x)(1+x^2)(1+x^3)\cdots = \prod_{j\geq 1}(1+x^j). \tag{3.38}$$

14. We claim that $p_{odd}(n) = p_d(n)$ for all n. We can prove this by showing that the two sequences have identical generating functions, that is, by showing that $\sum_{n\geq 0} p_{odd}(n)x^n = \sum_{n\geq 0} p_d(n)x^n$. By the results of the two previous exercises, this is equivalent to showing that

$$\prod_{\substack{i\geq 1\\ i\ odd}} \frac{1}{(1-x^i)} = \prod_{j\geq 1}(1+x^j). \tag{3.39}$$

In order to prove the last inequality, let us multiply both sides by $\prod_{j\geq 1}(1-x^j)$. Then on the left-hand side we will have

$$\frac{\prod_{j\geq 1}(1-x^j)}{\prod_{\substack{i\geq 1\\ i\ odd}}(1-x^i)} = \prod_{\substack{i\geq 1\\ i\ even}} (1-x^i),$$

whereas on the right-hand side we will have

$$\prod_{i\geq 1}(1-x^j)(1+x^j) = \prod_{i\geq 1}(1-x^{2i}).$$

So the two sides of (3.39) are indeed equal as claimed.

15. The infinite product we are evaluating is quite similar to the generating function $\sum_{n\geq 0} p_d(n)x^n$ that we computed in the solution of Exercise 13. The difference lies in the *signs*. That is, a partition of n that consists of an *odd* number of distinct parts will contribute $-x^n$ to the generating function, and a partition of n that consists of an *even* number of distinct parts will contribute x^n to the generating function. By Euler's pentagonal number theorem (Theorem 2.28), these terms will cancel each other except when n is pentagonal. See that theorem for the precise details. Those details yield that

$$A(x) = \prod_{i\geq 1}(1-x^i) = \sum_{-\infty}^{\infty}(-1)^n x^{n(3n+1)/2} = 1 - x - x^2 + x^5 + x^7 - x^{12} + \cdots .$$

16. Recall the identity $B(x) = \sum_{n\geq 0} p(n)x^n = \prod_{i\geq 1}\frac{1}{1-x^i}$. Comparing this with the result of the previous exercise, we see that $A(x)B(x) = 1$. This implies that $[x^n](A(x)B(x)) = 0$ if $n \geq 1$. Expressing this by the coefficients of $A(x)$ and $B(x)$, we get the identity

$$p(n) - p(n-1) - p(n-2) + p(n-5) + p(n-7) - \cdots = 0,$$

$$p(n) = \sum_{k=1}^{n}(-1)^{k+1}p\left(\frac{k(3k+1)}{2}\right) + \sum_{k=1}^{n}(-1)^{k+1}p\left(\frac{k(3k-1)}{2}\right),$$

where $p(m) = 0$ if $m < 0$.

17. Let $A(x) = \sum_{n\geq 0} a_n \frac{x^n}{n!}$ be the exponential generating function of the sequence a_n. Multiply both sides of the defining equation by $\frac{x^n}{n!}$ and sum over all $n \geq 1$ to get

$$\sum_{n\geq 1} a_n \frac{x^n}{n!} = \sum_{n\geq 1} n a_{n-1}\frac{x^n}{n!} + \sum_{n\geq 1}(n+1)x^n.$$

It is easy to recognize $A(x)$ on both sides. Note that the last term on the right-hand side is simply $\left(\frac{1}{1-x}\right)' - 1$. Therefore, the previous equation leads to

$$A(x) = xA(x) + \frac{1}{(1-x)^2} - 1,$$

$$A(x)(1-x) = \frac{1}{(1-x)^2} - 1,$$

$$A(x) = \frac{1}{(1-x)^3} - \frac{1}{1-x}.$$

All that is left to do is to find the coefficient of x^n on the right-hand side. In the second term of the right-hand side, it is 1. The first term is equal to

$$\frac{1}{2}\left(\frac{1}{1-x}\right)'' = \frac{1}{2}\left(\sum_{n\geq 0} x^n\right)'' = \frac{1}{2}\sum_{n\geq 2} n(n-1)x^{n-2}.$$

So the coefficient of x^n in the first term of the right-hand side is $\binom{n+2}{2}$, bringing the coefficient of x^n on the right-hand side to $\binom{n+2}{2} - 1$. Therefore,

$$a_n = n!\left(\binom{n+2}{2} - 1\right) = \frac{(n+2)!}{2} - n!.$$

18. Let $A(x) = \sum_{n\geq 0} a_n \frac{x^n}{n!}$ be the exponential generating function of the sequence. Multiplying both sides by $x^n/n!$, and summing over $n \geq 1$, we get

$$\sum_{n\geq 1} a_n \frac{x^n}{n!} = \sum_{n\geq 1} a_{n-1}\frac{x^n}{(n-1)!} + \sum_{n\geq 1}(-1)^n \frac{x^n}{n!},$$

$$A(x) - 1 = xA(x) + e^{-x} - 1.$$

This leads to

$$\begin{aligned} A(x) &= \frac{e^{-x}}{(1-x)} \\ &= \sum_{n\geq 0}\left(\sum_{k=0}^{n}\frac{(-1)^k}{k!}\right)x^n. \end{aligned}$$

Therefore, the coefficient of $x^n/n!$ in $A(x)$ for $n \geq 1$ is

$$a_n = \sum_{k=0}^{n}(-1)^k \frac{n!}{k!}.$$

19. First, choose the president of the conference committee in one of n ways. The remaining $n-1$ people can be sent to any of the three committees as long as the appropriations committee has an even number of members and the business committee has an odd number of members.

Let us look at the case of even n first. If n is even, then $n-1$ is odd. Let a, b, and c denote the number of members of the three committees. First, we claim that exactly half of the cases in which $a+b$ is odd are good. Indeed, in this case there is a bijection between the set of good assignments and bad assignments. This bijection simply swaps all senators from appropriations to business and vice versa. So if a were odd and b were even, and the assignment was therefore bad, then our bijection turns into one in which a is even and b is odd, which is a good assignment. The converse is also true.

Therefore, our task is now reduced to proving that there are exactly $(3^{n-1}+1)/2$ words of length $n-1$ consisting of letters A, B, and C in which $a+b$ is odd. And that is true, since, if w is such a word, with c letters C, then there are $\binom{n-1}{c}2^{n-1-c}$ possibilities for c. Summing over all even c, we get

$$\sum_{\substack{c=0 \\ c \text{ even}}}^{n-2} \binom{n-1}{c} 2^{n-1-c} = \frac{(2+1)^{n-1}+(2-1)^n}{2}.$$

The case of odd n is very similar and is therefore left to the reader.

20. Consider an allowed arrangement of an n-page thesis, and remove the last page. Then there are two possibilities:

    (a) The last chapter is still properly arranged. Then there are b_{n-1} possibilities for the remaining part of the thesis and two possibilities for the removed page, resulting in $2b_{n-1}$ possibilities for the original thesis.

    (b) The last chapter is no longer properly arranged. That means that the last chapter no longer contains pages of both types. Let us assume that the last chapter consisted of i pages, then we have b_{n-i} possibilities for the chapters other than the last one and two possibilities for the last chapter (all pages are of one kind, the last page is the other kind). This yields $2\sum_{j=0}^{n-2} b_j$ more possibilities.

    Now that our recurrence relation is proved; it is routine to prove that $b_n = 2\cdot 3^{n-2}$ by induction.

21. (a) We simply have to split $[n]$ into three nonempty blocks, then carry out the trivial task on each block. The trivial task can be carried out in exactly one way. Therefore, we have

    $$A(x) = B(x) = C(x) = \sum_{n\geq 1} \frac{x^n}{n!} = e^x - 1.$$

    So the product formula implies that the exponential generating function for the number of ways to carry out the composite task is

    $$\begin{aligned} S_3(x) &= A(x)B(x)C(x) \\ &= (e^x-1)^3 \\ &= e^{3x} - 3e^{2x} + 3e^x - 1. \end{aligned}$$

(b) It follows from the result of the previous exercise and the identity $e^{kx} = \sum_{n\geq 0} k^n \frac{x^n}{n!}$ that

$$OS(n,3) = 3^n - 3 \cdot 2^n + 3$$

for positive n. Therefore, $S(n,3) = (3^{n-1} - 2^n + 1)/2$ for positive n, agreeing with our earlier results.

22. We can visualize the question as follows: The number d_n is the number of lattice paths from $(0,0)$ to (n,n) using steps $(1,0)$, $(0,1)$, and, when on the main diagonal, $(1,1)$.

Let p be such a lattice path that does not start with a $(1,1)$ step. The number of such paths is $d_n - d_{n-1}$, and their generating function is $D(x)(1-x)$. Such paths can be decomposed into two parts, the part before they first touch the main diagonal and the part after that. The first parts are enumerated by the generating function $2xC(x)$, where $C(x)$ is the generating function of the Catalan numbers that we computed in Example 3.16. The reader may want to read that example again for further justification of the last sentence. Indeed, the first part of these paths is a northeastern lattice path that is either below the diagonal or above it, explaining the factor 2. The second parts are enumerated by $D(x)$ itself. Using the product formula, we get

$$D(x)(1-x) - 1 = 2xC(x)D(x),$$

since this decomposition is only possible if $n \geq 1$. Expressing $D(x)$ from this, we get

$$D(x) = \frac{1}{1 - x - 2xC(x)} = \frac{1}{\sqrt{1-4x} - x} = \frac{\sqrt{1-4x} + x}{1 - 4x - x^2}.$$

23. In the language of lattice paths, M_n is the number of lattice paths from $(0,0)$ to $(n/2, n/2)$ with steps $(1,0)$, $(0,1)$, and $(1/2, 1/2)$ that never go above the diagonal $x = y$. Otherwise, this proof will be very similar to that of Example 3.16. The reader may want to reread that proof first.

To start with, we have $M_0 = 1$. Let $M(x) = \sum_{n\geq 0} M_n x^n$. Let us look at all series of $n \geq 1$ games ending in an aggregate tie that did not start with a tie game. Their number is $M_n - M_{n-1}$ for $n \geq 1$, and therefore their generating function is $M(x)(1-x) - 1$. We will now obtain this same generating function in another way, using the product formula. Let us call the first moment in which the score is a tie (after 0-0) the *critical moment.* The critical moment breaks our series of games into two parts, namely, the part before the critical moment and the part after the critical moment. The part after the critical moment is just a series like the original one, so the generating function of the relevant numbers is $M(x)$. The part before the critical moment is a series that is never a tie except at the beginning and at the end. By an argument analogous to that of Example 3.16, we see that the generating function of the relevant numbers is $M(x)x^2$ (the series starts with a win of A, and the critical moment arrives with a loss of A, so there are M_{i-2} choices for this part, where i is the length of the part). By the product formula, this leads to the functional equation

$$M(x)(1-x)-1 = M(x)x^2 \cdot M(x),$$

$$M(x)^2x^2 + M(x)(x-1)+1 = 0.$$

Solving this quadratic equation for $M(x)$, and verifying that the *negative* root is the combinatorially meaningful one, we get

$$M(x) = \frac{1-x-\sqrt{1-2x-3x^2}}{2x^2}.$$

The numbers M_n are called the *Motzkin numbers.* See [72] for more than a dozen interpretations of these numbers. Some additional interpretations can be found in [13].

24. In the language of lattice paths, r_n is the number of lattice paths from $(0,0)$ to (n,n) with steps $(1,0)$, $(0,1)$, and $(1,1)$ that never go above the main diagonal. Otherwise, this exercise is similar to the previous exercise, with some significant differences. Let $R(x)$ be the generating function we are looking for. Then $R(x)(1-x)-1$ is the generating function for the numbers of possible series that do not start with a tie game. Define the critical moment of a series as in the solution of the previous exercise. Then the product formula shows that

$$R(x)(1-x)-1 = xR(x) \cdot R(x),$$

since the part before the critical moment must start with a win of A, end with a win of B, and contain a series counted by r_{i-1} in between, where the critical moment arrives when the score is $i-i$. Solving the last equation for $R(x)$, we get

$$R(x) = \frac{1-x-\sqrt{1-6x+x^2}}{2x}. \quad (3.40)$$

The numbers r_n are called the *Schröder numbers.* See the remark at the end of the solution of the previous exercise to find out where to read more about these numbers.

25. We will use the compositional formula of ordinary generating functions. We partition the interval $[1,n]$ into an unspecified number of

intervals, then on each interval we carry out the task of tiling that interval *using one tile only.* There are two ways to do that for intervals of size 1, three ways to do it for intervals of size 2, and no ways to do it for larger intervals. Therefore, $A(x) = 2x + 3x^2$. Hence the compositional formula yields $F(x) = 1/(1-A(x)) = 1/(1-2x-3x^2)$. This leads to the formula $f(n) = \left(3^{n+1} + (-1)^n\right)/4$.

26. We will use the compositional formula of ordinary generating functions. We partition the interval $[1, n]$ into an unspecified number of intervals, then on each interval we carry out the task of partitioning that interval into one part of size $rk + 1$, for some nonnegative integer k. This is possible in one way if that interval had length $rk+1$, and in zero ways otherwise. Therefore, the inside generating function is $A(x) = x + x^{r+1} + x^{2r+1} + \cdots = x/(1 - x^r)$, so the compositional formula yields

$$H(x) = \frac{1}{1 - A(x)} = \frac{1 - x^r}{1 - x - x^r}.$$

In the special case of $r = 2$, we get

$$\frac{1 - x^2}{1 - x - x^2} = 1 + \frac{x}{1 - x - x^2},$$

which is the generating function of the Fibonacci numbers (shifted by 1).

27. Let $A_d(x)$ be the requested generating function. Then we have

$$A_d(x) = \prod_{i \geq 1} \left(1 + x^i + \cdots + x^{(d-1)i}\right) = \prod_{i \geq 1} \frac{1 - x^{di}}{1 - x^i}.$$

28. Let $B_d(x)$ be the requested generating function. Then we have

$$B_d(x) = \prod_{\substack{j \geq 1 \\ j \neq dk}} \frac{1}{1 - x^j}.$$

29. The equality $A_d(x) = B_d(x)$ holds as the factors $1 - x^j$ where $j = di$ cancel out in $A_d(x)$. Therefore, for all n and $d \geq 2$, the number of partitions of n in which each part occurs less than d times is equal to the number of partitions of n in which no part is divisible by d.

30. The moments at which the game was tied break up the sequence of goals into an undetermined number of segments. Each such segment starts with the home team scoring, and ends with the visiting team scoring. In between, there is a regular sequence of goals in

which the home team never trails. So the number of possible individual segments has generating function $xH(x)$, and therefore, the Compositional formula yields

$$H(x) = \frac{1}{1 - xH(x)},$$

which is equivalent to (3.19).

3.10 Supplementary exercises

1. For which formal power series $A(x)$ can one find a formal power series $B(x)$ so that $A(x)B(x) = 1$?
2. Verify that (3.10) is correct for all $n \geq 0$.
3. Find an explicit formula for a_n if $a_0 = 3$ and $a_n = 7a_{n-1} + 2$ for $n \geq 1$.
4. Find an explicit formula for a_n if $a_0 = 1$, $a_1 = 2$, and $a_n = 2a_{n-1} - a_{n-2}$ for $n \geq 2$.
5. A small country has a regressive taxation system for its police officers. An officer starts her career with a yearly salary of 40 units, and she has no taxes to pay during her first year of work. At the end of each year, she gets a five percent raise and a one-unit increase in taxes. How much will the after-tax income of this officer be after n years of work?
6. Find an explicit formula for a_n if $a_0 = 1$ and $a_n = (n+1)a_{n-1} + 3^n$ for $n \geq 1$.
7. Find an explicit formula for a_n if $a_0 = 1$, $a_1 = 2$, and

$$a_n = n(a_{n-1} + a_{n-2})$$

for $n \geq 2$.

8. Solve the recurrence relation $a_0 = 1$ and $a_n = \sum_{i=0}^{n-1}(i+1)a_i$ for $n \geq 1$.
9. Let $f(n)$ be the number of ways to pay n cents using pennies, nickels, dimes, and quarters. Find the ordinary generating function of the numbers $f(n)$.
10. Let $p(n)$ denote the number of partitions of the integer n, with $p(0) = 1$. Find the generating function $\sum_{n\geq 0} p(n)x^n$.
11. Let $p_k(n)$ denote the number of partitions of the integer n into at most k parts. Find the generating function $\sum_{n\geq 0} p_k(n)x^n$.

12. Find the ordinary generating function for the number of partitions of n in which no part occurs more than three times.
13. Let S be the set of all positive integers that can be obtained from 1 by repeatedly applying the operations $f(x) = 2x - 1$ and $g(x) = 4x + 1$ in any order. Let

$$s_n = |S \cap [2^n]|.$$

Find the ordinary generating function $S(x) = \sum_{n \geq 0} s_n x^n$.
14. Find a combinatorial proof for the result of Exercise 3.21.
15. Let h_n be the number of partitions of $[n]$ in which no block contains more than three elements. Find the exponential generating function of the sequence $\{h_n\}$.
16. Let a_n be the number of partitions of $[n]$ in which each block has odd size. Find the exponential generating function of the sequence $\{a_n\}$.
17. Deduce a formula for the number of all compositions of n using Theorem 3.25.
18. The semester of a college consists of n days. In how many ways can we separate the semester into sessions if each session has to consist of at least five days?
19. Find a proof for the result of Example 3.26 showing that, for $n \geq 1$, we have $b_n = (4^n - 1)/3$. Do not use generating functions.
20. The solution we provided for Exercise 21 uses the product formula. Solve the same exercise using the compositional formula.
21. Let k be a fixed positive integer. Find a closed form for the generating function $\sum_{n \geq 0} S(n,k) \frac{x^n}{n!}$, then use that result to deduce a formula for $S(n,k)$.
22. Find the exponential generating function for the numbers of all ordered partitions of $[n]$. Ordered partitions are defined in Exercise 21.
23. Assume that in Example 3.37 each table can order red wine, white wine, both, or neither. How many possibilities are there now for the seating arrangements and wine orders?
24. Find a combinatorial proof for the result proved in Example 3.38.
25. In how many different ways can we cut a convex $(n+2)$-gon into triangles by $n-1$ noncrossing diagonals?

In the following four exercises, the reader may want to translate the problem into the language of lattice paths.

26. + Two grandmasters played a series of chess games, according to the scoring system of Exercise 22. The final score was $n - n$, and a reporter pointed out that tie games only occurred when the current

aggregate score was *not* a tie. Let f_n be the number of ways this could happen, and set $f_0 = 1$. Find the ordinary generating function of the numbers f_n.

27. Solve Exercise 22 if one of the grandmasters, say A, has never trailed during the series.

28. Solve Exercise 23 if grandmaster A might have trailed during the series.

29. Solve Exercise 24 if grandmaster A might have trailed during the series.

30. Find an explicit formula for a_n if $a_0 = 1$, and $a_n = n^2 a_{n-1} + n!^2$ if $n \geq 1$.

31. Let k be a fixed positive integer. Let $f_k(n)$ be the number of s of n into parts that are not larger than k. Find the closed form of the ordinary generating function $F_k(x) = \sum_{n\geq 0} f_k(n)x^n$.

32. Let k be a fixed positive integer, and let $g_k(n)$ be the number of compositions of n with no parts less than k. Find the closed form of the ordinary generating function $G_k(x) = \sum_{n\geq 0} g_k(n)x^n$.

33. Let $h(n)$ be the number of ways to tile a $2 \times n$ rectangle using 1×2 and 2×1 rectangles. Find the ordinary generating function of the numbers $h(n)$.

Part II

Topics

4

Counting permutations

Two buses full of tourists arrived at a large rest area. The n tourists who traveled on the first bus entered the self-service food court one by one (we denote them by elements of $[n]$ in the order in which they get off the bus) and they formed a line at each of the k different concessions that were open. Nobody passed anybody, so the person who got off the bus first got in line first, the person who got off the bus second got in line second, and so on. The n tourists who traveled on the second bus entered a full-service restaurant, where they took their seats around each of k identical circular tables. In how many ways could tourists on each bus proceed?

Both questions asked above are fundamental counting problems of combinatorics. They are both very closely related to permutations, that is, arrangements of the elements of $[n]$ in a line that use each element exactly once. The answers to both questions are also closely related to set partitions, more precisely to the Stirling numbers of the second kind.

4.1 Eulerian numbers

Let us first consider the tourists on the first bus, who formed lines at each of k counters. If we denote the n tourists by the elements of $[n]$ in the order in which they arrived at the food court, then the k lines at the concessions will correspond to k increasing subsequences. (For example, if $n = 6$ and $k = 2$, and the first, fourth, and fifth people go to the first concession, and the second, third, and sixth people go to the second concession, then we get the increasing subsequences 145 and 236.) If we write these k-increasing subsequences one after another in a specified order (say in the order the concessions are located), we get a permutation $p_1p_2 \cdots p_n$ of the set $[n]$. In this permutation, there are *at most* $k-1$ entries p_i that are followed by an entry p_{i+1} satisfying $p_i > p_{i+1}$. Indeed, that can only happen if p_{i+1} starts a new increasing subsequence (coming from a line in front of a new concession), and that happens $k-1$ times.

This way of decomposing permutations into increasing subsequences of consecutive elements is our main topic in this section. Therefore, we make the concept formal.

DOI: 10.1201/9781003304272-4

Definition 4.1 *If k is the smallest integer so that the permutation $p = p_1p_2 \cdots p_n$ can be decomposed into k increasing subsequences of consecutive entries, then we say that p has k* ascending runs, *or, if there is no danger of confusion,* runs.

Example 4.2 *The permutation $p = 245169378$ has three ascending runs, namely, 245, 169, and 378.*

So the tourists of the first bus form a permutation with at most k ascending runs.

Sometimes it is easier to work not with the ascending runs themselves, but with the breakpoints between them. Clearly, a new ascending run must start every time a larger element is followed by a smaller one. This is again a basic phenomenon, and it has its own name.

Definition 4.3 *Let $p = p_1p_2 \cdots p_n$ be a permutation. We say that $i \in [n-1]$ is a* descent *of p if $p_i > p_{i+1}$. If $i \in [n-1]$ is not a descent, then it is called an* ascent.

Note that the descents are the *positions* of p, not its entries.

Example 4.4 *The permutation $p = 245169378$ has two descents, $i = 3$ and $i = 6$. Therefore, p has six ascents.*

At this point, the reader should stop and verify that p has $k-1$ descents if and only if p has k ascending runs. Figure 4.1, showing the permutation of Example 4.4, should help the reader visualize this fact.

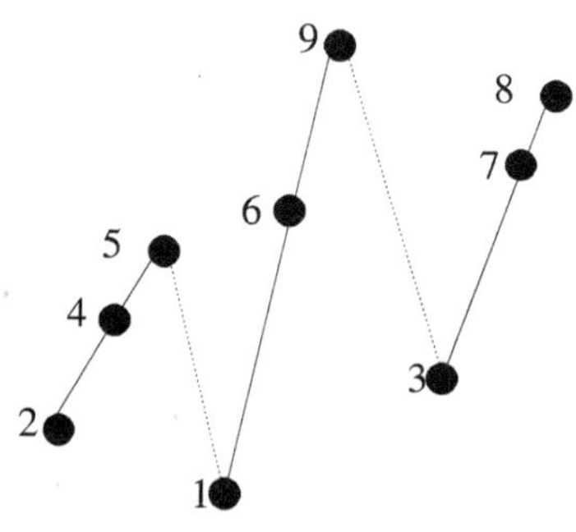

Figure 4.1
Ascending runs and descents in a permutation.

For the rest of this book, permutations of the set $[n]$ will be called n-permutations. If we could tell how many n-permutations there are with exactly k ascending runs, we could also tell how many n-permutations there are with *at most* k ascending runs by the addition principle. It turns out that this is a good way to go, because counting permutations with exactly k ascending runs is easier than counting permutations with at most k ascending runs. Therefore, it is the former to which we give a name.

Definition 4.5 *Let k and n be positive integers satisfying $k \leq n$. Then the number of n-permutations having exactly k ascending runs is denoted by $A(n,k)$ and is called an* Eulerian number.

Example 4.6 *There are four 3-permutations with exactly two ascending runs. They are 132, 213, 231, and 312. Therefore, $A(3,2) = 4$.*

Example 4.7 *For all n, there is one permutation (the increasing one) with one ascending run, and there is one permutation (the decreasing one) with n ascending runs. Therefore, $A(n,1) = A(n,n) = 1$.*

We can extend the definition of $A(n,k)$ for other nonnegative integers by setting $A(n,0) = 0$ and $A(n,k) = 0$ for $k > n$.

An explicit formula for Eulerian numbers has been known since the late nineteenth century. However, we present a relatively recent and remarkably simple proof that was communicated to this author by Richard Stanley in 2003.

Theorem 4.8 *For all nonnegative integers n and k satisfying $k \leq n$, the Eulerian numbers are obtained by the explicit formula*

$$A(n,k) = \sum_{i=0}^{k} (-1)^i \binom{n+1}{i} (k-i)^n. \tag{4.1}$$

Before we start the proof, let us warm up by looking at the special cases of $k = 1$ and $k = 2$. If $k = 1$, then the permutation must have zero descents, so it must be the increasing permutation, implying that $A(n,1) = 1$. The reader can verify that (4.1) holds in this special case. If $k = 2$, then our permutations have two ascending runs. That means that, in order to obtain these permutations, we need to split $[n]$ into nonempty sets X and Y (the future ascending runs) in one of $2^n - 2$ ways, and then order these two sets increasingly, and finally concatenate the obtained strings. We have to be careful, however. If the largest entry of X is smaller than the smallest entry of Y (in other words, when $X = [j]$ and $Y = \{j+1, j+2, \cdots, n\}$), then the obtained permutation will be $12 \cdots n$, which has only one run. This happens in $n-1$ cases, showing that $A(n,k) = 2^n - 2 - (n-1) = 2^n - n - 1$, which is in concordance with (4.1) and also with Example 4.6.

Now we start seeing why proving (4.1) is not easy. As k grows, there could be more and more situations when, in certain circumstances, two substrings that are supposed to form two ascending runs form only one, like X and Y did above. This is a serious problem, because, on one hand, we get fewer ascending runs than we wanted, and, on the other hand, some permutations are obtained more than once. Fortunately, our new friend the inclusion-exclusion principle, as well as an ingenious argument, will help us out.

Proof: (of Theorem 4.8) Let us say that the k concessions are separated by $k-1$ fences, and the tourists just fill the spaces between the fences, as in

$$p = 245|136||79|8. \tag{4.2}$$

Every tourist is free to go to any of the k concessions, which creates k^n possibilities. However, many of these will not be good for our purposes since they will result in a permutation with fewer than k runs. There are two reasons for this that could happen.

(a) There could be empty concessions (no entries between two fences), and

(b) it could happen that the entry immediately preceding a fence is smaller than the following entry.

The reader is invited to verify the easy statements that if neither of (a) and (b) occurs, then we get an n-permutation with exactly k runs, and, crucially, we get each such permutation exactly once.

Now we are going to count the obtained permutations in which at least one of (a) and (b) happens, that is, in which something goes wrong. This may look like a daunting task, since we have to keep track of two properties, not just one. Fortunately, we are able to do that simultaneously, thanks to the definition that follows.

Let p be an n-permutation with $k-1$ fences inserted among its entries (possibly at the beginning or the end). We will call this structure, that is, the permutation and the fences together, an *arrangement*. We say that a fence f is *removable* if it satisfies both of the following requirements:

1. If we remove f, we still have a permutation whose entries increase between any two consecutive fences, and
2. the fence f is not immediately followed by another fence.

Note that, for instance, a fence at the very end of a permutation is always removable. For example, in (4.2), the third fence is removable, and the others are not.

If an arrangement has no removable fences, then the corresponding permutation has k ascending runs, and we are happy. If an arrangement does have at least one removable fence, then the corresponding permutation has less than k ascending runs. Using the inclusion-exclusion principle, we will now enumerate these permutations.

Let us call the $n-1$ spaces between two consecutive entries of our permutations and the space at the beginning and at the end of them *positions*. So there are $n+1$ positions, and we will number them from left to right. Let $S \subseteq [n+1]$, and let A_S be the set of arrangements in which there is a removable fence in each position that belongs to S. Note that while there might be more than one fence in any given position, only one of them (the last one) can be removable.

Fortunately, the size of A_S is easy to determine.

Proposition 4.9 *Let $S \subseteq [n+1]$ so that $|S| = i \leq k-1$. Then*

$$|A_S| = (k-i)^n.$$

Proof: First put down $k-i-1$ fences, and line up the n entries to get an arrangement with at most $k-i$ ascending runs. This can be done in $(k-i)^n$ ways. Now insert the remaining i fences in those positions belonging to S. (If there are already fences there, just put the new fence on the right of the old ones in the same position.) This assures that we get removable fences in the positions that belong to S. Moreover, we will get each arrangement with the required property exactly once. Indeed, if a is an arrangement with the required property, then, if we remove the last fence from each position of a that belongs to S, we get the unique original arrangement that leads to a. ♢

Note that we did not need any specific property of S besides its size, implying that

$$R_i = \sum_{\substack{S\subseteq[n+1]\\ |S|=i}} |A_S| = \binom{n+1}{i}(k-i)^n. \tag{4.3}$$

Now we are ready to use the inclusion-exclusion principle. With a slight abuse of notation, we write A_i for the set of arrangements that have a removable fence in position i. Then Theorem 2.38 yields

$$\begin{aligned} |A_1 \cup A_2 \cup \cdots \cup A_{n+1}| &= R_1 - R_2 + R_3 - \cdots + (-1)^k R_{k-1} \\ &= \sum_{i=1}^{k-1}(-1)^{i+1}\binom{n+1}{i}(k-i)^n. \end{aligned}$$

Note that $R_i = 0$ if $i \geq k$ since there are altogether only $k-1$ fences, so there cannot be more than $k-1$ removable fences.

Recalling that the total number of arrangements we could have is k^n, the subtraction principle shows that the number of arrangements with no removable fences is

$$A(n,k) = \sum_{i=0}^{k-1}(-1)^i\binom{n+1}{i}(k-i)^n, \tag{4.4}$$

which was to be proved. ♢

In many aspects, Eulerian numbers behave similarly to binomial coefficients, and in many other aspects, they are closely related to Stirling numbers of the second kind.

An example for the former is *symmetry*, that is, the property, that for any fixed n and any $k \leq n$, the equality $A(n,k) = A(n, n+1-k)$ holds. The reader should try to prove this identity first, then check our two solutions given in Exercise 1.

Just like binomial coefficients and Stirling numbers of the second kind, Eulerian numbers also satisfy a triangular recurrence.

Theorem 4.10 *Let k and n be nonnegative integers satisfying $k \leq n$. Then*

$$A(n,k) = kA(n-1,k) + (n-k+1)A(n-1,k-1).$$

Proof: The left-hand side clearly counts n-permutations with $k-1$ descents; all we need to show is that the right-hand side counts the same.

Let p be a permutation counted by $A(n,k)$. Let p' be the permutation obtained from p by omitting the entry n. Then there are two cases, namely, that p' has as many descents as p, or it has one less. We will show that the two terms of the right-hand side count permutations that fall into these respective cases.

(Case 1) When the omission of the entry n leaves the number of descents of p unchanged.

This happens when n is at the end of p, or when n is inserted right after the ith entry of p' where i is a descent of p'. (Try $\ldots 3n1 \ldots$.) In either case, we are left with a permutation p' that is of length $n-1$ and has $k-1$ descents. There are, by definition, $A(n-1,k)$ such permutations. Each of them is obtained this way from k different permutations p. Indeed, to get these permutations, we can insert n into any of the $k-1$ descents of p', or we can put it to the end of p'. Therefore, $kA(n-1,k)$ permutations fall into this case.

(Case 2) When the omission of the entry n causes the number of descents of p to decrease by 1.

This happens when n is the first entry of p, or when n is inserted right after the ith entry of p' where i is an ascent of p'. (Try $\ldots 1n3 \ldots$.) In either case, we are left with a permutation p' that is of length $n-1$ and has $k-2$ descents. There are $A(n-1,k-1)$ such permutations. Each of them is obtained this way from $n-k+1$ different permutations p. Indeed, to get these permutations, we can insert n into any of the $n-k$ ascents of p', or we can put it to the front of p'. By the product formula, this explains the second term of the right-hand side.

We have shown that the two sides count the elements of the same set; therefore, they have to be equal. $\diamond$

Theorem 4.10 enables us to prove that the Eulerian numbers occur in many other contexts. We show one example for that here, and leave some others to the exercises.

Example 4.11 *A city has n different hotels, labeled from 1 to n, where 1 denotes the cheapest hotel and n denotes the most expensive one. On each of n consecutive days, one new, mathematically inclined tourist arrives in the city. These tourists arrive in increasing order of spending power. That is, the tourist who arrives on day i can afford to go to any hotels labeled by an element of $[i]$.*

Let $H(n,k)$ be the number of ways the tourists can make their selections so that at the end there will be exactly k hotels that host at least one of our n tourists. Then $H(n,k) = A(n,k)$.

For instance, if $n = 2$, then either both tourists go to hotel 1, or the first goes to 1 and the second goes to 2, yielding $H(2,1) = H(2,2) = 1$, in concordance with $A(2,1) = A(2,2) = 1$. The reader is invited to verify the result for $n = 3$. Note that it is not surprising that the total number of possibilities is $\sum_{k=1}^{n} A(n,k) = n!$, since tourist k has k possibilities.

Solution: (of Example 4.11) We prove the statement by induction on n, the initial case of $n = 1$ being trivial. Assume that we know the statement for $n-1$, and prove it for n. We claim that

$$H(n,k) = kH(n-1,k) + (n-k+1)H(n-1,k-1). \tag{4.5}$$

The left-hand side is, by definition, the number of all possible selections made by the tourists in which k hotels get at least one customer from this group. We claim that the right-hand side counts the objects. Indeed, there are two different ways that a situation counted by the left-hand side could occur, namely, the last tourist either went to a hotel where there was already someone from the group, or not. In the first case, the first $n-1$ tourists split into k hotels in one of $H(n-1,k)$ ways, then the last tourist chooses one of these k hotels. This explains the first term of the right-hand side. In the second case, the first $n-1$ tourists split into $k-1$ hotels in one of $H(n-1,k-1)$ ways, and then the last tourist chooses one of the remaining $n-k+1$ hotels. This explains the second term of the right-hand side, and therefore proves (4.5).

By our induction hypothesis, we know that $A(n-1,k) = H(n-1,k)$, and also that $A(n-1,k-1) = H(n-1,k-1)$. Looking at Theorem 4.10, we see that this means that $A(n,k)$ and $H(n,k)$ are obtained from identical numbers, by identical operations. Therefore, they must be equal. ◊

Note that we proved Theorem 4.10 by a particular kind of induction, that is, we proved that two arrays of numbers have to be identical *because they start the same way* and *they satisfy the same recursive relation.* We will see further applications of this proof technique soon.

The following concept ties descents to another interesting permutation statistic that will be discussed in the following section. A *desarrangement* is a permutation whose first ascent occurs in an even position. We make a single exception to this rule. If n is even, then we will also consider the *decreasing* permutation $n(n-1)\cdots 21$ a desarrangement, as if it had an ascent at its last position n. So 213, 43215 and 4321 are all desarrangements, but 12, 321, and 52134 are not. What interests us here is the *number* of desarrangements of length n.

The term "desarrangements" was coined by Michelle Wachs, to honor her co-author Jacques Désarmenien in an amusing way. The editors of the journal in which the term was first used originally wanted to change it to "disarrangements," which *is* an English word, but when they got the joke, they left "desarrangements" unchanged.

Lemma 4.12 *The number of desarrangements of length n is*

$$J(n) = n!\sum_{i=0}^{n} \frac{(-1)^i}{i!}.$$

Proof: Let n be fixed, and let b_i be the number of n-permutations whose first ascent occurs in position $2i$. In such a permutation, there are $2i$ different possibilities for the relative order of the first $2i+1$ entries. Indeed, the entry in position $2i+1$ can be anything but the smallest of these $2i+1$ entries, but once that entry is chosen the remaining $2i$ entries must all be in decreasing order. Therefore, for $i < n/2$, the ratio of permutations enumerated by b_i to all n-permutations is $\frac{2i}{(2i+1)!}$; consequently

$$b_i = \frac{2i}{(2i+1)!}n! = \left(\frac{1}{(2i)!} - \frac{1}{(2i+1)!}\right) n!.$$

Taking the sum of all possible b_i, we get

$$J(n) = \sum_{i=1}^{\lfloor (n-1)/2 \rfloor} n!\left(\frac{1}{(2i)!} - \frac{1}{(2i+1)!}\right) = n!\left(\frac{1}{2!} - \frac{1}{3!} + \frac{1}{4!} - \cdots\right).$$

This is equivalent to our claim, since our claim was that

$$J(n) = n!\left(1 - 1 + \frac{1}{2} - \frac{1}{6} + \cdots\right).$$

What we proved is the same, without the first two terms that cancel each other anyway. ◊

Finally, we mention that, instead of simply counting permutations with a given number of descents, we could count permutations whose *set* of descents is a given set S, or whose set of descents is contained in a given set S. The reader is invited to look at Exercises 2–6 for results of this kind.

Quick Check

1. Let $n = 2k+1$. Prove that the Eulerian number $A(n, k+1)$ is even.
2. Give a simple proof of the fact that $A(2n, n+1) \geq n!^2$.
3. Find a direct proof for the formula $A(n, 2) = 2^n - n - 1$ without using the general formula for $A(n, k)$.

4.2 The cycle structure of permutations

So far, we have viewed permutations as linear orders of objects. This is not the only way to view them, which makes them even more interesting.

4.2.1 Stirling numbers of the first kind

Let us now turn to the problem of the second tourist bus. Recall that the n tourists traveling on that bus chose to go to a full-service restaurant, where they sat down around k identical tables. Two seating arrangements are considered identical if each person has the same left neighbor in both of them. In how many different ways can the tourists do this? (Note that we have considered this problem before, in Example 3.31, where we solved it using generating functions.)

At first sight, it is not even clear that this problem is related to permutations. Indeed, permutations are, based on what we have learned so far, linear orders. Here, the tables are circular. Furthermore, in permutations, the order of the objects certainly matters, whereas here the tables are all identical. However, we will show that this is very much a permutation enumeration problem; we just need to broaden our conception of permutations to understand that.

Let $p = p_1 p_2 \cdots p_n$ be a permutation. Then we can view p as a *function*, in fact, a bijection, from $[n]$ onto $[n]$, namely, the function defined by $p(i) = p_i$ for all $i \in [n]$. This way of looking at permutations has a few advantages on its own. One of them is that it provides for a natural way of defining the *product* of two permutations.

Definition 4.13 *Let f and g be two permutations, that is, two bijections from $[n]$ onto $[n]$. Then fg is the permutation defined by $fg(i) = g(f(i))$ for all $i \in [n]$.*

In other words, fg is the composition of f and g as functions.

Example 4.14 *Let $n = 6$, and let $f = 421563$, and let $g = 361524$. Then the reader is invited to verify that $fg = 563241$, while $gf = 134625$.*

You may wonder why we set $fg(i) = g(f(i))$ as opposed to $fg(i) = f(g(i))$. The reason for this is that we want to apply the functions in the order in which they are multiplied together. In $g(f(i))$, we *first* apply f to i, and *then* we apply g to $f(i)$.

We would like to stress that, as the above example shows, fg and gf are in general not identical; in other words, multiplication of permutations is not a commutative operation.

Another useful property of permutations as functions is that we can now talk about the *inverse* of a permutation. The inverse of the permutation p, denoted by p^{-1}, is just the inverse of the bijection p. That is, if $p(i) = j$, then $p^{-1}(j) = i$. Equivalently, we have $pp^{-1} = p^{-1}p = 123\cdots n$, the increasing permutation, which we also call the *identity permutation.*

The fact that we are able to multiply permutations together and take their inverses, along with some other important properties, opens new avenues of studying permutations. Many of these aspects belong to *group theory,* where the set of all n-permutations is called the *symmetric group* of degree n and is denoted S_n. Though the theory of permutation groups is beyond the scope of this book, we will sometimes use the notation S_n for the set of all n-permutations.

As we are now able to multiply n-permutations together, we can in particular talk about *powers* of the permutation p. The computation of powers of p is even easier than that of arbitrary products. Indeed, we only need to draw the *short diagram* of p and watch what happens to each element of $[n]$ if p is repeatedly applied. (The short diagram of a function $f : [n] \rightarrow [n]$ consists of n dots, representing the elements of $[n]$, and n arrows, representing the action of f on $[n]$.)

Example 4.15 *Let $p = 321645$, and let us take a closer look at permutations p^k, with $k = 1, 2, \cdots$. Using the short diagram of p shown in Figure 4.2, we find the following:*

- $p = 321645$,
- $p^2 = 123564$,
- $p^3 = 321456$,
- $p^4 = 123645$,
- $p^5 = 321564$,
- $p^6 = 123456$,
- $p^7 = 321645 = p$, *and so on.*

We see that the entries 1 and 3 keep being permuted among each other, the entry 2 keeps being fixed, and the entries 4, 5, and 6 keep being cyclically permuted among themselves.

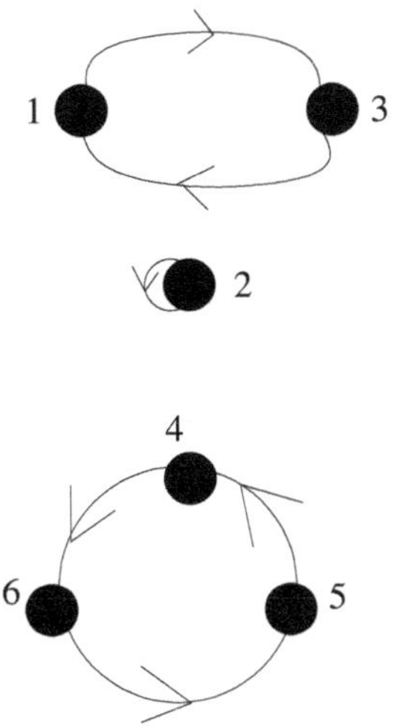

Figure 4.2
The short diagram of $p = 321645$.

In general, if p is an n-permutation, and $i \in [n]$, then there is a positive integer k so that $p^k(i) = i$. Indeed, by the pigeonhole principle, there exist two positive integers t and u so that $t > u$ and $p^t(i) = p^u(i)$ (since there are only n possibilities for $p^t(i)$, but an infinite number of possibilities for t). This implies (why?) that $p^{t-u}(i) = i$, so our claim is proved by setting $k = t - u$.

Let us now choose the *smallest* positive k so that $p^k(i) = i$. Then, if we draw the diagram of p, the entry i will lie on a circle C of length k. As p is a bijection, this circle must be *disjoint* from the rest of the cycles in the short diagram of p. Therefore, if $j \in C$, then $p(j) \in C$, so p permutes the entries of C among each other. Furthermore, $p(j)$ is the left neighbor of j in C, so p permutes the entries of C among each other *cyclically*. Therefore, C is called the *cycle* of p containing i.

If C has fewer than n entries, then let us choose an entry i' that is not part of C. Repeat the above procedure, with i' playing the role of i, to find the cycle C' of p containing i'. Then C' and C are disjoint since p is a bijection. If $C \cup C' = [n]$, then stop, otherwise continue this way until the entire set $[n]$ has been decomposed into disjoint cycles. The obtained set $\{C, C', \cdots\}$ of cycles is called the *cycle decomposition* of p.

The reader may feel that we are cheating as we talk about *the* cycle decomposition while we have not even proved yet that each permutation p has a *unique* cycle decomposition. In order to allay those concerns, we point out that, if we know p, then we know $p(i)$ for all $i \in [n]$, and therefore we can draw the entire diagram of p.

If we want to write a permutation in a way that makes it easy to see what its cycles are, we can put entries in the same cycle within parentheses in the order in which they appear in the cycle.

Example 4.16 *Using cycle notation, the permutation* $p = 321645$ *can be written as* $(31)(2)(645)$.

Hopefully, you are now asking why did I not write $(13)(2)(456)$ or maybe $(564)(2)(21)$. Those notations, and other notations, are also acceptable as long as they all tell correctly what the image of each entry is. (For instance, (456) cannot be replaced by (465), since 4 goes to 5 in the first cycle, while in the second one, 4 goes to 6.) Certainly, the order among the cycles can be changed in any way, and each cycle can be started anywhere. However, for the sake of clarity, it is certainly advantageous to define a *canonical* cycle notation of permutations, which is unique for each permutation.

Definition 4.17 *Let* p *be a permutation. Then the* canonical cycle notation *of* p *consists of writing each cycle of* p *with its largest entry first, then ordering the cycles in increasing order of their largest entries.*

Example 4.18 *The canonical cycle notation of the permutation* p *of Example 4.16 is* $(2)(31)(645)$.

By now the reader might have noticed where we are trying to go with all this. We have shown that a permutation can be conceived not only as a linear order, but also as a set of cycles. This is precisely what we need for the question that started this section, that is, how many ways are there for n tourists to sit down around k identical circular tables. The number that answers this question is so important that it has its own name.

Definition 4.19 *The number of* n*-permutations having exactly* k *cycles is denoted by* $c(n,k)$ *and is called a* signless Stirling number of the first kind.

We promise to explain by the end of the section why the mysterious adjective "signless" is used here. The name "Stirling numbers" may sound a little less surprising. Indeed, the Stirling numbers of the second kind were the numbers of partitions of $[n]$ into k blocks, and now we hear that the (signless) Stirling numbers of the first kind are the numbers of permutations of $[n]$ having k cycles. The connection between these two arrays of numbers is actually a lot stronger than that, as we will explain later in this section as well.

Note that the number of all n-permutations to any number of cycles is, of course, equal to

$$n! = \sum_{k=1}^{n} c(n,k),$$

which demystifies the result of Example 3.31. Recall that, in that example, we found that the number of ways for n people to sit around an unspecified number of circular tables was $n!$. This looked mysterious since at that point in our studies we only knew permutations as linear orders, not as unions of cycles.

By now, the reader is a seasoned veteran of combinatorial definitions and therefore will not be surprised to know that we set $c(n,k) = 0$ for $k > n$, and $c(n,0) = 0$ for positive n, with $c(0,0) = 1$.

It is clear that $c(n,n) = 1$, since an n-permutation f can only have n cycles if each of those cycles is of length 1, that is, when $f(i) = i$ for all i. This happens when f is the identity map of $[n]$. The number $c(n,1)$ is a little bit more complex to determine, but we have essentially done it in Example 1.22. The reader is asked to try to prove that $c(n,1) = (n-1)!$ at this point, then compare her argument to ours in the mentioned example.

While an explicit formula for the numbers $c(n,k)$ in the general case exists, it is very complicated, and its proof is beyond the scope of this book. We will see that this is not a very serious problem, since we can find the numbers $c(n,k)$ we want remarkably fast by some other means. To that end, we first prove a triangular recurrence for the numbers $c(n,k)$.

Theorem 4.20 *For all positive integers n and k satisfying $k \leq n$, the recurrence relation*

$$c(n,k) = c(n-1,k-1) + (n-1)c(n-1,k) \tag{4.6}$$

holds.

We can verify this identity in the special case of $k = 1$. Then we know that $c(n,1) = (n-1)!$, while on the right-hand side, $c(n-1,0) = 0$ and $c(n-1,1) = (n-2)!$, showing that the identity indeed holds in this special case. Let us now turn to the general case.

Proof: (of Theorem 4.20) All we need to show is that the right-hand side of (4.6) also enumerates permutations of $[n]$ having k cycles. Let p be a permutation counted by $c(n,k)$. Then the entry n of p can either form a cycle by itself or can share a cycle with other entries. (Just as the last tourist entering the restaurant may either sit at a new table or at a table where there are some other tourists.) In the first case, the remaining $n-1$ entries form a permutation with $k-1$ cycles, which they can do in $c(n-1,k-1)$ ways. In the second case, the remaining entries form a permutation q having k cycles, which they can do in $c(n-1,k)$ ways. The entry n can be inserted into any existing cycle of q, before any element. Note that inserting q into the end of a cycle would not be different from inserting it into the beginning of that same cycle. All the permutations obtained this way will be different (why?), resulting in $(n-1)c(n-1,k)$ permutations. Combining the counts in the two cases completes the proof. ◊

In Exercise 35, the reader will be asked to provide a nongenerating function proof for the mysterious result of Example 3.37. Ideas like the one in the above proof can be useful when solving that exercise.

The following theorem provides a very simple way of determining the numbers $c(n,k)$. This is the method that we promised for computing these numbers.

Theorem 4.21 *For all positive integers n, the identity*

$$\sum_{k=1}^{n} c(n,k)x^k = (x+n-1)(x+n-2)\cdots(x+1)x \tag{4.7}$$

holds.

In other words, if we need the numbers $c(n,k)$ for some fixed n, all we need to do is to expand the product $(x+n-1)(x+n-2)\cdots(x+1)x$, which is a breeze for any software package, and then we can read off the numbers $c(n,k)$ as the coefficients of the obtained polynomial.

Example 4.22 *Let $n=3$. Then $(x+2)(x+1)x = x^3+3x^2+2x$, yielding $c(3,3)=1$, $c(3,2)=3$, and $c=(3,1)=2$. We know that these numbers are correct because there is one 3-permutation with three cycles (the identity, since all cycles must be 1-cycles), there are three 3-permutations with two cycles (one cycle has to be a 1-cycle, the other a 2-cycle—we choose the 1-cycle, then must put the remaining two entries in a 2-cycle), and there are two 3-permutations with one cycle (we can choose the image of 1, and the rest follows).*

Proof: (of Theorem 4.21) Note that the right-hand side can be written in the short form $(x+n-1)_n$. We are going to prove our claim by induction on n.

For $n=1$, our equation reduces to $x=x$, so our claim trivially holds. Now assume that the claim holds for $n-1$, that is,

$$\sum_{k=1}^{n-1} c(n-1,k)x^k = (x+n-2)_{n-1}.$$

In order to obtain $(x+n-1)_n$, that is, the right-hand side of (4.7), let us multiply both sides of this equation by $x+n-1$. We get

$$\sum_{k=1}^{n-1} c(n-1,k)x^{k+1} + (n-1)\sum_{k=1}^{n-1} c(n-1,k)x^k = (x+n-1)_n.$$

Note on the left-hand side that the coefficient of x^k is precisely $c(n-1,k-1)+(n-1)c(n-1,k)$, which, by Theorem 4.20, is equal to $c(n,k)$. (This is true even in the special case of $k=n$.) Therefore, the last equation reduces to

$$\sum_{k=1}^{n} c(n,k)x^k = (x+n-1)_n,$$

and completes the proof of our induction step. $\diamondsuit$

Note that (4.7) stays true if the sum on the left-hand side is taken from $k = 0$ instead of $k = 1$. In fact, that replacement will only change the value of the left-hand side when $n = 0$, since otherwise $c(n, 0) = 0$.

The time has come for us to reveal what connects the signless Stirling numbers of the first kind to the Stirling numbers of the second kind, and what the adjective "signless" means in the first place.

Recall from your studies in linear algebra that the set V of all polynomials with real coefficients forms a *vector space* over the field of real numbers, and that the infinite set B of polynomials $\{1, x, x^2, \cdots\}$ forms a *basis* of this vector space. In the very unlikely case that you forgot what that means, we remind you that it means that each element of V can be written uniquely as a linear combination of the elements of B, so that the coefficients of that linear combination are real numbers.

What formula (4.7) tells us, in this terminology, is that, when we write the polynomial $(x + n - 1)_n$ as a (unique) linear combination of the elements of B, then the *coefficients* of that linear combination are precisely the signless Stirling numbers of the first kind, in the right order.

Now let us revisit Exercise 11 of Chapter 2, which we are sure the reader has already solved. In that exercise, we proved that

$$x^n = \sum_{k=0}^{n} S(n, k)(x)_k. \tag{4.8}$$

In other words, if we write the elements of B as linear combinations of the elements of the basis $B' = \{1, (x)_1, (x)_2, \cdots\}$ of V, then the coefficients will be precisely the Stirling numbers of the second kind, and in the right order.

Formulae (4.7) and (4.8) do sound quite similar to each other, but one can see that they are not quite "inverses" of each other in the loose sense that one would "undo" what the other does. This is because, while one expresses the polynomials $(x + n - 1)_n$ in terms of the polynomials x^n, the other expresses the polynomials x^n not in terms of the polynomials $(x + n - 1)_n$, but rather in terms of the polynomials $(x)_k$. This will be corrected by the following definition and proposition.

Definition 4.23 *For all nonnegative integers n and k, set $s(n, k) = (-1)^{n-k}c(n, k)$. The numbers $s(n, k)$ are called the* Stirling numbers of the first kind.

This takes care of the adjective "signless" in the name of the numbers $c(n, k)$. This definition also lets us translate (4.7) in the following manner.

Proposition 4.24 *For all nonnegative integers n, we have*

$$\sum_{k=0}^{n} s(n, k)x^k = (x)_n. \tag{4.9}$$

Proof: In (4.7), replace x by $-x$, then multiply both sides by $(-1)^n$. $\diamondsuit$

Now we can revisit our statements about the bases B and B' of V and see that the Stirling numbers of the first and second kind have inverse roles. That is, if we write the elements of B as linear combinations of B', then the coefficients are the numbers $S(n,k)$. If we do the opposite, that is, we write the elements of B' as linear combinations of B, then the coefficients are the numbers $s(n,k)$. This is why these two arrays of numbers deserve to have such similar names. Exercise 9 will explore the established connection between these numbers further.

4.2.2 Permutations of a given type

Let us return to the tourists of the second bus, who sit down around some circular tables. However, this time we will have more precise restrictions on the way in which the tourists can sit around tables. Instead of simply specifying the number of tables the tourists must use, we specify the number of tables that have to be used by one tourist each, the number of tables that have to be used by two tourists each, and so on. This new scenario leads to the following definition.

Definition 4.25 *Let p be an n-permutation that has exactly a_i cycles of length i for all $i \in [n]$. Then the array $(a_1, a_2, \cdots, a_n)$ is called the* type *of p.*

Example 4.26 *Let $p = (21)(534)(7)(986)$. Then the type of p is the array* $(1, 1, 2, 0, 0, \cdots, 0)$.

In this terminology, what we are looking for is the number of permutations of a given type. We do not even have to say "the number of n-permutations of a given type," since the length of a permutation can easily be read off its type by taking the sum $\sum_{i=1}^{n} ia_i$.

It is significantly easier to find an explicit formula for this number than for the Stirling numbers $c(n,k)$, as is shown by the next theorem.

Theorem 4.27 *Let n be any positive integer, and let $a = (a_1, a_2, \cdots, a_n)$ be any array of nonnegative integers satisfying $\sum_{i=1}^{n} ia_i = n$. Then the number of permutations of type a is*

$$\frac{n!}{1^{a_1} \cdot 2^{a_2} \cdot \cdots \cdot n^{a_n} \cdot a_1! \cdot a_2! \cdot \cdots \cdot a_n!}. \tag{4.10}$$

Proof: Formula 4.10 suggests that we try to count all permutations of type a by the division principle. This can be done in the following way.

Let us write down the entries of $[n]$ in a line, in all possible ways. This can be done in $n!$ different ways. Now let us place pairs of parentheses between the entries so that our permutations are now written in cycle notation, and their first a_1 cycles are of length 1, the following a_2 cycles are of length 2, and so on. Note that this way we will get a permutation whose cycle lengths are nondecreasing from left to right.

Notice that a completely determines where we insert the parentheses. All permutations we obtain this way will have type a. However, they will not all be *distinct* permutations, as each n-permutation will be obtained many times. The question is, of course, how many times. In other words, how many ways are there to write a permutation of type a in cycle notation if the length of the cycles is nondecreasing from left to right.

If p is a permutation of type a with that nondecreasing property, then, for all i, the set of i-cycles of p can be permuted in $a_i!$ ways, and p will still have nondecreasing cycle lengths from left to right. Furthermore, each i-cycle can be written with any of its i elements in the first position. Therefore, by the product principle, the number of ways p can be written in cycle notation so that its cycle lengths are nondecreasing from left to right is $a_1!a_2!\cdots a_n!1^{a_1}2^{a_2}\cdots n^{a_n}$. Our statement is then proved by the division principle. ◊

Example 4.28 *The number of 6-permutations of type* $(1,1,1,0,0,0)$ *is*

$$\frac{6!}{1\cdot 1\cdot 1\cdot 1^1\cdot 2^1\cdot 3^1}=120.$$

Note that Theorem 4.27 provides quick solutions to several exercises that we had to solve on a case-by-case basis before. For instance, it shows that the number of n-permutations with one cycle only is $(n-1)!$ and that the number of $(2n)$-permutations in which each cycle is of length two is $\frac{(2n)!}{n!2^n}$.

So far we have seen two well-defined ways of writing a permutation p, namely, the so-called *one-line notation*, covered earlier, in which p is written as a linear order $p_1p_2\cdots p_n$ of the elements of $[n]$, and the *canonical cycle notation*, covered more recently, in which p is written as a list of cycles in which each cycle starts in its largest entry, and the cycles are written in increasing order of their first entries. It is natural to ask what the relation is between these two ways of writing p. A random check shows that similar-looking permutations can be actually quite different when written in these two notations. For instance, $(421)(53)\neq 42153$, since $(421)(53)=41523$. Similarly, $(654321)=612345$. So it is simply not true that, if we omit the parentheses from the canonical cycle notation of p, then we get p in the one-line notation. What *do* we get then? While that question does not have a very compact answer, it does turn out to be a very useful question. The following lemma explains this in detail.

Lemma 4.29 (transition lemma) *Let n be any positive integer, and let p be an n-permutation written in canonical cycle notation. Define $f(p)$ as the permutation obtained from p by omitting all parentheses and reading the entries of p in the one-line notation. Then the map $f: S_n\rightarrow S_n$ is a bijection.*

In other words, while it is not simple to describe what we get if we omit the parentheses of p, at least we can be sure that we get each permutation exactly once! Let us convince ourselves of this first by looking at the special case of $n = 3$.

Example 4.30 *Let $n = 3$. Then the action of f is as follows:*

- $f((1)(2)(3)) = 123$,
- $f((21)(3)) = 213$,
- $f((2)(31) = 231$,
- $f((1)(32)) = 132$,
- $f((312)) = 312$, *and*
- $f((321)) = 321$.

Proof: (of Lemma 4.29) It follows from the definition of f that f maps into S_n. Therefore, the statement will be proved if we can show that f has an inverse. So let q be an n-permutation written in the one-line notation. We will show that there is exactly one $p \in S_n$, so that $f(p) = q$.

In order to find the unique p whose image is q, it would suffice to know which entries of p start a new cycle. The crucial observation is the following: An entry of p starts a new cycle if and only if it is larger than all entries on its left. Indeed, if an entry x has that property, then it must start a new cycle, otherwise the cycle containing x would not start with its largest entry, contradicting the assumption that p is written in canonical cycle form. Conversely, if an entry y of p is the first (leftmost) element in its cycle, then it is larger than all entries on its left. Indeed, by the canonical structure of p, each cycle on the left of y starts with its largest entry, and the sequence of these cycle-starting entries is increasing.

Therefore, we can find the one and only permutation p satisfying $f(p) = q$ by going from left to right in q and starting a new cycle each time we find an entry that is larger than all entries on its left. The obtained permutation p is in canonical cycle notation, and since we do not change the left-to-right order of the entries at all, $f(p) = q$ holds. By the previous paragraph, the permutation obtained this way is the only permutation p for which $f(p) = q$, proving that f has an inverse and is therefore a bijection. ◊

Note that the canonical cycle representation of a permutation was defined precisely so that the leftmost entry of each cycle could be identified as a left-to-right maximum. Other canonical representations could have been chosen, but they would not have this useful property.

If q is a permutation written in one-line notation, then an entry of q that is larger than all entries on its left is called a *left-to-right maximum*. For instance, the leftmost entry of q is always a left-to-right maximum (as there is nothing on its left), and so it is the largest entry. The proof of Lemma 4.29 said that we can find $f^{-1}(q)$ by starting a new cycle with each left-to-right maximum of q.

Example 4.31 *Let* $p = (312)(4)(75)(86)$. *Then* $f(p) = 31247586$. *The left-to-right maxima of* 31247586 *are 3, 4, 7, and 8, and we can indeed recover p by starting a new cycle at each of these entries of* $f(p)$.

The transition lemma is useful since it allows us to translate enumerative results on permutations written in canonical cycle notation into enumerative results on permutations written in one-line notation. The following is a classic example of this.

Example 4.32 *A discus-throwing competition had n participants, each of whom got only one try. There were no ties. At any point in time, the person currently in first place was wearing a yellow hat, and when someone else took the lead, then that new leader took the yellow hat.*

While it is clear that the competition could have $n!$ different endings, how many possible endings are there in which exactly k people wore the yellow hat?

Solution: In the language of permutations, this problem asks how many n-permutations are there with exactly k left-to-right maxima. Indeed, if we assign an element of $[n]$ to each participant—n to the winner, $n-1$ to the runner-up, and so on—then list these numbers in the order in which the athletes took their single try, then we get a permutation in which an entry is a left-to-right maximum if and only if at one point of time the corresponding athlete was the leader.

Let q be an n-permutation with k left-to-right maxima. Then the bijection f^{-1} (where f is the map defined in the transition lemma) turns q into a permutation p with k cycles. Conversely, if p has k cycles, then $f(p)$ has k left-to-right minima. Therefore, the number of n-permutations with k left-to-right minima is equal to the number of n-permutations with k cycles, that is, $c(n,k)$. ◊

Let us look at another striking application of the transition lemma. Now assume n people will sit down around a few circular tables, the number of tables not being fixed in advance. One person, called A, would really like to sit at the same table as B, but he cannot influence the seating in any way. How good are his chances to get his wish in a random seating arrangement?

Mathematically speaking, all we ask here is how many of the $n!$ permutations of length n contain the entries i and j in the same cycle. First, note that the answer will certainly not depend on what i and j are. Indeed, if a seating process is truly random, then any two people must have the same chance to get to the same table. Therefore, we might as well assume that $i = n-1$ and $j = n$, and ask in how many n-permutations do they get in the same cycle.

We claim that the permutation $p \in S_n$ contains entries $n-1$ and n in the same cycle if and only if in $f(p)$ the entry n is on the *left* of the entry $n-1$. Here f is the map defined in the transition lemma. Indeed, if p does have the property that $n-1$ and n are in the same cycle, then in the canonical cycle notation that cycle starts with n, so $f(p)$ contains n on the left of $n-1$. On the other hand, if n and $n-1$ are in different cycles, then they are both the leftmost entries of their cycles in the canonical cycle notation, and therefore $n-1$ is on the right of n. So our claim is proved. However, it happens that n precedes $n-1$ in exactly half of all n-permutations. Therefore, we proved the following interesting result.

Proposition 4.33 *Let n be any positive integer, and let i and j be two distinct elements of $[n]$. Then exactly half of all n-permutations contain i and j in the same cycle.*

So A has a fifty percent chance of getting to the same table as B.

The transition lemma has other striking applications, leading to fundamental enumeration results on permutations. Therefore, the reader is strongly urged to examine Exercises 10, 11, and 18—and try to solve them—before looking at the solutions we provided.

Quick Check

1. What is the number of permutations of length n that consist of one 1-cycle and one $(n-1)$-cycle?
2. How many permutations of length n are there in which the entries 1 and 2 are in the same cycle, but the entry 3 is in a different cycle?
3. How many permutations of length n are there in which the entry 1 is in a 2-cycle, and the entry 2 is also in a 2-cycle?

4.3 Cycle structure and exponential generating functions

If we think about it for a minute, we see why the exponential formula, learned in the previous chapter, is very useful for counting permutations. Indeed, a permutation is nothing other than a set $[n]$ split into an arbitrary number of nonempty blocks, on each of which a cycle is taken.

We will be particularly interested in the permutation analog of Theorem 3.33 that follows.

Theorem 4.34 *Let C be any set of positive integers, and let $g_C(n)$ be the number of permutations of length n whose cycle lengths are all elements of C. Then*

$$G_C(x) = \sum_{n\geq 0} g_C(n)\frac{x^n}{n!} = \exp\left(\sum_{n\in C}\frac{x^n}{n}\right).$$

Proof: The proof is similar, but not identical, to the proof of Theorem 3.33. Let us partition $[n]$ into blocks. For a block of size m, set $a(m) = 0$ if $m \notin C$, and set $a(m) = (m-1)!$ otherwise. In other words, $a(m)$ is the number of ways we can take an allowed cycle on an m-element block. Then

$$A(x) = \sum_{n\geq 1} a(n)\frac{x^n}{n!} = \sum_{n\in C}\frac{x^n}{n}$$

is the exponential generating function of the number of possibilities of the first task, and the statement is proved by the exponential formula. ♢

Let us start with a very basic example. Before we start, we remind the reader that $\frac{1}{1-x} = \sum_{n\geq 0} x^n$, and, integrating both sides, we get the equality $\ln(1-x)^{-1} = \sum_{n=1}^{\infty}\frac{x^n}{n}$.

Example 4.35 *Let C be the set of all positive integers. Then*

$$\begin{aligned} G_C(x) &= \exp\left(\sum_{n=1}^{\infty}\frac{x^n}{n}\right) \\ &= \exp\left(\ln(1-x)^{-1}\right) \\ &= \frac{1}{1-x} \\ &= \sum_{n\geq 0} x^n. \end{aligned}$$

So the coefficient of x^n in $G_C(x)$ is 1, and therefore the coefficient of $x^n/n!$ in $G_C(x)$ is $n!$. This is not surprising at all since this says that, if all cycle lengths are allowed, then there are $n!$ permutations of length n.

The following example shows a typical application of Theorem 4.34. A permutation p is called an *involution* if p^2 is the identity permutation. It goes without saying that p is an involution if and only if all cycles in p are of length 1 or 2. Let us use our new technique to compute the number of involutions of length n.

Example 4.36 *The number of all involutions of length n is*

$$f(n) = \sum_{i=0}^{\lfloor n/2\rfloor}(2i-1)!!\binom{n}{2i},$$

where we set $(-1)!! = 1$.

Solution: We use Theorem 4.34, with $C = \{1, 2\}$. Then we have

$$\begin{aligned} G_C(x) &= \exp\left(x + \frac{x^2}{2}\right) \\ &= \left(\sum_{k\geq 0} \frac{x^k}{k!}\right) \cdot \left(\sum_{i\geq 0} \frac{x^{2i}}{i!2^i}\right). \end{aligned}$$

Note that the coefficient of x^n on the right-hand side is equal to $\sum_{i=0}^{\lfloor n/2 \rfloor} \frac{1}{(n-2i)! \cdot i! \cdot 2^i}$. The result now follows if we observe that $i!2^i = \frac{(2i)!}{(2i-1)!!}$. ◊

With our fresh knowledge, we re-prove the formula for the number of derangements that we proved in Section 2.4.3.4. Recall that derangements are permutations without 1-cycles (or *fixed points*). This time, we will not need to resort to the somewhat cumbersome computation using the principle of inclusion-exclusion.

Indeed, set $C = \{\mathbf{P} - 1\}$. Then Theorem 4.34 yields

$$\begin{aligned} G_C(x) &= \exp\left(\sum_{n\geq 2} \frac{x^n}{n}\right) \\ &= \exp(\log(1-x)^{-1} - x) \\ &= \frac{e^{-x}}{1-x}. \end{aligned}$$

In other words, we get that

$$G_C(x) = \sum_{n=1}^{\infty} \left(\sum_{i=0}^{n} \frac{(-1)^i}{i!}\right) \cdot x^n,$$

so the coefficient of $x^n/n!$ in $G_C(x)$ is

$$n! \cdot \left(\sum_{i=0}^{n} \frac{(-1)^i}{i!}\right), \tag{4.11}$$

just as we have proved in Section 2.4.3.4. Note that, as n goes to infinity, the sum $\sum_{i=0}^{n} \frac{(-1)^i}{i!}$ converges to e^{-1}. That is, for large n, more than one-third of all n-permutations are fixed point free, that is, derangements.

Comparing (4.11) with the result of Lemma 4.12, we find the surprising fact that, for any $n \geq 1$, the number of derangements of length n is equal to the number of desarrangements of length n. This is very interesting as derangements are defined using cycles of permutations, whereas desarrangements are defined using the one-line notation. Recall that a permutation p is called a desarrangement if the first ascent of p occurs in an even position, or if p is the decreasing n-permutation and n is even.

Such a nice identity certainly asks for a bijective proof. Such a proof is reasonably easy to find once we find the right modification of canonical cycle notation.

Let p be any derangement of length n. Let us write p in cycle notation so that each cycle contains its smallest entry in its *second* position, and so that the cycles are ordered in decreasing order of their smallest entries. Define $f(p)$ as the permutation in one-line notation that is obtained from p by omitting all parentheses.

Example 4.37 *If* $p = (324)(51)$, *then* $f(p) = 32451$. *If* $p = (43)(215)$, *then* $f(p) = 43215$.

Theorem 4.38 *The "parenthesis omitting map" f defined above is a bijection from the set D_n of all derangements of length n onto the set J_n of all desarrangements of length n.*

Proof: First we show that f indeed maps into J_n, that is, that $f(p)$ is always a desarrangement. The first cycle C of p contains its smallest entry x in its second position, so, if C contains a third entry y, then $y > x$, and $f(p)$ has its first ascent in the second position, and therefore $f(p)$ is a desarrangement. If C only contains two entries, and the first entry of the second cycle C' is larger than x, then we are done, otherwise repeat the argument applied to C for C'. Continuing this way, we will stop and find the first ascent of $f(p)$ in an even position as soon as we find a cycle with more than two entries or a cycle whose first entry is larger than the last (second) entry of the previous cycle. The only case in which neither of these scenarios occurs is when all cycles are of length two, and the entries in them get smaller and smaller. In that case, we must have $n = 2k$ and $p = (2k\ 2k-1)(2k-2\ 2k-3)\cdots(21)$, and so $f(p)$ is the decreasing permutation of length $2n$, which is a desarrangement.

Now we show that $f : D_n \to J_n$ is a bijection by proving that it has an inverse. Let $q \in J_n$, and set $q = q_1 q_2 \cdots q_n$ in the one-line notation. We are going to find the unique $p \in D_n$ satisfying $f(p) = q$. It follows from the definition of f that, if $q_i = 1$, then the last cycle of p must be $(q_{i-1}\ q_i \cdots q_n)$, as 1 is always the smallest entry of the last cycle. Since q is a desarrangement, it could not happen that $i = 3$ (as that would mean that the first ascent of q is either in the first or in the third position), so the string $q' = q_1 q_2 \cdots q_{i-2}$, when not empty, is at least of length 2. Therefore, we can repeat the same argument for q', that is, we can find its last cycle by starting it at the entry preceding the smallest entry. Iterating this procedure until all of q was broken into cycles, we get the (unique) preimage of q. Uniqueness follows from the fact that f does not change the left-to-right order of the entries, and the cycles cannot start anywhere else without violating our cleverly chosen conditions.

Therefore, f is a bijection as claimed. ◇

The above example of derangements, while a classic one, was about a special case when the set C of allowed cycle lengths was just one element short of all positive integers. In order to increase the reader's appreciation of the exponential formula as a tool of permutation enumeration, we present an example that concerns an *infinite* set C.

Example 4.39 *The number of n-permutations in which each cycle length is even is zero if n is odd and is*

$$\mathbf{even}(2m) = 1^2 \cdot 3^2 \cdots (2m-1)^2 = (1 \cdot 3 \cdots (2m-1))^2 = (2m-1)!!^2$$

if $n = 2m$.

Solution: We will apply Theorem 4.34, with C being the set of all even positive integers. Then Theorem 4.34 implies

$$G_C(x) = \exp\left(\sum_{m=1}^{\infty} \frac{x^{2m}}{2m}\right).$$

Now note that the right-hand side looks very similar to the formal power series $\exp(\ln(1-x)^{-1})$, only x is replaced by x^2 and then the argument of exp is divided by 2. Therefore, we have

$$G_C(x) = \exp\left(\frac{1}{2}\ln(1-x^2)^{-1}\right) = \sqrt{\frac{1}{1-x^2}}.$$

Therefore, we will get the numbers $\mathbf{even}(2m)$ as the coefficients of $\frac{x^{2m}}{(2m)!}$ in $\sqrt{\frac{1}{1-x^2}}$. We see without any computation that $\mathbf{even}(2m+1) = 0$, since the sum of even integers cannot be odd. To compute $\mathbf{even}(2m)$, we will use the binomial theorem.

$$\begin{aligned} G_C(x) = \sqrt{\frac{1}{1-x^2}} &= (1-x^2)^{-1/2} \\ &= \sum_{m=1}^{\infty} \binom{-1/2}{m} (-1)^m x^{2m} \\ &= \sum_{m=1}^{\infty} (-1)^m \frac{(-1/2)\cdot(-3/2)\cdots(-(2m-1)/2)}{m!} x^{2m} \\ &= \frac{(2m-1)!!}{m!2^m} x^{2m}. \end{aligned}$$

So the coefficient of x^{2m} in $G_C(x)$ is $\frac{(2m-1)!!}{m!2^m}$; therefore, the coefficient of $x^{2m}/(2m)!$ is $(2m)!$ times that. In other words, it is indeed $(2m-1)!!^2$ as claimed. ◇

Not that we need this extra piece of evidence, but note that the coefficient of $x^{2m+1}/(2m+1)!$ in $G_C(x)$ is indeed 0 since $G_C(x)$ is a power series in x^2. This shows again that $\mathbf{even}(2m+1) = 0$ for all m.

The nice formula we have just proved for $\mathbf{even}(2m)$ certainly calls for a combinatorial proof, and we are going to present one. Let p be a permutation counted by $\mathbf{even}(2m)$, and think about p as a bijection from $[2m]$ onto $[2m]$. Then $p(1)$ can be anything but 1 since that would create a 1-cycle, which is not allowed because of its odd length. So we have $2m-1$ possibilities for $p(1)$. Once $p(1)$ is chosen, $p(p(1))$ can again be anything except for $p(1)$, so we have $2m-1$ choices for $p(p(1))$ as well. If 1 and $p(1)$ form a 2-cycle, then we are done by induction on m, otherwise we continue. In that case, $p(p(p(1)))$ can be anything but $1, p(1)$, and $p(p(1)))$, so we have $2m-3$ choices for $p(p(p(1)))$. By similar argument, we have $2m-3$ choices for $p^4(1)$ since $p^i(1)$ is not allowed for $i = 1, 2, 3$. Then if $p^4(1) = 1$, we are done by induction, otherwise we continue the same way. In each step, we see that we have $2m - 2i + 1$ possibilities for each of $p^{2i-1}(1)$ and $p^{2i}(1)$ (since 1 is a possibility for the latter). Therefore, our claim is proved.

Quick Check

1. Let C be the set of all positive integers divisible by 3. Find the closed form of $G_C(x)$.
2. Let $h(n)$ be the number of permutations of length n that contain exactly one 1-cycle. Find the exponential generating function of the numbers $h(n)$.
3. Let $h(n)$ be the number of permutations of length n that contain exactly one cycle that is longer than 1. Find the exponential generating function of the numbers $h(n)$.

4.4 Inversions

In Section 4.1, we studied descents of permutations. To that end, we wrote permutations in one-line notation, then studied pairs of *consecutive entries* that were in the *wrong* order, that is, the larger one preceded the smaller one. Now we will do the same, relaxing the requirement that the entries in the wrong order be consecutive.

Definition 4.40 *Let $p = p_1p_2\cdots p_n$ be a permutation. Then we say that the pair (p_i, p_j) is an* inversion *of p if $i < j$, but $p_i > p_j$. The number of inversions of p will be denoted by $i(p)$.*

If a pair (p_i, p_j) is not an inversion, then it is called, less than shockingly, a *non-inversion*.

Example 4.41 *The permutation 31524 has four inversions. These are* $(3,1)$, $(3,2)$, $(5,2)$, *and* $(5,4)$.

The number of n-permutations with k inversions will be denoted by $b(n,k)$. Note that if we read p backwards, then every inversion becomes a non-inversion, and vice versa. So there are as many n-permutations with k inversions as there are with $\binom{n}{2} - k$ inversions.

The number of inversions of the permutation p keeps being interesting even when we look at permutations as products of cycles. In order to prove a simple result in that direction, we define the *full diagram* of a function $f : [n] \rightarrow [n]$. This diagram consists of $2n$ points, n for the domain of f, and n for the range of f, and both of these n-element subsets are arranged vertically, facing each other. If $f(i) = j$, then an arrow goes from point i of the first set to point j of the second set.

Example 4.42 *Figure 4.3 shows the full diagram of the permutation* 24153.

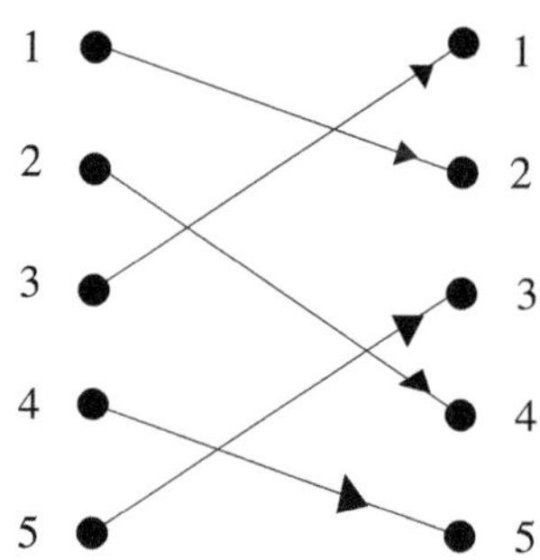

Figure 4.3
The full diagram of 24153.

The following proposition connects the notion of inversions, which was defined using the one-line notations, to the concept of inverse permutations, which was defined considering permutations as functions.

Proposition 4.43 *For every permutation* p, *we have* $i(p) = i(p^{-1})$.

Proof: If we draw the full diagram of the function $p : [n] \rightarrow [n]$, then $i(p)$ is precisely the number of crossings in this diagram. (The reader should think about this.) However, the full diagram of p^{-1} is the same as that of p, with the arrows reversed. Therefore, the number of crossings in the full diagram of p^{-1} is the same as that in the full diagram of p, proving our statement. $\diamondsuit$

The following notion turns out to be useful in many parts of mathematics outside combinatorics.

Definition 4.44 *A permutation p is called* even *if $i(p)$ is even. Similarly, p is called* odd *if $i(p)$ is odd.*

One feels that "even" and "odd" should be properties that occur equally often. The following proposition shows that this is indeed the case.

Proposition 4.45 *Let $n \geq 2$. Then the number of even (equivalently, odd) n-permutations is $n!/2$.*

Proof: Let p be any n-permutation, and let p' be the permutation obtained from p by swapping its first two entries. Then the difference between $i(p)$ and $i(p')$ is plus or minus 1, so they are of different parity. Repeating this argument for all n-permutations, we see that S_n can be split into two subsets of equal size, one consisting of even permutations, the other one consisting of odd permutations. ◊

It turns out that, if we know the parity of the permutations p and q, then we can figure out that of pq and qp. One way to do this is through the following very useful representation of permutations by matrices.

Definition 4.46 *Let $p \in S_n$, with $p = p_1p_2 \cdots p_n$. The* permutation matrix *of p is the $n \times n$ matrix A_p defined by*

$$A_p(i,j) = \begin{cases} 1 \text{ if } p_i = j, \\ 0 \text{ otherwise.} \end{cases}$$

In other words, each row and column of A_p contains exactly one 1 and $n-1$ zeros. To figure out which entry of row i is equal to 1, just find the value of the ith *position* in p. To figure out which entry of column j is equal to 1, find the *entry* that is in position j of p, that is, p_j.

Example 4.47 *If $p = 3241 = (2)(431)$, then*

$$A_p = \begin{pmatrix} 0 & 0 & 1 & 0 \\ 0 & 1 & 0 & 0 \\ 0 & 0 & 0 & 1 \\ 1 & 0 & 0 & 0 \end{pmatrix}.$$

The crucial property of representing permutations by the matrices A_p is that it is a *homomorphism*, that is, that these matrices can be multiplied together as the corresponding permutations.

Lemma 4.48 *Let $p = p_1p_2 \cdots p_n$ and $q = q_1q_2 \cdots q_n$ be n-permutations. Then*

$$A_pA_q = A_{pq}.$$

Proof: We know from the definition of permutation matrices that A_p, A_q, and A_{pq} are all 0-1 matrices. It is easy to see that so is A_pA_q, since the entries of that matrix can be obtained as dot products of a row of A_p and a column of A_q and are therefore equal to either 0 or 1. It is therefore sufficient to prove that $x = A_{pq}(i,j) = 1$ if and only if $y = (A_pA_q)(i,j) = 1$.

It is a direct consequence of the definition of matrix multiplication that $y = A_pA_q(i,j) = \sum_{k=1}^{n} A_p(i,k)A_q(k,j)$. That is, $y = 1$ if and only if there is a $k \in [n]$ so that both $A_p(i,k) = 1$ and $A_q(k,j) = 1$ hold. Considering p and q as functions, this is equivalent to $p(i) = k$ and $q(k) = j$, meaning that $q(p(i)) = (pq)(i) = j$, which by definition is equivalent to $x = A_{pq}(i,j) = 1$. ◊

Many important pieces of information about p can easily be read off A_p.

Proposition 4.49 *The permutation p is odd if and only if* $\det A_p = -1$. *Similarly, the permutation p is even if and only if* $\det A_p = 1$. *In other words,* $\det A_p = (-1)^{i(p)}$.

Proof: We prove the statement by induction on n. For $n = 1$ and $n = 2$, the statement is true. Assume the statement is true for $n-1$ and let $p = p_1p_2\cdots p_n$. Let $p_j = 1$. Then it follows from the expansion theorem of determinants that

$$\det A_p = (-1)^{j-1} \det A_{p'},$$

where p' is the $(n-1)$-permutation obtained from p by deleting the entry 1 and relabeling.

Indeed, let us expand A_p with respect to its first column, and we get the above identity. Finally, note that $i(p) = i(p') + (j-1)$, and then our statement follows from the induction hypothesis applied to p'. ◊

Corollary 4.50 *Figure 4.4 shows what happens if we multiply together permutations of various parities.*

	EVEN	ODD
EVEN	EVEN	ODD
ODD	ODD	EVEN

Figure 4.4
The parity of the product of two permutations.

In particular, the product of even permutations is always even, so the set of even permutations is closed under multiplication. Similarly, the inverse of an even permutation is always even. Therefore, the set of all even n-permutations is interesting on its own, and it is called the *alternating group of degree n*, denoted by A_n.

We point out that the parity of a permutation is often called the *sign* of the permutation. The reason for this is that, as we have just seen, permutations of various parities multiply together like real numbers of various signs.

A permutation that simply interchanges two elements of $[n]$ and leaves all other elements fixed is called a *transposition.* It turns out that transpositions are all odd permutations.

Proposition 4.51 *If p is a transposition, then $i(p)$ is odd.*

Proof: First assume that $p = (i\ i+1)$, that is, p interchanges two *consecutive* elements of $[n]$. Then p has one inversion, and the statement is true. Now assume that $p = (i, i+k)$, that is, there are $k-1$ elements between the two entries that p moves. We claim that $i(p) = 2k-1$. Indeed, each of the $k-1$ elements located between i and $i+k$ is a part of two inversions, one with i, one with $i+k$. Since $(i, i+k)$ forms the last inversion, $i(p)$ is odd. $\Diamond$

The following lemma may sound a little bit counterintuitive.

Lemma 4.52 *The permutation $a = (a_1\ a_2 \cdots a_n)$ that consists of an n-cycle is an even permutation if and only if n is odd.*

Proof: For the purposes of this proof, we will *omit* all 1-cycles from the cycle notation of a permutation. For instance, the 6-permutation (1)(32)(54)(6) would be written as (32)(54). This does not cause any ambiguity, since entries that are not shown in the cycle notation must be fixed points.

We prove the statement by induction on n. The initial cases of $n = 1$ and $n = 2$ are true because the permutation (x) has an even number of inversions (zero), while the permutation $(y\ z)$ has an odd number of inversions, as we have seen in Proposition 4.51.

Now let us assume the statement is true for all positive integers less than n, and let $a = (a_1\ a_2 \cdots a_n)$. The crucial observation is that

$$a = (a_{n-1}\ a_n)(a_1\ a_2 \cdots\ a_{n-1}).$$

Setting $q = (a_{n-1}\ a_n)$ and $a' = (a_1\ a_2 \cdots\ a_{n-1})$, the identity $p = qp'$ holds. Taking determinants, and using Proposition 4.49, we see that

$$\det A_a = \det A_q \cdot \det A_{a'} = -\det A_{a'},$$

since $\det A_q = -1$ by Proposition 4.51. So the parity of a is the opposite of that of a', and our induction step, and therefore our proof, is complete. $\Diamond$

The reader should spend a moment justifying the following corollary of Lemma 4.52.

Corollary 4.53 *The permutation a is even if and only if it has an even number of even cycles.*

4.4.1 Counting permutations with respect to inversions

While an explicit formula does exist for the numbers $b(n, k)$, it is quite complicated and somewhat cumbersome to prove. On the other hand, for any fixed n, the generating function $I_n(x)$ that counts n-permutations according to their number of inversions has a remarkably simple form.

Theorem 4.54 *For all positive integers $n \geq 2$,*

$$I_n(x) = \sum_{p \in S_n} x^{i(p)} = \sum_{k=0}^{\binom{n}{2}} b(n,k)x^k = (1+x)(1+x+x^2)\cdots(1+\cdots+x^{n-1}).$$

Example 4.55 *For $n = 3$, we have one permutation with no inversions, 123; we have two permutations with one inversion, 132 and 213; we have two permutations with two inversions, 231 and 312; and one permutation with three inversions, 321. This yields*

$$I_3(x) = 1 + 2x + 2x^2 + x^3 = (1+x)(1+x+x^2).$$

Proof: (of Theorem 4.54) Let $d(n, k)$ be the coefficient of x^k in $I_n(x)$. All we have to show is that, for all n and all k, we have $d(n, k) = b(n, k)$. We are going to prove this by induction on n, the initial case of $n = 2$ being trivial. We claim that both the numbers $b(n, k)$ and $d(n, k)$ satisfy the same horizontal recursive relation, namely,

$$b(n,k) = b(n-1,k) + b(n-1,k-1) + \cdots + b(n-1,m) \tag{4.12}$$

and

$$d(n,k) = d(n-1,k) + d(n-1,k-1) + \cdots + d(n-1,m), \tag{4.13}$$

where $m = \max(0, k-n+1)$. In other words, m is the larger of the numbers 0 and $k - n + 1$.

First we prove that (4.12) holds. The left-hand side is the number of n-permutations with k inversions. The right-hand side is the same, counted with respect to the number of inversions n is part of. That is, $b(n-1, k)$ is the number of permutations of length $n - 1$ with k inversions. If we insert n to the last position of such a permutation, we get a permutation counted by $b(n, k)$ since no new inversions are created. Similarly, $b(n-1, k-1)$ is the number of permutations of length $n-1$ with $k-1$ inversions; if we insert n into the next-to-last position of such a permutation, we create one new inversion, and we get a permutation that is again enumerated by $b(n,k)$. In general, for $0 \leq i \leq \min(k, n-1)$, the number $b(n-1, k-i)$ is the number of permutations of length n with k inversions in which n is in position $n-i$. Indeed, in such permutations, n is part of i inversions. This proves (4.12).

It is even easier to prove that (4.13) holds. By the definition of $I_n(x)$, we have

$$I_n(x) = I_{n-1}(x)(1 + x + \cdots + x^{n-1}).$$

Equating coefficients of x^k on both sides, (4.13) follows.

As the numbers $b(n,k)$ and $d(n,k)$ are obtained from the identical numbers $b(n-1,i)$ and $d(n-1,i)$ using identical operations, they must be equal. ◊

Just like Eulerian numbers, the numbers $b(n,k)$ also occur in many different contexts. One of the most famous notions leading to the same numbers is that of the *major index* of a permutation. The major index was introduced by Percy MacMahon, who called it the *greater index*. It was later called the major index because of MacMahon's rank in the British Army.

Definition 4.56 *Let $p = p_1p_2\cdots p_n$ be an n permutation, and let $maj(p)$ be the* sum *of all descents of p. Then $maj(p)$ is called the* major index *of p.*

Example 4.57 *Because $p = 31452$ has two descents, in positions 1 and 4, the major index of p is $1 + 4 = 5$.*

Theorem 4.58 *For every positive integer n, there are $b(n,k)$ permutations of length n having major index k. In other words, the number of n-permutations that have k inversions is the same as the number of n-permutations that have major index k.*

Example 4.59 *Let $n = 3$. Then there is one permutation with major index 0, namely, 123; there are two permutations with major index 1, namely, 213 and 312; there are two permutations with major index 2, namely, 231 and 132; and there is one permutation with major index 3, namely, 321. This agrees with the numbers we obtained when we enumerated 3-permutations according to the number of their inversions, in Example 4.55.*

Theorem 4.58 was first proved by MacMahon [55]. A combinatorial proof, in a more general setup, was given by Dominique Foata and Marcel-Paul Schützenberger in [33]. We will present a simple recursive proof using generating functions, which uses ideas from [81].

Proof: (of Theorem 4.58) Our statement is equivalent to saying that the generating function of n-permutations according to their major index is equal to the generating function $I_n(x)$ of permutations according to their number of inversions. In other words, if $m(n,i)$ denotes the number of all n-permutations with major index i, then it suffices to show that

$$\begin{aligned}\sum_{p\in S_n} x^{maj(p)} &= \sum_{i=0}^{\binom{n}{2}} m(n,i)x^i \\ &= (1+x)(1+x+x^2)\cdots(1+x+\cdots+x^{n-1}) \\ &= I_n(x).\end{aligned}$$

We prove the statement by induction on n, the initial case of $n=1$ being obvious. Assume now that we know that the statement is true for $n-1$. Let p be a permutation of length $n-1$ that has k descents. Let these descents be at positions $i_1, i_2, \cdots, i_k$. Furthermore, introduce the notations $i_0 = 0$ and $i_{k+1} = n-1$.

Let us now insert the entry n immediately after position i_j of p, for $j \in [0, k+1]$. Let $p_{<j>}$ be the obtained permutation. The crucial observation is the following proposition.

Proposition 4.60 *For all $j \in [0,k]$, the equality*

$$maj(p_{<j>}) = maj(p_{<j+1>}) + 1$$

holds.

That is, as we move n from one descent of p to the next, the major index will decrease by one.

Example 4.61 *Let $p = 53412$. Then*

- $p_{<0>} = 653412$, *with major index* $1+2+4=7$,
- $p_{<1>} = 563412$, *with major index* $2+4=6$,
- $p_{<2>} = 534612$, *with major index* $1+4=5$, *and*
- $p_{<3>} = 534126$, *with major index* $1+3=4$.

Proof: (of Proposition 4.60) When we move n from i_j to i_{j+1}, we do not change the descents on the left of i_j or on the right of i_{j+1}. So changes can only occur at i_j (the contribution of the descent to the major index will decrease by one) and at i_{j+1} (there will be no change). Therefore, the major index of our permutation will indeed decrease by one. ◊

After this result, it is not surprising that something similar will happen for the ascents of the permutation. Let $a_1, a_2, \cdots, a_{n-2-k}$ be the ascents of p, and let $(p^{<j>})$ denote the n-permutation obtained from p by inserting n immediately after position a_j.

Proposition 4.62 *For all $j \in [n-2-k]$, the equality*

$$maj(p^{<j>}) = maj(p^{<j+1>}) - 1$$

holds.

Example 4.63 *Let $p = 53412$. Then*

- $p^{<1>} = 536412$, *with major index* $1+3+4 = 8$, *and*
- $p^{<2>} = 534162$, *with major index* $1+3+5 = 9$.

Proof: (of Proposition 4.62). We will use a symmetry argument. If $p \in S_{n-1}$ has $d(p)$ descents, and p^c denotes the complement of p, then it is obvious that $maj(p) + maj(p^c) = d(p) \cdot (n-1)$, which only depends on $d(p)$. Applying this fact to $maj(p^{<j>})$ and $maj(p^{<j+1>})$, we see that it suffices to prove that

$$maj(p^{<j>^r}) = maj(p^{<j+1>^r}) + 1.$$

However, this identity follows from Proposition 4.60, since $p^{<j>^r}$ and $p^{<j+1>^r}$ are nothing else but p^r, with the entry n inserted after two consecutive *descents*. $\diamondsuit$

Example 4.64 *Keeping the notation of Example 4.63,*

$$maj(p) + maj(p^c) = 8.$$

In order to prove that $maj(p^{<1>}) + 1 = maj(p^{<2>})$, it suffices to show that $maj(p^{<1>^c}) = maj(p^{<2>^c}) + 1$. This is just the identity

$$maj(163254) = maj(132654) + 1,$$

which follows from Proposition 4.60.

Considering our running example, the permutation $p = 53412$ (which has major index 4), we see that inserting 6 into all possible positions resulted in permutations having major index 4, 5, 6, 7, 8, and 9, that is, each integer from the interval $[maj(p), maj(p) + n - 1]$ occurs once. We claim that this always happens.

Lemma 4.65 *Inserting n into each position of the $(n-1)$-permutation p, we will get n permutations of length n, so that the set of major indices of these permutations is the interval $[maj(p), maj(p) + n - 1]$.*

Proof: We already know that inserting n immediately after the descents of p or at either end will result in major indices forming an interval, and that inserting n immediately after the ascents of p results in major indices forming another interval. In order to show that the union of these two intervals is another interval, note that $maj(p_{<k>}) = maj(p)$, since adding n to the end of p does not change any descents. Therefore, by Proposition 4.60, the major indices of the permutations $maj(p_{<i>})$, where i goes from $k+1$ to 0, are $maj(p), maj(p)+1, \cdots, maj(p)+k+1$.

On the other hand, inserting n after the rightmost ascent of p increases the major index of p by $n-1$. Indeed, if this ascent is in position m, then inserting n there will create a new descent there and will increase the major index by m. Furthermore, m was the rightmost ascent, so the remaining $n-1-m$ positions of p were all descents. They all get shifted by one because of the insertion of n, and they will therefore contribute $n-1-m$ more to the major index. Therefore, $maj(p^{<n-2-k>}) = maj(p)+n-1$. This implies, by Proposition 4.62, that the major indices of the permutations $maj(p^{<i>})$, where i goes from $n-2-k$ to 1, are $maj(p)+n-1, maj(p)+n-2, \cdots, maj(p)+k+2$. ◊

The proof of Theorem 4.58 is now immediate. Using our induction hypothesis and the result of Lemma 4.65, we get that

$$\begin{aligned}\sum_{p\in S_n} x^{maj(p)} &= \left(\sum_{p\in S_{n-1}} x^{maj(p)}\right)\cdot(1+x+\cdots+x^{n-1})\\ &= I_{n-1}(x)(1+x+\cdots+x^{n-1}) = I_n(x).\end{aligned}$$

◊

We mention that the concept of inversions can be generalized in several different ways. On one hand, an inversion is just two entries of a permutation that are in a prescribed order, that is, the larger one comes first. We could instead look at three, or four, or k entries of a permutation that are in a prescribed order. This leads to the theory of pattern-avoiding permutations. The reader should take a look at Exercise 32 for the precise definition, and at Exercises 33 and 34 for further information about this relatively recent and rapidly developing topic.

On the other hand, just as we defined inversions of permutations, we could define inversions of permutations of *multisets*. This leads to the theory of Gaussian coefficients, named after Carl Friedrich Gauss. See Exercises 29 and 30 to explore this direction.

Quick Check

1. Let $n \geq 3$. Prove that the number of permutations $p = p_1p_2\cdots p_n$ for which $i(p) = k$ and $p_1 < p_2$ is equal to the number of permutations for which $i(p) = k$ and $p_{n-1} < p_n$.
2. Let p as in the previous question, and let $2 \leq j \leq n-1$. Prove that the number of permutations p for which $i(p) = k$ and $p_{j-1} > p_j <$

p_{j+1} is equal to the number of permutations p for which $i(p) = k$ and $p_{j-1} < p_j > p_{j+1}$.

3. Let n be a positive integer for which $\binom{n}{2}$ is even. Prove that the number $b(n, \binom{n}{2}/2)$ is even.

4.5 Advanced applications of generating functions to permutation enumeration

Let us consider a few counting problems where generating functions are still very useful, but we have to use them in ways that we have not seen before.

4.5.1 The combinatorial meaning of the derivative

In Theorem 4.21, we proved the interesting formula

$$\sum_{k=1}^{n} c(n,k)x^k = x(x+1)\cdots(x+n-1) \tag{4.14}$$

for any fixed positive integer n.

Now let us take the *derivative* of both sides with respect to x. We get the identity

$$\sum_{k=1}^{n} k \cdot c(n,k)x^{k-1} = \sum_{i=0}^{n-1} \frac{x(x+1)\cdots(x+n-1)}{x+i}.$$

Note that, differentiating the right-hand side of (4.14), we have simply used the product rule, which was easy to do since the derivative of any individual factor $(x+i)$ is 1.

Now let us substitute $x = 1$ in the last displayed equation, to get

$$\sum_{k=1}^{n} k \cdot c(n,k) = \sum_{j=1}^{n} \frac{n!}{j}. \tag{4.15}$$

The crucial observation is that the expression on the left-hand side, $\sum_{k=1}^{n} k \cdot c(n,k)$, is simply the number of *all* cycles in all permutations of length n. Indeed, permutations counted by $c(n,k)$ contribute k cycles to that total. So formula (4.15) is in fact a formula for the total number of cycles in all permutations of length n. Equivalently, if we divide both sides of (4.15) by $n!$, we get that the *average* number of cycles in permutations of length n is $\sum_{j=1}^{n} \frac{1}{n}$. This number is often called the nth *harmonic number*. While this result can be proved without using generating functions (see Exercise 18), the idea of taking the derivative of a generating function and then finding its value at $x = 1$ is often useful in similar situations.

4.5.2 Multivariate generating functions

In this section, our goal is to combine *all* signless Stirling numbers of the first kind in one single generating function. That is, our goal is to find a generating function whose coefficients consist of the numbers $c(n,k)$ for every n and every k, as long as $k \leq n$, and both n and k are nonnegative integers. This can be done if we introduce the *two-variable* generating function

$$H(x,t) = \sum_{n\geq 0}\sum_{k\geq 0} c(n,k)t^k \frac{x^n}{n!}. \tag{4.16}$$

How can we find a closed formula for $H(x,t)$? Note that $c(n,k)t^k$ is the number of permutations of length n in which each cycle is colored by one of the colors $\{1,2,\cdots,t\}$. Now we can apply the exponential formula, since what we need to count is to partition $[n]$ into nonempty blocks, then put a cycle on each block and color that cycle by one of t colors. The number of ways to proceed on one cycle of length m is $a_m = (m-1)!t$, so we obtain the inside generating function

$$A(x) = A(x,t) = \sum_{m\geq 1} t(m-1)! \cdot \frac{x^m}{m!} = t\ln\left(\frac{1}{1-x}\right).$$

Therefore, by the exponential formula, we get

$$H(x,t) = \exp\left(A(x,t)\right) = \exp\left(t \cdot \ln\left(\frac{1}{1-x}\right)\right) = (1-x)^{-t}. \tag{4.17}$$

So the remarkably simple formula at the far right-hand side contains *all* signless Stirling numbers of the first kind as coefficients. This result is useful for many purposes, one of which is the easy generation of Stirling numbers. Typing

```
mtaylor[(1-x)^-t,[x,t],20]
```

into Maple will produce all the numbers $c(n,k)/n!$ for all pairs (k,n) satisfying $k+n<20$.

Example 4.66 *Let $n = 2m$. Find the number of r_n permutations of length n in which each cycle length is even and there are an even number of cycles.*

Solution: Similarly to the argument above, let us partition $[n]$ into nonempty blocks, then take a cycle on each block, and color that cycle with one of t colors. As we also require that each block be of even size, the number of ways to proceed on each block of size n is $a_n = (n-1)!t$ if n is even, and $a_n = 0$ if n is odd. So,

$$A(x) = A(x,t) = \sum_{m\geq 1} t\frac{x^{2m}}{2m} = -t\frac{1}{2}\ln(1-x^2),$$

as we have computed in Example 4.39. Therefore, the generating function for the combined task is

$$H(x) = H(x,t) = \exp(A(x,t) = \left(\sqrt{\frac{1}{1-x^2}}\right)^{-t}. \qquad (4.18)$$

Note that the combinatorial meaning of $H(x,t)$ is that

$$H(x,t) = \sum_{n,k} h(n,k)\frac{x^n}{n!}t^k,$$

where $h(n,k)$ is the number of permutations of length n in which all cycle lengths are even, and there are a total of k cycles. Setting $t = -1$ and using (4.18), we get

$$H(x,-1) = \sum_{n,k} h(n,k)\frac{x^n}{n!}(-1)^k = \sqrt{1-x^2}.$$

Note that $\sum_{n,k} h(n,k)\frac{x^n}{n!}(-1)^k$ is just the number of our permutations with an even number of cycles minus the number of our permutations with an odd number of cycles, which we will denote by $r_n - s_n$. Using the binomial theorem, the last displayed formula yields

$$r_n - s_n = (2m-1)!! \cdot (2m-3)!!,$$

where $n = 2m$. We have seen in Example 4.39 that the total number of permutations of length $n = 2m$ with all cycle lengths even is

$$r_n + s_n = (2m-1)!!^2.$$

Note that we could also get this result by setting $t = 1$ in (4.18). The last two displayed equations then yield that

$$r_n = (m-1) \cdot (2m-1)!! \cdot (2m-3)!!,$$

and, for the number of permutations of length $n = 2m$ in which all cycle lengths are even and there are an odd number of cycles,

$$s_n = m \cdot (2m-1)!! \cdot (2m-3)!!.$$

Note the interesting fact that $s_n > r_n$ holds for all even n, that is, for all n for which these numbers are not 0. ◇

Supplementary Exercises 44–47 contain further advanced applications of the exponential formula to permutation enumeration.

Quick Check

1. Use the compositional formula of exponential generating functions to find the total number of cycles of all permutations of length n.
2. Compute the value of $\sum_{k=1}^{n} c(n,k)k(k-1)$.
3. Find the closed form of the multivariate generating function

$$V(x,t) = \sum_{n\geq 0}\sum_{k\geq 0} s(n,k)t^k \frac{x^n}{n!}.$$

4.6 Notes

The book *Combinatorics of Permutations* [13] is a natural follow-up to this chapter for the interested reader.

Just for the sake of completeness, we close this chapter by presenting the explicit formula by Schlömilch for the Stirling numbers of the first kind. A proof can be found in [13].

Theorem 4.67 *For all positive integers k and n satisfying $k \leq n$, we have*

$$s(n,k) = \sum_{0\leq h\leq n-k} (-1)^h \binom{n-1+h}{n-k+h}\binom{2n-k}{n-k-h} S(n-k+h,h)$$

$$= \sum_{0\leq i\leq h\leq n-k} (-1)^{j+h}\binom{h}{j}\binom{n-1+h}{n-k+h}\binom{2n-k}{n-k-h}\frac{(h-j)^{n-k+h}}{h!}.$$

As we mentioned earlier, desarrangements were introduced by Jacques Désarmenien and Michelle Wachs in [25]. That paper contains generalizations of the concept and of the results we presented on the topic.

4.7 Chapter review

(A) Explicit formulae

1. Eulerian numbers. [They count n-permutations with k runs.]

$$A(n,k) = \sum_{i=0}^{k} (-1)^i (k-i)^n \binom{n+1}{i}.$$

2. Number of permutations of type $(a_1, a_2, \cdots, a_n)$.

$$\frac{n!}{1^{a_1}2^{a_2}\cdots n^{a_n}a_1!a_2!\cdots a_n!}.$$

Here $\sum_{i=1}^{n} ia_i = n$.

(B) Recurrence relations.

1. Eulerian numbers.

$$A(n,k) = kA(n-1,k) + (n-k+1)A(n-1,k-1).$$

2. Signless Stirling numbers of the first kind.
[They count n-permutations with k cycles.]

$$c(n,k) = c(n-1,k-1) + (n-1)c(n-1,k).$$

(C) Generating functions

1. Permutations of fixed size n.
   (a) Signless Stirling numbers of the first kind.

$$\sum_{k=0}^{n} c(n,k)x^k = (x+n-1)_n.$$

   (b) Stirling numbers of the first kind.

$$\sum_{k=0}^{n} s(n,k)x^k = (x)_n.$$

   (c) Permutations according to the number of their inversions.

$$\sum_{p\in S_n} x^{i(p)} = I_n(x) = (1+x)(1+x+x^2)\cdots(1+x+\cdots+x^{n-1}).$$

   (d) Permutations according to their major index.

$$\sum_{p\in S_n} x^{maj(p)} = I_n(x).$$

2. Permutations of variable size n.
   (a) Permutations of length n with cycle lengths from the set C.

$$G_C(x) = \exp\left(\sum_{n\in C}\frac{x^n}{n}\right).$$

4.8 Exercises

1. Prove that, for all positive integers k and n satisfying $k \leq n$, we have $A(n,k) = A(n, n+1-k)$. Try to give two different solutions.
2. Let $S \subseteq [n-1]$, and let $\alpha(S)$ denote the number of n-permutations whose descent set is *contained* in S.

   Find the one-element set $\{i\} \subseteq [n-1]$ for which $\alpha(\{i\})$ is maximal.
3. Let $n = 10$ and let $S = [3, 7]$. Compute $\alpha(\{3, 7\})$.
4. Let $S = \{i_1, i_2, \cdots, i_k\} \subseteq [n-1]$. Find an explicit formula for $\alpha(S)$.
5. How many 10-permutations are there whose set of descents is *equal* to the set $\{3, 7\}$?
6. Let $S = \{i_1, i_2, \cdots, i_k\} \subseteq [n-1]$. Let $\beta(S)$ denote the number of n-permutations whose set of descents is *equal* to S. Find a formula for $\beta(S)$. Your answer may contain the α function.
7. Let
$$M(n,k) = \sum_{k=1}^{n} A(n,k).$$
   That is, $M(n,k)$ is the number of n-permutations with at most k ascending runs. Prove, by constructing an appropriate surjection, that
$$M(n,k) \leq k!S(n,k).$$
8. The *order* of the permutation p is the smallest positive integer k for which $p^k = id$.
   (a) Prove that, if p and q have the same type, then their order is the same.
   (b) How can we figure out that order given just the type of p and q?
9. Let S be the infinite matrix whose rows are indexed by the non-negative integers and whose row k consists of the Stirling numbers $S(n,0), S(n,1), \cdots$. Let s be the similarly defined infinite matrix, but with the numbers $s(n,k)$ replacing the numbers $S(n,k)$. Find a matrix identity satisfied by the matrices S and s.
10. An airplane has n seats, and each of them is sold for a particular flight. Passengers board the plane one by one, and the first $n-1$ passengers sit at *random* places, not necessarily their assigned seats. When the last passenger boards the plane, he goes to his assigned seat. If his assigned seat is free, then he takes it. If he finds someone in his assigned seat, then he asks that person to move. Then that person goes to her assigned seat, and acts similarly. This continues until the person currently moving finds her assigned seat empty, and

so nobody else is asked to move. Let $k \in n$. What is the probability that (counting the last passenger) exactly k people had to move because of the arrival of the last passenger?

11. What is the average number of fixed points of all n-permutations?

12. How many 6-permutations are there whose fourth power is the identity permutation?

13. How many 6-permutations are there whose sixth power is the identity permutation?

14. How many 12-permutations are there whose cube contains exactly two 3-cycles?

15. How many even 6-permutations have order four? The order of a permutation is defined in Exercise 8.

16. Find an explicit formula for the exponential generating function for the number of n-permutations whose third power is the identity permutation.

17. Find an explicit formula for the exponential generating function for the number of n-permutations whose order is six.

18. Solve this exercise *without* using generating functions.

    (a) Let $H(n)$ be the average number of cycles of an n-permutation. Prove that

    $$H(n) = \frac{n-1}{n} H(n-1) + \frac{1}{n}(H(n-1)+1).$$

    (b) Find an explicit formula for $H(n)$.

19. Prove that, if $n \geq 1$, then

    $$n! = \sum_{k=0}^{n} \binom{n}{k} D(k),$$

    where $D(k)$ denotes the number of derangements of length k and we set $D(0) = 1$.

20. Prove, by a combinatorial argument, that, for all $n \geq 2$, we have

    $$D(n) = (n-1)(D(n-1) + D(n-2)).$$

21. (a) Find the exponential generating function for the numbers $f_k(n)$ of permutations of length n in which each cycle length is divisible by k.

    (b) Find a formula for the numbers $f_k(n)$ defined in part (a).

    (c) Find a combinatorial proof for the result of part (b).

22. Let p be a prime number. For what $k \in [p]$ will $c(p,k)$ be divisible by p?

23. (Wilson's theorem) Let p be a prime number. Prove that $(p-1)!+1$ is divisible by p.

24. Compute $b(10,3)$.

25. Prove that, if $n-1 \geq k$, then

$$b(n,k) = b(n,k-1) + b(n-1,k). \tag{4.19}$$

26. Modify the recursive relation of the previous exercise for the case when $n-1 < k$.

27. Let $k \leq n-1$.

    (a) How many n-permutations are there for which $i(p)$ is divisible by k?

    (b) How many n-permutations are there for which $i(p) \equiv r$ for some r, modulo k?

28. Let $Inv(n,k)$ be the set of n-permutations having exactly k inversions. Let

$$a_i = |\{p = p_1p_2\cdots p_n \in Inv(n,k) | p_i > p_{i+1}\}|.$$

Prove that a_i is independent of i.

29. Let $k \leq n$ be fixed positive integers. Let K be the multiset that consists of k copies of 1 and $n-k$ copies of 2. Let p be a permutation of K, and define an inversion of p as a pair formed by a 2 and a 1 so that the 2 precedes the 1. For instance, 221 has two inversions, while 1212 has one. We know that altogether, K has $\binom{n}{k}$ permutations, but let us now count them according to their inversions. In particular, define

$$\begin{bmatrix}\mathbf{n}\\\mathbf{k}\end{bmatrix} = \sum_p q^{i(p)}.$$

Here, $i(p)$ is the number of inversions of p, and the sum ranges over all permutations of K. The polynomials $\begin{bmatrix}\mathbf{n}\\\mathbf{k}\end{bmatrix}$ are called *Gaussian coefficients*, or q-binomial coefficients, or *Gaussian polynomials*.

    (a) Compute $\begin{bmatrix}\mathbf{4}\\\mathbf{2}\end{bmatrix}$, $\begin{bmatrix}\mathbf{5}\\\mathbf{1}\end{bmatrix}$, and $\begin{bmatrix}\mathbf{5}\\\mathbf{3}\end{bmatrix}$.

    (b) Prove that $\begin{bmatrix}\mathbf{n}\\\mathbf{k}\end{bmatrix} = \begin{bmatrix}\mathbf{n}\\\mathbf{n-k}\end{bmatrix}$.

30. Prove that

$$\begin{bmatrix}\mathbf{n}\\\mathbf{k}\end{bmatrix} = q^{n-k} \cdot \begin{bmatrix}\mathbf{n-1}\\\mathbf{k-1}\end{bmatrix} + \begin{bmatrix}\mathbf{n-1}\\\mathbf{k}\end{bmatrix}. \tag{4.20}$$

31. Let us enumerate integer partitions whose Ferrers shape fits within an $m \times n$ rectangle according to their sizes. That is, let

$$p(m, n, q) = \sum_a q^{|a|},$$

where a ranges over all integer partitions that have at most m rows and at most n columns, and $|a|$ denotes the integer of which a is a partition. For instance, $p(2, 2, q) = 1 + q + 2q^2 + q^3 + q^4$.

Prove that $p(m, n, q) = \left[{\mathbf{m+n} \atop \mathbf{m}} \right]$.

32. We say that a permutation *avoids* the pattern 132 if it does not have three elements that relate to each other the same way as 1, 3, and 2. That is, if $p = p_1p_2 \cdots p_n$, then p is 132-avoiding if there are no three indices $i < j < k$ so that $p_j > p_k > p_i$. That is, there are no three entries in p among which the leftmost is the smallest and the one in the middle is the largest, just as in 132. For instance, 42351 is 132-avoiding, but 35241 is not, for the three entries 3, 5, and 4. So we will say that 35241 *contains* 132.

Prove that the number of 132-avoiding n-permutations is the nth Catalan number.

33. Let q be any permutation of length k, and define q-avoiding permutations in an analogous way to 132-avoiding permutations, with q playing the role of 132. Let $S_n(q)$ be the number of n-permutations avoiding the pattern q. For what permutations q does the result of the previous exercise immediately imply that $S_n(q) = C_n$?

34. Prove that, if the n-permutation $p = p_1p_2 \cdots p_n$ contains a 312-pattern, then it must contain a 312-pattern in which entries playing the role of the entries 3 and 1 of the 312-pattern are *consecutive entries* in p.

35. Find a non-generating function proof for the result of Example 3.37.

36. A *regular tetrahedron* is a solid with four vertices, six edges, and three faces, so that each edge is of the same length. See Figure 4.5 for an illustration.

    (a) Find the number of all symmetries of a regular tetrahedron.
    (b) Find the number of all symmetries of a regular tetrahedron that can be obtained by a series of rotations.
    (c) Symmetries of a regular tetrahedron are permutations of its vertex set. Does the sign (or parity) of such a permutation tell us whether the corresponding symmetry can be obtained by a series of rotations?

37. (+) We have seen how to compute the number of all cycles in all permutations of length n in Section 4.5.1, and also, in Supplementary Exercise 16. Let us consider the following generalization of that

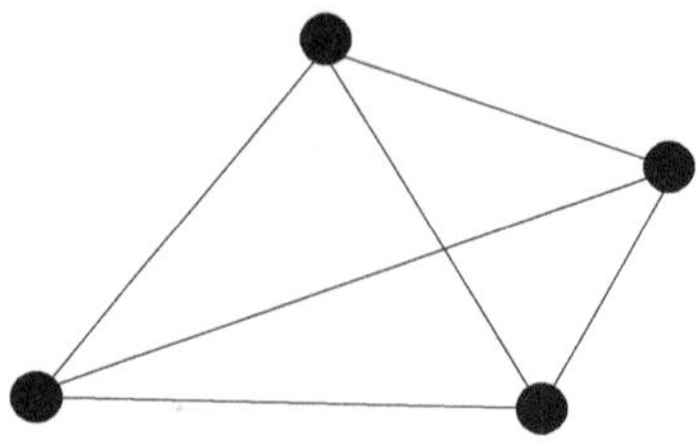

Figure 4.5
A regular tetrahedron.

problem. Take all permutations of length n, and in each of them, label each cycle with a positive divisor of the length of that cycle. Let $h(n)$ be the number of all these labeled cycles. Find the exponential generating function of the sequence of the numbers $h(n)$. The first four values of the sequence $h(n)$ are 1, 3, 11, and 59.

(In other words, $h(n)$ counts ordered triples (p, C, d) so that p is a permutation of length n, while C is a cycle of p, and d is a positive divisor of the length of C.)

4.9 Solutions to exercises

1. Recall that $A(n, k)$ is the number of n-permutations with $k-1$ descents. If $p = p_1p_2\cdots p_n$ has $k-1$ descents, then its reverse, $p^r = p_np_{n-1}\cdots p_1$, has $k-1$ ascents, so it has $(n-1)-(k-1) = n-k$ descents. In other words, "taking reverses" is a bijection from the set of permutations counted by $A(n, k)$ into the set of permutations counted by $A(n, n+1-k)$.

   Alternatively, instead of taking reverses, we can take *complements.* That is, for $p = p_1p_2\cdots p_n$, define p^c as the permutation whose ith entry is $n+1-p_i$. If p has $k-1$ descents, then its complement p^c has $k-1$ ascents, and the proof is completed as above.

2. In a permutation counted by $\alpha(\{i\})$, the first i entries must form an increasing sequence, and the last $n-i$ entries have to form an increasing sequence. Therefore, once we know the set of the first i entries, we know the permutation. This yields $\alpha(\{i\}) = \binom{n}{i}$. It is easy to prove computationally that for fixed n the maximum of $\binom{n}{i}$ is taken for $i = \lfloor n/2 \rfloor$ (if you do not believe it, see Proposition 6.39).

3. Let $\pi = \{B_1, B_2, B_3\}$ be any ordered partition of $[10]$ in which the

blocks B_1, B_2, B_3 have respective sizes 3, 4, and 3. Let us order the elements of each block increasingly, and write down the elements of the blocks starting with B_1, following with B_2, and then ending with B_3. Clearly, we get a permutation whose descent set is contained in $\{3,7\}$ as each block is ordered increasingly. It is also clear that we obtain each such permutation exactly once. So all we have to do is to find the number of such partitions π. That is not difficult, because we have $\binom{10}{3}$ choices for B_1, then $\binom{7}{4}$ choices for B_2, and because our hands are tied when we choose B_3. Therefore, by the product principle, we have

$$\alpha(\{3,7\}) = \binom{10}{3} \cdot \binom{7}{4}.$$

4. Assume the s_i are listed in increasing order. Using the ideas of the solution of the previous exercise, we get that

$$\begin{aligned}\alpha(S) &= \binom{n}{s_1}\binom{n-s_1}{s_2-s_1}\binom{n-s_1-s_2}{s_3-s_2}\cdots\binom{n-s_1-\cdots-s_{k-1}}{s_k-s_{k-1}} \\ &= \binom{n}{s_1, s_2-s_1, \cdots, s_k-s_{k-1}, n-s_k}.\end{aligned}$$

5. In Exercise 3, we have computed the number of permutations of length 10 whose descent set was contained in $\{3,7\}$. That means that the descent set of those permutations was one of the following sets: $\{3,7\}$ or $\{3\}$ or $\{7\}$ or $\emptyset$. Following the line of the previous exercise, for $S \subseteq [10]$ we define $\beta(S)$ to be the number of permutations whose set of descents is equal to S. Then we claim that

$$\beta(\{3,7\}) = \alpha(\{3,7\}) - \alpha(\{3\}) - \alpha(\{7\}) + \alpha(\{\emptyset\}).$$

Indeed, the second and the third terms on the right-hand side subtract the number of permutations whose descent set is *strictly* contained in $\{3,7\}$, and in doing this, they subtract the number of permutations with no descents twice. We have computed $\alpha(\{3,7\})$ in the previous exercise. By the same method, we see that $\alpha(\{3\}) = \binom{10}{3}$, and $\alpha(\{7\}) = \binom{10}{7} = \binom{10}{3}$. As there is only one permutation with no descent, we get that

$$\beta(\{3,7\}) = \binom{10}{3} \cdot \binom{7}{4} - 2 \cdot \binom{10}{3} + 1.$$

6. Using the ideas of the previous exercise, we get

$$\beta(S) = \sum_{T \subseteq S} \alpha(S)(-1)^{|S-T|}.$$

7. Take any of the $S(n,k)$ partitions of $[n]$ into k blocks. Order the elements within each block increasingly, then order the *set* of k blocks in any of $k!$ ways. This way, we get $S(n,k)k!$ permutations, and they all have at most k ascending runs. We will obtain some of them several times; this is why only inequality, not equality, is true.
8. (a) By part (b), the order of a permutation is the least common multiple of all cycle lengths. If two permutations have the same type, then their cycle lengths are pairwise equal, as are their least common multiples.
   (b) If the length of a cycle of a permutation p is t, then for any i, the permutation p^{ti} will contain the entries of that cycle as fixed points. So if m is a multiple of all cycle lengths of p, then $p^m = id$. Therefore, the order of p is the *least common multiple* of all cycle lengths of p.
9. We claim that $Ss = sS = I$, where I is the identity matrix. At this point, the reader may reproach us for the fact that we have not even defined the product of two infinite matrices. Therefore, we will say that we multiply two infinite matrices X and Y very much as if they were finite, that is, the (i,j)-entry of XY is the dot product of row i of X and column j of Y, as long as that dot product is defined. That dot product is the infinite sum $\sum_k x_{ik}y_{kj}$, which is certainly defined if all but a finite number of its summands are equal to 0. If, for some i and j, this dot product is not defined (infinite), then we say that XY is not defined. In our examples, both sS and Ss are defined because $S(n,k) = s(n,k) = 0$ for $k > n$.

   Returning to the proof of our statements, we claim that s and S are inverses of each other. This follows from the fact that the columns of S are in fact the elements of B written in the basis B', and the columns of s are the elements of B' written in the basis B.

   So S is the *change of basis matrix* from B' to B, and s is the change of basis matrix from B to B'. That is, if p is any polynomial in V, and $\mathbf{p_C}$ denotes p written in basis C, then we have

   $$\mathbf{p_{B'}} = s\mathbf{p_B},$$

   and

   $$\mathbf{p_B} = S\mathbf{p_{B'}}.$$

   This proves that S and s are indeed inverses of each other.
10. The way the n passengers sit down in their assigned seats determines a permutation of $[n]$. A passenger will have to move because of the arrival of the last passenger if and only if she is in the same cycle as the last passenger. Therefore, our task is to determine the probability of a randomly selected cycle of a random permutation being exactly k long.

Since all entries of $[n]$ have the same role in this problem, we might as well answer this question for the cycle containing the maximum entry n. Using the transition lemma, we see that the length of the cycle containing n is k if and only if n is the $(n+1-k)$th entry of $f(p)$. Indeed, n always starts the last cycle of $f(p)$. The probability of n being in any given position is, of course, $1/n$, so that is the probability that the cycle containing n is of length k.

11. Let a_n be the average we are looking for. We claim that $a_n = 1$ for all n, and we will prove this by induction on n.

    For $n = 1$, the statement is obvious. Assume now that $a_{n-1} = 1$. It is clear that there are $(n-1)!$ permutations of length n in which 1 is a fixed point. By induction, the remaining part of the permutation, which is a permutation on the set $\{2, 3, \cdots, n\}$, has one fixed point on average. So permutations in which 1 is a fixed point have two fixed points on average.

    On the other hand, if $k > 1$, then we know from the previous exercise that there are again $(n-1)!$ permutations of length n in which 1 is in a cycle of length k. Elements of this cycle will not be fixed points. Therefore, all fixed points of the permutation will come from the rest of the permutation. The rest of the permutation is a permutation on $n-k$ elements. Therefore, by induction, it has on average one fixed point, *as long as it is not the empty permutation*, that is, as long as $k < n$. If $k = n$, then the remaining part of the permutation has no fixed points.

    Therefore, adding for the three cases—that is, when the length of the cycle containing the entry 1 is one, when it is k, where $1 < k < n$, and when it is n—we get that

$$a_n = \frac{(n-1)! \cdot (1+1) + (n-2)(n-1)! \cdot 1 + (n-1)! \cdot 0}{n!} = \frac{n!}{n!} = 1.$$

12. For a permutation p to satisfy $p^4 = id$, the length of each cycle in p has to be a divisor of 4, that is, 4, 2, or 1. As we are looking at 6-permutations, we have to consider partitions of 6 into parts that are 4, 2, or 1. These are

    - $4+2$, that is, a 4-cycle and a 2-cycle. By Theorem 4.27, there are $\frac{6!}{4 \cdot 2} = 90$ permutations of this type.
    - $4+1+1$. There are $\frac{6!}{4 \cdot 1 \cdot 1 \cdot 2} = 90$ permutations of this type.
    - $2+2+2$. There are $\frac{6!}{2^3 \cdot 3!} = 15$ permutations of this type.
    - $2+2+1+1$. There are $\frac{6!}{2^2 \cdot 2^2} = 45$ permutations of this type.
    - $2+1+1+1+1$. There are $\frac{6!}{2 \cdot 4!} = 15$ permutations of this type.

- $1+1+1+1+1+1$. There is obviously one permutation of this type.

Therefore, there are $90+90+15+45+15+1 = 256$ permutations of length six whose fourth power is the identity permutation.

We point out, for future reference, that the last four types of permutations are involutions. Therefore, there are 76 involutions of length six.

13. For a permutation p to satisfy $p^6 = id$, the length of each cycle in p has to be a divisor of 6, that is, 6, 3, 2, or 1. As we are looking at 6-permutations, we have to consider partitions of 6 into parts that are 6, 3, 2, or 1. These are

    - 6, that is, when p consists of one 6-cycle. Then by Theorem 4.27, there are $\frac{6!}{6} = 120$ permutations of this type.
    - $3+3$. There are $\frac{6!}{3^2 \cdot 2} = 40$ permutations of this type.
    - $3+2+1$. There are $\frac{6!}{3 \cdot 2} = 120$ permutations of this type.
    - $3+1+1+1$. There are $\frac{6!}{3 \cdot 3!} = 40$ permutations of this type.
    - Involutions. We know from the previous exercise that there are 76 of them.

    Therefore, the number of 6-permutations whose sixth power is the identity is $120+40+120+40+76 = 396$.

14. What happens to the cycles of p when we take p^3? That depends on the length of the cycle.

    (a) If the length of a cycle was relatively prime to 3, then it will not change. In particular, it will still be relatively prime to 3.
    (b) If, however, the length of the cycle was $3r$, then the cycle will split into three cycles of length r.

    Note that it is only in the second case that p^3 can have a cycle whose length is divisible by three, namely, when r is divisible by three. But then p^3 will immediately have three of those r-cycles. So the number of cycles of p^3 that are of length $3i$ is always a multiple of three. Therefore, it cannot be two.

15. Such permutations must contain a 4-cycle. Since they are even, the rest of their entries must be fixed points. So Theorem 4.27 shows that their number is $\frac{6!}{2! \cdot 4} = 90$.

16. In such permutations, each cycle length is one or three. Therefore, Theorem 4.34 implies that

$$G_C(x) = \exp\left(x + \frac{x^3}{3}\right).$$

17. In such permutations, each cycle length is at 1, 2, 3, or 6. However, either both the cycle lengths 2 and 3 have to occur, or the cycle length 6 has to occur. In other words, we have to exclude permutations whose third or second power is the identity permutation. The only permutation p so that $p^2 = p^3 = id$ is the identity permutation itself. Therefore, by the inclusion-exclusion principle, the generating function we are looking for is

$$\exp\left(x + \frac{x^2}{2} + \frac{x^3}{3} + \frac{x^6}{6}\right) - \exp\left(x + \frac{x^2}{2}\right)$$
$$- \exp\left(x + \frac{x^2}{2}\right) + \exp x.$$

18. (a) Let p be a permutation of length $n-1$, written in cyclic notation taken not on the set $[n]$, but on the set $\{2, 3, \cdots, n\}$. Now insert 1 into any of the following n positions of p: either to the front of p, forming a new cycle by itself or after any entry x of p, in the same cycle as x. In the first case, the number of cycles grows by one, otherwise it does not change. As the first case occurs in $1/n$ of all times, the formula is proved.
    (b) We claim that
$$H(n) = \sum_{i=1}^{n} \frac{1}{i}.$$
    This is trivial to prove for $n = 1$, and is routine by induction on n, and by the recursive formula of part (a) after that.

19. Let us write the right-hand side as $\sum_{k=0}^{n} \binom{n}{n-k} D(k)$. Then the right-hand side counts all n-permutations according to their fixed points. Indeed, if an n-permutation has exactly $n - k$ fixed points, then we have $\binom{n}{n-k}$ choices for the set of these fixed points. The permutation of the entries that are not fixed must be a derangement on k entries.

20. We show that the right-hand side also counts all derangements of length n. First, let p be any derangement of length $n - 1$. Then the entry n can be inserted into $n - 1$ different positions of p, namely, after each entry x in the same cycle as x. This way we get $(n-1)D(n-1)$ derangements, in each of which the entry n is in a cycle of length at least three. So we are still missing those derangements of length n in which the entry n is in a 2-cycle. In these, we have $n - 2$ choices for the entry y sharing a cycle with n, and then we have $D(n-2)$ choices for the rest of the permutation. This gives us the missing $(n-2)D(n-2)$ derangements. A little thought shows that we have counted all derangements of length n, so the proof is complete.

21. (a) We use Theorem 4.34, with C being the set of all positive integers divisible by k. Then we get

$$\begin{aligned} G_C(x) &= \exp\left(\sum_{k\geq 1} \frac{x^{kn}}{n}\right) \\ &= \exp\left(\frac{1}{k}\ln(1-x^k)^{-1}\right) \\ &= \left(\frac{1}{1-x^k}\right)^{1/k}. \end{aligned}$$

If the reader wants more details about this computation, she should read Example 4.39 and the paragraphs preceding it. That example is a special case of this exercise, namely, the one with $k=2$.

(b) Again, similarly to Example 4.39, we need to find the coefficient of $x^n/n!$ in $(1-x^k)^{-1/k}$. Obviously, $f_k(n)=0$ if n is not divisible by k. Otherwise, let $n=rk$. We then find that this number is

$$f_k(n) = 1^2 \cdot 2 \cdots (k-1)\cdot(k+1)^2 \cdots (2k-1)\cdot(2k+1)^2 \cdots (rk-1).$$

In other words, $f_k(n)$ is a product of n terms, just as $n!$, but the terms that are divisible by k are missing and are replaced by the integer that is $k-1$ smaller.

(c) This is similar to the proof given in the text for the special case of $k=2$. That is, assume $p=rk$. If p is enumerated by $f_k(n)$, then $p(1)$ cannot be 1, but it can be anything else, that is, we have $rk-1$ choices for $p(1)$. Then $p(p(1))$ can be anything but 1 and $p(1)$, so we have $rk-2$ choices for it. This trend continues until $p^k(1)$. (That is, if $i<k$, then $p^i(1)$ cannot be any of 1 and $p^j(1)$ for $j<i$, yielding $rk-i$ choices.) Then p^i can be 1 as well, giving us $rk-k+1$ choices. Continuing this argument as in the text, the proof follows.

22. We claim that $c(p,k)$ is divisible by p except for $k=1$ and $k=p$. It is obvious that these two cases are indeed exceptions. Otherwise, $c(p,k)$ can be obtained by adding the numbers of all permutations of all types that have k cycles. By Theorem 4.27, all these numbers are fractions in which the numerator is divisible by p and the denominator is not. Therefore, since p is prime, all these numbers are divisible by p, and so is their sum.

23. Recall that $(p-1)!=c(p,1)$. By the result of the previous exercise,

we have that

$$\begin{aligned} p! = \sum_{k=1}^{p} c(p,k) &\equiv 0 \ (mod\ p) \\ c(p,1) + c(p,p) &\equiv 0 \ (mod\ p) \\ (p-1)! + 1 &\equiv 0 \ (mod\ p), \end{aligned}$$

which was to be proved.

24. By Theorem 4.54, we know that $b(n,k)$ is the coefficient of x^3 in $I_{10}(x)$. In other words, $b(n,k)$ is the number of weak compositions of 3 into nine parts, so that the first part is at most 1, and the second part is at most 2. There are altogether $\binom{11}{3}$ weak compositions of 3 into nine parts. One of them violates the condition that the second part is at most 2, and nine of them violate the condition that the first part is at most 1. No weak composition of 3 into nine parts violates both conditions. Therefore,

$$b(10,3) = \binom{11}{3} - 1 - 9 = 165 - 1 - 9 = 155.$$

25. The left-hand side counts all n-permutations with k inversions. The second term of the right-hand side counts those permutations with this property whose last entry is n. Finally, the first term of the right-hand side counts those n-permutations with k inversions in which n is not the last entry. Indeed, in these permutations, n can be interchanged with the entry immediately following it, decreasing the number of inversions by 1, to $k-1$. This way, we will get an n-permutation with $k-1$ inversions in which n is not in the first position; but n could not be there anyway, since that would result in n inversions, which is not allowed.

26. Following the line of thinking of the solution of the previous exercise, we see that we need an additional term on the right-hand side, namely, the number of n-permutations with $k-1$ inversions in which n is in the first position. If n is in the first position, then it is part of $n-1$ inversions, so the rest of the permutation must contain $k-n$ inversions, which can happen in $b(n-1,k-n)$ ways. Therefore, we have

$$b(n,k) = b(n-1,k-n) + b(n,k-1) + b(n-1,k).$$

27. The answer to both questions is $n!/k$. To see this, one only has to consider the term $(1 + x + \cdots + x^{k-1})$ of $I_n(x)$.

28. We claim that $a_i = a_{i+1}$, and it is obvious that then all a_i have to be equal. We can clearly disregard permutations in which both i and $i+1$ are descents, as well as permutations in which neither one is a descent. So all we have to show is that there are as many permutations in $Inv(n,k)$ in which $p_i > p_{i+1} < p_{i+2}$ (set A) as there are permutations in $Inv(n,k)$ in which $p_i < p_{i+1} > p_{i+2}$ (set B). This is easy to do bijectively. Let $p \in A$. Leave the entries on the left of p_i and on the right of p_{i+2} unchanged. Then rearrange the entries p_i, p_{i+1}, p_{i+2} by first reversing them, and then taking complements within this three-element set. [In the terminology of Exercise 32, if the pattern of these three entries was 312, turn it into 231, and if it was 213, turn it into 132.] Because the number of total inversions of the permutation did not change, the new permutation is still in $Inv(n,k)$, and it is obviously in B. The map is bijective, since taking reverse complements is a bijective operation.

29. (a) For $\begin{bmatrix} \mathbf{4} \\ \mathbf{2} \end{bmatrix}$, we count the permutations 1122, 1212, 1221, 2112, 2121, and 2211. Therefore,

$$\begin{bmatrix} \mathbf{4} \\ \mathbf{2} \end{bmatrix} = 1 + q + 2q^2 + q^3 + q^4.$$

For $\begin{bmatrix} \mathbf{5} \\ \mathbf{1} \end{bmatrix}$, we must count the permutations 12222, 21222, 22122, 22212, and 22221. Therefore,

$$\begin{bmatrix} \mathbf{5} \\ \mathbf{1} \end{bmatrix} = 1 + q + q^2 + q^3 + q^4.$$

Finally, for $\begin{bmatrix} \mathbf{5} \\ \mathbf{3} \end{bmatrix}$, we must count the permutations 11122, 11212, 11221, 12112, 12121, 12211, 21112, 21121, 21211, and 22111. Therefore,

$$\begin{bmatrix} \mathbf{5} \\ \mathbf{3} \end{bmatrix} = 1 + q + 2q^2 + 2q^3 + 2q^4 + q^5 + 1.$$

(b) Let p be a permutation of the multiset K. Then reverse p, and turn all 1s into 2s, and vice versa. Then the new permutation p' is a permutation of the multiset consisting of k copies of 2 and $n-k$ copies of 1, so it is a permutation counted by $\begin{bmatrix} \mathbf{n} \\ \mathbf{n-k} \end{bmatrix}$. Furthermore, $i(p) = i'(p)$, proving our claim.

30. The term $\begin{bmatrix} \mathbf{n-1} \\ \mathbf{k} \end{bmatrix}$ enumerates permutations that end in a 2, and the term $q^{n-k}\begin{bmatrix} \mathbf{n-1} \\ \mathbf{k-1} \end{bmatrix}$ enumerates permutations that end in a 1.

31. It is easy to prove that $p(m,n,q)$ satisfies the same recursive relation as $\begin{bmatrix} \mathbf{m+n} \\ \mathbf{m} \end{bmatrix}$. Indeed, if the Ferrers shape of a partition fits into an $m \times n$ rectangle, then there are two possibilities: Either the partition has

at most $m-1$ parts, and then its Ferrers shape fits even into an $(m-1)\times n$ rectangle, or the partition has m parts, and then, after removing the first column of its Ferrers shape, its remaining shape fits into an $m\times(n-1)$ rectangle. These two cases correspond to the two summands in the recursive relation satisfied by the Gaussian coefficients.

32. We prove the statement by induction on n, the initial case of $n=1$ being trivial. We will say that there is one permutation of length 0 that avoids 132.

    Now assume that we know the statement for all nonnegative integers less than n. Suppose we have a 132-avoiding n-permutation in which the entry n is in the ith position. Then it is clear that any entry to the left of n must be larger than any entry to the right of n, otherwise the two entries violating this condition and the entry n would form a 132-pattern. Moreover, by our induction hypothesis, there are C_{i-1} possibilities for the substring of entries to the left of n, and C_{n-i} possibilities for that to the right of n. Summing over all allowed i, we get the following recursion:

$$C_n=\sum_{i=0}^{n-1}C_{i-1}C_{n-i}, \tag{4.21}$$

    and we know from (3.25) that this is the recursion of the Catalan numbers.

33. If the n-permutation p contains the pattern q, then the reverse of p will contain the reverse of q. This implies $S_n(132)=S_n(231)$. Similarly, if p contains the pattern q, then the complement of p will contain the complement of q. Therefore, $S_n(132)=S_n(213)$. Finally, $S_n(213)=S_n(312)$ by taking reverses. So all four patterns 132, 213, 231, 312 are avoided by C_n permutations of length n.

    This is not even the end of the story, since we also have $S_n(123)=S_n(321)=C_n$. The latter is somewhat harder to prove (Supplementary Exercise 39).

34. Let p_i, p_j, and p_k form a 312-pattern in p. If there are several 312-patterns in p, choose the one for which $j-i$ is minimal. In that case, we claim that $j-i=1$ must hold, meaning that p_i and p_j are consecutive entries of p. Indeed, assume p_t is located between p_i and p_j. Then p_t cannot be larger than p_k, since then $p_tp_jp_k$ would form a 312-pattern, contradicting the minimality of $j-i$. Similarly, p_t cannot be smaller than p_k, since then again $p_ip_tp_k$ would form a 312-pattern. So p_t simply cannot exist, meaning that p_i and p_j are consecutive.

35. We prove the statement by induction on n, the case of $n=1$ being

obvious. Now assume that the statement is true for $n-1$, and prove it for n. Say $n-1$ tourists are already sitting at tables, and wine has been served to those tables. Now the last tourist comes in. He can either sit at a new table, and get red wine or white wine, or sit at one of the tables with tourists, choosing his left neighbor in one of $n-1$ ways. Since the first scenario yields $2n!$ possible outcomes, and the second scenario yields $(n-1)n!$ possible outcomes, then altogether we have $(n+1)n! = (n+1)!$ possible outcomes, as claimed.

Note that we have essentially used the fact that the last tourist can take $n+1$ different courses of action.

36. (a) Since all permutations of the vertices define a symmetry, the regular tetrahedron has 24 symmetries.

    (b) There are 12 symmetries that can be obtained by a series of rotations. To see this, take an axis that is perpendicular to one of the faces and contains the fourth vertex. There are two rotations around that axis, by 120 and 240 degrees. There are four axes like this, leading to $4 \cdot 2 = 8$ symmetries. Following a rotation by a rotation around another axis gives us a symmetry that interchanges the elements of two pairs of vertices. There are three possible pairs of pairs, so there are three such symmetries. Finally, the identity is certainly a rotation. So there are $8+3+1 = 12$ symmetries that can be obtained by a series of rotations.

    (c) Part (b) shows that the symmetries that can be obtained by a series of rotations have cycle types $(4,0,0,0)$, $(1,0,1,0)$, or $(0,2,0,0)$, so they are even permutations. Since there are 12 of them, they are *all* even permutations of S_4. So a symmetry of a regular tetrahedron can be obtained by rotations if and only if it is an even permutation.

37. We claim that

$$H(x) = \prod_{i \geq 1} \left(\frac{1}{1-x^i} \right)^{1/i}.$$

Indeed, to get an object that we are enumerating, partition $[n]$ into blocks, index them by positive integers i, and then on block i, take a permutation whose cycles are *labeled* i, and have length divisible by i. The generating function for the number of ways to do this on block i is then $(1-x^i)^{-1/i}$, (see Exercise) for details, and the proof is immediate by the Product formula. This result and its generalizations appeared in [1].

4.10 Supplementary exercises

1. Prove that $A(n,k) \geq k^{n-k}$.
2. Compute $\sum_{p \in S_n} i(p)$.
3. Let $d(p)$ be the number of descents of the permutation p. Compute $\sum_{p \in S_n} d(p)$.
4. How many 10-permutations are there whose set of descents is *contained* in $\{2, 5, 8\}$?
5. How many 10-permutations are there whose set of descents is *equal to* $\{2, 5, 8\}$?
6. Let n be any positive integer. Prove that there are at least $(n-1)/2$ permutations of length n that have the same number of descents, but all distinct numbers of inversions. Give two different proofs, namely,

   (a) one without actually presenting these permutations, and

   (b) one explicitly presenting these permutations.
7. There are n participants at a long-jump competition, each of whom gets only one try. After the jump of each participant, the officials write down the current rank of that participant. (Assume there are no ties.) This will lead to a sequence of n numbers (one for each participant). Let us call this sequence S.

   (a) Prove that we can determine the final ranking of the contest if we are given S.

   (b) It is clear that the contest can have $n!$ different final rankings, but in how many of them will S contain exactly k distinct numbers?

   (c) It is clear that the sum of the n numbers in S will be at least n and at most $\binom{n+1}{2}$, but for how many lists S will it be exactly m?
8. Prove that, for any positive integers n, there exists at least one k so that $A(n,k) \geq (n-1)!$.
9. Prove that multiplication of permutations is an associative operation. That is, prove that, for any positive integers n and any three n-permutations p, q, and r, we have $(pq)r = p(qr)$.
10. Find a formula for $b(n,1)$ and $b(n,2)$.
11. How many 10-permutations have exactly four inversions?
12. Let $n \geq 2$ be a fixed positive integer, and let x be any real or complex number for which there exists a positive integer m, with $1 < m \leq n$ so that $x^m = 1$. What is the value of $\sum_{k=0}^{\binom{n}{2}} b(n,k)x^k$?

13. Let $n \geq 3$. How many n-permutations contain the entries 1, 2, and 3 in three distinct cycles?

14. Let $n \geq 3$. How many n-permutations contain the entries 1, 2, and 3 in the same cycle?

15. Find a simple formula for $c(n+1,2)$.

16. Find a formula for the total number of cycles in all n-permutations. Try to find a solution that uses exponential generating functions. Compare this result with the result of the previous exercise, and find a direct explanation of what you found.

17. Xavier has computed the square of a 30-permutation and got a permutation with exactly five 2-cycles. Prove that he made a mistake.

18. How many permutations of order two are there in the alternating group A_5? See Exercise 8 for the definition of the order of a permutation.

19. How many permutations of order six are there in the symmetric group S_6? How many of them are in A_6?

20. Is it true that, if a permutation is of odd order, then it is an even permutation?

21. Let $a(n)$ be the number of fixed point–free involutions of length n. Find the exponential generating function of the numbers $a(n)$. Then find a formula for the numbers $a(n)$.

22. Prove, by a generating function argument, that

$$(n-1)(D(n-1)+D(n-2)) = D(n).$$

23. Prove that, for all $n \geq 1$, we have $D(n+1) = (n+1)D(n)+(-1)^{n+1}$.

24. Find a proof for the result of Example 4.36 without the use of generating functions.

25. Let $E(n)$ be the number of n-permutations with exactly one fixed point. Prove that, for all $n \geq 1$, we have $|E(n) - D(n)| = 1$.

26. Let n be even. Which number is larger, the number $\mathbf{even}(n)$ of n-permutations in which each cycle length is even, or the number $\mathbf{odd}(n)$ of n-permutations in which each cycle length is odd? Answer this question in two different ways, namely,

    (a) by a computational argument, proving a formula for $ODD(n)$, and

    (b) by a combinatorial argument.

27. Find a closed formula for $\sum_{k=1}^{n} c(n,k)2^k$.

28. Let $I(n)$ be the number of involutions of length n, and let $D_2(n)$ be the number of n-permutations in which each cycle is longer than two. Prove that

$$\sum_{k=0}^{n} I(k)D_2(n-k) = n!.$$

29. For what positive integers n is it true that the number $I(n)$ of all involutions of length n is even? Find a combinatorial argument.

30. Prove that p and p^{-1} have the same type.

31. Let p and q be two n-permutations. We say that p and q are *conjugate permutations* if there exists a permutation $g \in S_n$ so that $g^{-1}pg = q$.

    (a) Prove that "being conjugates" is an equivalence relation, without using the result of part (b).

    (b) Prove that p and q are conjugate permutations if and only if they are of the same type.

32. The previous exercise shows that "being conjugates" is an equivalence relation on S_n, so this relation creates equivalence classes. Let us call these classes *conjugacy classes*. How many conjugacy classes does S_n have?

33. Let C be a conjugacy class in S_n. Prove that the number of permutations in C is a divisor of $n!$.

34. Prove that every n-permutation can be obtained as a product of transpositions.

35. A transposition is called *adjacent* if it interchanges to consecutive entries, that is, if it is of the form $(i\ i+1)$. Prove that all n-permutations can be obtained as products of adjacent transpositions.

36. Prove that $\begin{bmatrix} \mathbf{n} \\ \mathbf{k} \end{bmatrix}$ is always a symmetric polynomial (that is, show that the ith coefficient of this polynomial from the left is the same as its ith coefficient from the right, for any i).

37. Let us enumerate all k-element subsets of $[n]$ according to their sums as follows. If T is a k-element subset of $[n]$, then let $s(T)$ be the sum of all elements of T. Define

$$s_{n,k}(x) = \sum_{T} q^{s(T)},$$

where T ranges over all k-element subsets of $[n]$. Prove the identity $s_{n,k}(x) = \begin{bmatrix} \mathbf{n} \\ \mathbf{k} \end{bmatrix}$.

38. (Basic knowledge of finite fields required.) Let F be a finite field of q elements. Let V be an n-dimensional vector space over q. Prove

that the number of k-dimensional subspaces of V is $\left[\begin{smallmatrix}\mathbf{n}\\ \mathbf{k}\end{smallmatrix}\right]$. Try to give a solution in which the formula for the number of these subspaces is *deduced*, not *verified*.

39. Prove that, for all positive integers n, the identity

$$S_n(123) = S_n(321) = C_n$$

holds.

40. Prove that, for all patterns q, the identity $S_n(q) = S_n(q^{-1})$ holds.

41. For two patterns p and q, let $S_n(p, q)$ denote the number of n-permutations avoiding both patterns p and q.

Find $S_n(p, q)$ for each pair of patterns (p, q) where both p and q are of length three.

42. Find a formula for the number of n-permutations p that do not contain a 321-pattern in which the first two entries are consecutive entries of p.

43. Let $k > n/2$. Find the number of permutations of length n whose longest cycle is of length k.

44. Use the exponential formula to prove that, if $n \geq 2$, then there are as many permutations of length n with an even number of even cycles as permutations of length n with an odd number of even cycles.

45. Find a closed formula for the bivariate generating function

$$\sum_{n\geq 0}\sum_{k\geq 0} s(n,k)t^k \frac{x^n}{n!}.$$

46. Deduce Theorem 4.21 from the more general result of formula (4.17).

47. Substitute $t = -1$ in formula (4.17) and explain the combinatorial meaning of the obtained formula.

5

Counting graphs

A rural area of a remote county has n farms, no two of which are connected by roads. The owners want to build some roads between some of the farms, and they are interested in all solutions. What are their choices?

Figure 5.1 shows some plans the farmers may consider, for $n = 6$.

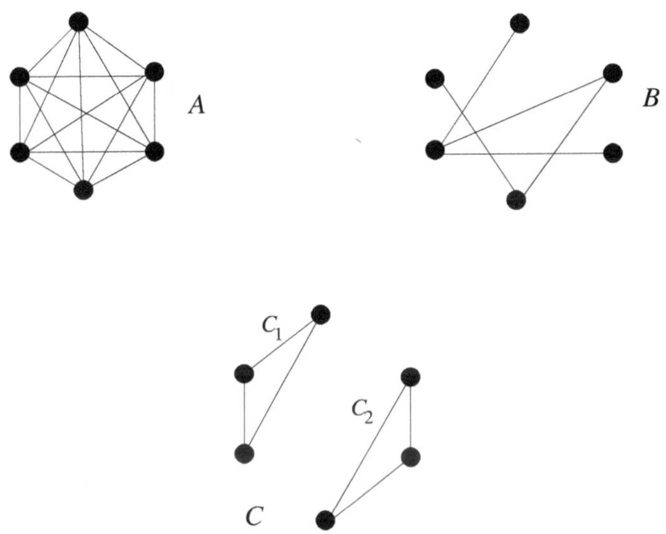

Figure 5.1
Possible road systems among six farms.

Of the three plans shown in Figure 5.1, plan A provides a very comfortable system of roads (each farm is connected to each farm by a direct road), but this system may be prohibitively expensive. Plan C is much cheaper, but it has the big drawback that it does not connect each farm to each farm, not even by indirect routes. In particular, there is no way to go from any of the three farms in C_1 to any of the three farms in C_2. Finally, plan B may be the golden mean. It connects every farm to every farm by some route, and contains no redundancy. Indeed, the reader is invited to verify that omitting any of the roads from plan B would destroy the property that every farm can be reached from every farm.

DOI: 10.1201/9781003304272-5

The model we used to represent the farms and the roads between them in the above discussion, that is, dots and lines connecting them, is called a *graph*, and it is probably the most widely used model in combinatorics. The dots are called the *vertices* of the graph, while the lines are called the *edges*. If all the edges corresponded to one-way roads, we would talk about a *directed* graph, and if only some of the edges corresponded to one-way roads, then we would talk about a *partially directed* graph. Unless otherwise specified, we will assume that our graphs are *undirected* (see Figure 5.2).

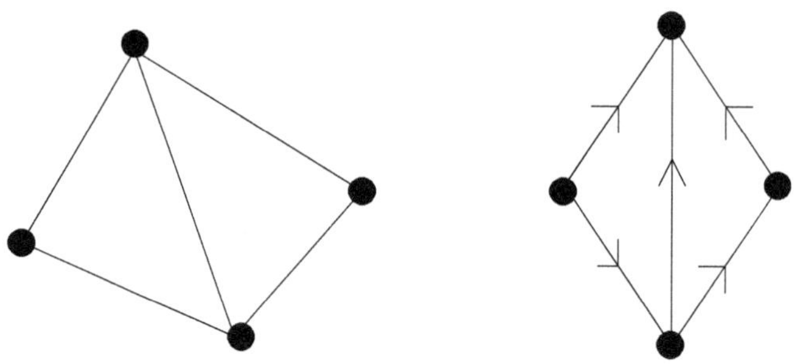

Figure 5.2
An undirected and a directed graph.

If an edge connects a vertex to itself, then that edge is called a *loop*, and if there are several edges between the same pair of vertices, then those edges are called *multiple edges*. See Figure 5.3 for examples. If a graph has no loops, and no multiple edges, then it is called a *simple graph*. For our enumeration purposes, loops and multiple edges would mean more pain than gain; therefore, *all graphs in this chapter will be simple graphs unless otherwise stated.*

Figure 5.3
A vertex with a loop, and a pair of vertices with multiple edges.

If every pair of vertices is connected by an edge, as in graph A in Figure 5.1, then the graph is called *complete*. A series $e_1e_2\cdots e_k$ of edges that lead from a vertex to another one is called a *walk*, and if a walk does not go through any vertex twice, then it is called a *path.* In a walk or path, the endpoint of e_i must be the same as the starting point of e_{i+1} (see Figure 5.4).

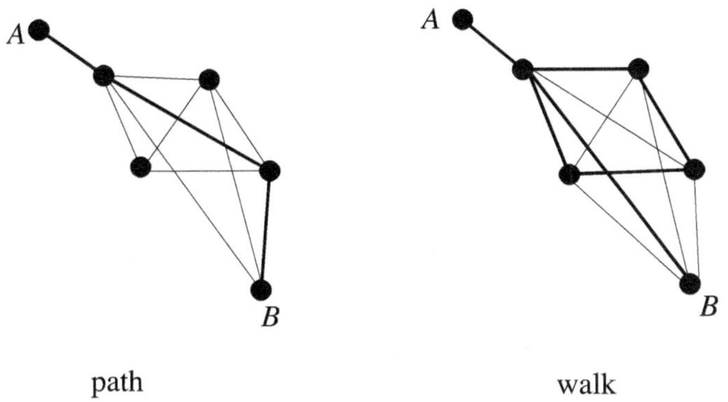

Figure 5.4
A path and a walk between A and B.

If there is a walk between every pair of vertices in a graph, then that graph is called *connected.* The reader is urged to prove that we can, in that case, also find a *path* from x to y for each choice of x and y. (A solution is provided in Exercise 2.) So in Figure 5.1, graphs A and B are connected, but C is not. If a graph is not connected, then its maximal connected subgraphs are called its *connected components.* So the connected components of C in Figure 5.1 are C' and C''.

In a graph, the number of edges adjacent to the vertex V is called the *degree* of V. The following proposition presents a simple but useful result on vertex degrees.

Proposition 5.1 *Let $d_1, d_2, \cdots, d_n$ be the degrees of the vertices of a graph G on n vertices that has e edges. Then we have*

$$d_1 + d_2 + \cdots + d_n = 2e.$$

Example 5.2 *In the graph shown in Figure 5.5, the degrees of the vertices are 1, 2, 2, 3, and 4. The sum of these degrees is 12, and, indeed, the graph has six edges.*

Solution: We claim that both sides count all *endpoints* of all edges of G. The left-hand side counts them by the vertices (a vertex of degree d_i is the endpoint of d_i edges), and the right-hand side counts them by the edges (each edge has two endpoints). ◊

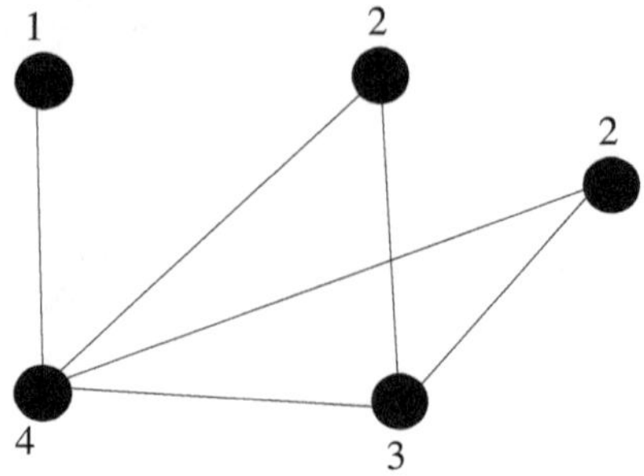

Figure 5.5
Degrees in a graph.

5.1 Trees and forests

In this section, we will count trees and other graphs whose vertex set is the set $[n]$.

5.1.1 Trees

While being connected is certainly a very desirable property when building any sort of network, it has its own costs. Therefore, it is a natural direction of research to study graphs that are connected but have no "redundancy." To make that notion more precise, we say that a graph G is *minimally connected* if G is connected, but if we remove any edge of G, then G will no longer be connected. As we have mentioned, graph B in Figure 5.1 is minimally connected.

Minimally connected graphs play a central role in graph theory. In order to understand their importance, we prove the following lemma, which provides three equivalent definitions for the same notion. Note how simple and natural all three of them are. We define a *cycle* as a walk whose starting point is the same as its endpoint, but which otherwise has no repeated vertices. For instance, C_1 and C_2 in Figure 5.1 are both cycles.

Lemma 5.3 *Let G be a connected simple graph on n vertices. Then the following are equivalent.*

(i) The graph G is minimally connected.

(ii) There are no cycles in G.

(iii) The graph G has exactly $n-1$ edges.

Proof:

(i)$\Rightarrow$ (ii) Assume there is a cycle C in G. Then G cannot be minimally connected since any one edge e of C can be omitted, and the obtained graph G' is still connected. Indeed, if a path uv used the edge e, then there would be a walk from u to v in which the edge e is replaced by the set edges of C that are different from e.

(ii)$\Rightarrow$ (iii) Pick any vertex x in G and start walking in some direction, never revisiting a vertex. As there is no cycle in G, eventually we will get stuck, meaning that we will hit a vertex of degree 1. This means that a connected simple graph with no cycles contains a vertex of degree 1. Removing such a vertex (and the only edge adjacent to it) from G, we get a graph $G*$ with one less vertex and one less edge, and the statement is proved by induction on n.

(iii) $\Rightarrow$ (i) This statement says that, if a simple graph on n vertices is connected and has $n-1$ edges, then it is minimally connected, that is, a graph on n vertices and $n-2$ edges cannot be connected. Let us assume that this is not true, and let H be a counterexample with a minimum number of vertices. The reader is invited to check that H must have more than three vertices for this. As H has $n-2$ edges, there has to be a vertex y of degree 1 in H, otherwise H would need to have at least n edges (since the sum of the degrees of its vertices would be at least $2n$). Removing y from H, we get an even smaller counterexample for our statement, which is a contradiction.

$\diamond$

Graphs that satisfy the equivalent conditions mentioned in Lemma 5.3 are of central importance in graph theory. Therefore, they have their own name.

Definition 5.4 *A simple graph that is minimally connected is called a* tree.

Equivalently, a simple graph that is connected and cycle-free is called a tree. Equivalently, a simple graph of n vertices that is connected and has $n-1$ edges is called a tree.

A graph whose connected components are all trees is called, not surprisingly, a *forest.* Trees can look quite different from each other. See Figure 5.6 for some examples.

In the proof of Lemma 5.3, we spoke about vertices of degree 1 in trees. Such vertices will be called *leaves.* So our argument in the second part of that proof showed that each tree has at least one leaf. This statement can be significantly refined. See Supplementary Exercises 4 and 5 for some questions in that direction.

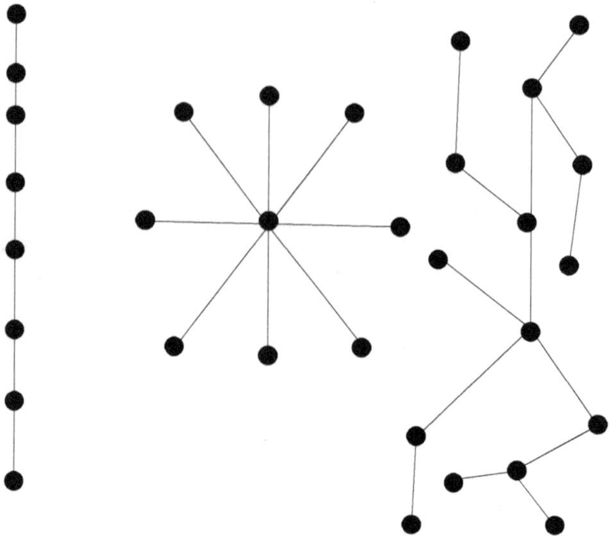

Figure 5.6
Three trees.

5.1.2 The notion of graph isomorphisms

In the rest of this chapter, we would like to count all kinds of graphs. First of all, however, we will have to agree on the definition of two graphs being different. There are many different ways to do this, since we can choose to be more demanding or less demanding when we declare what conditions two graphs have to satisfy if they are to be called identical.

For the purposes of this section, and most of this chapter, the way graphs are *drawn* does not matter. That is, we consider two graphs of labeled vertices *identical* (or *identical as labeled graphs*) if they have the same set of edges, that is, when the number of edges between i and j is the same in both graphs for all pairs of vertices i and j. See Figure 5.7 for an example.

On the other hand, the two graphs shown in Figure 5.8 are different as labeled graphs, since G has an edge between vertices 3 and 4, and H does not.

Nevertheless, the two graphs in Figure 5.8 look quite similar. In order to make this notion more precise, we define *isomorphism* of graphs.

Definition 5.5 *Let G and H be two graphs. We say that G and H are* isomorphic *if there exists a bijection from the vertex set $V(G)$ of G onto the vertex set $V(H)$ of H so that the number of edges between a and b in G is equal to the number of edges between $f(a)$ and $f(b)$ in H for any $a, b \in V(G)$. In the case of simple graphs, that means that there is an edge between $f(a)$ and $f(b)$ if and only if there is an edge between a and b.*

In that case, we call f an isomorphism *from G into H.*

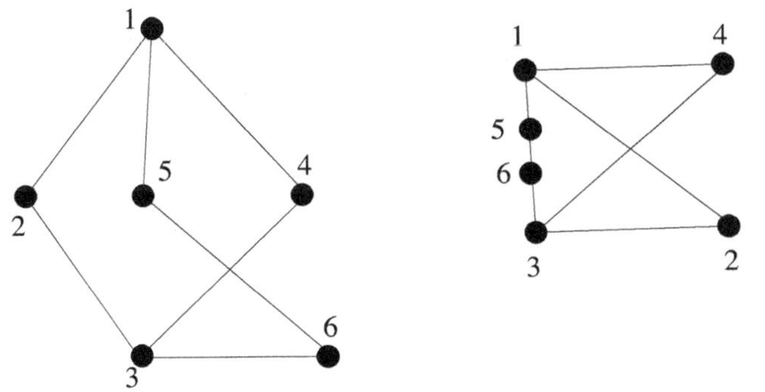

Figure 5.7
These two graphs are identical as labeled graphs, even if they look different.

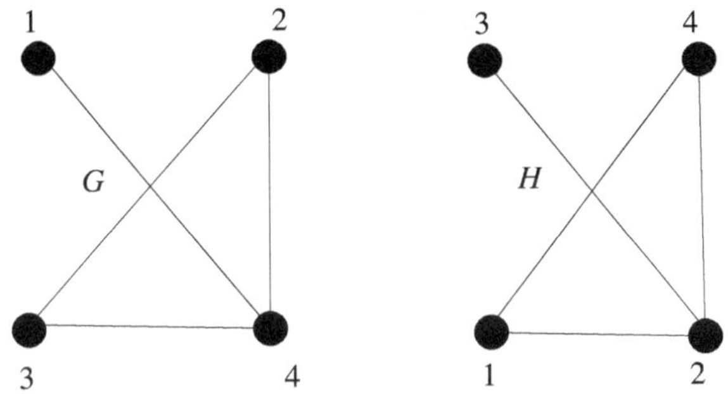

Figure 5.8
These graphs are different, but look similar.

The graphs G and H shown in Figure 5.8 are isomorphic. An isomorphism from G into H is given by $f(1) = 3$, $f(2) = 1$, $f(3) = 4$, and $f(4) = 2$. If two graphs are isomorphic, they are often called *identical as unlabeled graphs*, since, if we omit their labels, we get two graphs that are not distinguishable.

The reader is asked to verify that graph isomorphism is an equivalence relation, creating equivalence classes, or *isomorphism classes*, of graphs. We can then ask how many isomorphism classes of graphs (or trees, forests, etc.) exist if the graphs are to have n vertices. Another way of asking the same question would be to ask how many different *unlabeled* graphs exist on n vertices. One could also ask for the size of the largest, or smallest, isomorphism class.

In order to be able to determine the sizes of isomorphism classes of graphs, we introduce some new notions. An *automorphism* of a graph G is simply a graph isomorphism $f : G \to G$. The set of all automorphisms of G is denoted by $Aut(G)$. Readers who are familiar with group theory will notice that $Aut(G)$ is actually a *group*. We will define groups in Chapter 8. It is not necessary to know the definition of a group for the present discussion.

At this point, the reader should verify that graph G in Figure 5.8 has two automorphisms. Indeed, an automorphism takes an edge to an edge; therefore, if $f \in Aut(G)$, then the degree of v is equal to that of $f(v)$. Therefore, any $f \in Aut(G)$ must satisfy $f(1) = 3$ and $f(4) = 2$. However, both $f(2) = 4$ and $f(3) = 1$, and $f(2) = 1$ and $f(3) = 4$ are possible ways to complete the automorphism.

We would like to point out that, while we defined automorphism for graphs on *labeled* vertices, the number of automorphisms of G will not change if we permute the labels of the vertices. Therefore, we may talk about the number of automorphisms of a graph on *unlabeled* vertices. When we do that, we just mean "take any labeling, then use the definition of $Aut(G)$ for labeled graphs."

Now let J be a graph on n *unlabeled* vertices. Then we define $\ell(J)$ as the number of possible ways to bijectively label J by the elements of $[n]$ so that all obtained graphs are different. (Again, two graphs are different if their edge sets are different.)

Let us call the graph on n vertices that has no edges the *empty graph*, and let us recall that the simple graph on n vertices in which each pair of vertices is connected by an edge is called the *complete graph* on n vertices. Let us denote the latter by K_n.

If J is the empty graph, or if J is the complete graph on any number of vertices, then $\ell(J) = 1$. For the graph G in Figure 5.8 (after removing the labels) we easily see that $\ell(G) = 12$, since we can choose the label of the vertex with degree 3 in four ways, then we can choose the label of the vertex with degree 1 in three ways, and we have no more choices to make since interchanging the labels of the remaining two vertices does not change the set of edges of G.

Example 5.6 *If J is the graph shown in Figure 5.9, then $\ell(J) = 6!/2$ since first we can choose the label of the only vertex that has degree 4 in six ways, then we can choose the label of the only vertex that has degree 3 in five ways, then we can choose the label of the only vertex that has degree 2 in four ways. Finally, we choose the label of the degree-1 vertex connected to the degree-3 vertex in three ways, after which we have no more choices to make. Indeed, switching the labels of the remaining two vertices does not create a new graph, since it does not change the set of edges.*

The last few examples suggest that the more automorphisms a graph on vertex set $[n]$ has, the fewer labelings it has. Indeed, if G is the complete graph or the empty graph, then $Aut(G)$ has $n!$ elements, whereas $\ell(G) = 1$. On the other hand, we have seen graphs (in Example 5.6 and in Figure 5.8) that have only two automorphisms, but $n!/2$ labelings. This trend is true in general.

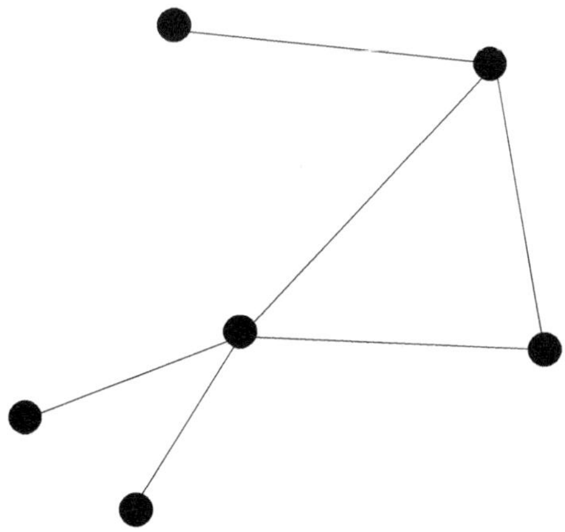

Figure 5.9
This graph has 360 different labelings.

Theorem 5.7 *For any graph H on vertex set $[n]$,*

$$|Aut(H)| \cdot \ell(H) = n!.$$

Proof: Consider the set S of all $n!$ possible ways to assign the elements of $[n]$ bijectively to the vertices of H. Let us say that two elements of S are equivalent if they are identical as labelings of H.

This equivalence relation creates equivalence classes on S. The *number* of these classes is $\ell(H)$, since two assignments are in different classes if and only if they are different as labelings.

On the other hand, the *size* of each equivalence class is $|Aut(H)|$. Indeed, within the same class, all graphs have the same set of edges, meaning that there is a natural isomorphism between any two elements of any given class. For any given labeled graph H, there are $|Aut(H)|$ ways in which that is possible, proving our claim. $\diamondsuit$

Example 5.8 *Let H be the graph shown at the top of Figure 5.10. Then $\ell(H) = 3$, $|Aut(H)| = 2$, and the equivalence classes are also shown in Figure 5.10.*

It is often easier to count automorphisms than labelings, in which case the previous theorem helps us to determine the latter using the former. Below is a typical application.

Example 5.9 *Let H be the graph of the cube, that is, the graph whose vertices are the vertices of a cube and whose edges are the edges of the cube. Then H has 840 labelings.*

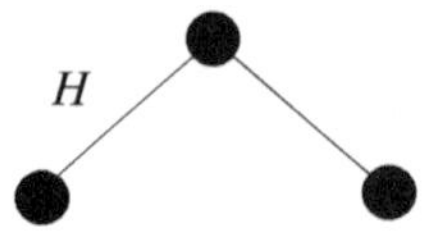

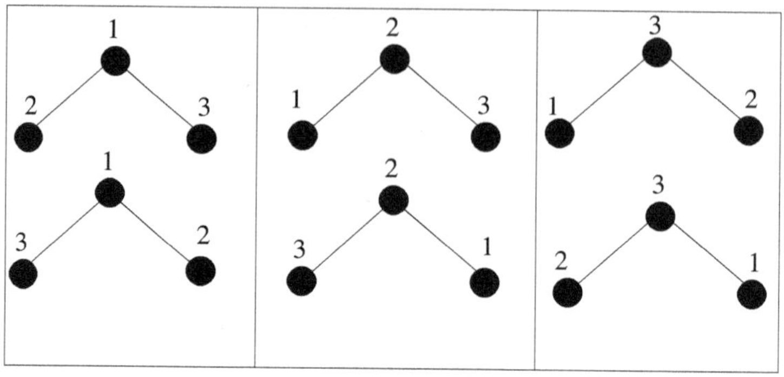

Figure 5.10
Three equivalence classes of labelings, each of size $|Aut(H)|$.

Solution: Because the cube has eight vertices, it suffices to show that H has 48 automorphisms. Our claim will then follow from Theorem 5.7, since $8!/48 = 840$.

Let $f \in Aut(H)$, and let A be a vertex of the cube. Then we have eight choices for $f(A)$. Once $f(A)$ is chosen, the neighbors of A in G must be mapped into the neighbors of $f(A)$ by f. Since A has three neighbors, this is possible in $3! = 6$ ways. Once the images of the neighbors are chosen, there are no more choices to make (why?). Therefore, we have $|Aut(H)| = 8 \cdot 6 = 48$. $\diamondsuit$

Enumerating unlabeled graphs with no additional structure is usually a difficult task. We will only confront these kinds of problems at the level of exercises. However, in the upcoming subsections, we will discuss interesting enumeration problems in which the vertices are unlabeled, but where there is some additional structure on our graphs.

5.1.3 Counting trees on labeled vertices

Since the end of the nineteenth century, combinatorialists have been fascinated by the task of enumerating trees having vertex set $[n]$. Taking a look at the numbers T_n of trees on $[n]$ for small values of n (see Figure 5.11 and Supplementary Exercise 7), we find that there is one tree on $[1]$, there is one tree on $[2]$, there are three trees on $[3]$, and there are 16 trees on $[4]$. If we look further, we find that $T_5 = 125$. This suggests the amazingly simple formula

$T_n = n^{n-2}$ for these numbers. Surprisingly, this actually turns out to be the *correct* formula.

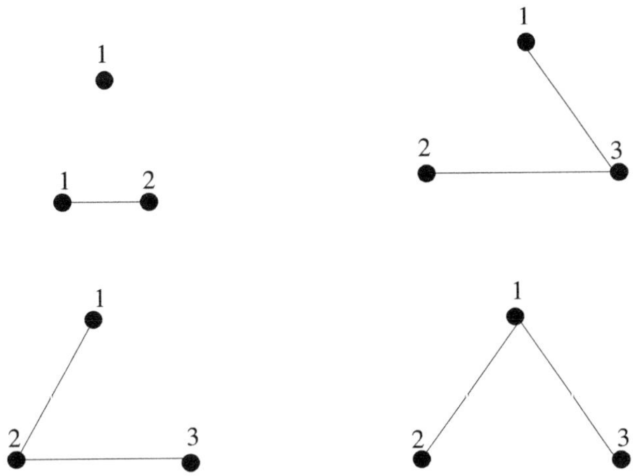

Figure 5.11
Trees on $[n]$, for $n \leq 3$.

Theorem 5.10 (Cayley's formula) *For all positive integers n, the number of all trees on vertex set $[n]$ is n^{n-2}.*

The first proof of this result was published in [23], and mathematicians have never since stopped admiring the surprising elegance of this formula. Indeed, who would have thought that such an extremely simple function would turn out to count the elements of a set as diverse as that of all trees on $[n]$? The great interest in this result gave rise to more than a dozen proofs for it, the most elegant of which may well be the following argument due to André Joyal [47].

Proof: We need to prove that $T_n = n^{n-2}$, which is the number of all *functions* from $[n-2]$ to $[n]$. This is certainly equivalent to proving the identity

$$n^2 T_n = n^n. \tag{5.1}$$

Here the right-hand side is the number of all functions from $[n]$ into $[n]$. The left-hand side, on the other hand, is equal to the number of all trees on $[n]$ in which we select two special vertices, called Start and End (which may be identical). Let us call these trees *doubly rooted trees*. See Figure 5.12 for an example.

Consequently, formula (5.1), and therefore Theorem 5.10, will be proved if we present a bijection G from the set A of all functions from $[n]$ into $[n]$ onto

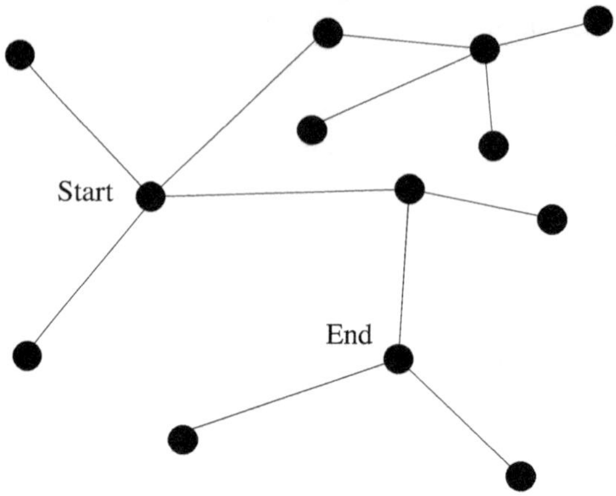

Figure 5.12
A doubly rooted tree.

the set B of all doubly rooted trees on $[n]$. In what follows, we will construct such a bijection.

Let $f \in A$, and draw the short diagram of f, that is, represent the elements of $[n]$ by vertices and draw an arrow from x to y if $f(x) = y$. This creates two kinds of vertices, namely, those that are in a directed cycle and those that are not. Let C and N, respectively, denote these two subsets of $[n]$.

Now we start creating the doubly rooted tree $G(f)$. First, note that f acts as a *permutation* on C. Write down the elements of C in a line according to that permutation as follows: If $C = \{c_1, c_2, \cdots, c_k\}$ so that $c_1 < c_2 < \cdots < c_k$, then write the elements of C in the order $f(c_1), f(c_2), \cdots, f(c_k)$. In other words, write the permutation defined by f on C in one-line notation. Call the vertex $f(c_1)$ Start, and the vertex $f(c_k)$ End, and create a path with vertices $f(c_1), f(c_2), \cdots, f(c_k)$. Note that so far we have defined a graph with k vertices and $k-1$ edges.

If $x \in N$, then simply connect x to $f(x)$, just as in the short diagram of f, except that no arrow is needed here since we know the elements of N are mapped *toward* the elements of C. This will define $n-k$ more edges. Therefore, we now have a graph on $[n]$ that has $n-1$ edges and has two vertices (called Start and End, respectively). This is the graph that we want to call $G(f)$. In order to justify that name, we must prove that $G(f)$ is indeed a doubly rooted tree. It remains to prove that $G(f)$ is connected. This is true, however, since, in the short diagram of f, each directed path starting at any $x \in N$ must reach a vertex of C at some point (there is no other way it could end). So indeed, $G(f) \in B$ for all $f \in A$.

Example 5.11 *Let $n = 9$, and let $f : [9] \to [9]$ be defined by $f(i) = 10 - i$ if i is odd; $f(2) = f(4) = 5$; $f(6) = 8$; and $f(8) = 9$. Then $C = \{1, 3, 5, 7, 9\}$, and $G(f)$ is the doubly rooted tree shown in Figure 5.13.*

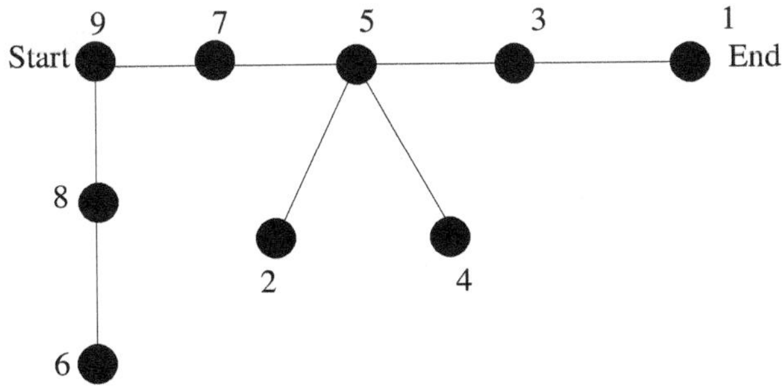

Figure 5.13
The doubly rooted tree $G(f)$.

In order to show that $G : A \to B$ is a bijection, we prove that it has an inverse. Let $t \in B$. Then there is a unique path p from Start to End in t (this is not difficult, but see Exercise 3 if you are doubtful). To find $f = G^{-1}(t)$, just put the vertices along p into C, and put all the other vertices to N. If $x \in N$, then define $f(x)$ as the unique neighbor of x in t that is closer to p than x. (Again, it is not hard to see that x has only one such neighbor; see Exercise 4 for a proof.) For the vertices $x \in C$, we define f so that the ith vertex of the Start-End path is the image of the ith smallest element of C. It is a direct consequence of the definition of G that this way we will get an $f \in A$ satisfying $G(f) = t$, and that this f is the only preimage of t under G. Therefore, G is a bijection and the proof is complete. ◇

Example 5.12 *(for finding $G^{-1}(t)$) If $n = 6$ and t is the doubly rooted tree shown in Figure 5.14, then the Start-End path shows that $C = \{2, 5, 6\}$ and that $f(2) = 5$, $f(5) = 6$, and $f(6) = 2$. Considering the remaining vertices, we see $f(3) = f(4) = 1$ and $f(1) = 6$.*

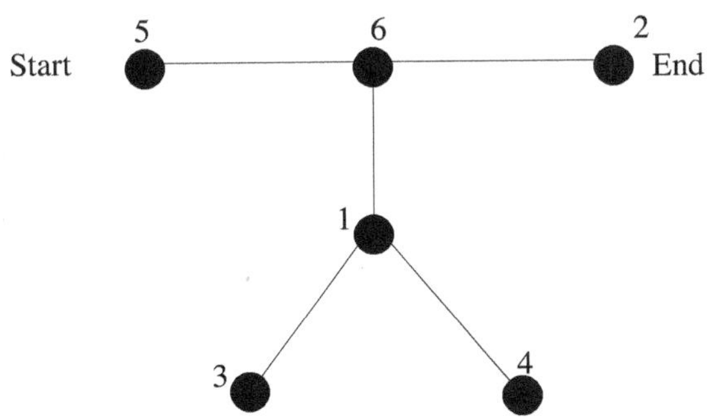

Figure 5.14
We compute the preimage of this tree.

Let us now try to count not simply all trees on vertex set $[n]$, but all trees on vertex set $[n]$ that have a given *degree sequence*. The degree sequence of a tree on $[n]$ is the sequence $(d_1, d_2, \cdots, d_n)$, where d_i is the degree of vertex i.

Figure 5.11 shows that there is one degree sequence, (1), for $n = 1$; there is one degree sequence, $(1, 1)$, for $n = 2$; and there are three degree sequences, $(2, 1, 1)$, $(1, 2, 1)$, and $(1, 1, 2)$ for $n = 3$. For each of these degree sequences s, there exists exactly one tree with degree sequence s. For $n = 4$, the situation is more complicated. While there are 16 trees on vertex set $[4]$, there are only 10 degree sequences. Indeed, all degree sequences must consist of either a 3 and three 1s (four possibilities) or two 2s and two 1s (six possibilities). Therefore, some sequences must belong to more than one tree. It is not hard to see that this happens precisely for the degree sequences of the second type, which will each belong to two trees. See Figure 5.15 for an example.

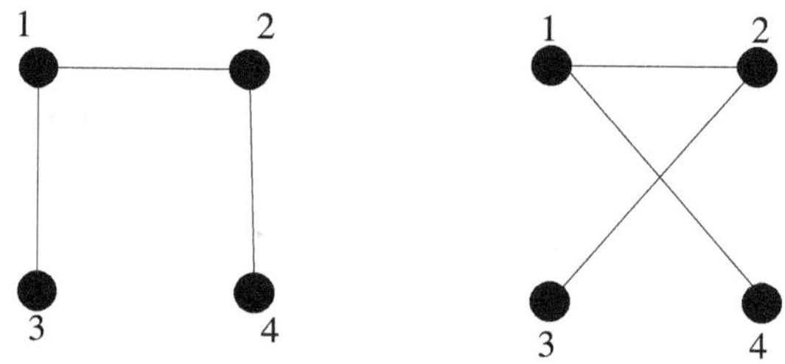

Figure 5.15
Different trees, same degree sequence.

A convenient way to count objects according to several parameters is by introducing a generating function in *several* variables. It is natural to start with a generating function in which there are n variables, the ith of which is responsible for taking the degree d_i of the vertex $i \in [n]$ of our trees into account. More precisely, define

$$F_n(x_1, x_2, \cdots, x_n) = \sum_{T \in \mathbf{T}(n)} x_1^{d_1} x_2^{d_2} \cdots x_n^{d_n}, \tag{5.2}$$

where $\mathbf{T}$(n) is the set of all trees on vertex set $[n]$, and $(d_1, d_2, \cdots, d_n)$ is the degree sequence of $T \in \mathbf{T}(n)$. (To be perfectly proper, we should write $(d_1(T), d_2(T), \cdots, d_n(T))$, but we do not want to upset the reader with excessive notation.)

The examples that we computed for small values of n show that

1. $F_1(x_1) = 1$,
2. $F_2(x_1, x_2) = x_1 x_2$, and
3. $F_3(x_1, x_2, x_3) = x_1^2 x_2 x_3 + x_1 x_2^2 x_3 + x_1 x_2 x_3^2 = (x_1 + x_2 + x_3) x_1 x_2 x_3$.

The next function, $F_4(x_1, x_2, x_3, x_4)$, has 16 summands. In order to find some pattern in the sequence of functions F_n, we look at F_2 and F_3 and see that they are all divisible by $x_1 \cdots x_n$. This is not surprising, since there are no isolated vertices in a tree with at least two vertices. This shows that $F_4(x_1, x_2, x_3, x_4)$ is divisible by $x_1x_2x_3x_4$. Now we only have to look at what is left after we divide $F_4(x_1, x_2, x_3, x_4)$ by $x_1x_2x_3x_4$. Looking at all possible degree sequences of trees on [4], we see that what is left is the sum of the squares of all variables and twice the product of any pair of variables. Therefore,

$$F_4(x_1, x_2, x_3, x_4) = (x_1 + x_2 + x_3 + x_4)^2 x_1x_2x_3x_4.$$

Now this starts getting interesting. Based on the computed functions F_4, it seems that we should have

$$F_5(x_1, x_2, x_3, x_4, x_5) = (x_1 + x_2 + x_3 + x_4 + x_5)^3 x_1x_2x_3x_4x_5. \tag{5.3}$$

Indeed, a tree on [5] must have a degree sequence identical to one of the trees shown in Figure 5.16.

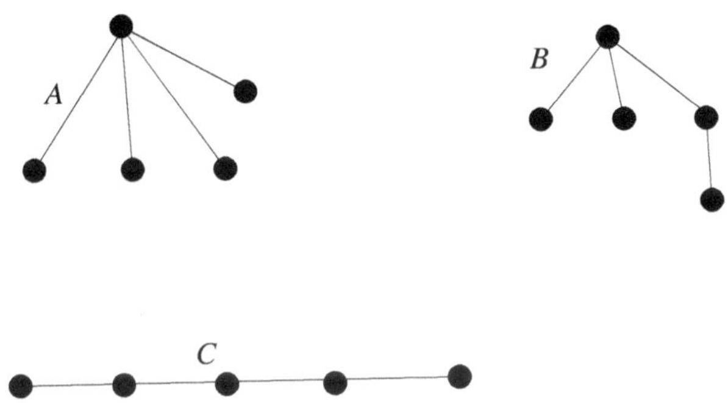

Figure 5.16
Trees on five vertices with different degree sequences.

Trees of type A (in the notation of Figure 5.16) can be labeled in five different ways, and they correspond to the five monomials that look like $x_1^4x_2x_3x_4x_5$, that is, monomials of degree 5 in which one variable has exponent 4, and the rest have exponent 1. Trees of type B can be labeled in $5 \cdot 4 \cdot 3$ different ways, corresponding to the 60 monomials of degree 5 in which one variable is of exponent 3 and one is of exponent 2. Finally, trees of type C can be labeled in $5!/2 = 60$ different ways, corresponding to the 60 monomials of degree 5 in which there are three variables of exponent 2. It is a direct consequence of the multinomial theorem that these 125 monomials together form $(x_1 + x_2 + x_3 + x_4 + x_5)^3 x_1x_2x_3x_4x_5$, confirming our conjecture.

After all this evidence, we do not think that the reader will be surprised by the following general theorem.

Theorem 5.13 *Let $n \geq 2$. Then the number of trees on vertex set $[n]$ having degree sequence $(d_1, d_2, \cdots, d_n)$ is equal to the coefficient $\binom{n-2}{d_1, d_2, \cdots, d_n}$ of $x_1^{d_1} x_2^{d_2} \cdots x_n^{d_n}$ in the expression*

$$(x_1 + x_2 + \cdots + x_n)^{n-2} x_1 x_2 \cdots x_n.$$

Equivalently,

$$F_n(x_1, x_2, \cdots, x_n) = (x_1 + x_2 + \cdots + x_n)^{n-2} x_1 x_2 \cdots x_n,$$

where F_n is defined in (5.2).

Proof: We prove our statement by presenting a bijection f from the set $\mathbf{T}(n)$ into the set B_n of all $(n-2)$-element sequences formed from the elements of $[n]$ that has the following additional property: If the degree of vertex i in $T \in \mathbf{T}(n)$ is k, then $f(T)$ will contain the element i exactly $k-1$ times.

A little thought shows that this will prove the theorem. Indeed, trees with the same degree sequence $(d_1, d_2, \cdots, d_n)$ will all map to sequences in which i appears $d_i - 1$ times, and the number of such sequences is $\binom{n-2}{d_1, d_2, \cdots, d_n}$. In order to define $f(T)$, we will decompose our tree $T \in \mathbf{T}(n)$ from the outside in, the way we would peel an onion.

Let us first cut off all leaves of T, and write down the unique neighbor of each of these leaves in increasing order of the *leaves*. Call the remaining tree T_1. Then repeat the same procedure. That is, cut off the leaves of T_1, write down their unique neighbors in increasing order of the leaves (continuing the sequence obtained in the previous step), and call the remaining tree T_2. Stop when the encoding sequence has length $n-2$, then define $f(T)$ to be that sequence.

Example 5.14 *Let T be the tree shown in Figure 5.17.*

When we pass from T to T_1, we cut off the leaves 2, 5, 6, 7, and 10. Therefore, we write down their neighbors in this order, obtaining 94393. When passing from T_1 to T_2, we cut off the leaves 3, 4, and 9, so we continue the previous sequence by writing down the neighbors of these leaves in increasing order of the leaves, obtaining 94393881. This sequence now has length 8, so we stop and define $f(T) = 94393881$.

It follows from the definitions that $f(T) \in B_n$. We need to show that $f : \mathbf{T}(n) \rightarrow B_n$ is a bijection.

To that end, first note that the leaves of T are precisely the elements of $[n]$ that do not appear in $f(T)$. Therefore, given $f(T)$, we can at least immediately identify the leaves of T. We can simply read off their unique neighbors as the elements at the beginning of $f(T)$. That is, if we find that T has k leaves, then the first k elements of $f(T)$ are the unique neighbors of these leaves when the leaves are listed in increasing order.

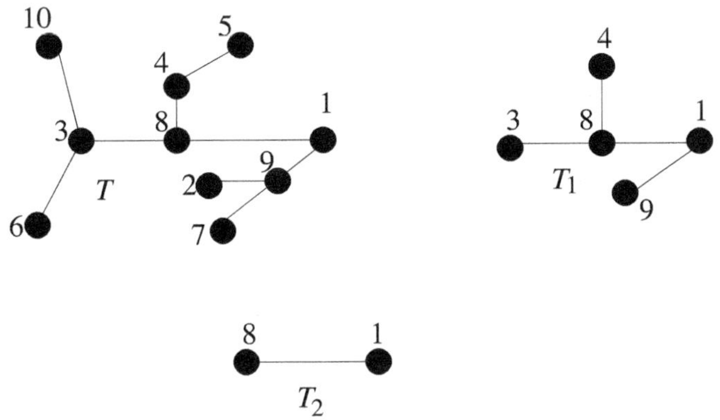

Figure 5.17
Decomposing T and constructing $f(T)$.

Second, the m leaves of T_1 are those elements of $[n]$ that do occur among the first k elements of B, but never after. Indeed, these vertices are connected to some leaves of T, but when those leaves of T are gone, the vertices themselves become leaves of T_1.

Next, the leaves of T_2 are those elements of $[n]$ that do occur among the first $k+m$ elements of B, but never after, and are not leaves of T_1 (and so on).

Continue in this manner until all of $f(T)$ has been used, and a graph with $n-2$ edges has been created. When that point is reached, there are still two elements of $[n]$ that have not played the role of leaves in the procedure so far. (Indeed, each leaf is cut off from the tree only once, so the $n-2$ elements of $f(T)$ correspond to $n-2$ distinct leaves.) The last edge of T connects these two vertices.

Since the procedure described above is deterministic, we conclude that, for any $B \in B_n$, there is at most one $T \in \mathbf{T}(n)$ so that $f(T) = B$. We still have to show that such a T indeed exists, that is, that the graph $f^{-1}(B)$ which we construct from B is always a tree. As that graph has $n-1$ edges, it suffices to show that it does not contain a cycle. Each of the first $n-2$ edges connects two vertices of two different kinds, an "earlier" and a "later" vertex, where "earlier" and "later" refer to the rightmost occurrence of that vertex in B. Since each "earlier" vertex is connected to only one "later" vertex, the subgraph after $n-2$ steps is still cycle free, that is, it is disconnected since it has only $n-2$ edges. The two vertices x and y that have not played the role of leaves so far must be in different components. However, then connecting x and y by the last edge will not create a cycle either. This completes the proof. $\diamondsuit$

Example 5.15 *Let $n = 9$ and let $B = 2311454$. Then the leaves of $T = f^{-1}(B)$ are the elements of $[9]$ not occurring in B, that is, 6, 7, 8, and 9. In*

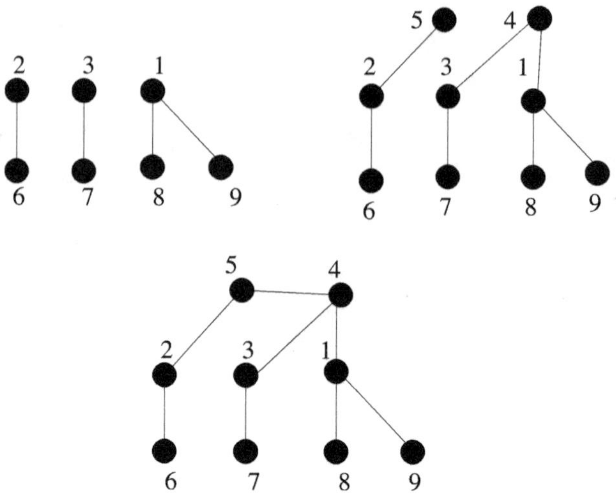

Figure 5.18
The three stages of finding $f^{-1}(2311454)$.

particular, the unique neighbor of 6 is 2, that of 7 is 3, and that of 8 and 9 is 1.

The leaves of T_1 are those elements of $[9]$ that occur among the first four elements of B, but never after, that is, 1, 2, and 3. Their unique neighbors in T_1 are the next three elements of B, that is, 4, 5, and 4. This gives us seven edges of T. The last edge is the one that connects the two remaining vertices, 4 and 5. See Figure 5.18 for an illustration.

Note that, if we substitute $x_i = 1$ for all i in the previous theorem, we get that $\sum_{T\in\mathbf{T}(n)} 1 = |\mathbf{T}(n)| = n^{n-2}$, providing yet another proof of Cayley's formula. (In case the reader still had doubts that Cayley's formula was true...)

5.1.4 Forests

A *forest* is a graph whose connected components are all trees. If a forest has n vertices and k components, then it is easy to see that it must have $n-k$ edges. Indeed, each component is a tree, which means the number of edges in each component is the number of vertices in that same component minus one.

A *rooted forest* is a forest in which each component is a rooted tree, that is, a tree with a distinguished vertex. The following simple proposition connects rooted forests to trees in a natural way.

Proposition 5.16 *The number of rooted forests on vertex set $[n]$ is*

$$(n+1)^{n-1}.$$

Proof: We present a bijection f from the set $\mathbf{R}(n)$ of rooted forests on vertex set $[n]$ onto the set $\mathbf{T}(n+1)$ of trees on vertex set $[n+1]$. Let $r \in \mathbf{R}(n)$, and let $f(r)$ be the tree obtained from r by adding the vertex $n+1$ and connecting it to the root of each component of r by an edge. Then $f(r) \in \mathbf{T}(n+1)$. It is easy to see that f is a bijection. Indeed, if $t \in \mathbf{T}(n+1)$, then we get the unique preimage of t under f by removing the vertex $n+1$—which will split t into components—and making the former neighbors of $n+1$ the roots of these newly obtained components. $\diamondsuit$

See Figure 5.19 for an example of the bijection f just constructed.

A spectacular argument by Jim Pitman [59] proves a much stronger result by providing an explicit formula for the number of rooted forests on $[n]$ with a given number of components. Because a rooted forest with one component is a rooted tree, Cayley's formula will follow as a special case of Pitman's result.

Before we can present Pitman's argument, we need some definitions. First, in this argument, we will look at our rooted forests as *directed graphs.* A directed graph is a graph in which each edge is directed toward one of its endpoints (that is, has an arrow on it). See Figure 5.20 for some directed graphs. Our rooted forests will be directed in a special way, namely, each edge will be directed *away* from the root. (See Figure 5.20 again.) Second, if F and G are both rooted forests, then we say that F contains G if F contains G as a directed graph. (This means that each directed edge of G is also a directed edge of F.) Finally, if $F_1, F_2, \cdots, F_k$ are rooted forests on $[n]$ so that the forest F_i has i components for all $i \in [k]$ and F_i contains F_{i+1}, then we say that $F_1, F_2, \cdots, F_k$ is a *refining sequence.* In other words, a refining sequence starts with a rooted tree on $[n]$, then its subsequent members are obtained by omitting an edge from the preceding member.

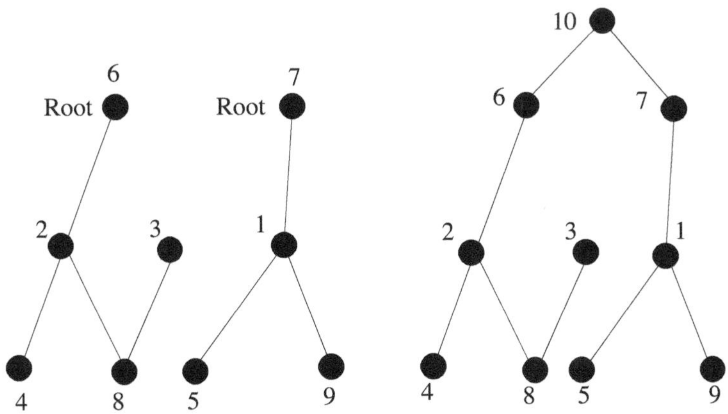

Figure 5.19
From a rooted forest on $[n]$ to a tree on $[n+1]$.

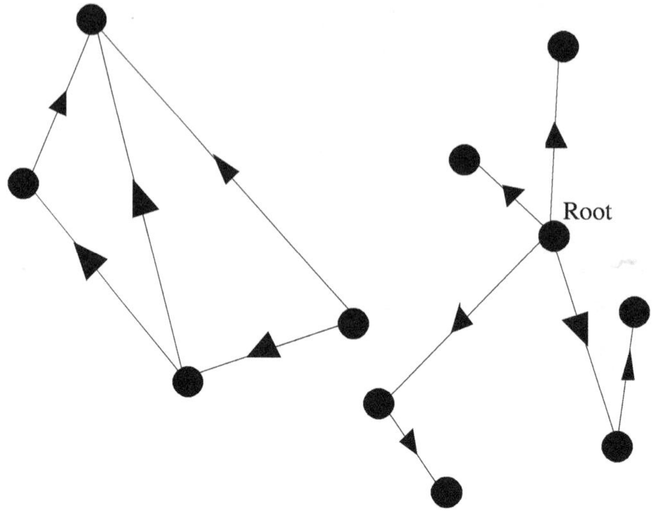

Figure 5.20
On the left, a generic directed graph. On the right, a rooted tree whose edges are directed away from the root.

Having announced all necessary definitions, we can finally start counting. The first step is the following lemma, which counts refining sequences by their last members.

Lemma 5.17 *Let F_k be a rooted forest on $[n]$ having k components. Let $R(F_k)$ be the number of refining sequences ending in F_k. Then*

$$R(F_k) = n^{k-1}(k-1)!.$$

Proof: We are going to prove this by induction on k. If $k = 1$, then there is one refining sequence ending in (and consisting of) F_1, and the statement is trivially true. Now assume we know the statement for $k-1$, and prove it for k. Given F_k, how many different rooted forests can play the role of F_{k-1}? Note that in a rooted forest, every vertex has in-degree 1, except for the roots of the components. Therefore, when we get F_{k-1} from F_k, we must add one edge e that can start anywhere but must end in one of the k roots in F_k. That is, we have $n(k-1)$ choices for F_{k-1} because e must end in a component different from the one in which it starts. Once we choose F_{k-1}, the induction hypothesis yields that there are $n^{k-2}(k-2)!$ ways to complete the refining sequence. Therefore, our induction step is proved by the product principle. $\diamond$

The advantage of the concept of refining sequences is that they are very easy to count by their length. This will be helpful in the proof of the following lemma.

Lemma 5.18 *Let F_k be a rooted forest on $[n]$ having k components, and let $T(F_k)$ be the number of rooted trees on $[n]$ containing F_k. Then*

$$T(F_k) = n^{k-1}.$$

Proof: If F_1 is a rooted tree containing F_k, then F_1 has $k-1$ more edges than F_k. A refining sequence starting at F_1 and ending in F_k will remove these $k-1$ edges in some order, so there are $(k-1)!$ such refining sequences. On the other hand, we know from Lemma 5.17 that there are $n^{k-1}(k-1)!$ refining sequences ending in F_k. Therefore, by the division principle, there must be n^{k-1} rooted trees where these sequences can start, that is, n^{k-1} rooted trees that contain F_k. ◊

We are now ready to announce our main result in this subsection.

Theorem 5.19 *The number of rooted forests on $[n]$ with k components is*

$$\frac{n^{n-1}(n-1)!}{n^{k-1}(k-1)!(n-k)!} = \binom{n}{k} k n^{n-1-k} = \binom{n-1}{k-1} n^{n-k}.$$

Proof: Let us call a refining sequence of length n *complete* since it is as long as a refining sequence can be. These sequences end in the empty forest F_n. The number of complete refining sequences is $R(F_n) = n^{n-1}(n-1)!$, as can be obtained from Lemma 5.17, setting $k = n$. Let us now count complete refining sequences again, this time by their kth element F_k.

Let $K(F_k)$ be the number of complete refining sequences whose kth member is F_k. We then claim that

$$K(F_k) = n^{k-1}(k-1)!(n-k)!. \tag{5.4}$$

Indeed, by Lemma 5.17, there are $n^{k-1}(k-1)!$ possibilities for the initial part $F_1, F_2, \cdots, F_k$; then there are $(n-k)!$ possible orders to remove the remaining $n-k$ edges.

Note that $K(F_k)$ depends only on k, not on F_k. Therefore, each rooted forest on $[n]$ having k components is in the kth place of the same number of complete refining sequences. On the other hand, only one rooted forest with k components is part of each complete refining sequence. So, by the division principle, the number of these forests is $R(F_n)/K(F_k)$ as claimed. ◊

The strength of this result can be seen if we simply set $k = 1$. Then we get the number of rooted forests on $[n]$ having one component; in other words, we get the number of rooted trees on $[n]$. This number, not surprisingly, turns out to be n^{n-1}. Applying the division principle (removing the root), we get another proof of Cayley's formula.

Quick Check

1. Let P_1 and P_2 be two paths of maximum length in a tree T. Prove that P_1 and P_2 have at least one vertex in common.
2. What is the total number of leaves in all trees on vertex set $[n]$?
3. Find the number of all trees on vertex set $[n]$ in which exactly two vertices have degree higher than 1. sequence h_n.

5.2 Graphs and functions

We have seen that the number of all trees on vertex set $[n]$ is equal to the number of all functions $f : [n-2] \rightarrow [n]$. This is not the only identity connecting the enumeration of functions to the enumeration of graphs.

All functions in this section will be functions $f : [n] \rightarrow [n]$. We will refer to this fact by saying that they are functions *on* $[n]$.

5.2.1 Acyclic functions

A function $f : [n] \rightarrow [n]$ is called *acyclic* if its short diagram does not contain cycles except for loops. Loops must be allowed, since the short diagram of f contains n edges, and so the short diagram cannot be cycle free.

Example 5.20 *Let $n = 6$. Then the function $f : [n] \rightarrow [n]$ defined by $f(i) = i+1$ if i is not divisible by 3, and $f(i) = i$ otherwise, is acyclic, as we can see in Figure 5.21.*

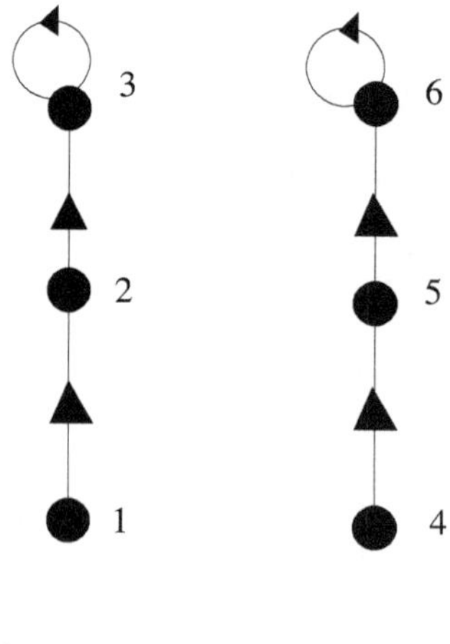

Figure 5.21
An acyclic function on $[6]$.

The following enumerative result connects acyclic functions to trees and forests.

Theorem 5.21 *The number of acyclic functions on* $[n]$ *is* $(n+1)^{n-1}$.

Without this result, it may not even seem immediately obvious that the inequality $(n+1)^{n-1} \leq n^n$ holds for all positive n, but our theorem implies that inequality.

Proof: We present a very natural bijection H from the set of acyclic functions on $[n]$ onto that of rooted forests on $[n]$. If f is an acyclic function on $[n]$, then we claim that the short diagram of f is a forest (disregarding the arrows except for those on the loops) with one vertex in each component having a loop around it. Indeed, since f is acyclic, any path starting at any point of $[n]$ in the short diagram must end in a loop. So each component has a loop. No component can have two loops since that would imply that a vertex has out-degree of at least 2. (Look at the path connecting the two loops in the short diagram.) Therefore, we can simply set $H(f)$ to be the undirected short diagram of f with the vertices with loops being the roots of the components.

Conversely, if F is a rooted forest, it is straightforward to find $H^{-1}(F)$ as follows: If x is the root of a component, set $f(x) = x$. Otherwise, set $f(x) = y$, where y is the unique neighbor of x that is closer to the root of its component than x. This shows that H has an inverse, so it is a bijection. Therefore, there are as many acyclic functions on $[n]$ as rooted forests, and the statement is proved. $\diamondsuit$

5.2.2 Parking functions

A shopping center has n parking spots along a one-way street. One morning, exactly n cars will arrive at the shopping center at various times. Once a car is parked, it does not leave before all other cars arrive. Each of the cars has a favorite parking spot, and a rather short-sighted strategy about it. namely, each car first goes to its favorite spot. If it is free, the car will take it; if it is not, it will go to the next spot. Then, if the next spot is not free, the car will go to the next spot, and so on, until it finds a free spot. If no spots are free after (and including) the favorite spot of a car, then that car will not have a spot, and we will say that the parking process was unsuccessful. If each car finds a spot, then we will say that the parking process was successful.

The question for us is what set of parking preferences will lead to a successful parking process? Let us denote the set of the n cars by the elements of $[n]$, and let us denote the n parking spots, in the order they follow along the one-way street, also by the elements of $[n]$. Let $f(i)$ be the favorite spot of car i; then $f : [n] \rightarrow [n]$ is a function. If the parking procedure is successful, then we will say that f is a *parking function* on $[n]$.

Example 5.22 *If f is a permutation, then f is a parking function since each car will find its favorite spot empty. The function $f(i) = 1$ for all $i \in [n]$ is also a parking function, but the function $f(i) = 2$ for all $i \in [n]$ is not, since the car arriving last will have no space to park.*

Two questions are natural to ask at this point. First, looking at a function $f : [n] \to [n]$, how can we tell whether f is a parking function or not? Second, how many parking functions are there on the set $[n]$?

When we look at the first question, some necessary conditions will easily come to mind. For instance, there could be at most one car whose favorite spot is the last spot (that is, there can be at most one i so that $f(i) = n$), since cars that go to the last spot first will not get a second chance. Similarly, there could be at most two cars whose favorite spots are from the set $\{n-1, n\}$.

While it is not difficult to extend this line of thought into a general necessary condition, it is somewhat surprising that the same set of conditions will then be sufficient as well. This is the content of the next lemma.

Lemma 5.23 *The function $f : [n] \to [n]$ is a parking function if and only if, for all $k \in [n]$, there are at most k values $i \in [n]$ so that $f(i) \geq n - k + 1$.*

In other words, f is a parking function if and only if, for all $k \in [n]$, at most k cars have a favorite parking spot that is one of the k last spots.

Proof: No car will attempt to park before its favored spot. Therefore, if more than k cars favor a spot from the last k spots, then there are more than k cars that will only accept a spot that is among the last k spots. Therefore, there will be at least one car which will not be able to park.

Conversely, assume there is at least one car that is not able to park. Let C be the first such car (chronologically). That means that when C drives up to its favorite spot i, then all spots $i, i+1, \cdots, n$ are taken. It could be that there are further spots immediately preceding i that are also taken. Let j be the smallest index so that all spots numbered $j, j+1, \cdots, n$ are taken at the moment C arrives at the parking lot. Then we must have $j \geq 2$ since C has not parked yet, so we know, by the definition of j, that $j-1$ is a free spot. Therefore, no car favoring a spot numbered $j-1$ or less had to take a spot numbered j or more so far. That means that the $n-j+1$ spots $j, j+1, \cdots, n$ are favored by at least $n-j+2$ cars, namely, the $n-j+1$ cars who are parking at these spots when C arrives, and C itself. In other words, our condition is violated for $k = n - j + 1$. ◊

In particular, note that the sufficient and necessary condition for f being a parking function does not depend on the order in which the cars arrive at the lot.

Let us now turn to our second natural question, that of enumerating parking functions on $[n]$. For shortness, let us write $(f(1), f(2), \cdots, f(n))$ for the function f. By Lemma 5.23, if f is a parking function, then so are all functions obtained by permuting the values of f. We will call these functions *permutations* of f. So $(1, 2, 1)$ is a permutation of $(1, 1, 2)$.

Let $P(n)$ be the number of parking functions on $[n]$. Then $P(1) = 1$, and $P(2) = 3$ (for the functions $(1, 1)$, $(1, 2)$, and $(2, 1)$). One checks easily that $P(3) = 16$ (for the functions $(1, 2, 3)$, $(1, 1, 3)$, $(1, 1, 2)$, $(1, 2, 2)$, $(1, 1, 1)$, and their permutations).

If the sequence 1, 3, 16 reminds the reader of something recently learned, then the reader is right.

Theorem 5.24 *For all positive integers n, $P(n) = (n + 1)^{n-1}$.*

The result is certainly very interesting for its elegance. Besides, who would have thought that parking functions would be so directly related to trees?

While it is possible to prove Theorem 5.24 by constructing a bijection from the set of all trees on $[n + 1]$ onto the set of all parking functions on $[n]$ or from the set of all forests on $[n]$ onto that of all parking functions on $[n]$, we will first prove Theorem 5.24 by a particularly elegant and direct argument due to Pollack, but published by Foata and Riordan in [32].

Proof: First let us reorganize the parking lot so that it becomes *circular* and has $n+1$ spots. We still have only n cars, but they prefer one of $n+1$ spots, not just n. Therefore, the total number of possible parking preferences is $(n+1)^n$.

When a car reaches spot $n + 1$ and cannot park, it simply continues to spot 1, then, if needed, to spot 2, and so on. This way, since we still have only n cars, at the end of the procedure each car will have a spot, and therefore one spot will be left empty.

We claim that each of the $n + 1$ spots is equally likely to be left empty. Indeed, if a particular set of preferences leaves spot i empty, then adding 1 to each parking preference will shift the entire parking procedure by one notch, leaving spot $i+1$ empty. (Here $(n+1)+1$ is interpreted as 1.) So indeed, each spot is left empty $1/(n + 1)$ of the time. Now note that, if spot $n + 1$ is left empty, then it was never used during the procedure, so only spots belonging to $[n]$ were needed. However, that means that we actually dealt with a *parking function*. Therefore, the number of parking functions is $\frac{(n+1)^n}{n+1} = (n + 1)^{n-1}$ as claimed. ◊

While the above proof is certainly beautiful, it may well leave the reader curious about why the number of parking functions on $[n]$ is equal to the number of rooted forests on $[n]$. Below we show that the reason is that both objects have the same kind of recursive structure.

In particular, we will *combinatorially* explain the recurrence relation

$$P(n + 1) = \sum_{i=0}^{n} \binom{n}{i} (i + 1)P(i)P(n - i), \tag{5.5}$$

both for rooted forests on $[n+1]$ and for parking functions on $[n+1]$. The key word here is *combinatorially*, since the above identity could be proved purely by algebraic manipulations, but that would not give us the insight we are looking for.

First, let F be a rooted forest on $[n+1]$. Let F_1 be the component of F that contains the vertex 1. Say F_1 has $i+1$ vertices. Then we have $\binom{n}{i}$ ways to choose the vertices other than 1 for F_1, and we have $P(i)$ ways to choose a tree on these $i+1$ vertices. Once that is done, any of these $i+1$ vertices can be the root of F_1. Finally, we have $P(n-i)$ ways to choose a rooted forest on the remaining $n-i$ vertices. Therefore, by the product principle, we indeed have $\binom{n}{i}(i+1)P(i)P(n-i)$ rooted forests on $[n+1]$ in which the vertex 1 is in a component of size $i+1$. Summing over all allowed i, we get (5.5).

Second, let f be a parking function on $[n+1]$. Say that when the last car arrives, spot $i+1$ is free for some $i \in [0,n]$. That means that there is a set S of i cars who parked in the first i spots, so their parking preferences formed a parking function on $[i]$. If we subtract $i+1$ from the parking preferences of the remaining $n-i$ cars, then they form a parking function on $[n-i]$. (Note that none of these $n-i$ cars could have had a parking preference in $[i+1]$ since the spot $i+1$ would not be free.) Finally, f is a parking function, so $f(n+1) \leq i+1$. That is, we have $i+1$ possibilities for $f(n+1)$. Since we have $\binom{n}{i}$ possibilities for S, formula (5.5) is proved again by the product principle and by summing over all allowed i.

Quick Check

1. How many parking functions $f : [n] \to [n]$ are there that satisfy $\sum_{i=1}^{n} f(i) = \binom{n+1}{2}$?
2. How many parking functions $f : [n] \to [n]$ are there that satisfy $\sum_{i=1}^{n} f(i) = \binom{n+1}{2} - 1$?
3. What is the number of functions $f : [n] \to [n]$ such that the short diagram of f contains exactly one cycle, and that cycle is of length k?

5.3 When the vertices are not freely labeled

There are some situations when we restrict the labels of the vertices of trees in some natural way. In this section, we consider some of the ensuing enumeration problems.

5.3.1 Rooted plane trees

So far in our graphical enumeration problems, we have paid little attention to the way in which our graphs were drawn. Now we are going to change this. Imagine, for instance, that we are drawing the family tree of all descendants of a person. It is then natural to list the children of people in that tree in some order that is easy to remember. For instance, we can always draw the nodes corresponding to siblings so that reading them from left to right we proceed from the older siblings toward the younger siblings. If we do this, we will sometimes get trees that are isomorphic in the sense of Definition 5.5, but that nevertheless describe different family situations.

Example 5.25 *The graph on the left of Figure 5.22 shows a family in which A has three children, the oldest of whom also has three children. The graph on the right of Figure 5.22 shows a family in which A has three children, the oldest of whom has only one child.*

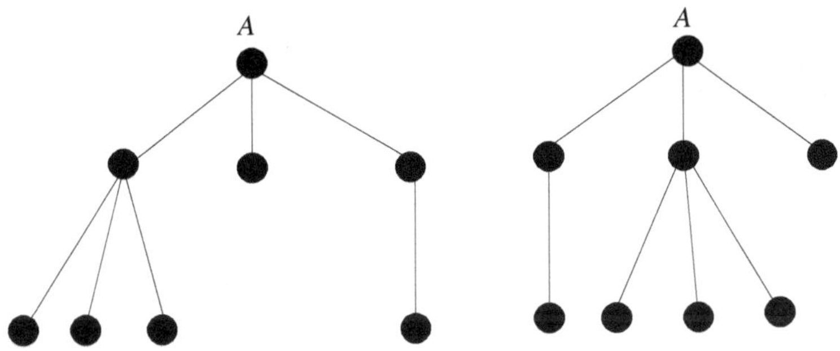

Figure 5.22
Different families, isomorphic graphs.

This phenomenon shows that, in order to model family ties and other structures in which there is an *order* among nodes at the same level, we need to refine our notion of rooted trees, or at least our notion of *isomorphisms* of rooted trees. Roughly speaking, there will have to be an order among the nodes that are at equal distance from the root, and this order will be expressed by positioning these nodes from left to right. That is, the way in which we *draw* the tree in the plane *matters*. The following definitions make this notion more formal.

Definition 5.26 *A* rooted plane tree *is a tree drawn in the plane with a distinguished vertex R, called the root, so that the neighbors $n_1, n_2, \cdots, n_k$ of R, from left to right, in this order, are all roots of smaller plane trees $T_1, T_2, \cdots, T_k$.*

Note that, if we know the *ordered* k-tuple $(T_1, T_2, \cdots, T_k)$ of subtrees of T, then we know T itself, since we can simply connect these subtrees to a root R. Now we can define isomorphism of plane trees.

Definition 5.27 *Two rooted plane trees* $T = (T_1, T_2, \cdots, T_k)$ *and* $U = (U_1, U_2, \cdots, U_m)$ *are called* isomorphic as rooted plane trees *if* $k = m$, *and if* T_i *is isomorphic to* U_i *as rooted plane trees for all* $i \in [k]$.

In other words, it is not enough for the subtrees of the two trees to be pairwise isomorphic to each other; they also have to be in the same order. Therefore, the two trees shown in Figure 5.22 are *not* isomorphic as rooted plane trees, even if they are isomorphic as trees.

How many rooted plane trees are there on $n+1$ unlabeled vertices? This is a more natural question than asking the same for labeled vertices. Indeed, once we know the answer for unlabeled vertices, we can label them in any way, implying that the number of rooted plane trees on $[n+1]$ is just $(n+1)!$ times the number obtained in the unlabeled case.

The following theorem provides the answer to this question, and it shows the first graph theoretical occurrence of the Catalan numbers in our studies. Recall that this fascinating sequence of integers was discussed in Example 1.34, and from a different viewpoint, in Example 3.16.

Theorem 5.28 *The number of rooted plane trees on* $n+1$ *unlabeled vertices is* $c_n = \binom{2n}{n}/(n+1)$.

Proof: We present a bijection f from the set A_n of all rooted plane trees on $n+1$ vertices onto the set B_n of all northeastern lattice paths from $(0,0)$ to (n,n) that never go above the main diagonal. From Example 1.34, we know that $|B_n| = c_n$, so the proof of our claim will follow.

Our bijection will be defined recursively. That is, we assume that f has already been defined for all positive integers less than n. We can do this since, for $n = 1$, the statement is trivially true, with the only rooted plane tree on two vertices corresponding to the only subdiagonal lattice path from $(0,0)$ to $(1,1)$. Let $t \in A_n$. There are two simple cases.

1. When the root of t has more than one child, then let $t_1, t_2, \cdots, t_k$ be the subtrees of t, from left to right. Add a new root to the top of each of these trees to get the new plane trees $t'_1, t'_2, \cdots, t'_k$. Crucially, all these trees are smaller than t. Then define $f(t)$ as the concatenation of the images of the t_i by this same algorithm (that is, by the version of f defined for integers smaller than n). Then $f(t) \in B_n$ since the trees t'_i together have as many nonroot vertices as t does.
2. When the root of t has only one child, then the above algorithm will not work since $t'_1 = t$ would hold. Therefore, we define $f(t)$ differently. Let us remove the root of t to get the plane tree t'. Take

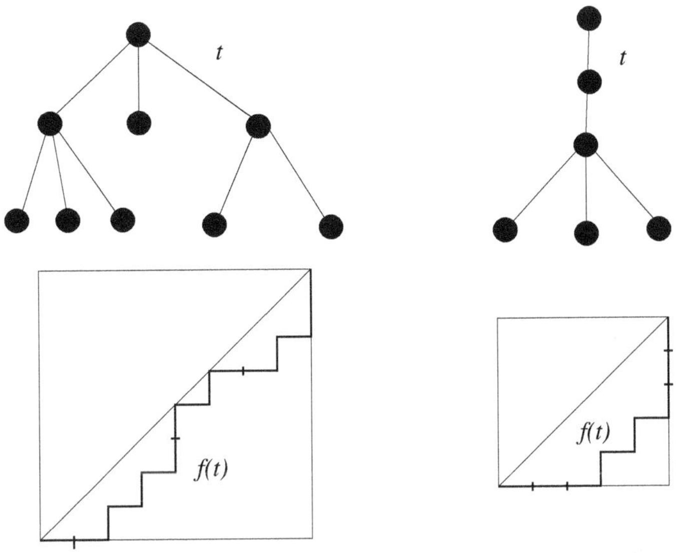

Figure 5.23
The two cases of bijection $f : A_n \to B_n$.

$f(t')$, which is a northeastern lattice path from $(0, 0)$ to $(n-1, n-1)$ that never goes above the main diagonal, and translate it one notch to the east. Then $f(t')$ becomes a path from $(1, 0)$ to $(n, n-1)$. Now prepend it with an east step and add a north step to its end, and call the obtained path $f(t)$.

It follows from the definition we just gave for f that $f(t) \in B_n$. So all we have to show is that f is a bijection. We will do that by proving that, if $p \in B_n$, then there is exactly one $t \in A_n$ so that $f(t) = p$. After the recursive definition of f, it should not come as a surprise that this proof will be by induction. Again, for $n = 1$, the statement is true. Assume we know the statement for all positive integers less than n. There are two cases, and they are both illustrated in Figure 5.23.

(a) If p does touch the diagonal $x = y$, then it follows from the definition of f that it can only be the image of a tree t that belongs to case 1. The sizes of the subtrees t_i must then be the distances between consecutive points in which p touches the main diagonal. Finally, once the sizes of the t_i are known, the trees themselves are unique by our induction hypothesis.

(b) If p does not touch the main diagonal, then it can only be the image of a tree that belongs to case 2. Then cutting of the first and last step of p and translating the remaining path one notch to the west, we get a path p', which by our induction hypothesis has a unique preimage t'. Adding a new root on the top of t', we get the unique tree t satisfying $f(t) = p$.

So f is indeed a bijection, completing the proof. $\diamondsuit$

As we mentioned, it follows from Theorem 5.28 that the number of rooted plane trees on vertex set $[n+1]$ is $(n+1)!C_n$. Let us take the time to digest this result. We know from Cayley's formula that the number of all rooted trees on $[n+1]$ is $(n+1)^n$. Using Stirling's formula (1.13), we see that the number of rooted plane trees on $[n+1]$ satisfies

$$(n+1)!C_n = \frac{(2n)!}{n!} \simeq \left(\frac{4}{e}\right)^n \cdot n^n \frac{1}{\sqrt{\pi n}}.$$

As n grows, this number will grow somewhat faster than the number $(n+1)^n$ of all rooted trees on $[n+1]$, since $4 > e \cdot \frac{n+1}{n}$ for $n > 2$. So for large n, there are more rooted plane trees on $[n+1]$ than there are rooted trees on $[n+1]$. This is how it should be, since there exist pairs of trees on $[n+1]$ that are identical as rooted trees, but not identical as rooted plane trees. See Figure 5.24 for an example.

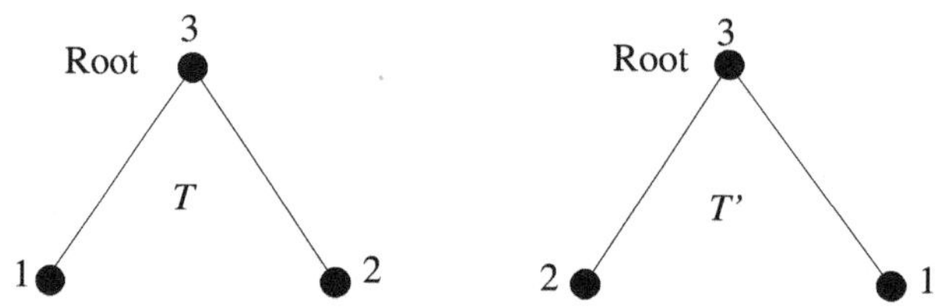

Figure 5.24
Rooted trees T and T' are identical as rooted trees, but not as rooted plane trees.

Now let us call a vertex of a rooted plane tree that is not a leaf and not the root an *internal node*. Let us look at all rooted plane trees on $n+1$ unlabeled vertices and count them according to their number of internal nodes. For instance, for $n = 4$, there are $c_4 = 14$ such trees; one of them has no internal nodes, six of them have one internal node, six of them have two internal nodes, and one of them has three internal nodes. This is a special case of the following theorem.

Theorem 5.29 *For a fixed positive integer n and a fixed nonnegative integer $k \leq n-1$, let $N(n,k)$ be the number of rooted plane trees on $n+1$ unlabeled vertices that have k internal nodes. Then we have*

$$N(n,k) = N(n, n-1-k).$$

Equivalently, there are as many rooted plane trees on $n+1$ unlabeled vertices that have k internal nodes as there are such trees with k leaves.

This is an interesting statement since it does not seem obvious how we could turn internal nodes into leaves and vice versa.

The proof of Theorem 5.29 will be given in the solution of Exercise 23. We suggest that the reader try to solve that exercise *after* reading the next subsection. Until then, we mention that the numbers $N(n, k)$ are called the *Narayana numbers*, which are obtained by the formula

$$N(n,k) = \frac{1}{n}\binom{n}{k}\binom{n}{k+1}. \tag{5.6}$$

See [13] or [72] for a proof of this fact. Note that, in accordance with Theorem 5.29, the sequence of Narayana numbers is symmetric in k for any fixed n.

5.3.2 Decreasing binary trees

There are several natural ways to bijectively represent *permutations* by trees. In this subsection, we will discuss one of them, leaving another representation to the exercises.

Let p be an n-permutation written in one-line notation. Let us introduce the notation $p = LnR$, where L is the string of entries on the left of n, and R is the string of entries on the right of n. So it is possible that one of L and R is empty. We would like to represent p by a certain kind of rooted plane tree $T(p)$ as follows: The root of $T(p)$ will be labeled n, the largest entry of p. The root will have at most two children, the left child and its subtree corresponding to L, and the right child and its subtree corresponding to R.

In other words, $T(p)$ is constructed recursively, by taking the trees $T(L)$ and $T(R)$ and joining them under a new root n. If one of L and R is empty, then the new root will only have one child, but that child will still be a left child or a right child, depending on whether its subtree encodes L or R. The trees $T(L)$ and $T(R)$ are constructed by this same rule, that is, we first take their largest entry, make it the root of the tree to be constructed, then look at their left and right, and construct the subtrees corresponding to the substrings on the left and on the right recursively.

Example 5.30 *If $p = 4527613$, then $T(p)$ is the tree shown in Figure 5.25.*

As we mentioned, the tree $T(p)$ is a special kind of labeled plane tree. First, each of its vertices has at most two children. Second, if a vertex has only one child, then the direction (left or right) of that child still matters. Indeed, if the child is on the left of the parent, then in p the entry corresponding to the child is on the left of the entry corresponding to the parent. Similarly, if the child is on the right of the parent, then in p the entry corresponding to the child is on the right of the entry corresponding to the parent. Third, it follows from

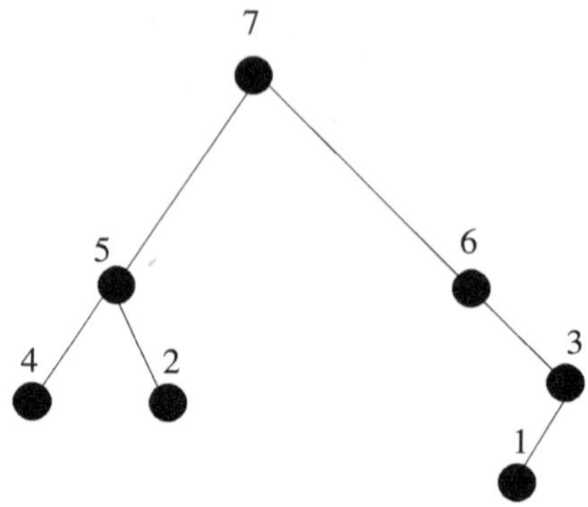

Figure 5.25
The tree $T(p)$ of $p = 4527613$.

the recursive structure of $T(p)$ that the label of every child is smaller than the label of its parent. All these observations warrant the following definition.

Definition 5.31 *A* decreasing binary tree *on vertex set* $[n]$ *is a tree with root vertex* n *and an ordered pair* (T_L, T_R) *of subtrees (the left subtree and the right subtree), which together have vertex set* $[n-1]$ *and in which the label of each child is smaller than that of its parent.*

The following proposition shows how close decreasing binary trees are to permutations.

Proposition 5.32 *The map* f *described above is a bijection from the set* S_n *of all* n*-permutations into the set* $BT(n)$ *of decreasing binary trees on vertex set* $[n]$.

In particular, this means that the number of decreasing binary trees on vertex set $[n]$ is $n!$.

Proof: All we have to show is that f has an inverse, that is, that given $T \in BT(n)$, we can find a unique $p \in S_n$ so that $f(p) = T$. Again, because of the recursive definition of f, we prove this by induction on n. For $n = 1$, the statement is true. Now let us assume that the statement is true for all positive integers less than n. Take the left subtree T_L and the right subtree T_R of T. The sizes of T_L and T_R determine the unique place where the entry n must be in any permutation p satisfying $f(p) = T$. That is, the entry n has to be in position $|L|+1$. Each of T_L and T_R have less than n vertices, so the induction hypothesis applies to them. Therefore, there is a unique permutation L of the labels of T_L so that $f(L) = T_L$, and there is a unique permutation R of the labels of T_R so that $f(R) = T_R$. Consequently, the only permutation for which $f(p) = T$ must be LnR, and that permutation does map into T. $\diamond$

Note that, when we compute $f^{-1}(T)$ for some $T \in BT(n)$, we always obtain the next entry of the inverse image by finding the leftmost branch of the tree and going to the "deepest" vertex on that branch. In the theory of tree traversals, this is called a *left-to-right depth-first search.*

Proposition 5.32 shows that the number of decreasing binary trees on n vertices is $n!$. The interesting problem here is enumerating *unlabeled* binary plane trees on n vertices, that is, decreasing binary trees from which the labels are omitted. The number of these will be smaller than $n!$, since there are many different decreasing binary trees that become identical once their labels are omitted. For instance, the reader is asked to verify that the trees associated with permutations 1243, 1342, and 2341 by our bijection f constructed above differ only in their labels.

Again, due to the recursive structure of our trees, our counting argument will be recursive. Let us say that there is one binary plane tree on zero unlabeled vertices.

Theorem 5.33 *Let n be any nonnegative integer. The number of all binary plane trees on n unlabeled vertices is $c_n = \binom{2n}{n}/(n+1)$, the nth Catalan number.*

Before we prove the theorem, let us think about it for a while. The theorem says that, in the unlabeled case, there are as many binary plane trees on n vertices as there are rooted plane trees on $n+1$ vertices. This may sound surprising, since $n+1$ is more than n, and also since in a rooted plane tree a parent can have any number of children, whereas in a binary plane tree a parent can have at most two children. What balances this out is the fact that in binary plane trees the direction of an only child matters, while in rooted plane trees, it does not. See Figure 5.26 for an example.

Proof: Let p_n be the number of binary plane trees on n vertices. We want to prove that $p_n = c_n$ for all n.

We prove the statement by induction on n. For $n = 0$, the statement is true (since $p_0 = 1$), and for $n = 1$, the statement is true as well. Now let us assume that we know the statement for all nonnegative integers less than n, and prove it for n. Let $T \in BT(n)$. Then T has a left subtree with i vertices and a right subtree with $n-1-i$ vertices (the last vertex is the root). Both of these subtrees are binary subtrees on at most $n-1$ vertices; therefore, by the induction hypothesis, there are p_i choices for the left subtree, and p_{n-1-i} choices for the right subtree. Using the product principle, there are $p_i p_{n-1-i}$ choices for the entire tree for this particular i, and using the addition principle for all possible i, we get that there are altogether $\sum_{i=0}^{n-1} p_i p_{n-1-i}$ possibilities for T. We have pointed out after Example 3.16 that

$$c_n = \sum_{i=0}^{n-1} c_i c_{n-1-i}.$$

That is, the numbers p_n and c_n satisfy the same recurrence relation for $n \geq 1$. Since $p_0 = c_0 = 1$, this proves our claim. $\diamondsuit$

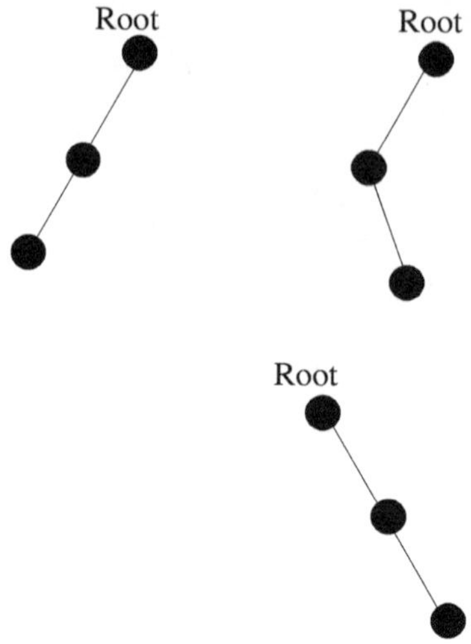

Figure 5.26
These four trees are all different as binary plane trees, but all identical as rooted plane trees.

A very useful property of binary plane trees and decreasing binary trees is a certain *symmetry*. Let us call an edge of a binary plane tree a *left edge* if it connects a node to its left child, and a *right edge* if it connects a node to its right child.

Proposition 5.34 *For any fixed positive integer n and any nonnegative integer $k \leq n-1$, there are as many binary plane trees on n vertices with k left edges as there are with k right edges.*

Proof: Simply reflect a binary plane tree through a vertical axis, and left edges will turn into right edges, and vice versa. See Figure 5.27. $\Diamond$

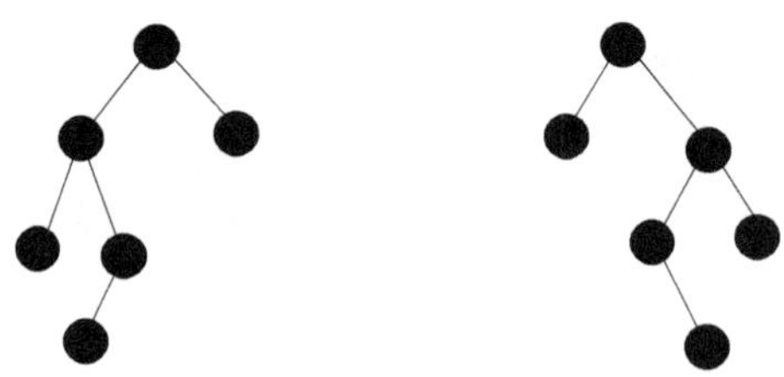

Figure 5.27
A binary plane tree and its reflected image.

The reader is probably wondering what the big deal is, since reflecting a tree through an axis is not all that interesting. What is interesting, however, is the fact that this symmetry can be translated into all 150 interpretations of Catalan numbers, and in many of those interpretations, it will prove symmetries that are far less obvious than the one proved in Proposition 5.34.

For instance, in Exercises 21 and 22 we show how to use Proposition 5.34 to prove the much less obvious fact that the number of 132-avoiding n-permutations with k descents is the same as the number of 132-avoiding n-permutations with k ascents. In case you do not remember, 132-avoiding permutations were defined in Exercise 32.

Then, Exercises 18 and 20 connect the topic of the previous subsection, rooted plane trees, to these objects as well. Therefore, the reader solving all these exercises can conclude in Exercise 23 that there are as many rooted plane trees on $n+1$ vertices with k internal nodes as there are with k leaves. This is what we claimed in Theorem 5.29.

In general, a similar symmetry can be found for each interpretation of the Catalan numbers. Furthermore, for each interpretation of the Catalan numbers, we can find a natural partition of the set of objects counted by the Catalan numbers so that the sizes of the blocks of the partition are given by the Narayana numbers, defined by (5.6). The interested reader should consult [72].

Quick Check

1. What is the number of rooted plane trees on $n+1$ unlabeled vertices in which no nonroot vertex has degree higher than 1?
2. What is the number of decreasing binary trees on vertex set $[n]$ in which no vertex has two children?
3. What is the number of binary plane trees on n vertices that have exactly two leaves?

5.4 Graphs on colored vertices

Let us assume that representatives of four countries convene for a series of negotiations. The negotiations will consist of a round of one-on-one conversations, each occurring between two people coming from different countries.

If we want to represent these negotiations by a graph, then the vertices will correspond to the negotiators, and there will be an edge between two vertices if the two corresponding people talked to each other. However, the graph we get will have a special structure. That is, there will be no edges between vertices corresponding to people from the same country. See Figure 5.28 for an example.

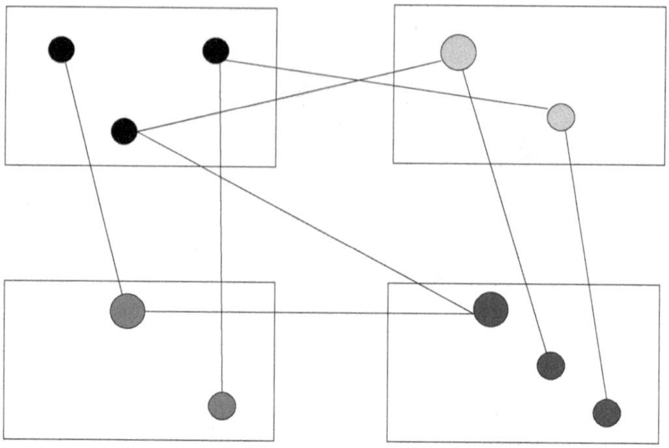

Figure 5.28
Diplomats from the same country will not negotiate with each other.

This example shows that sometimes we want to study graphs in which we do *not* want edges between certain subsets of vertices. Another example of this is communication towers and the frequencies on which they broadcast. We can represent communication towers by vertices of a graph and join two vertices by an edge if the corresponding towers are closer to each other than a specified distance d. Towers that are closer to each other than d should not broadcast on the same frequency. Therefore, if we represent frequencies by colors, then the distribution of frequencies to the towers is equivalent to coloring the vertices of the graph so that adjacent vertices are of different colors.

The idea of colorings is an easy way to visualize such requirements, that is, to keep track of pairs of vertices that cannot be adjacent. In the example of the negotiators, we would color vertices corresponding to representatives of the same country with the same color (or we would color vertices corresponding to communication towers built close to each other with the same color) and then we would not connect vertices of the same color by edges.

Definition 5.35 *Let G be a graph. We say that a coloring of the vertices of G is* proper *if adjacent vertices are of different colors.*

There are two types of questions we are going to look at in this section:

1. Given a *graph* G and a number n of colors, how many proper colorings does G have using only those n colors?
2. Given a *colored vertex set*, how many graphs are there on that vertex set so that the coloring is proper?

In the next chapter, we will look at other related problems. One of them will be to determine the smallest integer k so that there exists a proper coloring of G with k colors.

5.4.1 Chromatic polynomials

Let G be a graph on vertex set $[m]$, and let $\chi_G(n)$ be the number of proper colorings of G using only the colors $1, 2, \cdots, n$. Note that we do not have to use all these colors.

Example 5.36 *If G is the complete graph on vertex set $[m]$, then it is not difficult to see that $\chi_G(n) = (n)_m$. Indeed, there are n choices for the color of the first vertex, then $n-1$ choices for the color of the second vertex, and so on.*

Example 5.37 *If G is the empty graph on vertex set $[m]$, then $\chi_G(n) = n^m$ since any vertex can get any color.*

Example 5.38 *If G is the graph shown in Figure 5.29, then we compute $\chi_G(n) = n(n-1)(n-2)^2$. Indeed, we first need two different colors for A and*

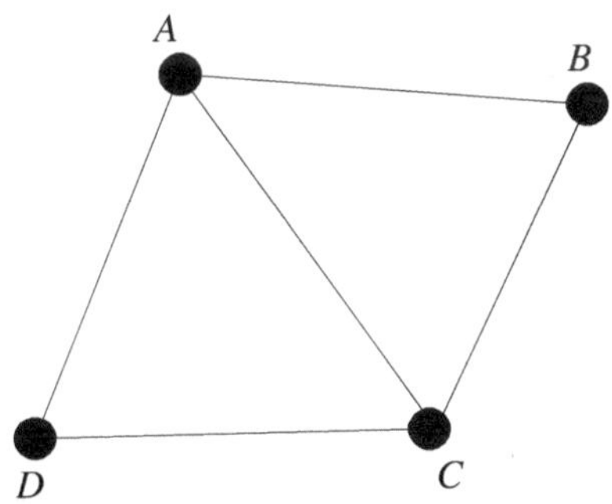

Figure 5.29
For this graph, $\chi_G(n) = n(n-1)(n-2)^2$.

C, but then the colors of B and D do not have to be different from each other.

The three examples above seem to suggest that $\chi_G(n)$ is always a *polynomial* function of n. The following theorem shows that this is indeed the case.

Theorem 5.39 *For any graph G, the function $\chi_G(n)$ is a polynomial function of n (for nonnegative values of n). The degree of this polynomial is equal to the number m of vertices of G.*

Proof: No matter how large n is, no proper coloring of G uses more than m colors. So if $m < n$, then we first have to choose the m colors that will actually be used, then color G with those colors.

For $i \in [m]$, let r_i denote the number of ways to color the vertices of G with colors from $[i]$ *actually using each of those i colors.* The argument of the previous paragraph then shows that

$$\chi_G(n) = \sum_{i=1}^{m} \binom{n}{i} r_i.$$

Since the r_i are constants and the $\binom{n}{i}$ are polynomials of n, the right-hand side is the sum of a constant number of polynomials of n, so it is a polynomial of n. The degree of this polynomial is equal to the highest degree of all summands. Therefore, it is equal to the degree of $\binom{n}{m}$, which is m. $\diamondsuit$

Therefore, we can call the function $\chi_G(n)$ the *chromatic polynomial* of G.

A crucial property of $\chi_G(n)$ is that it is possible to compute it recursively. In order to facilitate that computation, for a graph G and an edge e of G, we define G_e to be the graph G with e omitted, and G^e to be the graph in which the edge e is *contracted* to a vertex. That is, the two endpoints of i are identified, and the obtained new vertex is adjacent to all vertices that were adjacent to at least one of its two predecessors. See Figure 5.30 for an example.

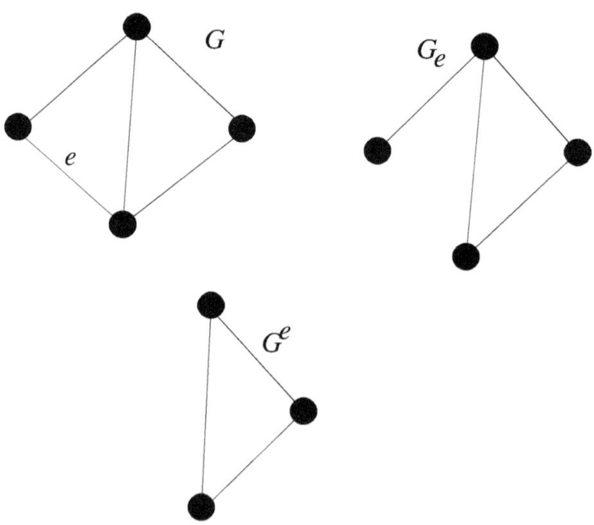

Figure 5.30
Transformations of G.

Proposition 5.40 *For all nonnegative integers n, the following holds:*

(1) $\chi_{K_1}(n) = n$, where K_1 is the one-vertex graph with no edges;

(2) $\chi(G+H)(n) = \chi_G(n) \cdot \chi_H(n)$, where $G+H$ is the union of the disjoint graphs G and H; and

(3) $\chi_G(n) = \chi_{G_e}(n) - \chi_{G^e}(n)$.

Example 5.41 *If G is the graph shown in Figure 5.30, then $\chi_G(n) = n(n-1)(n-2)^2$, while $\chi_{G_e}(n) = n(n-1)^2(n-2)$, and $\chi_{G^e}(n) = n(n-1)(n-2)$, verifying part 3 of Proposition 5.40.*

Proof: (of Proposition 5.40) Proofs of part 1 and part 2 of the proposition are left to the reader. For part 3, note that a proper coloring of G_e is a proper coloring of G if and only if the two endpoints of e are of different colors. If they are the same color, we get a coloring of G^e by giving their common color to the vertex obtained by contracting e. ◊

Now that we know that $\chi_G(n)$ is a polynomial function, we may substitute *negative* values into this polynomial as well. You could say, Yes, we can, but why would we? To compute the number of ways to color a graph with -5 colors? Interestingly, as shown by Richard Stanley [73], it turns out that $\chi_G(n)$ does have a combinatorial meaning, even for negative values of n.

In order to find this combinatorial meaning, we define an *acyclic orientation* of a graph. An acyclic orientation of G is an assignment of a direction to each edge of G so that no directed cycles are formed.

Example 5.42 *The orientation of G on the left-hand side of Figure 5.31 is acyclic. The orientation of G on the right-hand side is not acyclic since the edges connecting the black vertices form a directed cycle.*

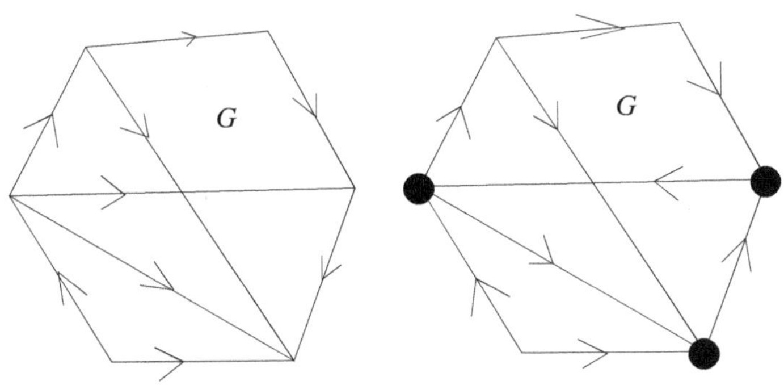

Figure 5.31
An acyclic, and a nonacyclic, orientation.

Definition 5.43 *For any graph G, let $\bar{\chi}_G(n)$ be the number of pairs (f, A) so that A is an acyclic orientation of G and f is a coloring of the vertices of G using only colors from $[n]$ so that, if there is an edge from u to v in G, then $f(u) \geq f(v)$.*

Example 5.44 *Let G be a triangle. Then G has six acyclic orientations. Let A be the orientation shown in Figure 5.32. Then the colors x, y, and z must be chosen so that $1 \leq z \leq y \leq x \leq n$, which can be done in $\binom{n+2}{3}$ ways. By symmetry, this means that $\bar{\chi}_G(n) = 6\binom{n+2}{3} = (n+2)(n+1)n$.*

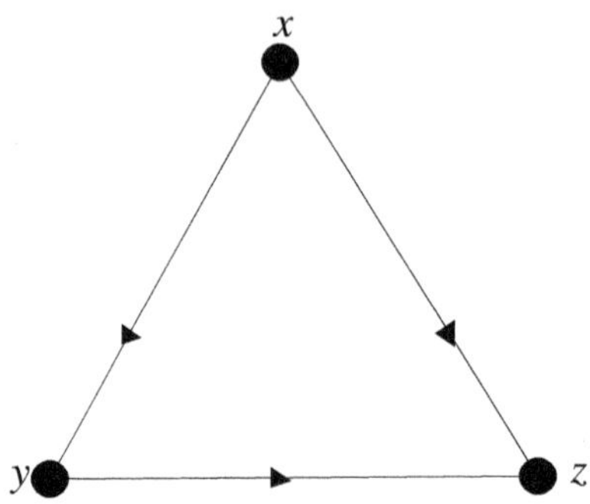

Figure 5.32
There are $\binom{n+2}{3}$ ways to label this directed graph.

Example 5.45 *Generalizing the technique used in the previous example, we see that, if G is the complete graph on m vertices, then $\bar{\chi}_G(n) = \binom{n+m-1}{m}$.*

Just as in Example 5.37, if G is the empty graph on m vertices, then $\bar{\chi}_G(n) = n^m$. Comparing the values of $\bar{\chi}_G(n)$ and $\chi_G(n)$ in the cases when we computed both, we see that there is a rather close connection between them. This is the content of the next theorem.

Theorem 5.46 *For all graphs G on vertex set $[m]$ and all nonnegative integers n,*

$$\bar{\chi}_G(n) = (-1)^m \chi_G(-n).$$

The reader is urged to verify this equality for the examples in which we computed the values of $\bar{\chi}_G(n)$ and $\chi_G(n)$.

Note that the right-hand side is meaningful because we have extended the domain of χ_G to all integers. As a consequence of the theorem, we will see that the values of $\chi_G(n)$ for $n = 0, -1, -2, \cdots$ alternate in sign, since the left-hand side is always positive.

Proof: (of Theorem 5.46) We claim that the function $\bar{\chi}_G$ satisfies recurrence relations that are identical to those satisfied by the function $\chi_G(-n)$. If we can prove this, the theorem will follow by induction.

The following proposition lists the mentioned recurrence relations.

Proposition 5.47 *Keep the notation of Proposition 5.40. Then, for all non-negative integers n,*

(1) $\bar{\chi}_{K_1}(n) = n$;

(2) $\bar{\chi}(G + H)(n) = \bar{\chi}_G(n) \cdot \bar{\chi}_H(n)$; *and*

(3) $\bar{\chi}_G(n) = \bar{\chi}_{G_e}(n) + \bar{\chi}_{G^e}(n)$.

Proof: Part 1 and part 2 are left to the reader. We are going to prove part 3 as follows: Let e be an edge of G connecting vertices u and v. Let (f, A) be a pair counted by $\bar{\chi}_{G_e}(n)$, or, as we will call these pairs for the rest of this proof, a *good* pair. We will now orient e in both possible ways, once from u to v (call the new orientation A_1), and once from v to u (call the new orientation A_2). We will then prove that one of these orientations will turn (f, A) into a good pair (that is, at least one of the pairs (f, A_1) and (f, A_2) will not violate the constraint that $f(x) \geq f(y)$ if there is an edge from x to y). Then, we will prove that *both* orientations are good for exactly $\bar{\chi}_{G^e}(n)$ original (f, A) pairs. This will prove our claim.

There are three possibilities for the colors of u and v:

1. When $f(u) > f(v)$, then the reader can check that (f, A_2) is not good. On the other hand, we claim that (f, A_1) is. In order to prove this, all we need is to show that A_1 is acyclic. This is true, since if A_1 contained a directed cycle C, that cycle would contain the edge uv. That is a contradiction, however, since for any vertex x from which there is a directed path to u, we have $f(x) \geq f(u)$, and for any vertex x to which there is a directed path from y, we have $f(x) \leq f(v)$, contradicting $f(u) > f(v)$.
2. When $f(u) < f(v)$, the argument is the same, reversing the roles of u and v.
3. When $f(u) = f(v)$, then we claim that at least one of A_1 and A_2 is acyclic, that is, at least one of (f, A_1) and (f, A_2) is good. Let us assume that the opposite is true. Then A_1 contains a directed cycle $uva_1a_2 \cdots u$ and A_2 contains a directed cycle $vub_1b_2 \cdots v$. See Figure 5.33. However, this means that

$$ub_1b_2 \cdots va_1a_2 \cdots u$$

is a directed cycle in A, which is a contradiction.

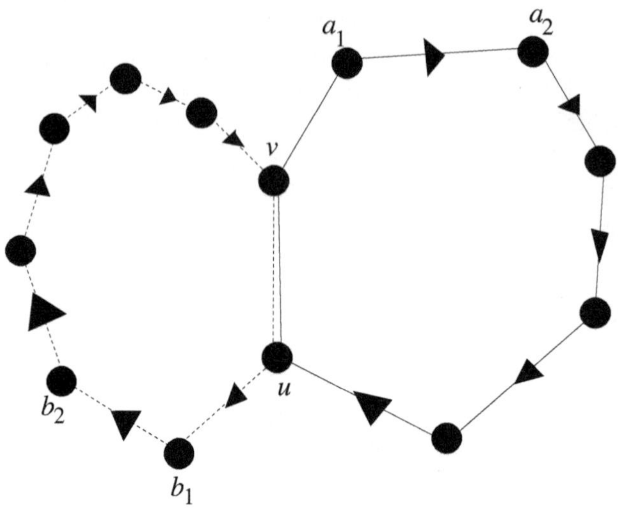

Figure 5.33
If both A_1 and A_2 contain a directed cycle, then so does A.

So at least one of (f, A_1) and (f, A_2) is good. How many pairs (f, A) are there so that they are *both* good? We will bijectively prove that their number is precisely $\bar{\chi}_{G^e}(n)$. Indeed, let (f, A) be a pair so that both of (f, A_1) and (f, A_2) are good. By our analysis above, this implies that we have $f(u) = f(v)$, where u and v are the vertices of the edge e. Set $x = f(u) = f(v)$, and let UV be the vertex of G^e into which the uv edge is contracted. Note that, since the colors of u and v agree, there is no vertex t in G^e so that the orientation of the edge between u and t is not the same as the orientation of the edge between v and t. If there were such a vertex t, then either A_1 or A_2 would contain a directed cycle. See Figure 5.34.

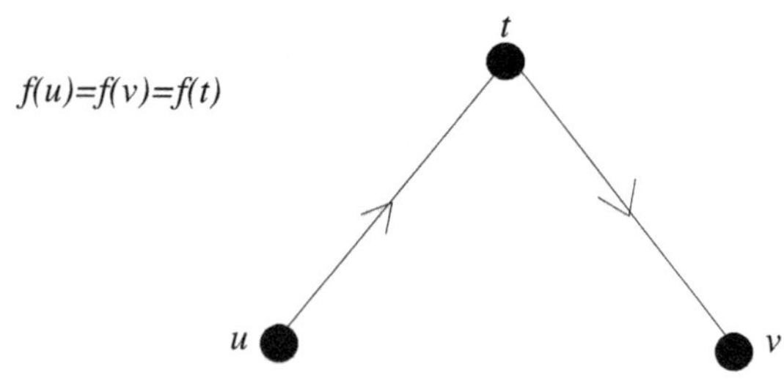

Figure 5.34
If there is a directed path from u to v (or back), then both (f, A_1) and (f, A_2) cannot be good.

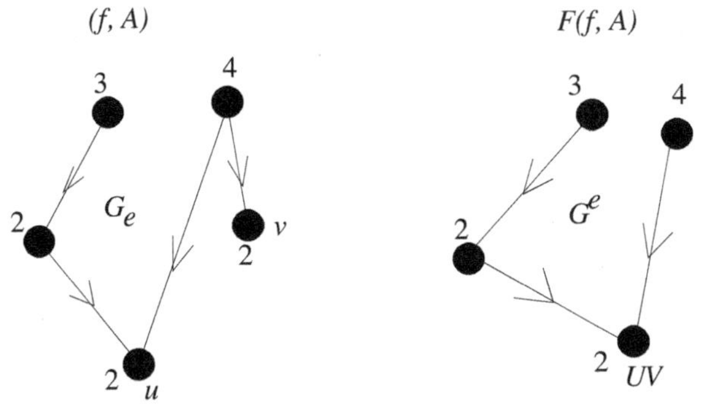

Figure 5.35
How the bijection F works.

Define $F(f, A)$ as the pair (f', A'), where A' is the acyclic orientation of G^e induced by A (all edges inherit their orientation; edges adjacent to UV inherit the orientation of the relevant edges adjacent to u and v), and f' is the coloring induced by f (all vertices inherit their color; UV gets color x). See Figure 5.35.

Then the reader is invited to verify that F is a bijection from the set of good pairs into the set of pairs (g, B) counted by $\bar{\chi}_{G^e}(n)$, so part 3 is proved, completing the proof of our proposition.

◊

The statement of Theorem 5.46 is now immediate by induction on the number of edges of G, and by comparing the recurrence relations satisfied by $\chi(G)$ and $\bar{\chi}(G)$ that we have proved in Propositions 5.40 and 5.47. ◊

In the special case when $n = 1$, there is exactly one allowed coloring for each acyclic orientation, which leads to the following interesting fact.

Corollary 5.48 *For all graphs G on vertex set $[m]$, the number of acyclic orientations of G is $(-1)^m \chi_G(-1)$.*

We note that the phenomenon discussed here occurs quite frequently in combinatorics. That is, it happens often that we define a function f as the function enumerating certain combinatorial objects in terms of the combinatorial, and therefore nonnegative, parameter n. Then it turns out that $f(n)$ is a polynomial function of n, and therefore we can talk about $f(x)$ for negative values of x as well, once the coefficients of f are known. And the values of $f(-n)$ often prove to have some natural combinatorial meaning, usually somehow reciprocating the original meaning of $f(n)$.

5.4.2 Colored graphs

Coloring the vertices or edges of graphs leads to two very different families of interesting combinatorial problems. Here we will explore the former, while leaving the latter to the exercises.

5.4.2.1 Counting all k-colored graphs

Let us now count all graphs on vertex set $[n]$ that have a given number of vertices of each color, say n_1 vertices of color 1, n_2 vertices of color 2, and so on, with $n_1 + n_2 + \cdots + n_k = n$. First, we have to decide which vertices get which colors. This can be done in

$$\binom{n}{n_1, n_2, \cdots, n_k}$$

ways. Once we know the color of each vertex, we must decide, for each pair of vertices of different colors, whether we put an edge between those two vertices or not. The number of such pairs is the number of all pairs minus the number of monochromatic pairs, that is,

$$\binom{n}{2} - \sum_{i=1}^{k} \binom{n_i}{2} = \frac{n^2 - n - \sum_{i=1}^k n_i^2 + \sum_{i=1}^k n_i}{2} = \frac{n^2 - \sum_{i=1}^k n_i^2}{2}.$$

Therefore, the number of properly colored graphs on vertex set $[n]$ in which n_i vertices have color i is

$$\binom{n}{n_1, n_2, \cdots, n_k} 2^{\frac{n^2 - \sum_{i=1}^k n_i^2}{2}}.$$

5.4.2.2 Counting colored trees

The following is one of the many variations of Cayley's formula counting trees.

Theorem 5.49 **(Scoins' theorem)** *The number of properly bicolored trees on $n+m$ vertices, n of which are red and labeled bijectively by the elements of $[n]$ and m of which are blue and labeled bijectively by the elements of $[m]$, is*

$$n^{m-1}m^{n-1}.$$

Note that a graph whose vertices can be properly bicolored is often called a *bipartite graph*. We will study these graphs in more detail in the following chapter.

Example 5.50 *Let $n = 3$ and $m = 2$. Then $n^{m-1}m^{n-1} = 3 \cdot 4 = 12$. Indeed, there are 12 trees with the given parameters; six are isomorphic to the tree shown in the left-hand side of Figure 5.36, and six are isomorphic to the tree shown in the right-hand side of Figure 5.36.*

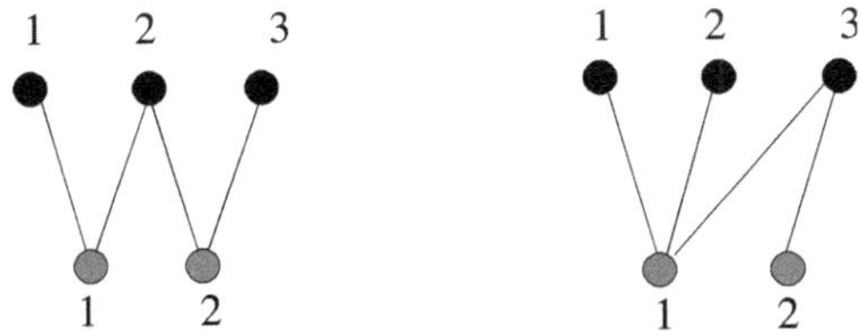

Figure 5.36
Two nonisomorphic trees on the same vertex set.

The original proof of Theorem 5.49 by Scoins [66] was computational, and used a sophisticated tool called the *Lagrange inversion formula.* We are going to give an elegant combinatorial proof, which is an extension of the argument by André Joyal we presented in the proof of Theorem 5.10. A similar proof is given in [72].

Proof: (of Theorem 5.49) Instead of counting bicolored trees, let us count *doubly rooted* bicolored trees, that is, bicolored trees in which there are two distinguished vertices (the roots), namely, Start, which is always a blue vertex, and End, which is always a red vertex. Our statement is equivalent to the statement that the number of doubly rooted trees with the given parameters is $n^m m^n$.

We will prove this new statement bijectively. Let A be the set of all pairs of functions (f, g), where $f : [n] \rightarrow [m]$, and $g : [m] \rightarrow [n]$. Let B be the set of all doubly rooted bicolored trees on red vertex set $[n]$ and blue vertex set $[m]$. We will construct a bijection $F : A \rightarrow B$. Since the size of A is $m^n \cdot n^m$, this will prove our theorem.

Let $(f, g) \in A$, and represent both f and g within the same directed graph $G_{f,g}$. That is, take a graph on red vertex set $[n]$ and blue vertex set $[m]$, and, if $f(i) = j$, then add an edge from the red vertex i to the blue vertex j. Similarly, if $g(a) = b$, then add an edge from the blue vertex a to the red vertex b.

Example 5.51 *Let* $n = 5$ *and* $m = 4$*, and let* $f(1) = f(2) = 1$*,* $f(3) = 3$*, and* $f(4) = f(5) = 4$*. Furthermore, let* $g(1) = 3$*,* $g(2) = g(4) = 2$*, and* $g(3) = 4$*. Then the graph* $G_{f,g}$ *is shown in Figure 5.37.*

This procedure will create two kinds of vertices in our graph $G_{f,g}$, those that are part of a directed cycle, and those that are not. Let C and N, respectively, denote these two subsets of our vertex set. Note that $|C|$ is even since our graph is bipartite. Let $C = \{c_1, c_2, c_3, \cdots, c_{2k}\}$, where $c_1 < c_3 < \cdots < c_{2k-1}$ are the *red vertices* in C and $c_2 < c_4 < \cdots < c_{2k}$ are the *blue vertices* in C. There are as many red vertices in C as there are blue vertices, since each cycle must contain the same number of vertices of each color.

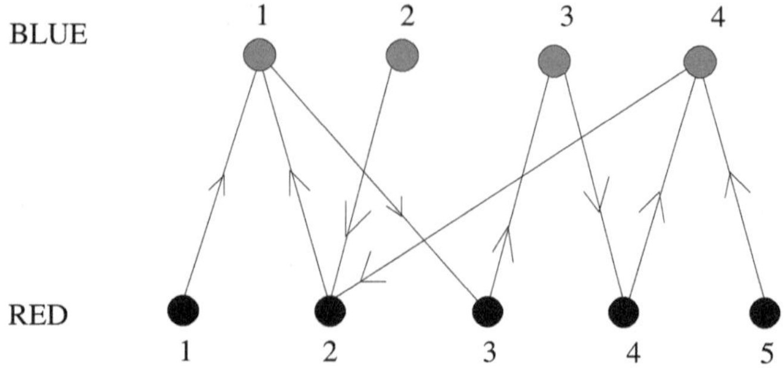

Figure 5.37
The bipartite graph $G_{f,g}$.

Example 5.52 *If $G_{f,g}$ is the graph shown in Figure 5.37, then one sees easily that $C = \{R2, B1, R3, B3, R4, B4\}$ and $N = \{R1, R5, B2\}$.*

Now we are ready to define the doubly rooted bicolored tree $F(f, g)$. First, let us take the path of length $2k$ whose vertices are $f(c_1)$, $g(c_2)$, $f(c_3)$, $g(c_4)$, and so on, ending with $g(c_{2k})$. Let us call the vertex $f(c_1)$ Start and the vertex $g(c_{2k})$ End. Note that the Start-End path is properly colored. Finally, if $x \in N$, then let us connect x to its image (either $f(x)$ or $g(x)$, whichever is defined) by an edge. This procedure results in a graph having $n+m-1$ edges and $n+m$ vertices. This graph is connected since we can get to the Start-End path from any vertex. Therefore, our graph is indeed a doubly rooted tree, and we call this tree $F(f, g)$.

Example 5.53 *If f and g are as in Example 5.51, and therefore C and N are as in Example 5.52, then $F(f, g)$ is the doubly rooted bicolored tree shown in Figure 5.38.*

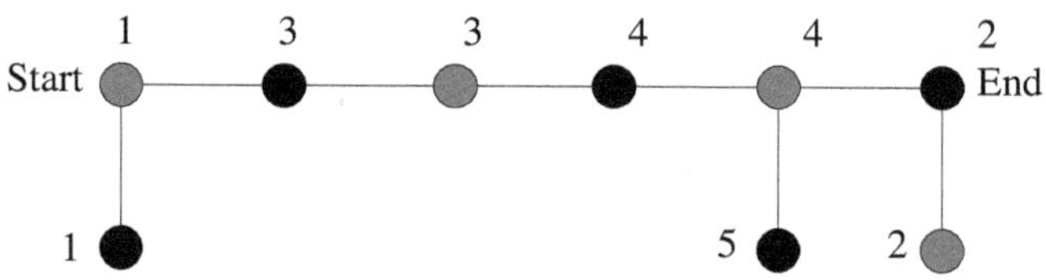

Figure 5.38
The doubly rooted tree $F(f, g)$.

All that is left to prove is that $F : A \to B$ is a bijection. We can do this by showing that each doubly rooted tree $T \in B$ has exactly one preimage under F. Let $T \in B$, then the Start-End path tells us what the sets C and N must be in a preimage of T and defines the values of f and g for the elements of C. For the elements of N, the value of $f(x)$ or $g(x)$ (whichever is meaningful) is the neighbor of x toward the Start-End path on the unique path from x to the Start-End path. ◊

Theorem 5.49 can be generalized to an arbitrary number of colors, as the following theorem shows.

Theorem 5.54 *Let $n_1+n_2+\cdots+n_k = n$, where the n_i are positive integers. Then the number of properly k-colored trees on n vertices, n_i of which are of color i and are bijectively labeled by the elements of $[n_i]$ (for all i), is*

$$n^{k-2}\prod_{i=1}^{k}(n-n_i)^{n_i-1}.$$

Note that this formula implies not only Theorem 5.49, but also Theorem 5.10, since we can choose $n_i = 1$ for all i. See [53] for a proof.

Quick Check

1. Find the chromatic polynomial of a path of n vertices.
2. Is it true that all trees on n vertices have the same chromatic polynomial?
3. What is the number of acyclic orientations of the complete graph K_n?

5.5 Graphs and generating functions

It is high time that we applied the powerful counting techniques that we learned in the previous chapter to enumerate various kinds of trees. We start by exploring a classic family of trees.

5.5.1 Trees counted by Cayley's formula

Rooted trees on $[n]$ have a nice recursive structure. If we cut off the root R of such a tree, we get a rooted forest, that is, trees that are rooted at the vertex which was a neighbor of R. In other words, rooted forests are built up from rooted trees. This structure can be expressed by a simple functional equation. We know from Cayley's formula that the number of rooted trees on

$[n]$ is $t_n = n^{n-1}$. Let

$$T(x) = \sum_{n=1}^{\infty} t_n \frac{x^n}{n!} = \sum_{n=1}^{\infty} \frac{n^{n-1}}{n!} x^n.$$

We cannot find a closed form for this generating function since the convergence radius of the series is 0. However, as the following theorem shows, we can prove an interesting identity about it.

Theorem 5.55 *Let $T(x)$ be defined as above. Then we have*

$$T(x) = xe^{T(x)}. \tag{5.7}$$

Proof: By the exponential formula, the exponential generating function for the number of rooted forests is $e^{T(x)}$ since a rooted forest is just a collection of an arbitrary number of rooted trees. On the other hand, removing the root vertex of a rooted tree, we get a rooted forest. The exponential generating function for the number of ways $[n]$ can be a root vertex is x; indeed, there is one way for this to happen if $n = 1$, and zero ways otherwise. Therefore, by the product formula, we get $T(x) = xe^{T(x)}$ as claimed. $\diamondsuit$

Let us assume that we do not know in advance that $t_n = n^{n-1}$. Can we deduce that formula from (5.7)? It turns out that the answer to that question is in the affirmative. However, it involves the Lagrange inversion formula, which we will discuss only later in this Chapter, in Section 5.6. A key observation is that (5.7) yields

$$x = \frac{T(x)}{e^{T(x)}} = F(T(x)),$$

with $F(x) = x/e^x$. In other words, the composition of the functions x/e^x and $T(x)$ is the identity function. In other words, $T(x)$ is the *compositional inverse* of the power series x/e^x.

5.5.2 Rooted trees

We can often obtain the explicit form of a generating function counting some kind of rooted trees. This is our next subject.

5.5.2.1 Ordinary generating functions

Let b_n be the number of all rooted plane trees on n unlabeled vertices. These trees were defined in Definition 5.26. In Theorem 5.28, we proved that $b_n = c_{n-1}$ by an elegant bijective argument. Now we will re-prove that result by a more or less automatic argument that uses generating functions.

Let $B(x) = \sum_{n\geq 0} b_n x^n$ be the ordinary generating function of our sequence, where we set $b_0 = 0$. Note that removing the root of a rooted plane tree, we get a sequence of rooted plane trees. Therefore, by Theorem 3.25 and the product formula, we get

$$B(x) = x \cdot \frac{1}{1 - B(x)},$$

$$B(x)^2 - B(x) + x = 0.$$

This is a quadratic equation for $B(x)$, and therefore we can solve it as such. The solution is

$$B(x) = \frac{1 - \sqrt{1 - 4x}}{2}. \tag{5.8}$$

See the discussion after the very similar formula (3.20) to understand why the square root is taken with a negative sign.

The explicit formula that we have just computed for $B(x)$ can be used to compute b_n by the binomial theorem, and obtain the result $b_n = \binom{2(n-2)}{n-1}/n = c_{n-1}$ for $n \geq 1$ as claimed. We have provided all the details for the very similar power series $H(x) = B(x)/x$ following formula (3.20).

The reader is encouraged to use the same technique to compute the ordinary generating function for the numbers s_n of rooted plane trees on n unlabeled vertices in which every nonleaf vertex has *at least two* children. An answer is given in Exercise 31.

5.5.2.2 Exponential generating functions

Let us find the number of all decreasing *nonplane* 1-2 trees on vertex set $[n]$, These are trees whose vertices are bijectively labeled with the elements of n so that each vertex has a label that is smaller than that of its parent, and each vertex that is not a leaf has one or two children. However, the trees are *not* plane trees, that is, the children of a given vertex are not ordered linearly. See Figure 5.39 for the five such trees on vertex set $[4]$.

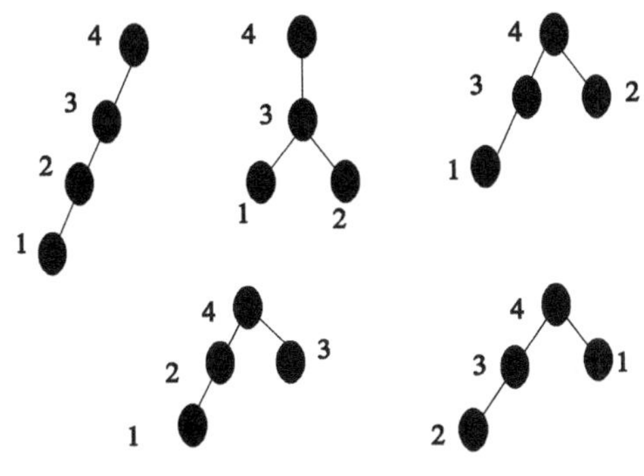

Figure 5.39
The five rooted 1-2 trees on vertex set $[4]$.

Let u_n be the number of such trees on vertex set $[n]$, set $u_0 = 1$, and let $U(x) = \sum_{n\geq 0} u_n \frac{x^n}{n!}$ be the exponential generating function of the sequence u_n. Then we have

$$U'(x) = \sum_{n\geq 1} u_n \frac{x^{n-1}}{(n-1)!}. \tag{5.9}$$

The key observation is the following. If we *remove the root* of a decreasing nonplane tree on vertex set $[n]$, we get a forest of two components (one of which may be empty) on vertex set $[n-1]$. The number of such forests is equal to u_n, since the root that we removed was always the vertex n. So, $U'(x)$ is the exponential generating function counting these forests. Each component of these forests is a decreasing nonplane tree, and such trees are counted by the generating function $U(x)$. Note that, if we partition $[n-1]$ into two (possibly empty) blocks, and then take a decreasing nonplane tree on each block and make them subtrees of the root n, then we get each decreasing nonplane tree on $[n]$ *twice*, except for the empty tree, which we only get once. Therefore, the product formula yields the differential equation.

$$2U'(x) = U(x)^2 + 1. \tag{5.10}$$

By definition, we have the initial condition $U(0) = 1$. We can solve (5.10) as a separable equation, and we get

$$U(x) = \tan\left(\frac{x}{2} + \frac{\pi}{4}\right) = \tan x + \sec x. \tag{5.11}$$

The numbers u_n are called the *Euler numbers*. They count numerous combinatorial objects, as we will see in the exercises. Note that $\sec x$ has only terms with even exponents, and $\tan x$ has only terms with odd exponents. Therefore, the numbers u_{2n} are sometimes called the *secant numbers*, and the numbers u_{2n+1} are sometimes called the *tangent numbers*.

The above application of derivatives is often useful. Let us take a second look at decreasing binary trees with it. We have seen in Proposition 5.32 that the number of such trees on vertex set $[n]$ is $n!$, but let us try to prove that fact by generating functions.

Let $A_n(x) = \sum_{n\geq 0} a_n \frac{x^n}{n!}$ be the generating function for the numbers a_n of decreasing binary trees on vertex set $[n]$. Removing the root, we get an *ordered* pair of two decreasing binary trees so that their vertex set is $[n-1]$. These ordered pairs have generating function $A'(x) = \sum_{n\geq 1} a_n \frac{x^{n-1}}{(n-1)!}$, since there is a natural (and trivial) bijection between the set of such pairs, and decreasing binary trees on vertex set $[n]$. Therefore, the product formula implies that

$$A'(x) = A(x)^2,$$

with initial condition $A(0) = 1$. We can solve this differential equation since it is a separable equation. We obtain the result $A(x) = 1/(1-x)$, so indeed, $a_n = n!$ as expected.

Supplementary Exercises 35, 36, and 38 illustrate further applications of the derivative of a generating function.

5.5.2.3 An application of multivariate generating functions

Let us return to the generating function $B(x)$ of Section 5.5.2.1 that counts rooted plane trees. What if we want to know how many children the root has in such a tree on n unlabeled vertices, on average? As the number of such trees is $b_n = c_{n-1} = \binom{2(n-1)}{(n-1)}/n$, it suffices to find the *total* number of all root-children in all c_{n-1} such trees.

To that end, we introduce the generating function

$$H(x,y) = \sum_{n\geq 1}\sum_{k\geq 0} h(n,k)x^n y^k,$$

where $h(n,k)$ is the number of rooted plane trees on n unlabeled vertices in which the root has k children. It is then easy to see that

$$H(x,y) = x + xyB(x) + xy^2B(x)^2 + \cdots = \frac{x}{1-yB(x)}.$$

Indeed, if such a tree has more than one vertex, then, removing its root, we get a sequence of k trees, and these trees contribute the summand $xy^kB(x)^k$ to $H(x,y)$.

The following computations could be a bit tedious when carried out by hand, but can be done very easily by a software package like Maple or Mathematica. Substituting the explicit form of $B(x)$ from (5.8) into the last displayed equation, we get that

$$H(x,y) = \frac{x}{1-y\left(\frac{1}{2}-\frac{1}{2}\sqrt{1-4x}\right)}. \tag{5.12}$$

The crucial idea is the following. If we differentiate $H(x,y)$ with respect to y, and then we set $y=1$, we get

$$\frac{\partial}{\partial y}H(x,y)\Big|_{y=1} = \sum_{n\geq 0}\left(\sum_{k\geq 0} kh(n,k)\right)x^n. \tag{5.13}$$

In other words, the *coefficient* of x^n in $\frac{\partial}{\partial y}H(x,y)\Big|_{y=1}$ is $\sum_k k\cdot h(n,k)$, that is, the *total number of the children* of the root in *all* rooted plane trees on n vertices. Indeed, the summand $k\cdot h(n,k)$ counts these children in all $h(n,k)$ trees of size n in which the root has k children.

Therefore, all we need to do is to compute $[x^n]\frac{\partial}{\partial y}H(x,y)\Big|_{y=1}$ using the closed form of $H(x,y)$ that we have computed in (5.12). Differentiating that form of $H(x,y)$ with respect to y, we get

$$\frac{\partial}{\partial y}H(x,y) = \frac{x\left(\frac{1}{2}-\frac{1}{2}\sqrt{1-4x}\right)}{\left(1-y\left(\frac{1}{2}-\frac{1}{2}\sqrt{1-4x}\right)\right)^2}. \tag{5.14}$$

Now let us take the value of the derivative we just computed at $y = 1$, to get

$$\frac{\partial}{\partial y} H(x, y)\Big|_{y=1} = \frac{1 - \sqrt{1-4x}}{2x} + \frac{\sqrt{1-4x}}{2} - \frac{3}{2}. \tag{5.15}$$

Fortunately, the first two summands of the right-hand side are familiar, as we computed their power series forms in (3.21) and (3.24). Indeed, those formulas show that $\frac{1-\sqrt{1-4x}}{2x} = \sum_{n\geq 0} c_n x^n$, while $\sqrt{1-4x} = 1 - 2\sum_{n\geq 1} c_{n-1}x^n$, where $c_n = \binom{2n}{n}/(n+1)$ is the nth Catalan number.

Therefore,

$$\frac{\partial}{\partial y} H(x, y)\Big|_{y=1} = \sum_{n\geq 2} (c_n - c_{n-1})\, x^n.$$

Comparing this with (5.13), we see that the total number of all children whose parent is the root in all rooted plane trees of size n is $c_n - c_{n-1}$. On the other hand, we have seen in Section 5.5.2.1 that the *number* of these trees is c_{n-1}. Therefore, on average, the root of a rooted plane tree on n vertices has

$$\frac{c_n - c_{n-1}}{c_{n-1}} = \frac{c_n}{c_{n-1}} - 1 = \frac{4n-2}{n+1} - 1 = \frac{3n-3}{n+1}$$

children, for $n \geq 1$. So as n goes to infinity, the average number of children converges to 3.

5.5.3 Connected graphs

In the previous subsections, we have studied the number of *minimally* connected graphs, or trees. In this section, we look at the number of *all* connected graphs on vertex set $[n]$. This is a very large class of graphs, containing all trees which have only $n-1$ edges, and also containing many graphs with more edges. Just as the number of edges of a connected graph can be anything from $n-1$ to $\binom{n}{2}$, the *structure* of a connected graph can also be a lot more varied than that of a tree. A tree was characterized as a connected acyclic graph, while a connected graph can contain any number of cycles.

All these difficulties show that it may be too much to ask for a compact, explicit formula for the number $C(n)$ of all connected graphs on $[n]$. However, we will get reasonably close to that goal by finding a way to compute the exponential generating function $C(x) = \sum_{n\geq 0} C(n)\frac{x^n}{n!}$ of the numbers $C(n)$.

To begin, let $G(n)$ be the number of all *graphs* on vertex set $[n]$. Then $G(n) = 2^{\binom{n}{2}}$ since each 2-subset of $[n]$ may or may not form an edge. Let $G(x)$ be the exponential generating function of the numbers $G(n)$, that is,

$$G(x) = \sum_{n\geq 0} G(n)\frac{x^n}{n!} = \sum_{n\geq 0} \frac{2^{\binom{n}{2}}}{n!} x^n = 1 + x + x^2 + \frac{8}{3!}x^3 + \frac{64}{4!}x^4 + \cdots.$$

The crucial observation is that every graph is just the union of its connected components. Therefore, the generating functions $C(x)$ and $G(x)$ are connected through the exponential formula, yielding

$$G(x) = e^{C(x)},$$

where we set $C(0) = 0$. Therefore, we have

$$C(x) = \ln(G(x)).$$

This equation enables us to easily compute the first few coefficients of $C(x)$, that is, the numbers of connected graphs on $[n]$ for small values of n.

Indeed, most mathematical software packages can compute the logarithm of a formal power series. For instance, to get the first six values of $C(n)$, we can type the following in Maple:

```
with(powerseries):
t:=powpoly(1+x+x^2+(8/6)*x^3+(64/24)*x^4+(1024/120)*x^5+
(32768/720)*x^6,x):
s:=powlog(t),
tpsform(s,x,7);
```

Hitting return, and bringing the coefficient of x^n to the form in which its denominator is $n!$, we get that

$$C(x) = x + \frac{1}{2!}x^2 + \frac{4}{3!}x^3 + \frac{38}{4!}x^4 + \frac{728}{5!}x^5 + \frac{26704}{6!}x^6 + \cdots .$$

As the first k coefficients of $\ln(F(x))$ only depend on the first k coefficients of $F(x)$, the numbers 1, 1, 4, 38, 728, and 26704 are indeed the first six values of $C(n)$.

5.5.4 Eulerian graphs

A newspaper carrier has to deliver the morning paper to houses on nine streets in a subdivision. He can start anywhere, as long as he finishes at the same point where he started. He wants to be as efficient as possible and therefore does not want to go through any streets twice. Can he do this and still deliver all papers if the layout of the subdivision is as shown in Figure 5.40?

After a while, the carrier realizes that he cannot do his job without walking through some streets twice (and therefore that he will not like this subdivision very much). He is interested, however, in finding a method to determine whether he can do the same in other subdivisions. To that end, he has to understand *why* he failed in this subdivision.

We will not surprise anyone by suggesting that we view the streets of the subdivision as the edges of a graph G, and the crossings as the vertices of G. The question is then whether there exists a walk in G that uses each edge of G exactly once and ends where it started. Such a walk will be called a *closed Eulerian walk*, and if G has a closed Eulerian walk, then we will say that G is *Eulerian*.

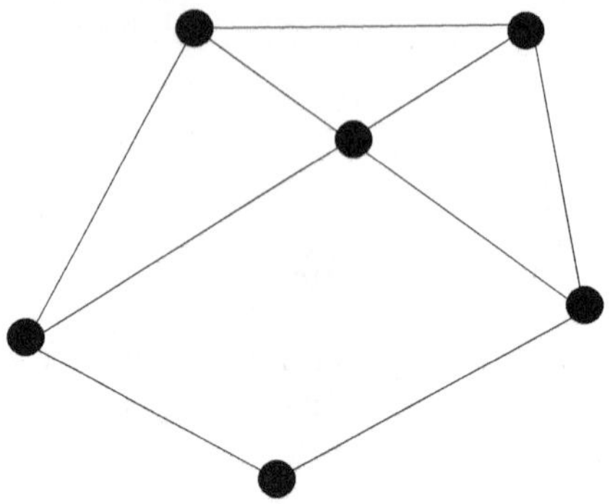

Figure 5.40
The nine streets of the subdivision.

Fortunately for all newspaper carriers of the world, there exists a simple characterization of Eulerian graphs. This theorem is valid for all graphs, not just simple graphs. This is the content of the next theorem, which was first proved by Euler.

Theorem 5.56 *Let G be a graph. Then G is Eulerian if and only if G is connected and all vertices of G have even degree.*

Proof: It is easy to see that the conditions are necessary. Indeed, if G is not connected, then walks starting in one component will never get to other components. Furthermore, every time we enter a vertex on a closed Eulerian walk w, we will have to leave that vertex using another, previously unused edge. Therefore, if G is Eulerian, all its vertices must have even degree. (Explain why this is true even for the vertex where w starts!)

We prove that the conditions are sufficient by induction on m, the number of edges in G. If $m = 2$, then G must be one of the graphs shown in Figure 5.41, and we are done.

Now assume we know the statement for all graphs having fewer than m edges, and let G have m edges, all even degrees, and be connected. Start walking at a vertex v of G using new edges in each step. Since G is finite, eventually we will get back to v, completing a closed walk C. If $C = G$, then C is a closed Eulerian walk, and we are done. If not, then remove the edges of C from G, to get the new graph G'.

Then G' still has only even degrees and fewer than m edges. So if G' is connected, then, by our induction hypothesis, it has a closed Eulerian walk w'. In that case, $C \cup w'$ is a closed Eulerian walk of G.

However, it can happen that G' is not connected, since the removal of C could split G into components. In that case, each connected component of G' has only even degrees, and therefore a closed Eulerian walk. Then the

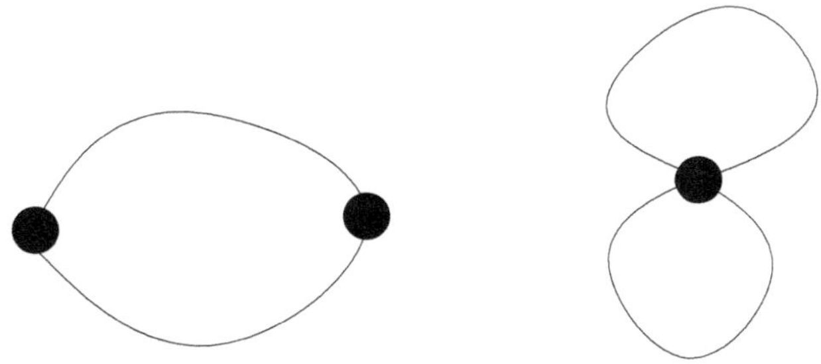

Figure 5.41
Connected graphs with two edges and all degrees even.

union of these closed Eulerian walks and C is a closed Eulerian walk of G. See Figure 5.42.
$\diamondsuit$

The enumeration of all simple Eulerian graphs on vertex set $[n]$ has some surprising results. First, let us consider graphs that are not necessarily connected, but each component of which is Eulerian. It then follows immediately from the previous theorem that the graphs we are looking for are precisely the simple graphs on $[n]$ in which each vertex has an even degree. Let us call these graphs *even* graphs.

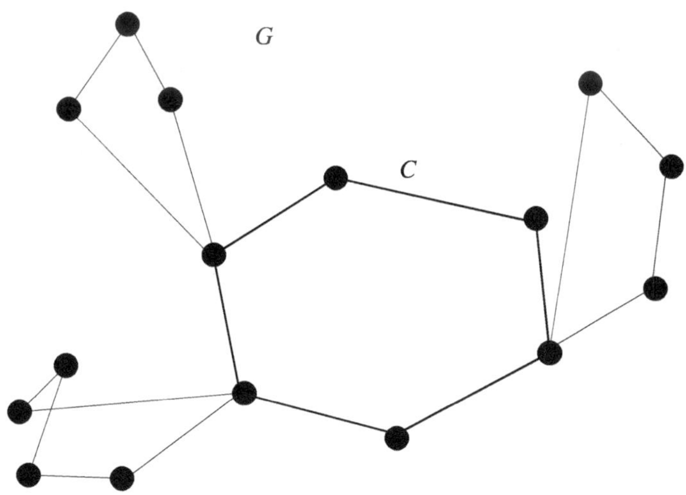

Figure 5.42
The closed walk C connects the closed Eulerian walks of the components of G'.

A little thought shows that there is one even graph on [2] (namely, the empty graph), there are two even graphs on [3] (namely, the empty graph and the complete graph), and there are eight even graphs on [4] (namely, the empty graph, four graphs consisting of a triangle, and three graphs consisting of a quadrilateral). Further investigation yields 64 even graphs on [5]. This starts looking suspicious, since the numbers we get, 1, 2, 8, and 64, are all powers of two. If we look at the exponent of two in these powers, we get the numbers 0, 1, 3, 6, which are precisely the numbers $\binom{i}{2}$ for $i = 1, 2, 3, 4$. Therefore, it seems that the number of even graphs on $[n]$ is $2^{\binom{n-1}{2}}$. This is very interesting because this suggests that there are as many *even* graphs on $[n]$ as there are *graphs* on $[n-1]$. It turns out that this is indeed the case.

Theorem 5.57 *For all positive integers n, the number of even graphs on vertex set $[n]$ is $2^{\binom{n-1}{2}}$.*

Proof: We prove the theorem by presenting a bijection from the set A of all graphs on $[n-1]$ onto the set B of all even graphs on $[n]$. Let $G \in A$. To define $f(G)$, add the new vertex n to G and connect it to all vertices of G that have *odd* degree. Call the obtained graph $f(G)$. See Figure 5.43 for an illustration. This operation will turn the degrees of all old vertices of G into even numbers. Furthermore, the degree of n will be even, as G must have an even number of vertices of odd degree. Therefore, $f(G) \in B$. The function $f : A \to B$ is a bijection, since the unique preimage of any graph $H \in B$ is obtained by omitting vertex n and the edges adjacent to it from B. $\diamond$

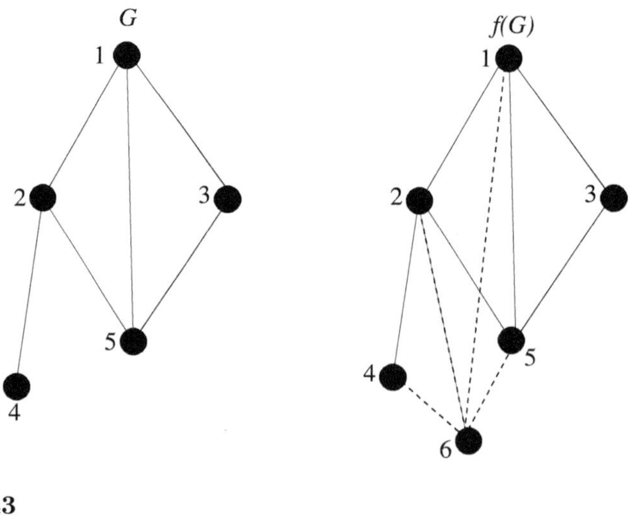

Figure 5.43
Turning a graph into an even graph.

The connection between even graphs and Eulerian graphs is the same as the connection between all simple graphs and connected simple graphs. That is, in both of these pairs, the *connected components* of the first class of graphs are precisely the graphs from the second class. Let us express this fact by generating functions.

Set $b_n = 2^{\binom{n-1}{2}}$ for $n \geq 1$, and $b_0 = 1$. Then the exponential generating function of the numbers of even graphs on $[n]$ is $B(x) = 1 + \sum_{n\geq 1} 2^{\binom{n-1}{2}} \frac{x^n}{n!}$. Let $Eu(x) = \sum_{n\geq 1} Eu(n)\frac{x^n}{n!}$, where $Eu(n)$ denotes the number of Eulerian graphs on vertex set $[n]$. Then the exponential formula yields

$$B(x) = e^{Eu(x)},$$

$$Eu(x) = \ln(B(x)).$$

We can now use a software package to compute the first few coefficients of $Eu(x)$ just as we did for the generating function of connected graphs. We get that

$$Eu(x) = x + \frac{1}{3!}x^3 + \frac{3}{4!}x^4 + \frac{38}{5!}x^5 + \frac{720}{6!} + \cdots.$$

So there are three Eulerian graphs on vertex set $[4]$, there are 38 Eulerian graphs on vertex set $[5]$, and 720 Eulerian graphs on vertex set $[6]$.

Quick Check

1. Find the exponential generating function for the number of simple graphs on vertex set $[n]$ in which every vertex has degree 2.
2. Find the exponential generating function for the number of simple graphs on vertex set $[n]$ in which every vertex has degree at most 2.
3. Let a_n be the number of rooted plane trees on $2n + 1$ unlabeled vertices in which every nonleaf vertex has two children. Find the ordinary generating function of the numbers a_n. Where have you seen that generating function before?

5.6 The Lagrange Inversion Formula

We have seen examples of trees varieties where *each* vertex of a tree is labeled, and of tree varieties where *no* vertex is labeled. Now we will consider trees where *only leaves* are labeled. A *k-phylogenetic tree* is a rooted non-plane trees whose *leaves* are bijectively labeled with the elements of the set $[n] = \{1, 2, \cdots, n\}$, and in which each non-leaf vertex has exactly k children. See Figure 5.44 for the set of all three 2-phylogenetic trees on label set $[3]$.

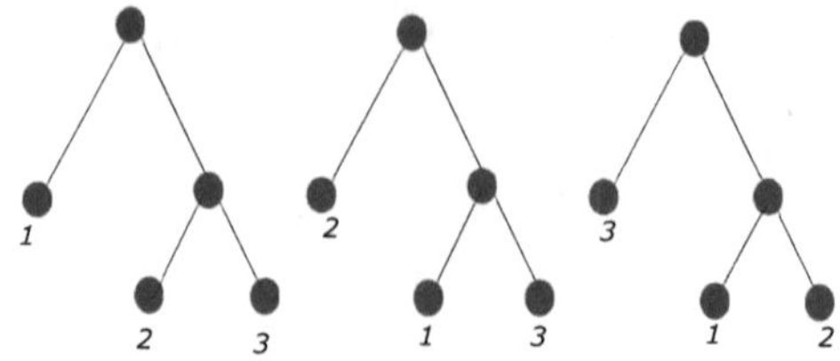

Figure 5.44
The three 2-phylogenetic trees on leaf set [3].

Let $t_{k,n}$ be the number of k-phylogenetic trees on leaf set $[n]$, and set $t_{k,0} = 0$. Let $T_k(x) = \sum_{n\geq 0} t_{k,n} \frac{x^n}{n!}$ be the exponential generating function of the sequence of these numbers.

Removing the root of such a tree, we get either the empty set, or an unordered set of k such trees, leading to the functional equation

$$T_k(x) = x + \frac{T_k^k(x)}{k!}. \tag{5.16}$$

The form

$$T_k(x) - \frac{T_k^k(x)}{k!} = x \tag{5.17}$$

will be particularly useful for our purposes.

If $k = 2$, then (5.16) is the quadratic equation $T_2(x) = x + \frac{T_2^2(x)}{2}$. Solving this equation, we get the formula

$$T_2(x) = 1 - \sqrt{1 - 2x},$$

which implies that

$$t_{2,n} = (2n-3)!!.$$

However, for $k > 2$, we need stronger techniques to find an explicit formula for t_n.

To that end, we first define the notion of the *compositional inverse* of a power series.

Definition 5.58 *Let $F(x)$ be a formal power series. We say that the formal power series $G(x)$ is the* compositional inverse *of $F(x)$ if $F(G(x)) = G(F(x)) = x$. In this case, we write $F^{\langle -1\rangle}(x) = G(x)$.*

Example 5.59 *If $F(x) = 3x + 2$, then $F^{\langle -1\rangle}(x) = (x-2)/3$.*

Solution: Indeed, $F\left(\frac{x-2}{3}\right) = (x-2)+2 = x$. Also, $F^{\langle -1\rangle}(3x+2) = 3x/3 = x$. $\Diamond$

Note that for the compositions $F(G(x))$ and $G(F(x))$ of *formal power series* to be defined, it suffices to assume that the constant term of F and the constant term of G are 0. In what follows, we will work with formal power series that satisfy that condition. We point out that if F has infinitely many nonzero coefficients and G has a nonzero constant term, then $F(G(x))$ is not defined.

Example 5.60 *If* $F(x) = \sum_{n\geq 1} x^n = x/(1-x)$, *then* $F^{\langle -1\rangle}(x) = \sum_{n\geq 0}(-1)^{n-1}x^n = x/(1+x)$.

Solution: We have

$$F\left(\frac{x}{1+x}\right) = \frac{(x/(1+x)}{1-[x/(1+x)]} = \frac{x}{1+x-x} = x. \tag{5.18}$$

◊

Example 5.61 *If* $F(x) = \ln(1+x)$, *then* $F^{\langle -1\rangle}(x) = e^x - 1$.

The following lemma shows that for power series with 0 as a constant term, it is easy to decide if their inverse exists.

Lemma 5.62 *Let* $F(x) = f_1x + f_2x^2 + \cdots$ *be a formal power series. Then* $F^{\langle -1\rangle}(x)$ *exists if and only if* $f_1 \neq 0$. *In that case,* $F^{\langle -1\rangle}(x)$ *is unique.*

Proof: If $f_1 = 0$, then $G(F(x))$ will have coefficient 0 for its linear term, so $G(F(x)) \neq x$. Now let us assume that $f_1 \neq 0$. Let us see if we can find a power series $G(x) = \sum_{n\geq 1} g_nx^n$ so that $G(F(x)) = x$. (The reader is invited to verify that such a power series $G(x)$ must have a 0 constant term.) Equating the coefficients of x^n for $n = 1, 2, 3\cdots$ in $G(F(x))$ and x, we get

$$G(F(x)) = g_1F(x) + g_2F^2(x) + g_3F^3(x) + \cdots,$$

which implies that $G(F(x)) = x$ if and only if

$$\begin{aligned} g_1f_1 &= 1, \\ g_1f_2 + g_2f_1^2 &= 0, \\ g_1f_3 + g_2f_1f_2 + g_3f_1^3 &= 0, \end{aligned}$$

and so on. We can solve the first equation for g_1 as $f_1 \neq 0$. After that, we can solve the second one for g_2, the third one for g_3, and so on. There will never be a problem of dividing by 0, since in the nth equation in the array above, the coefficient of g_n will always be $f_1^n \neq 0$. ◊

A *formal Laurent series* is a power series with finitely many negative exponents. That is, a formal Laurent series with complex coefficients is of the form $\sum_{i\geq m} a_ix^i$, where $m \in \mathbf{Z}$. We are now in a position to state and prove the main result of this section.

Theorem 5.63 *Let n and ℓ be positive integers, and let $F^{\langle -1\rangle}(x)$ be the compositional inverse of the power series $F(x)=\sum_{i\geq 1} f_i x^i$. Then*

$$n[x^n](F^{\langle -1\rangle}(x))^\ell = \ell[x^{n-\ell}]\left(\frac{x}{F(x)}\right)^n. \tag{5.19}$$

Note that in particular, the case of $\ell = 1$ gives us the coefficients of $F^{\langle -1\rangle}(x)$. This is the most frequently used special case of the Lagrange Inversion Formula.

The following proof is due to Joseph-Louis Lagrange himself. The interested reader can consult Chapter 5 of [72] for additional proofs.

Proof: Let

$$F^{\langle -1\rangle}(x)^\ell = \sum_{i\geq \ell} h_i x^i.$$

Substituting $F(x)$ for x on both sides of the above equality, we get

$$x^\ell = \sum_{i\geq \ell} h_i F(x)^i.$$

Note that the last displayed formula contains the coefficients of $F^{\langle -1\rangle}(x)^\ell$, but not $F^{\langle -1\rangle}(x)$ itself. This is a significant step toward expressing these coefficients in terms of $F(x)$.

Let us take the derivative of both sides to get

$$\ell x^{\ell-1} = \sum_{i\geq \ell} i h_i F(x)^{i-1} F'(x).$$

After rearranging, this yields

$$\frac{\ell x^{\ell-1}}{F(x)^n} = \sum_{i\geq \ell} i h_i F(x)^{i-1-n} F'(x). \tag{5.20}$$

Let us compare the coefficients of x^{-1} on both sides of (5.20). Note that if $i \neq n$, then

$$F(x)^{i-1-n}F'(x) = \frac{\left(F(x)^{i-n}\right)'}{i-n}.$$

In other words, if $i \neq n$, then $h_i F(x)^{i-1-n}F'(x)$ is a constant multiple of the *derivative of a Laurent series*, and as such, its coefficient for the x^{-1} term is 0. (Since the antiderivative of x^{-1} is not a power function.) Therefore, the only summand on the right-hand side of (5.20) that will contribute to the coefficient of x^{-1} is the summand indexed by $i = n$. That coefficient is

$$\begin{aligned}
[x^{-1}]nh_nF(x)^{-1}F'(x) &= [x^{-1}]nh_n\left(\frac{h_1 + 2h_2x + 3h_3x^2_+ \cdots}{h_1x + h_2x_2 + \cdots}\right)\\
&= [x^{-1}]nh_n\left(\frac{1}{x} + \cdots\right)\\
&= nh_n,
\end{aligned}$$

which is the expression on the left-hand side of (5.19), that was to be proved. Now returning to (5.20), notice that

$$[x^{-1}]\frac{\ell x^{\ell-1}}{F(x)^n} = [x^{n-\ell}]\frac{\ell x^{n\ell}}{F(x)^n},$$

which is the right-hand side of (5.19).

◊

Now we can return to our k-phylogenetic trees. Formula (5.17) shows that $F(T(x)) = x$, so $T_k(x) = F^{\langle -1\rangle}(x)$, where

$$F(x) = \left(x - \frac{x^k}{k!}\right)^{\langle -1\rangle}.$$

Therefore, we can compute the coefficients of $T_k(x)$ using Theorem 5.63.

That theorem yields

$$n[x^n]T_k^\ell(x) = \ell[x^{n-\ell}]\left(\frac{x}{x - \frac{x^k}{k!}}\right)^n.$$

From this, we compute

$$\begin{aligned}
[x^n]T_k^\ell(x) &= \frac{\ell}{n}[x^{n-\ell}]\left(1 - \frac{x^{k-1}}{k!}\right)^{-n} \\
&= \frac{\ell}{n}[x^{n-\ell}]\sum_{s\geq 0}\binom{-n}{s}\left(-\frac{x^{k-1}}{k!}\right)^s \\
&= \frac{\ell}{n}[x^{n-\ell}]\sum_{s\geq 0}\binom{n+s-1}{s}\frac{x^{s(k-1)}}{k!^s}.
\end{aligned}$$

So, setting $n-\ell = s(k-1)$, we have $n = s(k-1)+\ell$, and the last displayed chain of equalities implies that

$$[x^n]T_k^\ell(x) = \frac{\ell}{s(k-1)+\ell}\binom{ks+\ell-1}{s}\frac{1}{k!^s}. \tag{5.21}$$

Note that in particular, for $\ell = 1$, we get

$$[x^n]T_k(x) = \frac{1}{s(k-1)+1}\binom{ks}{s}\frac{1}{k!^s}. \tag{5.22}$$

The previous example was about nonplane trees, and used exponential generating functions. The next example shows that the Lagrange Inversion Formula can be used in situations in which ordinary generating functions appear.

Example 5.64 *How many plane rooted unlabeled trees are there on n vertices in which every vertex has either 0 or k children?*

Solution: Let $U_k(x)$ be the ordinary generating function counting such trees. That is, $U_k(x) = \sum_{n\geq 1} u_{n,k}x^n$, where $u_{n,k}$ is the number of such trees on n vertices. Then

$$U_k(x) = x + xU_k(x)^k = x\left(1 + U_k(x)^k\right), \tag{5.23}$$

since removing the root of such a tree that consists of more than one vertex results in a k-element sequence of such trees.

Rearranging (5.23), we get

$$\frac{U_k(x)}{1 + U_k(x)^k} = x,$$

so

$$U_k(x) = \left(\frac{x}{1+x^k}\right)^{\langle -1\rangle}.$$

Setting $\ell = 1$, the Lagrange Inversion Formula yields

$$\begin{aligned}
[x^n]U_k(x) &= \frac{1}{n}[x^{n-1}]\left(\frac{x}{(x/(1+x^k))}\right)^n \\
&= \frac{1}{n}[x^{n-1}]\left(1+x^k\right)^n \\
&= \frac{1}{n}[x^{n-1}]\sum_{s\geq 0}\binom{n}{s}\left(x^k\right)^s \\
&= \frac{1}{n}[x^{n-1}]\sum_{s\geq 0}\binom{n}{s}x^{ks}.
\end{aligned}$$

So $u_{n,k} = 0$ if $n-1$ is not divisible by k, and

$$u_{n,k} = \frac{1}{n}\binom{n}{s} \tag{5.24}$$

if $n = ks+1$.

Note that for $k = 2$, this means that there are no such trees on n vertices if n is odd, and there are $\frac{\binom{2m+1}{m}}{2m+1} = \frac{\binom{2m}{m}}{m+1}$ such trees if $n = 2m+1$. Compare this result with Theorem 5.28. ◊

When we discussed Theorem 5.55, we were not yet able to deduce a formula for the number of all rooted trees on vertex set $[n]$ from the equality $T(x) = xe^{T(x)}$. Now, with the Lagrange Inversion Formula at hand, we can achieve that.

Example 5.65 *The number of rooted trees on vertex set $[n]$ is n^{n-1}.*

Solution: Let t_n be the number of rooted trees on n vertices, with $t_0 = 0$, and let $T(x) = \sum_{n\geq 1} t_n \frac{x^n}{n!}$ be the exponential generating function of the sequence of the numbers t_n. As we mentioned in the proof of Theorem 5.55, a rooted tree naturally decomposes to its root (that has generating function x) and a set of rooted trees (that has generating function $e^{T(x)}$.) Therefore, the Product Formula yields

$$T(x) = xe^{T(x)}. \tag{5.25}$$

So $T(x)/e^{T(x)} = x$, and therefore,

$$T(x) = \left(\frac{x}{e^x}\right)^{\langle -1\rangle}.$$

Setting $\ell = 1$, the Lagrange Inversion Formula yields

$$\begin{aligned}
[x^n]T(x) &= \frac{1}{n}[x^{n-1}]e^{nx} \\
&= \frac{1}{n}[x^{n-1}]\sum_{m\geq 0}\frac{(nx)^m}{m!} \\
&= \frac{1}{n}\frac{n^{n-1}}{(n-1)!} \\
&= \frac{1}{n}\frac{n^n}{n!} \\
&= \frac{n^{n-1}}{n!}.
\end{aligned}$$

So $t_n = n^{n-1}$ as claimed. ◊

Quick Check Exercises 1 and 2, as well as Supplementary Exercise 39 cover applications of the Lagrange Inversion Formula with $\ell > 1$.

Quick Check

1. Use the Lagrange Inversion Formula to compute the number of rooted non-plane forests with two components on vertex set $[n]$.
2. Let $k \geq 2$ be a fixed integer. Use the Lagrange Inversion Formula to compute the number of ordered pairs of plane rooted unlabeled trees in which each vertex has either k or zero children.
3. Prove that the following statement is equivalent to the Lagrange Inversion Formula. If $x = f(x)/G(f(x))$, where G is a formal power series with a nonzero constant term, while f is a formal power series with a zero constant term, then

$$n[x^n]f(x)^k = k[x^{n-k}]G(x)^n. \tag{5.26}$$

5.7 Notes

An early classic on counting graphs is the book by Harary and Palmer, *Graphical Enumeration* [44]. Further uses of the exponential formula in graph theory can be found in *Enumerative Combinatorics, volume 2*, by Richard Stanley [72].

For three recent, high-level surveys, the reader may consult Chapters 4, 5, and 6 of the *Handbook of Enumerative Combinatorics* [18]. The first survey, by Michael Drmota, discusses the enumeration of trees, the second, by Gilles Schaeffer, covers the enumeration of planar maps, and the third, by Mark Noy, considers the enumeration of graphs in general.

Theorems 5.10, 5.49, and 5.54 are all special cases of a very general theorem, called the *matrix-tree theorem.* If G is a graph, the tree T is called a *spanning tree* of G if the vertex set of G is the same as that of T and the set of edges of G contains the set of edges of T. It is then a natural question to ask what the number $span(T)$ of spanning trees of G is.

Example 5.66 *The graph G shown in Figure 5.45 has three spanning trees, so $span(G) = 3$.*

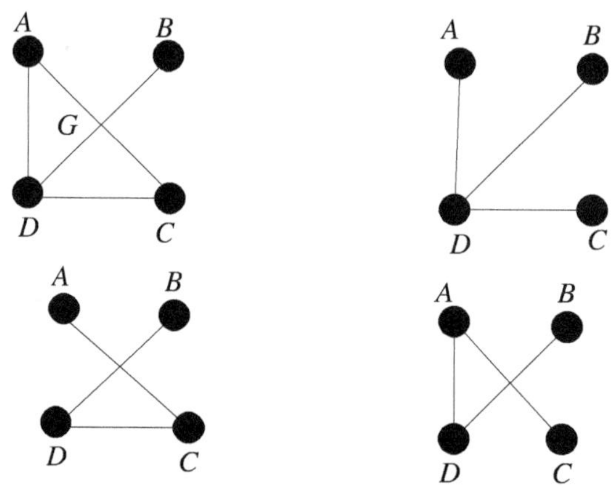

Figure 5.45
A graph and its three spanning trees.

In order to understand the matrix-tree theorem, the reader needs some background in linear algebra. If the reader has no such background, the reader can find the necessary definitions in any introductory textbook on linear algebra, such as [52]. For any graph G on vertex set $[n]$ that has no loops but may have multiple edges, we define the *Laplacian* of G as follows: The Laplacian

$\ell(G)$ of G is an $n \times n$ matrix whose entries are given by the rule

$$L(G)_{i,j} = \begin{cases} d_i \text{ if } i = j, \\ -m \text{ if there are } m \text{ edges between } i \text{ and } j. \end{cases}$$

The matrix-tree theorem uses $\ell(G)$ to determine the number of spanning trees of G as follows.

Theorem 5.67 (Matrix-tree theorem) *If G is a graph with no loops and $\ell(G)$ is its Laplacian, then for all i and j,*

1. $$span(G) = (-1)^{i+j} \det L_0(G),$$
   where $\ell_0(G)$ is just $\ell(G)$ with its ith row and jth column omitted, and also

2. $$span(G) = \frac{\lambda_1 \cdot \lambda_2 \cdot \cdots \cdot \lambda_{n-1}}{n},$$
   where $\lambda_1, \lambda_2, \cdots, \lambda_n$ are the eigenvalues of $\ell(G)$, with $\lambda_n = 0$.

Theorems 5.10, 5.49, and 5.54 are special cases of this theorem, namely, the special cases when $G = K_n$, or when in G any two vertices of different colors are connected by an edge. (In the following chapter, we will call those graphs *complete multipartite graphs.*)

For a proof and an introductory treatment of the matrix-tree theorem, the interested reader should consult *A Walk Through Combinatorics* [12]. A more advanced discussion can be found in Richard Stanley's above-mentioned book [72]. These books contain several alternative versions of the matrix-tree theorem.

We mention that a new class of parking functions, one in which cars are allowed to push another car out of a parking spot, has recently been introduced. These functions are called *Boston parking functions*, and the reader can find their precise definition in the paper [38] by F. Gilbey and L. Kalikow.

The Lagrange Inversion Formula has numerous generalizations. The following one is sometimes called the *Lagrange-Bürmann formula.*

Theorem 5.68 *Let G, and h be power series in variable x, and let us assume that $G(0) = 1$, and that*

$$f(x) = xG(f(x)). \tag{5.27}$$

Then there is a unique power series $f(x)$ that satisfies (5.27). Furthermore,

$$[x^n]h(f(x)) = \frac{1}{n}[f^{n-1}]\left(h'(f)G(f)^n\right). \tag{5.28}$$

See Theorem 5.1.1 in [84] for a direct proof that involves complex integration.

5.8 Chapter review

(A) Explicit formulae

1. Cayley's formula: The number of trees on $[n]$ is $T_n = n^{n-2}$.
2. Scoins' theorem: The number of properly colored trees on n labeled red vertices and m labeled blue vertices is $n^{m-1}m^{n-1}$.
3. The number of properly k-colored trees on n vertices, n_i of which are of color i and are bijectively labeled by the elements of $[n_i]$, is $n^{k-2}\prod_{i=1}^{k}(n-n_i)^{n_i-1}$.
4. Forests and functions:
   (a) The number of rooted forests on $[n]$,
   (b) the number of parking functions on $[n]$, and
   (c) the number of acyclic functions on $[n]$ is
$$P(n) = (n+1)^{n-1}.$$
5. The number of rooted plane trees on $n+1$ unlabeled vertices is
$$c_n = \frac{\binom{2n}{n}}{n+1}.$$
6. The number of binary plane trees on n vertices is c_n.
7. The Lagrange Inversion Formula: Let n and ℓ be positive integers, and let $F^{\langle -1\rangle}(x)$ be the compositional inverse of the power series $F(x) = \sum_{i\geq 1} f_i x^i$. Then
$$n[x^n](F^{\langle -1\rangle}(x))^{\ell} = \ell[x^{n-\ell}]\left(\frac{x}{F(x)}\right)^{n}. \tag{5.29}$$

(B) Other results

1. Chromatic polynomials:
   (a) The function $\chi_G(n)$ counts proper colorings of G using only colors from $[n]$. It is a polynomial function.
   (b) The number $\chi_G(-1)$ is equal to the number of acyclic orientations of G.
2. Generating functions:
   (a) Let $G(x) = \sum_{n\geq 0} \frac{2^{\binom{n}{2}}}{n!}x^n$, and let $C(x)$ be the exponential generating function for the numbers of all *connected* graphs on labeled vertices, with $C(0) = 0$. Then
$$C(x) = \ln(G(x)).$$

(b) Let $B(x) = 1 + \sum_{n\geq 1} 2^{\binom{n-1}{2}} \frac{x^n}{n!}$, and let

$$Eu(x) = \sum_{n\geq 1} Eu(n) \frac{x^n}{n!},$$

where $Eu(n)$ denotes the number of Eulerian graphs on vertex set $[n]$. Then

$$Eu(x) = \ln(B(x)).$$

5.9 Exercises

1. Prove that any finite simple graph with at least two vertices has two vertices that have the same degree.
2. Let G be any finite graph, and let x and y be two vertices of G. Prove that if there is a walk in G from x to y, then there is also a path from x to y in G.
3. Prove that the finite simple graph G is a tree if and only if, for any two distinct vertices x and y, there is exactly one path connecting x to y.
4. Let T be a tree, let P be a path in T, and let v be a vertex of T outside P. Prove that there is a vertex $w \in P$ that is closer to v than any other vertices of P.
5. Prove that there are at least 25 trees on 10 vertices so that no two of them are isomorphic.
6. + We say that the parking function f has no *like consecutive elements* if there is no i in the domain of f so that $f(i) = f(i+1)$. Find a formula for the number of parking functions on $[n]$ that have no like consecutive elements.
7. Let $f : [n] \rightarrow [n]$ be a parking function. Prove that the number of unsuccessful parking attempts in f is equal to $\binom{n+1}{2} - \sum_{j=1}^{n} f(j)$. An unsuccessful attempt is when a car tries to park in a spot but finds another car there.
8. Let us order the elements of the degree sequence of a graph in non-increasing order and call the obtained sequence the *ordered degree sequence* of the graph.

   How many ordered degree sequences are possible for a tree on n vertices? (Your answer can contain the function p denoting the number of partitions of an integer.)

9. + Let us generalize the notion of parking functions as follows: Let $1 \leq k \leq n$. We now have $n+1-k$ cars arriving at our parking lot, which still has n spots labeled 1 through n. Each car has a favorite spot, and the preferences are described by a function $f : [n+1-k] \rightarrow [n]$. One day, however, spots labeled less than k are closed for construction. If all of the $n+1-k$ cars can park using our usual parking process, then we say that f is a *k-shortened parking function* (a term coined by Catherine Yan). Note that 1-shortened parking functions are simply parking functions.

   (a) Find a sufficient and necessary condition for f being a k-shortened parking function.

   (b) Find a formula for the number of k-shortened parking functions $f : [n+1-k] \rightarrow [n]$.

10. Find a formula for the number $P(n,k)$ of parking functions on $[n]$ that contain exactly k values equal to 1.

11. + Let us call a parking function on $[n]$ *prime* if, by omitting a 1 from the parking function, we get a parking function on $[n-1]$. For instance, $1,1,2$ is a prime parking function, but $1,1,3$ is not. By convention, we say that the only parking function on $[1]$ is also prime.

    The rest of this exercise is meant to explain the name "prime parking functions" by showing how each parking function can be decomposed uniquely into a sequence of prime parking functions, the sum of whose sizes is n.

    To that end, prove that there exists a natural bijection from the set $\mathbf{P}(n)$ of all parking functions on $[n]$ onto the set $SPP(n)$ of strings

$$(p_1, p_2, \cdots, p_k, s_1, s_2, \cdots, s_k),$$

    where k ranges from 1 through n (so k is not fixed), each p_i is a prime parking function on $[a_i]$, and $\sum_{i=1}^{k} a_i = n$. Finally, $(s_1, s_2, \cdots, s_n)$ is a partition of the set $[n]$ so that block s_i has a_i elements. In other words, the bijection we are looking for will not only specify the prime parking functions into which our parking function is decomposed, but also their *location* within the original parking function.

12. Let $P(n,k)$ be the number of parking functions on $[n]$ containing exactly k values equal to 1, and let $PP(n,k)$ be the number of prime parking functions containing exactly k values equal to 1. Express $PP(n+1,k+1)$ by $P(n,k)$.

13. + Prove that we have $PP(n) = (n-1)^{n-1}$ for all $n \geq 1$.

14. Prove that

$$\sum_{n \geq 0} (n+1)^{n-1} \frac{x^n}{n!} = \frac{1}{1 - \sum_{n \geq 1} (n-1)^{n-1} \frac{x^n}{n!}}.$$

15. Find the number $u(n)$ of different unlabeled trees on n vertices for all $n \leq 6$.
16. For each of the three graphs shown in Figure 5.46, find the number of all automorphisms of that graph.

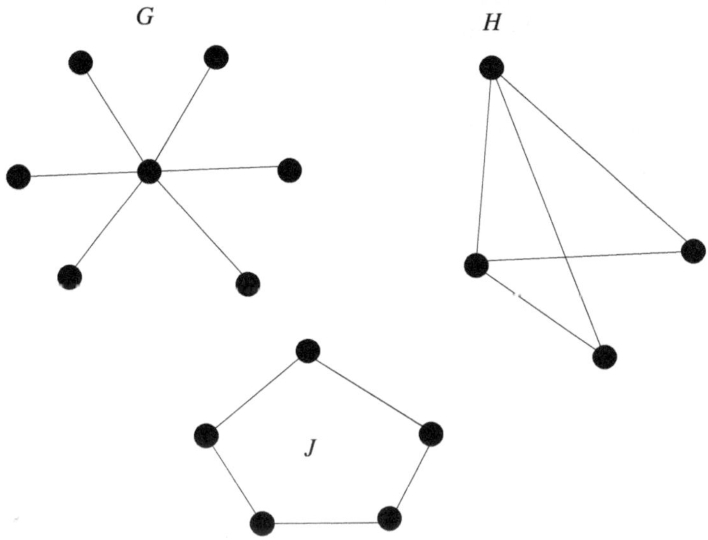

Figure 5.46
How many automorphisms do these graphs have?

17. Show an example of a graph on six vertices that has 720 labelings.
18. Let T be a rooted tree on $n+1$ unlabeled vertices. Let us now label the vertices of T by the elements of $[n+1]$ so that each label is used exactly once and the label of each vertex is smaller than the label of its parent. See Figure 5.47 for an example. We will say that trees labeled in this way are *decreasing* trees. Prove that the number of

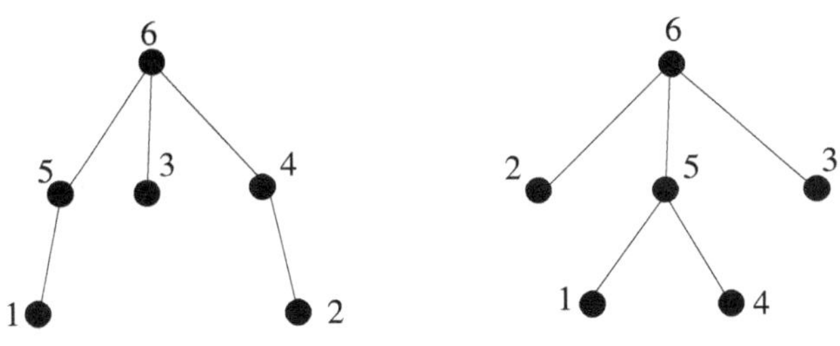

Figure 5.47
Decreasing trees on six vertices.

decreasing trees on $n+1$ vertices is $n!$.

19. Find a formula for the number of increasing trees on $n+1$ vertices in which the root has exactly k children.

20. (a) Find a bijection f from the set of 132-avoiding n-permutations, defined in Exercise 32 of Chapter 4, onto that of rooted plane trees on $n+1$ unlabeled vertices.

    (b) Give a bijection f as an answer to part (a) that turns the number of left-to-right minima of a permutation p into the number of leaves of the tree $f(p)$.

21. Pattern-avoiding permutations were defined in Exercise 32 of Chapter 4. Find a bijection from the set of all unlabeled binary plane trees on n vertices onto the set of all 132-avoiding n-permutations.

22. Use the result of the previous exercise to prove that the number of 132-avoiding n-permutations with k descents is the same as the number of 132-avoiding n-permutations with k ascents.

23. Compare the results of Exercises 20 and 22 and prove Theorem 5.29.

24. What is the number $a_{n,k}$ of acyclic functions on $[n]$ that have k fixed points?

25. Let $a_{n,k}$ be defined as in the previous exercise. Find a simple closed form for the generating function

$$A_n(x) = \sum_{k=1}^{n} a_{n,k} x^k.$$

26. Prove that, for all graphs G, the number of pairs (g, A) in which

    (a) A is an acyclic orientation of G, and

    (b) g is a coloring of the vertices of G using only colors from $[n]$ so that $g(u) > g(v)$ if there is an edge from u to v in A

    is equal to $\chi_G(n)$.

27. Let n and m be two integers with $n < m$, so that both n and m are roots of χ_G. Prove that, in that case, all integers in the interval $[n, m]$ are roots of χ_G.

28. + Prove that, if two graphs have identical chromatic polynomials, then they must have the same number of edges.

29. Let Co_n be the number of connected graphs on vertex set $[n]$.

    (a) Prove that the number of *rooted* graphs on vertex set $[n]$ is

$$\sum_{k=1}^{n} k \binom{n}{k} Co_k \cdot 2^{\binom{n-k}{2}}.$$

    A rooted graph is a graph in which one vertex is designated as the root.

(b) Prove that

$$Co_n = 2^{\binom{n}{2}} - \frac{1}{n}\sum_{k=1}^{n-1} k\binom{n}{k} Co_k \cdot 2^{\binom{n-k}{2}}.$$

30. Let G be any graph on vertex set $[n]$, with loops and multiple edges allowed. Let $A(G)$ be the $n \times n$ matrix whose (i, j)-entry is equal to the number of edges between i and j. Then $A(G)$ is called the *adjacency matrix* of G.

    Prove that the (i, j)-entry of $A(G)^k$ is equal to the number of *walks* of length k from i to j for any positive integers k.

31. Let $S(x) = \sum_{n\geq 1} s_n x^n$, where s_n is the number of rooted plane trees in which every nonleaf vertex has at least two children. Find a closed form for $S(x)$.

32. How many rooted plane trees are there on n unlabeled vertices in which each vertex has at most two children? Use the Lagrange Inversion Formula in your solution.

5.10 Solutions to exercises

1. Let G have n vertices. If G has an isolated vertex, then the maximum degree in G is $n-2$, meaning that the possible degrees of vertices are $0, 1, \cdots, n-2$. This is $n-1$ possibilities, so our claim follows by the pigeonhole principle.

   If G has no isolated vertices, then the possible degrees of vertices are $1, 2, \cdots, n-1$, which is again $n-1$ possibilities, and we conclude as above.

2. Let W be a walk from x to y. If W is a path, we are done. Otherwise, there is at least one vertex that occurs at least twice in W. Let z be the first such vertex. Then let us remove the part of W that is between the first and the second occurrences of z to get the new walk W', which still connects x and y. Now repeat the same procedure for W' instead of W, and keep doing this. The procedure will eventually have to stop, since each step decreases the number of edges in W. When the procedure stops, there are no more repeated vertices, so we have a path from x to y.

3. If G is a tree, then it is connected, so there is at least one path from x to y. If there were two, say p and p', then the symmetric difference of these two paths would contain a cycle. (In case you forgot, the symmetric difference of two sets is the set of elements

that belong to *exactly one* of the two sets.) Indeed, each component of this symmetric difference would be a graph in which each degree is two, and such a graph is a cycle.

Conversely, if G has the mentioned property, then G is connected and cycle free, since, if it had a cycle, there would be two paths between any two points of that cycle. Therefore, G is a tree.

4. Prove this by assuming the opposite, that is, that there are two vertices, x and y, that are closest to v in P, say at distance d. Then there would be two paths from v to x. One would be the shortest path of length d, and one would be the path of length $d+k$ that first goes from v to y, then from y to x along P. (Why is that a path?) This contradicts the property of trees we proved in the previous exercise.

5. There are 10^8 trees on 10 labeled vertices. Each of them can be isomorphic to at most $10! < 3.7 \cdot 10^6$ others (keep the tree fixed and permute the labels). Therefore, our claim follows by the pigeonhole principle.

6. This is similar to the proof of Theorem 5.24. We can again use the circular parking lot. The favorite spot of the first car can be any element of $[n+1]$, but there are only n possibilities (all but the favorite spot of the preceding car) for the favorite spot of each subsequent car. Therefore, there are $(n+1)n^{n-1}$ possible sets of parking preferences, and again, $1/(n+1)$ of them lead to parking functions. Therefore, the answer is n^{n-1}.

7. By induction on n. For $n = 1$ the statement is trivially true. Assume that we know the statement for $n-1$, and prove it for n. Let f be a parking function on $[n]$. Let i be the car that will end up parking in spot n if f governs the parking procedure.

   Then, omitting car i, its parking preference, and the last spot, we get a parking function f' on $[n-1]$, for which the induction hypothesis holds. That is, we know that the number of unsuccessful parking attempts for all cars excluding i is $\binom{n}{2} - \sum_{j \neq i} f(j)$. If we have $f(i) = a$, then car i had exactly $n-a$ unsuccessful parking attempts. Adding it to the previous expression, our claim is proved.

8. To solve this enumeration problem, we first characterize the possible ordered degree sequences. We claim that, for $n \geq 2$, a sequence $d_1 \geq d_2 \geq \cdots \geq d_n \geq 0$ is a possible sequence if and only if

   (a) $\sum_{i=1}^{n} d_i = 2n-2$, and

   (b) $d_{n-1} = d_n = 1$.

   On one hand, both conditions are necessary. Indeed, the first is equivalent to the graph having $n-1$ edges, and if the second one did not hold, we would have $d_i \geq 2$ for $i \leq n-1$, leading to $\sum_{i=1}^{n} d_i \geq 2n-1$, a contradiction since our tree has $n-1$ edges.

In order to prove that the two conditions together are sufficient, we use induction on n. If $n = 1$, then the only sequence allowed by the conditions is $d_1 = d_2 = 1$, and there is indeed a tree with that ordered degree sequence. Assume the statement is true for $n-1$, and prove it for n.

Let $S = d_1 \geq d_2 \geq \cdots \geq d_n$ be any allowed sequence with $n \geq 3$. Then there is a d_i in the sequence that is larger than 1. Choose the last such d_i. Now remove d_n from the sequence, and decrease d_i by 1. Then the obtained sequence S' satisfies the two criteria for $n-1$ (the sum of the elements is $2n-4$, and they are all positive, so the last two must be 1); therefore, the induction hypothesis applies. That is, there is a tree on $n-1$ vertices whose ordered degree sequence is S'. Adding a new vertex and connecting it to the vertex corresponding to d_i, we get a tree with ordered degree sequence S.

This completes the characterization of allowed sequences. For $n \geq 3$, their number is equal to the number of partitions $2n-4$ into $n-2$ parts (since the $(n-1)$st and nth parts must be 1).

9. (a) This is similar to the situation for parking functions. Roughly speaking, what causes a problem is when too many cars favor spots with high labels. The process will fail if more than one car wants the last spot, or in general, when more than j cars want a spot among the j last spots for any $j \in [n-k+1]$. (There are only $n-k+1$ cars, so we do not have to consider larger values of j.)

   We claim that, if, for any $j \in [n-k+1]$, there are at most j cars who want to park in one of the last j spots, then f is a k-shortened parking function. The reader is invited to prove this claim on her own. (Reviewing the proof of Lemma 5.23 may be helpful.) Note that the difference lies in the weaker condition on j. Indeed, here we only need that the condition holds for all $j \in [n-k+1]$, not all $j \in [n]$. Also note that an alternative description of the sufficient and necessary condition for f to be a k-shortened parking function is that the vector $(a_1, a_2, \cdots, a_{n-k+1})$ obtained by the nondecreasing rearrangement of the values $f(1), f(2), \cdots, f(n-k+1)$ is coordinate-wise smaller than or equal to the vector $(k, k+1, \cdots, n)$.

   (b) Consider again the circular parking lot of the proof of Theorem 5.24. We have $n+k-1$ cars, and they have parking preferences from $[n+1]$. Therefore, k spots will remain empty. If one of the empty spots is spot $n+1$, then f is a k-shortened parking function. We claim that that will happen in $k/(n+1)$

of all cases. Indeed, let $\mathbf{v} = (f(1), f(2), \cdots, f(n-k+1))$ be the vector of values of any function $f : [n-k+1] \rightarrow [n+1]$. For $i \in [n]$, define $\mathbf{v_i} = (f(1)+i, f(2)+i, \cdots, f(n-k+1)+i)$, where addition is meant modulo $n+1$. That is, the $\mathbf{v_i}$ are the circular translates of $\mathbf{v}$. Then exactly k of the $n+1$ vectors $\mathbf{v_i}$ will not contain $n+1$ as a coordinate since there are exactly k circular translations (rotations) that take $n+1$ into one of the k spots that do not appear in $\mathbf{v}$. This proves our claim.

As the number of all functions $f : [n-k+1] \rightarrow [n+1]$ is certainly $(n+1)^{n-k+1}$, the number of k-shortened parking functions $f : [n-k+1] \rightarrow [n]$ is $k(n+1)^{n-k}$.

10. First choose the k positions in which the k values equal to 1 will be. This can be done in $\binom{n}{k}$ ways. For the remaining $n-k$ positions, we need values so that the vector obtained by their nondecreasing rearrangement $(b_1, b_2, \cdots, b_{n-k})$ consists of coordinates larger than 1 and is coordinate-wise smaller than $(k+1, k+2, \cdots, n)$. Equivalently, the vector $(b_1 - 1, b_2 - 1, \cdots, b_{n-k} - 1)$ must consist of coordinates that are positive integers and must be coordinate-wise smaller than $(k, k+1, \cdots, n-1)$. In other words, it must be the nondecreasing rearrangement of a k-shortened parking function $f : [n-k] \rightarrow [n-1]$. We know from part (b) of the previous exercise that this is possible in kn^{n-1-k} ways. Therefore, we have

$$P(n,k) = \binom{n}{k} kn^{n-1-k} = \binom{n-1}{k-1} n^{n-k}.$$

11. Let p be a parking function on $[n]$. Find the smallest a_1 so that p contains only a_1 values that are elements of $[a_1]$, that is, the smallest a_1 for which the conditions of p being a parking function are satisfied in a tight way. If there is no such a_1, then p is prime, and the statement holds, with $k = 1$. Otherwise, there are a_1 cars whose parking preferences are within $[a_1]$, so their parking preferences form a parking function on $[a_1]$. Let s_1 be the set of these cars, and let p_1 be the parking function they determine. Then p_1 is prime, since a_1 was the *smallest* integer for which the parking function criteria held tightly, so a 1 can be omitted from p_1.

    Then iterate this procedure. That is, subtract a_1 from all remaining values of p to get a parking function p' on $[n-a_1]$. Look again for the smallest a_2 for which there are only a_2 values of p' that are elements of $[a_2]$, and so on. The procedure will stop when we get to a *prime* parking function; that function will be p_k.

    For instance, if $p = (1,7,1,6,2,1,4,7,7)$, then we have $a_1 = 5$. The values of p that are from $[5]$ are in the first, third, fifth, sixth, and seventh position, so $s_1 = \{1,3,5,6,7\}$ and $p_1 = (1,1,2,1,4)$. This leaves us with $p' = (2,1,2,2)$, which yields $a_2 = 1$. So $p_2 = (1)$ and $s_2 = \{4\}$, since the second value of p' was the fourth value of p. This in turn leaves us with $p'' = (1,1,1)$, which is a prime parking function. So the decomposition we obtain is

$$(1,1,2,1,4),(1),(1,1,1),\{1,3,5,6,7\},\{4\},\{2,8,9\}.$$

In order to recover the original parking function p from its decomposition $(p_1, p_2, \cdots, p_k, s_1, s_2, \cdots, s_k)$, simply put the values of p_i into the positions specified by s_i, then add $|s_1| + |s_2| + \cdots + |s_{i-1}|$ to each value of p_i.

12. Take a function counted by $P(n,k)$ and insert a 1 after any of its n values, or at the beginning. What we get is a prime parking function on $[n+1]$, containing $k+1$ values equal to one, and therefore a function counted by $PP(n+1, k+1)$. Each such function will be obtained $k+1$ times, since each of its entries equal to 1 can be the recently inserted 1. This shows

$$PP(n+1, k+1) = \frac{n+1}{k+1} P(n,k).$$

13. A prime parking function on $[n+1]$ must have at least two and at most $n+1$ values equal to 1. Using the result of the previous exercise and that of Exercise 10, we get that

$$\begin{aligned} PP(n+1) &= \sum_{k=1}^{n} PP(n+1, k+1) \\ &= \sum_{k=1}^{n} \frac{n+1}{k+1} P(n,k) \\ &= \sum_{k=1}^{n} \frac{n+1}{k+1} \cdot \binom{n-1}{k-1} n^{n-k} \\ &= \sum_{k=1}^{n} k \cdot \binom{n+1}{k+1} n^{n-k-1}. \end{aligned}$$

Dividing both sides by n^{n-1}, we get the identity

$$\frac{PP(n+1)}{n^{n-1}} = \sum_{k=1}^{n} \frac{k}{n^k} \cdot \binom{n+1}{k+1}.$$

Exercise 28 of Chapter 1 shows that the right-hand side is equal to n, and our result is proved.

14. We have seen that each parking function is decomposable into an *ordered* partition of prime parking functions. Therefore, the compositional formula (Theorem 3.36) applies, with $B(x) = \sum_{k \geq 0} k! \frac{x^k}{k!} = 1/(1-x)$.

15. For $n \leq 3$, there is only one unlabeled tree on n vertices, namely, the path on n vertices. For $n = 4$, we either have a vertex with degree three, and then we get a star, or we do not, and then we have a path. So $u(4) = 2$. If $n = 5$, then the maximum degree may be four (and then we get the star), or three (and then we get a star with an edge added to one of the leaves), or two (and then we get a path). So $u(5) = 3$. If $n = 6$, then the same holds for the cases when the maximum degree is five, or four, or two. However, when the maximum degree is three, there are several possibilities. Either there are two vertices with degree three, and then we get the tree shown on the left of Figure 5.48, or there is just one such vertex, leading to the trees on the right and at the bottom of Figure 5.48 (in one, the new vertices not connected to the vertex A of degree three are both at distance two from A; in the other one, one of them is at distance two from A, and the other one is at distance three). Therefore, $u(6) = 6$.

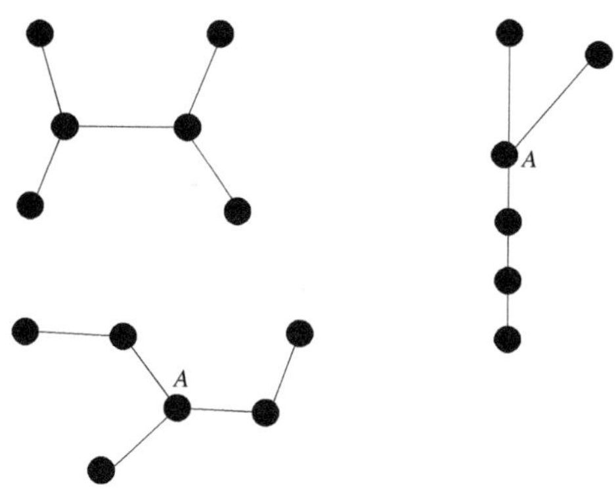

Figure 5.48
The three unlabeled trees on six vertices and maximum degree three.

16. We ask the reader to verify that the vertex at the center of the graph G has to be mapped into itself by every automorphism (which is the only vertex of degree six, and isomorphisms preserve degrees). Then any permutation of the remaining vertices is an automorphism. Therefore, G has $6! = 720$ automorphisms. In H, the two vertices of degree three can be interchanged, and the two vertices of degree two can be interchanged, so $|Aut(H)| = 4$. Finally, in J, let A and B be two neighboring vertices. Then $f(A)$ and $f(B)$ must be neighbors for any automorphisms f. We have five choices for the image of A, after which we have two choices for the image of B, and then our hands are tied. Therefore, J has 10 automorphisms.

17. By Theorem 5.7, this is equivalent to finding a graph on six vertices

that has only one automorphism, the trivial one. Figure 5.49 shows such a graph.

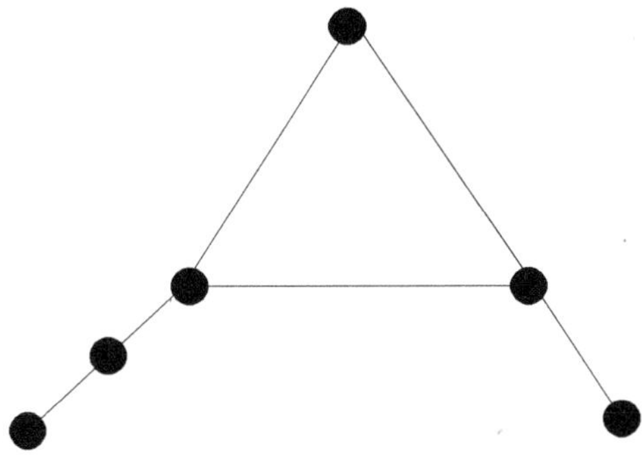

Figure 5.49
This graph has no nontrivial automorphisms.

18. We construct a bijection f from S_n into the set M_n of decreasing trees on $n+1$ vertices. Let $p \in S_n$ and let $p = p_1 p_2 \cdots p_n$. Each entry p_i will correspond to a vertex of $f(p)$. The unique parent of the vertex corresponding to p_i will be the vertex corresponding to p_j, where p_j is the *rightmost* entry on the left of p_i so that $p_j > p_i$. If there is no such entry p_j, then the parent of p_i is the extra vertex $n+1$. For instance, on the left of Figure 5.47, we see $f(34251)$, and on the right of that figure, we see $f(23514)$.

    To start with, $f(p) \in M_n$ since our definition results in a decreasing labeling of the obtained tree. Now we prove that f is a bijection by showing that it has an inverse. Let $M \in M_n$. Then the definition of f implies that the children of the root vertex $n+1$ must be the left-to-right maxima of p since they are precisely the entries that are larger than everything on their right, so it is for these entries that we will not find a p_j for a parent. So, if $f(p) = M$, then we can read off the set of left-to-right maxima of p from M. Because the left-to-right maxima of a permutation form an increasing sequence, we know their order as well. Say the left-to-right maxima are $m_1 < m_2 < \cdots < m_k$. Then the subtree of M that has m_i for its root uniquely describes the string of p starting in m_i and ending right before m_{i+1}, by iterated applications of this argument.

19. We have seen in the previous exercise that there is a bijection from the set of n-permutations with k left-to-right maxima and onto the set of our trees. Since the number of the first one is $c(n,k)$ by the transition lemma, this is the number of the latter as well.

20. (a) A 132-avoiding permutation p can be cut into blocks so that each entry of each given block is larger than all entries on the

right of that block. For instance, 78453612 can be cut into blocks as 78|4536|12. In general, the first cut comes immediately on the right of the maximal entry, then we proceed recursively to the right. If p ends with its maximal entry, then p is just one block.

An induction proof along the lines of the proof of Theorem 5.28 is now straightforward, with permutations ending in n corresponding to trees in which the root has only one child.

(b) The bijection of part (a) will work. This can be seen by induction, since the number of leaves of a rooted plane tree is just the sum of the numbers of leaves in the subtrees of the root. Similarly, the number of left-to-right minima of a 132-avoiding permutation is just the sum of the numbers of left-to-right minima in the blocks.

21. The definition of 132-avoiding permutations says that p is 132-avoiding if we cannot find three entries a, b, and c in p so that a is the leftmost of them, c is the rightmost of them, and b is the largest of them, while $a < c$.

This means that, in the binary plane tree $T(p)$ of p, all entries of $T(L)$ must be larger than all entries of $T(R)$. (Otherwise, if entries a and c violate this constraint, then anc is a 132-pattern.) Recursively, the same must hold for each node of $T(p)$. That is, for each node of $T(p)$, the labels of any vertex in the left subtree of that node must be larger than the label of any vertex in the right subtree of that node.

In other words, once the unlabeled tree $T(p)$ is given and we know that p is 132-avoiding, we can recover p by selecting the labels so that the conditions of the previous paragraph hold.

For a more explicit description of this argument, say that we have already defined the bijection f for binary plane trees on fewer than n vertices. Let T be a binary plane tree on n vertices, with left subtree T_L and right subtree T_R. Then we define $f(T) = f(T_L)'nf(T_R)$, where $f(T_L)'$ is obtained by adding $|T_R|$ to each entry of $f(T_L)$.

22. We claim that, for any n-permutation p, the number of descents of p is equal to the right edges of $T(p)$. This claim is easy to prove by induction (look at the two subtrees of $T(p)$). Then the claim of the exercise is straightforward, since $T(p)$ can be reflected through a vertical axis.

23. The statement will be proved if we can prove that there are as many 132-avoiding permutations with $k-1$ descents as there are such permutations with k left-to-right minima.

We claim that even more is true, that is, these two sets of permutations are not simply *equinumerous*, they are *identical*. In fact, we

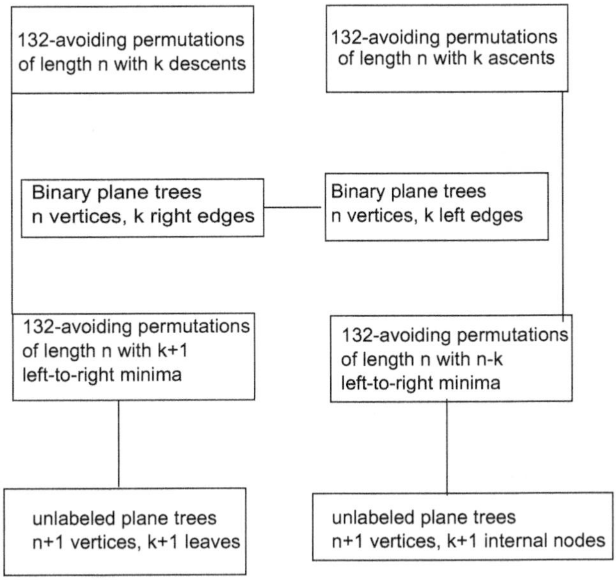

Figure 5.50
The bijections proved in Exercises 20–23.

claim that if $p = p_1p_2\cdots p_n$ is 132-avoiding, then p_i is a left-to-right minimum of p if and only if either $i = 1$ or $i-1$ is a descent of p. In other words, the left-to-right minima are precisely the ends of the descents (and of course the leftmost entry).

The "only if" part is trivial. To see the "if" part, assume the contrary, that is, that there exists $j < i$ so that $p_j < p_i$. Then $p_jp_{i-1}p_i$ is a 132-pattern, which is a contradiction.

Figure 5.50 shows all the bijections that were proved in the last several exercises. We point out again that the number of elements of each of these sets is the Narayana number $N(n,k) = \frac{1}{n}\binom{n}{k}\binom{n}{k+1}$. See the remark after Theorem 5.29 for references about these numbers.

24. These functions are in bijection with rooted forests on $[n]$ having k components. Therefore, by Theorem 5.19, their number is $a_{n,k} = \binom{n}{k}kn^{n-1-k}$.

25. It follows from the result of the previous exercise that we have $a_{n,k} = \binom{n-1}{k-1}n^{n-k}$. Therefore, by the binomial theorem,

$$\begin{aligned} A_n(x) &= \sum_{k=1}^{n}\binom{n-1}{k-1}n^{n-k}x^k = x\sum_{i=0}^{n-1}\binom{n-1}{i}n^{n-1-i}x^i \\ &= x(x+n)^{n-1}. \end{aligned}$$

26. If g is a proper coloring, then it defines a unique orientation A of G that satisfies the second criterion (edges point toward the smaller color), and that orientation is automatically acyclic.

    Conversely, if (g, A) satisfies the criteria, then g is a proper coloring since no edge has monochromatic vertices. For each g, there is one A so that (g, A) satisfies the criteria. This bijectively proves our claim.

27. First, n and m are nonnegative, since Theorem 5.46 implies that $\chi_G(n) \neq 0$ for $n < 0$. Indeed, $\bar{\chi}_G(n) \geq 1$ for $n > 0$, because, when we are looking for colorings counted by $\bar{\chi}_G(n) \neq 0$, we can always color all vertices with the same color.

    Second, if $\chi_G(m) = 0$, then there is no proper coloring of G using only colors from $[m]$, but then, of course, there is no proper coloring of G using only colors from $[i]$ where $i \leq m$.

28. We claim that the number of edges of the simple graph G on vertex set $[m]$ is equal to -1 times the coefficient of n^{m-1} (the term following the leading term) in χ_G. We will prove this statement by induction on the number of edges, using Proposition 5.40.

    If G is the empty graph, then $\chi_G(n) = n^m$, and the claim is true. Now let us assume that the claim is true for graphs with $k - 1$ edges, and let G have k edges. Then Proposition 5.40 says that $\chi_G(n) = \chi_{G_e}(n) - \chi_{G^e}(n)$. The term of degree $m - 1$ is the leading term in $\chi_{G^e}(n)$, and as such, its coefficient is 1 (Supplementary Exercise 26), and the term of degree $m - 1$ is the second term in $\chi_{G_e}(n)$, so, by our induction hypothesis, its coefficient is $k - 1$. So the coefficient of n^{m-1} on the left-hand side is $(k - 1) + 1 = k$, as claimed.

29. (a) The summand indexed by k counts the rooted graphs on $[n]$ in which the root is in a component of size k.

    (b) The summand indexed by k on the right-hand side is the number of rooted graphs on $[n]$ in which the root is in a component of size k. If $k < n$, then the graph is not connected. Dividing the sum by n will count *unrooted* graphs that are not connected, and then the statement is proved by the subtraction principle.

30. Use induction on k and the definition of matrix multiplication.

31. Just as we proceeded in Section 5.5.2.1, remove the root of such a tree, and notice that you either get the empty tree or a sequence of at least two trees. Therefore, we have

$$S(x) = x\left(1 + \sum_{n\geq 2} S(x)^n\right) = x + \frac{S(x)^2}{1 - S(x)}.$$

Solving this quadratic equation, we get that

$$S(x) = \frac{1 + x - \sqrt{1 - 6x + x^2}}{4}. \qquad (5.30)$$

So the numbers s_n are very close to the Schröder numbers that we have seen in (3.40). They are called the *little Schröder numbers*.

32. Let $H(x)$ be the ordinary generating function of these trees. Then we have $H(x) = x(1 + H(x) + H(x)^2)$. This can be solved in two ways, either with the Lagrange Inversion Formula, or by solving for $H(x)$ as a quadratic equation. Using the Lagrange Inversion Formula is faster.

    We obtain

$$x = \frac{H(x)}{1 + H(x) + H(x)^2},$$

so

$$H(x) = \left(\frac{x}{1 + x + x^2}\right)^{\langle -1 \rangle}.$$

So, using the Lagrange Inversion formula with $k = 1$, we get

$$h_n = \frac{1}{n}[x^{n-1}](1 + x + x^2)^n.$$

If we select x^2 from i parentheses, then we must select x from $n - 1 - 2i$ parentheses, which yields that

$$h_n = \sum_i \frac{1}{n}\binom{n}{i}\binom{n-i}{n-1-2i}.$$

The numbers h_n are called the *Motzkin numbers*.

5.11 Supplementary exercises

1. Is the statement of Exercise 1 true if we remove the restriction that the graph in question be simple?
2. True or False?
   (a) If we remove some edges of a tree (but keep all vertices), then we get a tree.
   (b) If we remove some edges of a forest (but keep all vertices), then we get a forest.
   (c) If we add some edges to a tree (but no new vertices), we never get a tree.
   (d) If we add some edges to a tree (but no new vertices), we never get a forest.

3. Nine people committed a serious crime. The police looked at their phone records and found, interestingly, that each of them had contact with either five or seven others. Prove that the police were wrong.

4. We have seen in the proof of Lemma 5.3 that each tree on n vertices has at least one leaf. Under what conditions on n is it true that each tree on n vertices has at least two leaves? How many trees on n vertices have exactly two leaves?

5. Let T be a tree in which the largest degree of any vertex is d. What can we say about the number of leaves in T?

6. For a finite simple graph G, let us define the *complement* of G as the graph $\bar{G}$ whose vertex set is the same as that of G and in which vertices x and y are adjacent in $\bar{G}$ if and only if they are *not* adjacent in G.

   Prove that, if G is any finite simple graph, then at least one of G and $\bar{G}$ has to be connected.

7. Draw all 16 trees on vertex set $[4]$.

8. Which isomorphism class of trees on $[n]$ is the largest?

9. Can two trees have the same degree sequence and still not be isomorphic?

10. Let us call a function $f : [n] \to [n]$ *strongly acyclic* on $[n]$ if the short diagram of f contains exactly one cycle of length 1 and no other cycles. Find the number of strongly acyclic functions on $[n]$.

11. Prove that, if a simple graph has more than one vertex, but less than six vertices, then it has a nontrivial automorphism.

12. Find $\ell(G)$ if G is the graph shown in Figure 5.51.

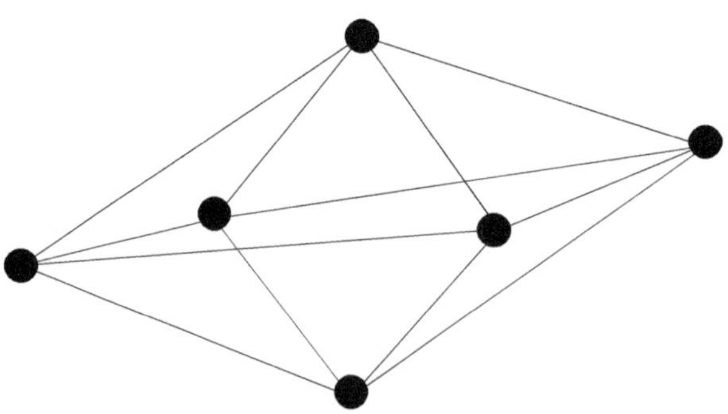

Figure 5.51
Find $\ell(G)$ for this graph G.

13. Find the number of automorphisms for the graphs G and H shown in Figure 5.52.

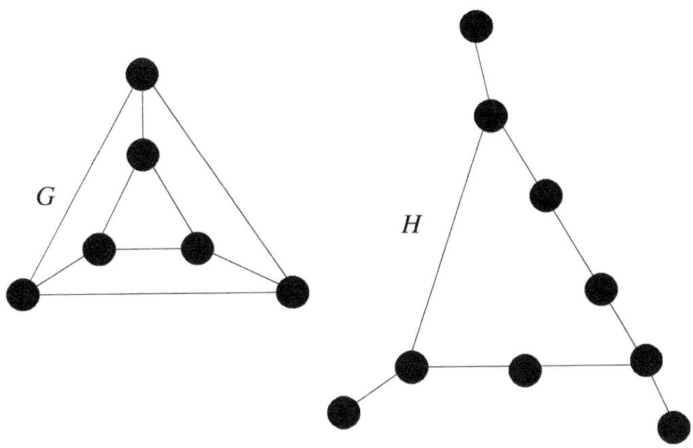

Figure 5.52
Find $|Aut(G)|$ and $|Aut(H)|$.

14. Let G be a simple graph that is not connected. At most how many edges can G have?
15. Let $n \leq m \leq 2n$. Find the number of all parking functions on $[n]$ so that $\sum_{i=1}^{n} f(i) = m$.
16. Prove formula (5.5) by replacing $P(n+1)$ by its numerical value, $(n+2)^n$, and then proving the obtained identity. Explain why your work proves Cayley's formula.
17. How many rooted plane trees are there on $n+1$ unlabeled vertices in which the root has exactly two children?
18. How many binary plane trees are there on n vertices with k right edges?
19. How many decreasing binary trees are there on n vertices with k right edges?
20. On average, how many components does a rooted forest on $[n]$ have?
21. Let $A(x) = \sum_{i=1}^{\binom{n}{2}} a_i x^i$, where a_i is the number of all graphs on vertex set $[n]$ that have exactly i edges. Find a closed formula for $A(x)$.
22. Let G be a graph with at least one edge. Prove that χ_G is not a symmetric polynomial, that is, if we reverse the sequence of coefficients of χ_G, we do not get the original sequence.
23. Let G be a cycle on m vertices. Find $\chi_G(n)$.

24. Let W_m be a cycle on m vertices with an extra vertex in the middle connected to all other vertices. Find $\chi_{W_m}(n)$.
25. Prove that, for any n, there exist n nonisomorphic graphs that have the same chromatic polynomial.
26. Prove that the leading coefficient of χ_G is always 1.
27. Can one find two graphs that do not have the same number of vertices but have the same chromatic polynomial?
28. Let d be the maximum degree of G. What can we say about the number of positive integer roots of χ_G?
29. Find $\chi(G)$ if G is the graph shown in Figure 5.53.

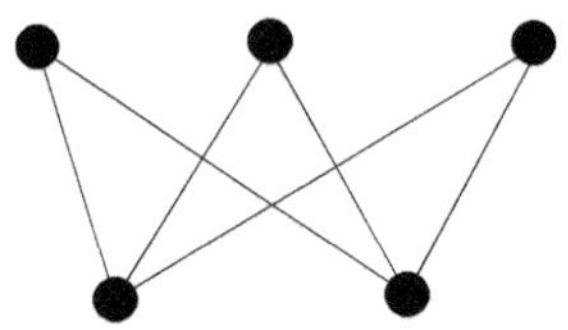

Figure 5.53
Find the chromatic polynomial of this graph.

30. Let $F(x)$ be the exponential generating function for the number of rooted forests on $[n]$. Find a functional equation satisfied by $F(x)$ similar to the one found for $R(x)$ in Theorem 5.55.
31. Find all 38 Eulerian graphs on vertex set $[5]$.
32. Extend the definition of the adjacency matrix of a graph given in Exercise 30 so that the statement of that exercise remains true for *directed graphs* as well.
33. State and prove the analog of Theorem 5.56 for directed graphs.
34. A *matching* in a graph is a set of pairwise vertex-disjoint edges. A *perfect matching* of G is a matching that covers all vertices of G. How many perfect matchings does the graph K_{2n} have?
35. Let $f(n)$ be the total number of leaves in all decreasing binary trees on vertex set $[n]$. Find an explicit formula for $f(n)$. Try to use generating functions.
36. Let us call a vertex in a tree an *almost-leaf* if it is not a leaf, but has at least one neighbor that is a leaf. Let $g(n)$ be the total number of almost-leaves in all decreasing binary trees on vertex set $[n]$. Find an explicit formula for $g(n)$. Try to use generating functions.
37. Decreasing trees were defined in Exercise 18. Let j be a leaf in a decreasing tree. What can we say about the entry j in the corresponding permutation?

38. Find a formula for the total number of leaves and the total number of almost-leaves in all decreasing trees on vertex set $\{0, 1, \cdots, n\}$.
39. Use the Lagrange Inversion Formula to find the number of rooted non-plane forests on vertex set $[n]$ that have ℓ components.
40. How many rooted plane trees are there on n unlabeled vertices in which each nonleaf vertex has either one child or three children?

6

Extremal combinatorics

6.1 Extremal graph theory

Real life is full of optimization problems, that is, we often have to find the best way to carry out a certain task. This often involves finding the graph that is best from a specific viewpoint. This is the topic of Extremal Graph Theory.

6.1.1 Bipartite graphs

In the previous chapter, we saw graphs whose vertices were colored so that there were no edges between vertices of the same color. If we have only one color, then such a coloring will exist only for empty graphs. So the first non-trivial case is when we have two colors. This case is very interesting on its own, which is why it has its own name.

Definition 6.1 *A graph G is called* bipartite *if its set of vertices has a proper 2-coloring.*

Bipartite graphs abound in everyday life. Imagine, for instance, a company that has some job openings and some candidates for them. A candidate may be qualified for more than one job, and a job may attract more than one candidate. The information relevant to this situation can be represented by a bipartite graph in which the red vertices are the jobs, the blue vertices are the candidates, and there is an edge between two vertices if the candidate associated with the blue vertex is qualified for the job associated to the red vertex. This definition assures that there will be no edges between vertices of the same color, so the obtained graph will be bipartite.

Bipartite graphs can be defined without resorting to colors, using the following fact.

Proposition 6.2 *A graph G is bipartite if and only if it does not contain a cycle of odd length.*

Proof: Let G be bipartite and let f be a proper 2-coloring of the vertices of G. If G had a cycle C of length $2k+1$, then C would contain at least $k+1$

DOI: 10.1201/9781003304272-6

vertices of the same color, so C would contain two adjacent vertices of the same color.

Conversely, assume that G has no odd cycles. Then start coloring the vertices of G at a vertex V, coloring it red, then coloring its neighbors blue, then coloring the neighbors of those red, and so on until all vertices of G are colored. This procedure will always lead to a proper 2-coloring, since vertices that are at even distance from V are red, and vertices that are at odd distance are blue. So if there were an edge between vertices A and B of the same color, then there would be an odd cycle in G. Indeed, take a shortest path p from V to A, and the shortest path p' from V to B. Note that A and B are at equal distances from V since they are neighbors and have the same color. If $p \cap p' = V$, then the union of path p, path p', and edge AB forms an odd cycle. If not, let W be a common vertex of p and p' that is as far away from V as possible. Then A and B are at equal distance from W, and so the part of p that is between A and W, the part of p' that is between B and W, and AB together form an odd cycle. See Figure 6.1 for an illustration. ◊

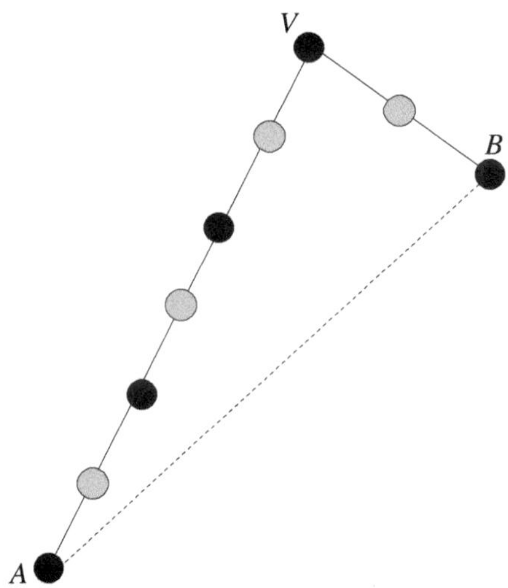

Figure 6.1
If there are no odd cycles, then our 2-coloring is proper.

Before we go any further, we should clarify what we mean by a *subgraph* of a graph, since this notion will be essential for the rest of this section. There are two different concepts as follows.

Definition 6.3 *Let G be a graph. We say that the graph H is a* subgraph *of G if the vertex set of H is a subset of the vertex set of G and the edge set of H is a subset of the edge set of G.*

Fair enough, you could say, but what else is there to say about defining subgraphs? The following definition—and the example after it—will hopefully answer that question.

Definition 6.4 *Let G be a graph. We say that the graph H is an* induced subgraph *of G if the vertex set $V(H)$ of H is a subset of the vertex set of G and if, when we take* all *edges of G that connect two vertices of $V(H)$, we get H itself.*

The following example should clarify the difference between these two notions.

Example 6.5 *Graph G in Figure 6.2 contains several subgraphs isomorphic to graph H. Indeed, any path of two edges is such a subgraph. However, G contains no* induced *subgraphs isomorphic to H since any 3-vertex induced subgraph of G is a copy of K_3.*

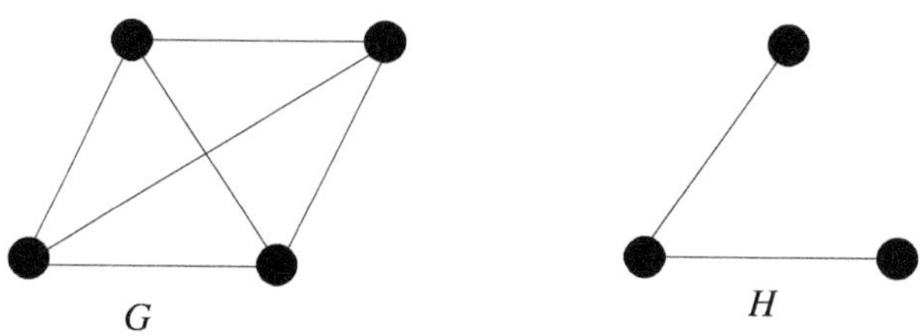

Figure 6.2
The graph G contains a copy of H as a subgraph, but not as an induced subgraph.

No matter which definition of subgraph graphs we use, we see that, if G is bipartite, then so is any subgraph, or induced subgraph, of G. Conversely, if we keep adding edges to G while not adding new vertices, then eventually G will lose the bipartite property, since the complete graph on n vertices is not bipartite if $n \geq 3$. The question is *when* this will happen.

On one hand, trees are certainly bipartite graphs since they do not have any cycles, even or odd. So a bipartite graph on n vertices can certainly have $n-1$ edges. On the other hand, since the complete graph K_n is not bipartite for $n \geq 3$, bipartite graphs on n vertices have fewer than $\binom{n}{2}$ edges. Where is the breaking point between $n-1$ and $\binom{n}{2}$? The following result shows that it is closer to $\binom{n}{2}$ than to $n-1$.

Theorem 6.6 *Let $n \geq 2$ and let G be a simple bipartite graph on n vertices. Then G has at most $a(n-a)$ edges, where $a = \lfloor n/2 \rfloor$.*

In other words, a bipartite graph on n vertices cannot have more than $n^2/4$ edges.

Proof: Assume G has a red vertices and $n-a$ blue vertices. Then the highest number of edges is achieved if all red vertices are adjacent to all blue vertices, which allows G to have $a(n-a)$ edges. By elementary calculus, the function $f(a) = a(n-a)$ has its maximum at $a = n/2$, and it increases when $0 < a < n/2$ and decreases when $a > n/2$. Because a has to be an integer, our claim follows. ◊

A few remarks are in order. First, the graph described in the proof above, that is, the bipartite graph in which there are a red vertices, $n-a$ blue vertices, and *all red vertices* are adjacent to *all blue vertices*, is called a *complete bipartite graph*, and is denoted by $K_{a,n-a}$. So, for instance, the graph shown in Figure 5.53 is $K_{2,3}$.

Second, we have proved a little bit more than the statement of Theorem 6.6. We have not only proved that the number of edges in such a bipartite graph is at most $a(n-a)$, with $a = \lfloor n/2 \rfloor$. We have also proved that this maximum can indeed be attained. In other words, the upper bound $a(n-a)$ cannot be further improved, since there are bipartite graphs on n vertices with exactly that many edges. In what follows, we will refer to this fact by saying that the upper bound $a(n-a)$ is *sharp*.

Third, we have also proved the fact that $a(n-a) \leq n^2/4$ is a direct consequence of the inequality between the *geometric* and *arithmetic mean*, that is, the inequality

$$\sqrt{xy} \leq \frac{x+y}{2}, \tag{6.1}$$

which holds for all nonnegative real numbers x and y. We will use this inequality many times. To prove this inequality, note that, after taking squares and rearranging, it becomes the inequality $(x-y)^2 \geq 0$. Because these steps are reversible, (6.1) follows. Now set $x = a$ and $y = n-a$ to get the inequality $\sqrt{a(n-a)} \leq n/2$.

Theorem 6.6 shows that, if a graph G on n vertices has more than $n^2/4$ edges, then G necessarily contains an *odd cycle* as a subgraph.

We claim that much more is true, namely, a graph with that many edges on that number of vertices will even contain a *triangle*—and therefore a very short odd cycle—as a subgraph.

Theorem 6.7 *If a graph G on n vertices contains more than $n^2/4$ edges, then it contains a triangle as a subgraph.*

We challenge the reader to prove this fact by induction, then check our solution in Exercise 1. An even stronger and more surprising fact (Supplementary Exercise 4) is that, if the graph G on $2n$ vertices has more than n^2 edges, so then G necessarily contains n triangles! Therefore, as long as G has no more than n^2 edges, G may contain zero triangles, but as soon as G has more than n^2 edges, it suddenly contains at least n triangles.

Proof: (of Theorem 6.7) Let G be a graph on n vertices that contains no triangles. Let us try to tweak the proof of Theorem 6.6 so that it gives us this stronger conclusion. In that proof, we knew that our graph was bipartite, so it had two color classes with no edges within them, so the edges had to be between vertices of different color classes.

Here, we do not know that G is bipartite. However, we can still look at the largest *induced subgraph* of G that is empty and call its vertex set X. That is, there are no edges between two vertices of X, and there is no induced subgraph of G with more than $|X|$ vertices having that property.

Let Y be the set of vertices of G that are not in X. Then $|X| + |Y| = n$.

Crucially, no vertex of G can have degree more than $|X|$. Indeed, no two neighbors of any vertex can be adjacent, since that would mean that G contains a triangle. So the neighbors of any vertex form an empty induced subgraph of G, and as such, there cannot be more than $|X|$ of them.

While we do not know that G is bipartite, we do know that its edges are either within Y or between X and Y. This means that all edges of G have at least one vertex in Y. So if we add the degrees of the vertices in Y, we count each edge of G at least once. Therefore, the number $E(G)$ of edges of G satisfies

$$\begin{aligned} E(G) &\leq \sum_{y\in Y} d_y \leq \sum_{y\in Y} |X| = |X||Y| \\ &\leq \left(\frac{|X|+|Y|}{2}\right)^2 = \frac{n^2}{4}. \end{aligned}$$

Here the inequality between the two lines follows from (6.1), the inequality between the geometric and arithmetic mean. ◊

Note that we used inequality (6.1) in the proof of Theorem 6.6 as well.

We point out that, in the above proof, in the step $E(G) \leq \sum_{y\in Y} d_y$, equality holds if and only if all edges of G have *exactly one* vertex in Y, that is, when G is bipartite with color classes X and Y. So it is again the complete bipartite graph that provides the highest number of edges without containing a triangle.

Now we know how many edges a graph on a given vertex set can have if it is not to contain a triangle as a subgraph. This question can be generalized in several natural directions. That is, a triangle is both a complete graph and a cycle. One can ask how many edges a graph on n vertices can have if it is not to contain the k-vertex complete graph K_k. One can also ask how many edges a graph on n vertices can have if it is not to contain the k-vertex cycle C_k. The latter can be further generalized by asking how many edges G can have if it is not to contain a cycle of length k *or less*.

6.1.2 Turán's theorem

In this subsection, we are going to look at the first of the generalizations we have just mentioned. That is, we will ask how many edges a graph G on n vertices can contain if it is not to contain K_k as a subgraph.

Our discussion of this problem will be made easier by the following definition.

Definition 6.8 *For a graph G, the* chromatic number *of G, denoted by $\chi(G)$, is the smallest positive integer n so that G has a proper coloring using colors from $[n]$ only.*

Example 6.9 *The chromatic number of any nonempty bipartite graph is two.*

Example 6.10 *The chromatic number of a pentagon is three.*

Solution: If G is a pentagon, then $\chi(G) > 2$, since the pentagon, being an odd cycle, is not bipartite. On the other hand, a proper 3-coloring of G is shown in Figure 6.3. ◊

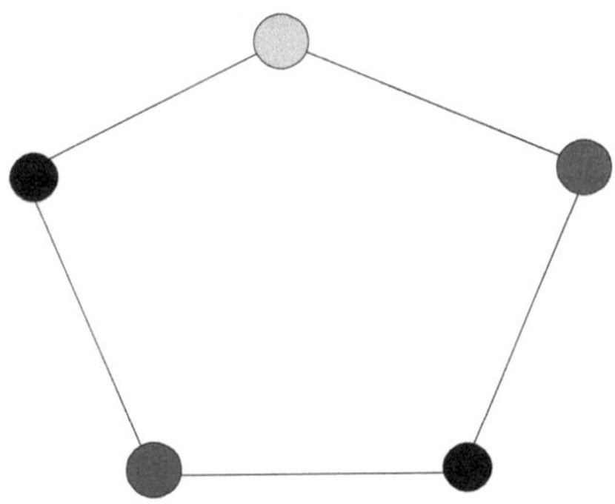

Figure 6.3
A proper 3-coloring of a pentagon.

Let us return to the question of determining the maximum number of edges a graph G can have without containing a copy of K_k. It would be easy to make a mistake by arguing as follows: Let us divide the vertices of G into $k-1$ (color) classes, then maximize the number of edges by connecting any two vertices that are not in the same class by an edge. Let there be no edges between vertices in the same class. This ensures that G contains no K_k. We then simply have to count how many edges are created and find what partition of vertices into $k-1$ parts results in the highest number of edges.

The problem with this argument is that we do not know in advance that this kind of construction will give us the highest number of edges. That is, it is possible that a graph has chromatic number k or higher, yet it does not contain K_k as a subgraph. For instance, in the case of $k = 3$, take a pentagon. As we saw in Example 6.10, its chromatic number is 3, yet it does not contain K_3. At this point, we do not know yet that a graph like that (that is, a graph with chromatic number k or higher but not containing K_k) will not have more edges than the graphs with chromatic number $k - 1$ we can construct in the way described in the previous paragraph.

Despite the fact that we exposed a gap in the previous argument, its final conclusion is correct, that is, the maximal number of edges will be attained by a graph of chromatic number $k-1$ in which any pair of vertices in different color classes is adjacent. Such a graph is called a *complete $(k-1)$-partite* graph. Furthermore, the distribution of vertices in the color classes will be as uniform as possible. That is, if $n = r(k-1) + q$, with $0 \leq q \leq k-2$, then in an optimal graph, there will be q color classes with $r+1$ vertices and $k-1-q$ color classes with r vertices. This is the content of the following classic theorem. Note that Theorem 6.7 is a special case of this result, namely, the case of $k = 3$.

Theorem 6.11 (Turán's theorem) *Let $n = r(k-1)+q$, with $0 \leq q \leq k-2$, and let $T(n, k-1)$ denote the complete $(k-1)$-partite graph on n vertices in which there are q color classes of size $r+1$ and $k-1-q$ color classes of size r.*

If a graph G on n vertices contains no copies of K_k, then it cannot have more edges than $T(n, k-1)$.

See Figure 6.4 for the graph $T(6, 3)$.

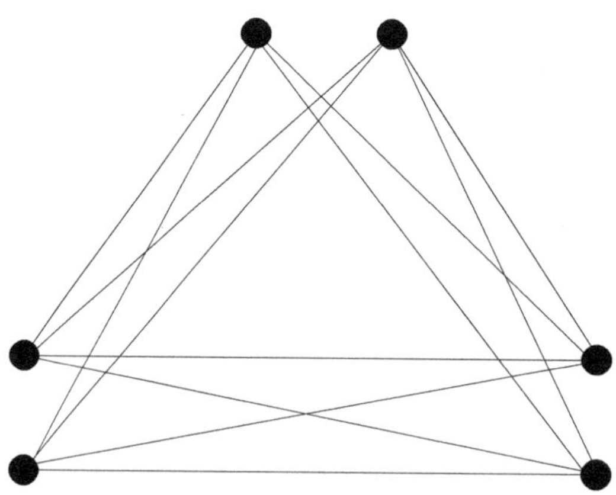

Figure 6.4
The Turán graph $T(6, 3)$.

Originally, Theorem 6.11 was proved in [82]. Nothing shows its importance more than the fact that it has been constantly re-proved ever since. The reader can find five different proofs of this theorem in *Proofs from THE BOOK* [2], a collection of beautiful proofs from many areas of mathematics as interpreted by Martin Aigner and Günther Ziegler. The proof we present uses very elementary tools.

Proof: (of Theorem 6.11) We will first remove the gap from the sketch of the proof above by showing that the optimal graph will indeed be a complete $(k-1)$-partite graph.

Let G be a graph on n vertices that does not contain K_k as a subgraph and that has a maximal number of edges among all graphs with this property. Let us say that vertices x and y of G are *similar* if they are *not adjacent* in G.

We claim that similarity of vertices is an *equivalence relation*, that is, it is reflexive, symmetric, and transitive. The first two of these properties are not difficult to prove (since G is a simple undirected graph). The interesting property is the third one, transitivity. In order to prove that vertex similarity is transitive, we need to show that, if x is similar to y, and y is similar to z, then x is similar to z.

Let us assume that this is not true, that is, that there are three vertices x, y, and z in G so that their induced subgraph has only one edge, the edge xz. Then there are two cases. Let d_x, d_y, d_z denote the degrees of the vertices x, y, and z.

1. When d_y is not maximal among these three degrees, we can assume that $d_y < d_x$. We will show that this contradicts the assumption that G has a maximum number of edges among n-vertex graphs avoiding K_k. Indeed, remove y (and all edges adjacent to it) from G, and add a new vertex x' that is connected to the same vertices as x. Then the number of edges of G grows, since the new vertex x' is adjacent to more edges than the old vertex y. Furthermore, the obtained graph G' still has n vertices and does not contain K_k as a subgraph. Indeed, if it did, then that copy K of K_k would have to contain x' (otherwise G would have contained K). Since x and x' are not adjacent and have the same set of neighbors, replacing x' by x in K would then yield a copy of K_k in G, a contradiction. So G' does not contain K_k as a subgraph, yet has more edges than G, which is a contradiction.
2. Now let us assume that d_y is maximal among the degrees of x, y, and z. Then remove both x and z, and add the new vertices y' and y'' having the same set of neighbors as y to G. Call the obtained graph G'. Then again, G' has more edges than G, since $d_{y'} = d_y \geq d_x$, $d_{y''} = d_y \geq d_z$, and the number of removed edges is $d_x + d_z - 1$, since the edge xz is counted by both d_x and d_z. Just as in the previous case, G' does not contain K_k as a subgraph (such a copy could contain only one of y, y', and y'', but then it might as well be y), which contradicts the maximality of G.

So similarity of vertices is indeed an equivalence relation. This means that this relation creates equivalence classes, and vertices that are in different classes are adjacent. If there are p equivalence classes, this is just the definition of a complete p-partite graph, with the color classes being the equivalence classes.

It is not difficult to see that, in an optimal graph, this p had better be $k-1$. Indeed, $p \geq k$ is impossible, since a complete p-partite graph contains K_p as a subgraph. On the other hand, if $p < k-1$, just split one of the color classes in half to increase the number of edges. (Okay, that splitting is not possible if all color classes are of size 1, but then G has at most $k-1$ vertices, and so G is complete because of its edge-maximal property.)

This means that G is indeed a complete $(k-1)$-partite graph. The last step is to show that the sizes of the color classes in G are indeed as balanced as possible. To see this, assume that there are two color classes whose sizes are equal to a and $a+b$, respectively, with $b \geq 2$. We claim that, in this case, the number of edges in G can be increased if we move one vertex from the class of size $a+b$ into the class of size a. Indeed, when we do that, there will be no change in the number of edges between the other classes of G or between the union of these two classes and the rest of the graph. The only change that occurs will be between these two classes, where the change will be

$$(a+b-1)(a+1)-(a+b)a = a+b-a-1 = b-1 \geq 1.$$

This completes the proof by showing that, if G is indeed optimal, then the difference between the sizes of two of its color classes cannot be more than 1. $\diamond$

6.1.3 Graphs excluding cycles

Let us now turn to the other generalization of Theorem 6.7. That is, let us try to find the highest number of edges a graph G on n vertices can have if it does not contain C_k, the cycle of length k.

Let us start with the next case, that is, the case of $k=4$. The following proposition is the key element of a simple but powerful method in this area.

Proposition 6.12 *Let G be a graph on n vertices whose degree sequence is $d_1, d_2, \cdots, d_n$, and assume G does not contain C_4 as a subgraph. Then*

$$\sum_{i=1}^{n} \binom{d_i}{2} \leq \binom{n}{2}.$$

Proof: For vertex i, there exist $\binom{d_i}{2}$ pairs of vertices so that i is adjacent to both of them. Let $P(i)$ be the set of these pairs. If we can find another vertex j so that $P(i) \cap P(j) \neq \emptyset$, then G contains C_4 since i, j, and any pair in $P(i) \cap P(j)$ form a copy of C_4. See Figure 6.5.

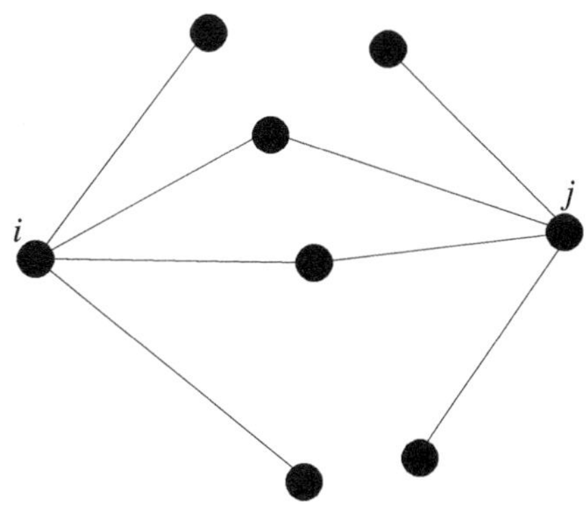

Figure 6.5
If $P(i)$ and $P(j)$ intersect, then G contains a cycle of length four.

Our claim now follows from the pigeonhole principle. ◊

What does this say about the *total* number of edges G can have if it is not to contain C_4? We want the sum of the degrees, $S = \sum_{i=1}^{n} d_i$, to be as large as possible without $T = \sum_{i=1}^{n} \binom{d_i}{2}$ becoming too large. We claim that, for a fixed value of S, the value of T will be the smallest if the degrees are as close to each other as possible. Indeed, assume that $d_1 - d_2 \geq 2$. Now decrease d_1 by one, to get d_1', and increase d_2 by one, to get d_2'. This operation does not change S, but it decreases T since

$$\begin{aligned}\binom{d_1'}{2} + \binom{d_2'}{2} &= \binom{d_1 - 1}{2} + \binom{d_2 + 1}{2} = \frac{d_1^2 + d_2^2 - 3d_1 + d_2 + 2}{2} \\ &= \binom{d_1}{2} + \binom{d_2}{2} - d_1 + d_2 + 1 \\ &< \binom{d_1}{2} + \binom{d_2}{2}.\end{aligned}$$

We can repeat this operation several times, as long as there are two degrees whose difference is at least two. So if we want G to contain $S/2$ edges, the minimal possible value of T that G can have is attained when all degrees are "as equal as possible." That is, when $S = na + b$, with $0 \leq b \leq n - 1$, then G has b vertices of degree $a + 1$ and $n - b$ vertices of degree a. Since $T \leq \binom{n}{2}$ must hold, the inequality

$$\begin{aligned}\sum_{i=1}^{n} \binom{d_i}{2} &= b\binom{a+1}{2} + (n-b)\binom{a}{2} \\ &= n\binom{a}{2} + ab \\ &\leq \binom{n}{2}\end{aligned}$$

must also hold.

In particular, comparing the last two lines, it has to be true that

$$\begin{aligned} n\binom{a}{2} &\leq \binom{n}{2}, \\ \binom{a}{2} &\leq \frac{n-1}{2}. \end{aligned}$$

That is, $a^2 - a \leq n - 1$, which implies $a - 1 \leq \sqrt{n-1}$. Since G has less than $n(a+1)/2 = n(a-1)/2 + n$ edges, we have proved the following theorem.

Theorem 6.13 *If the graph G has n vertices and does not contain C_4, then G cannot have more than*

$$\frac{n(\sqrt{n-1}+2)}{2}$$

edges.

Since Theorem 6.13 and its proof epitomize a wide class of theorems and proofs, it is a good time to take a break and look at some related details.

6.1.3.1 Convex functions and Jensen's inequality

In what follows, we are going to discuss two issues of a technical nature. While the reader is probably impatiently waiting for us to return to combinatorics, these minor digressions will be worth the time and effort.

The fact that we used concerning the sum $S = \sum_{i=1}^{n} d_i$—that is, the fact that, for a fixed $D = \sum_{i=1}^{n} \binom{d_i}{2}$, the value of S is the highest when the summands are all equal—is often referred to as the *convexity* of $\binom{x}{2}$. Indeed, the values of $\binom{x}{2}$, $x = 1, 2, 3, \cdots$ form a *convex* curve, that is, a curve *below* its chords (see Figure 6.6). This property will be used often in this section.

Let us make these notions more precise. We will say that a continuous function f is *convex* on an interval $[a, b]$ if its value at the midpoint of every subinterval of $[a, b]$ does not exceed the average of its values at the endpoints of the subinterval. That is, f is convex on $[a, b]$, if for any two points x and y in $[a, b]$, we have

$$f\left(\frac{x+y}{2}\right) \leq \frac{f(x)+f(y)}{2}.$$

Note that this definition does not require that x and y be integers. Nevertheless, the function $f(x) = \binom{x}{2}$ is convex on the interval $[1, \infty)$. See Exercise 3 for a proof of this (in fact, a stronger) statement.

Values of convex functions are often estimated using the following famous theorem by Jensen, proved in 1906 [46].

Theorem 6.14 (Jensen's inequality) *If the function f is convex on*

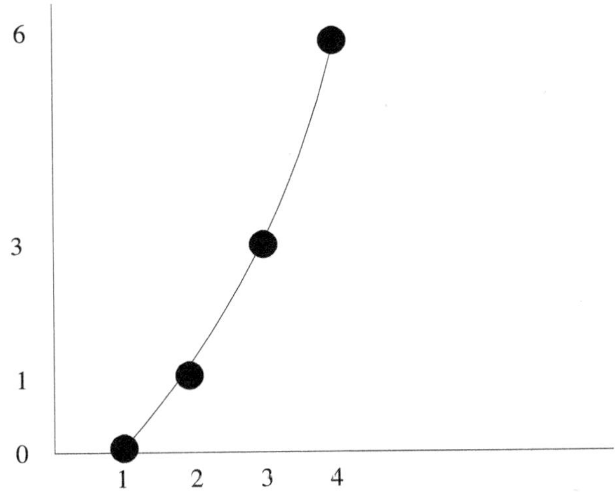

Figure 6.6
The convexity of $\binom{x}{2}$.

the interval $[a, b]$*, then, for any positive integer* n*, and for any* n *points* $x_1, x_2, \cdots, x_n$ *of* $[a, b]$*,*

$$f\left(\frac{x_1 + x_2 + \cdots + x_n}{n}\right) \leq \frac{f(x_1) + f(x_2) + \cdots + f(x_n)}{n}.$$

In other words, the convexity property, defined by just two points, implies the analogous property for n points. This will be very useful in proving the upcoming theorems in this section.

After reading the proof of Theorem 6.13, it is not difficult to prove Jensen's inequality. In Supplementary Exercise 6, we will ask the reader to do so.

6.1.3.2 Notation in approximate counting

In order to make it easier to discuss proofs like that of Theorem 6.13, we introduce the following notation.

Definition 6.15 *Let* $f : \mathbf{Z}^+ \rightarrow \mathbf{R}$ *be a function, and let* $g : \mathbf{Z}^+ \rightarrow \mathbf{R}$ *be another function. We say that* $f(n) = O(g(n))$ *(read "*f *is big* O *of* g*") if there exists a positive constant* c *so that*

$$f(n) \leq cg(n)$$

for all positive integers n*.*

That is, if $f(n) = O(g(n))$, then $f(n)$ is at most a constant times larger than $g(n)$, no matter what positive integer n is.

Example 6.16 *If* $f(n) = n + 10\sqrt{n}$*, and* $g(n) = n$*, then* $f(n) = O(g(n))$ *since we can choose* $c = 11$.

Example 6.17 *If* $f(n) = n \log n$*, and* $g(n) = 100n$*, then* $f(n) \neq O(g(n))$ *since no matter what constant* c *we choose, there exists a positive integer* n *so that* $\log n > 100c$.

In this terminology, the statement of Theorem 6.13 can be restated by saying that if $f_{C_4}(n)$ is the maximum number of edges a graph on n vertices can have without containing C_4, then $f_{C_4}(n) = O(n^{3/2})$.

At this point, the reader could object that simply saying that $f_{C_4}(n) = O(n^{3/2})$ is not a very precise statement. Indeed, there are many functions that are $O(n^{3/2})$, such as $f(n) = n$, $f(n) = \log n$, or $f(n) = 3$, but the values of these functions are very much smaller than the values of $f_{C_4}(n)$. In other words, it would be preferable to define classes of functions by *lower bounds* as well. The following definition takes care of that.

Definition 6.18 *Let* $f : \mathbf{Z}^+ \rightarrow \mathbf{R}$ *be a function, and let* $g : \mathbf{Z}^+ \rightarrow \mathbf{R}$ *be another function. We say that* $f(n) = \Omega(g(n))$ *(read "f is omega of g") if there exists a positive constant* c *so that*

$$f(n) \geq cg(n)$$

for all positive integers n.

Example 6.19 *If* $f(n) = n + 10\sqrt{n}$*, and* $g(n) = n$*, then* $f(n) = \Omega(g(n))$ *since we can choose* $c = 1$.

Finally, our last definition of this kind combines the two properties we have just defined.

Definition 6.20 *Let* $f : \mathbf{Z}^+ \rightarrow \mathbf{R}$ *be a function, and let* $g : \mathbf{Z}^+ \rightarrow \mathbf{R}$ *be another function. We say that* $f(n) = \Theta(g(n))$ *(read "f is theta of g") if* $f(n) = O(g(n))$ *and* $f(n) = \Omega(g(n))$.

Example 6.21 *It follows from Examples 6.16 and 6.19 that, if* $f(n) = n + 10\sqrt{n}$*, and* $g(n) = n$*, then* $f(n) = \Theta(g(n))$.

In other words, if $f(n) = \Theta(g(n))$, then $f(n)$ and $g(n)$ are roughly of the same size, that is, the ratio $f(n)/g(n)$ stays between certain constants.

The reader may now ask, with a hint of boredom in her voice, what the advantage of this terminology is, since it is still less precise than the language we used to describe our results before. Our answer is that, despite the relative lack of precision, this makes the description of our results much easier since we do not have to worry about precisely describing terms that are negligibly small compared to the other terms of the result. For instance, we do not have to worry about divisibility issues, so the result of Theorem 6.6 can be stated as follows: If the maximum number of edges a bipartite graph on n vertices can have is $h(n)$, then $h(n) = \Theta(n^2)$. Or, with a slight abuse of language, a bipartite graph on n vertices can have at most $\Theta(n^2)$ edges.

6.1.3.3 Refining the results on $f_{C_4}(n)$

Let us revisit Theorem 6.13 and the function $f_{C_4}(n)$ in light of our new terminology, and let us see if we can state and prove something stronger.

As far as the statement of Theorem 6.13 goes, which is a statement for *all* n, we claim that no improvement is possible.

Indeed, the inequality $\binom{a}{2} \leq \frac{n-1}{2}$ used in the proof of Theorem 6.13 is sharp, that is, it cannot be improved any further. A triangle is an example for this, with $a = 2$ and $n = 3$.

However, this is not the end of the story. At this point, one could imagine that this is the only example when the inequality $\binom{a}{2} \leq \frac{n-1}{2}$ is sharp, or that there are only a finite number of examples when this inequality is sharp. That could imply that, by choosing a large enough constant, we could prove that $f_{C_4}(n) = O(n^\alpha)$ for some positive real number $\alpha < 3/2$.

It turns out that this is *not* the case, though. The following sophisticated example of a graph with a lot of edges, but no C_4, was found by István Reiman [62]. Consider the graph in Figure 6.7.

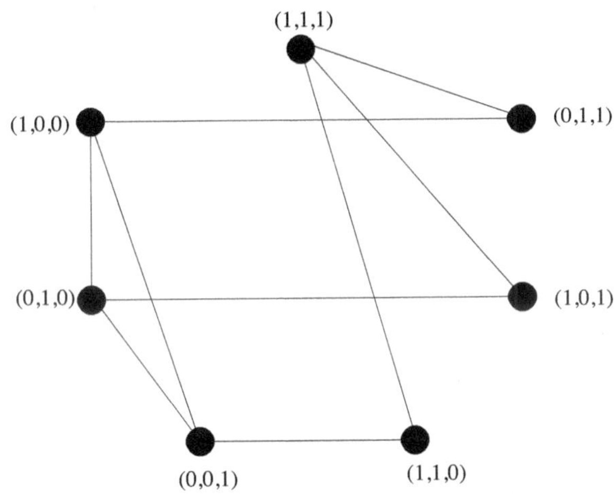

Figure 6.7
This graph contains no C_4.

In this graph, the vertices are labeled by the nonzero binary triples, that is, triples in which each digit is 0 or 1. Addition and multiplication are defined modulo 2. There is an edge between two vertices if their *dot product* is 0. That is, there is an edge between (a, b, c) and (x, y, z) if $ax + by + cz = 0$ or 2.

It turns out that this concept can be generalized in a sophisticated way. It is known (see [22], [62]) that, for all primes p, one can construct a graph on $n = p^2 + p + 1$ vertices with $p(p+1)^2/2$ edges that does not contain C_4. Since $p(p+1)^2/2 = \Theta(n^{3/2})$, this implies that the exponent of n in the upper bound $O(n^{3/2})$ cannot be improved. Indeed, there are infinitely many primes, so there are infinitely many numbers of the form $n = p^2+p+1$ for which such a graph will exist.

We can refine our result even further as follows. We know that $f_{C_4}(n) = \Theta(n^{3/2})$, meaning that there are constants c_1 and c_2 so that

$$c_1 n^{3/2} \leq f_{C_4}(n) \leq c_2 n^{3/2}.$$

This makes us wonder just how similar the function f_{C_4} is to the function $g(n) = n^{3/2}$, in other words, what can one say about the sequence $a_n = \frac{f_{C_4}(n)}{n^{3/2}}$. Assume now without proof that a_n is convergent (see [22] for a method to prove this). Then the limit of this sequence has to be equal to the limit of any of its infinite subsequences. This is true as well for the subsequence of elements a_m, where $m = p^2 + p + 1$ for some prime. It is not difficult to compute the limit of this subsequence and find that it is $1/2$. This proves the following theorem [22].

Theorem 6.22 *The equality*

$$\lim_{n\to\infty} \frac{f_{C_n}(n)}{n^{3/2}} = \frac{1}{2}$$

holds.

6.1.3.4 Avoiding longer cycles

What if we want to avoid longer cycles? Let $f_{C_k}(n)$ denote the maximum number of edges a graph on n vertices can have if it does not contain C_k as a subgraph. Interestingly, the cases of odd k and even k have to be treated differently. Indeed, the complete bipartite graph $K_{a,n-a}$ has $a(n-a)$ edges and avoids all odd cycles. For $a = \lfloor n/2 \rfloor$, this graph has $\Theta(n^2)$ edges, so $f_{C_k}(n) \geq \Theta(n^2)$.

On the other hand, for $k = 4$, we have seen in Theorem 6.13 that $f_{C_k}(n) = O(n^{3/2})$. In other words, it is significantly more difficult to avoid C_4 than to avoid C_k for any odd k. Furthermore, it is more difficult to find bounds for $f_{C_k}(n)$ in the case of even k than in the case of odd k, since there are no simple analogs of the complete bipartite graph for the case of even k. Nevertheless, there are very strong results concerning $f_{C_k}(n)$ for the even case as well.

Theorem 6.23 *Let $k \geq 4$ be an even positive integer. Then*

$$f_{C_k}(n) = O(n^{1+\frac{2}{k}}).$$

This theorem was first proved by Adrian Bondy and Miklós Simonovits in 1974 [19]. It was then generalized in [37] in the following manner. Let $C_{s,t}$ be the graph obtained by taking two vertices, x, and y, and connecting them by t vertex-disjoint paths of length s each. So $C_{s,t}$ has $2 + t(s-1)$ vertices and ts edges. See Figure 6.8 for an example.

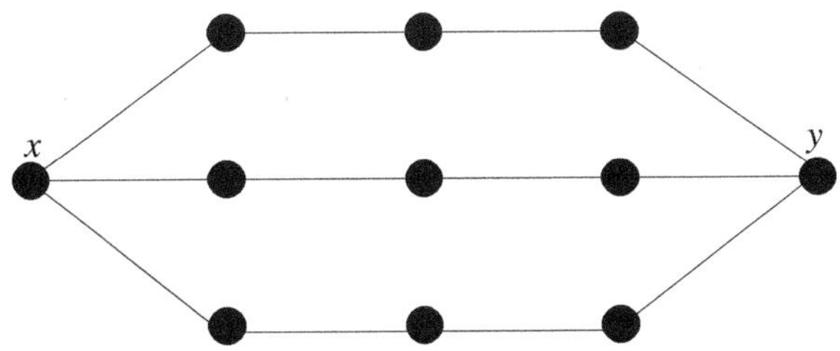

Figure 6.8
The graph $C_{3,4}$.

Theorem 6.24 *Let s and t be positive integers. Then*

$$f_{C_{s,t}}(n) = O(n^{1+\frac{1}{s}}).$$

Note that Theorem 6.24 indeed generalizes Theorem 6.23, since $C_{s,2} = C_{2s}$. We point out that it would be a mistake to say that the upper bound given by Theorem 6.24 does not depend on t. This is because Theorem 6.24 says that, *for all* positive integers s and t, there exists a constant c so that a graph avoiding $C_{s,t}$ as a subgraph cannot contain more than $cn^{1+\frac{1}{s}}$ edges. The theorem does not say that the constant c is the same for all choices of s and t. Therefore, the constant can (and does) depend on s and t; in other words, $c = c_{s,t}$. It is nevertheless interesting that it is only that part of the upper bound that depends on t.

While rigorous proofs of Theorems 6.23 and 6.24 are somewhat beyond the scope of this book, we do want to mention some of the ideas used in the proofs.

Let G be a graph on n vertices with no C_k subgraph for a fixed integer $k = 4l$, and assume that G has the highest number of edges among the graphs of this property. First, assume that G is *regular*, that is, all vertices have the same degree d. Then a vertex x has d neighbors, roughly d^2 second neighbors (vertices at distance two from x), and so on, and roughly d^l vertices at distance l from itself. Therefore, there are roughly d^{2l} *pairs of vertices* (a, b) so that both a and b are at a distance l from x. This holds for all vertices $y \in G$ instead of x, so, if $n < d^{2l}$, then the pigeonhole principle implies that there are two vertices x and y that both have the same (a, b)-pair. If the four paths between the pair (x, y) and (a, b) are vertex-disjoint, then they form a cycle of length $4l = k$, which is a contradiction. So, if we can somehow ensure that these paths are vertex disjoint, then the pigeonhole principle will force $n \geq d^{2l} = d^{k/2}$, or, in other words, $d \leq n^{2/k}$. This proves Theorem 6.23, since G has $nd/2$ edges.

Let us see what technical difficulties a rigorous proof would have to handle. It is easy to see that the assumption that G is regular is not a serious problem. Indeed, if G is not regular, then we can see, just as in the proof of Theorem

6.13, that the number of edges of G could be increased by making G more like a regular graph. (There are some divisibility concerns, but they are not important since we are only looking for an upper bound.) The condition that k is divisible by 4 can also be relaxed to the condition that k is even. Indeed, if $k = 4m + 2$, then one simply has to look at pairs (a, b) so that the distance between a and x is $m + 1$ and the distance between b and x is m.

A more serious concern is that the four paths between the pair (x, y) and (a, b) may not be vertex disjoint. The interested reader should consult [19] and [37] for this and other interesting details. We point out that, in the general case, it is not known whether Theorem 6.23 can be improved or not.

A very general answer is given by the following result of Paul Erdős and Miklós Simonovits [30].

Theorem 6.25 *Let H be a graph with chromatic number d. Then*

$$f_H(n) = f_{K_d}(n) + O(n^{2-c}),$$

where c is some positive constant.

If H is an odd cycle, then its chromatic number is 3. Therefore, the above theorem yields $f_H(n) = \Theta(n^2) + O(n^{2-c}) = \Theta(n^2)$. This proves the following corollary.

Corollary 6.26 *For all odd integers $k \geq 3$,*

$$f_{C_k}(n) = \Theta(n^2).$$

Recall that Exercise 4 showed that, if a graph on $2n$ vertices has enough edges to surely contain a triangle, then it has enough edges to contain n triangles as well. Corollary 6.26 now shows that, if G has enough edges to surely contain a triangle, then G has enough edges to surely contain C_k for any fixed odd k. (Or any k, for that matter, since we have seen in Theorem 6.23 that, for even k, the upper bounds are even lower.)

Note that Theorem 6.25 would not help us to prove good upper bounds for $f_{C_k}(n)$ in the case of even k since even cycles have chromatic number 2.

Miklós Simonovits even generalized Theorem 6.25 much further. The interested reader can consult [67] and [68] for details.

6.1.4 Graphs excluding complete bipartite graphs

Theorem 6.13 can be generalized in yet another way, one that is different from generalizations of Theorem 6.7. Indeed, note that C_4 is not just a circle, but also the complete bipartite graph $K_{2,2}$. Since finding good bounds on $f_{C_4}(n)$ was relatively simple, we might as well look for good upper bounds on $f_{K_{r,r}}(n)$, which we will abbreviate by $f_r(n)$.

Theorem 6.27 *[50] For all $r \geq 2$,*

$$f_r(n) = f_{K_{r,r}}(n) = O(n^{2-\frac{1}{r}}).$$

Proof: Let G be a graph on n vertices not containing $K_{r,r}$. Let us call the graph $K_{1,r}$ a *star*, let us call the vertices of a star that are in the color class of size r the *leaves* of the star, and let us call the remaining vertex the *center* of the star. See Figure 6.9. (We are not aspiring to get a good grade in astronomy here.) We will now count the stars in two different ways. First, we count the

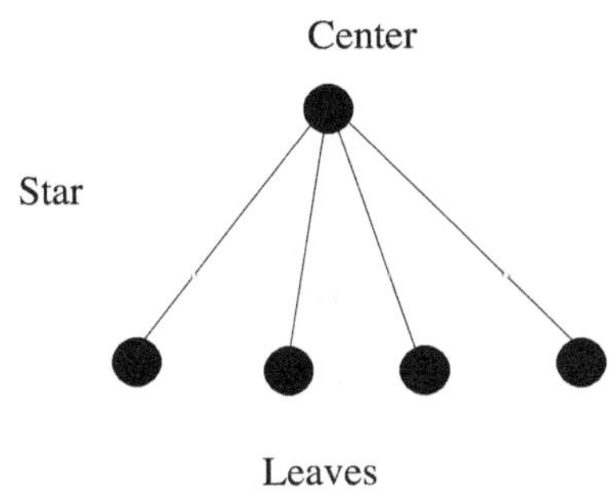

Figure 6.9
The anatomy of a star.

stars by their centers. Let $d_1, d_2, \cdots, d_n$ be the degrees of the vertices of the graph G. Then G contains $\sum_{i=1}^{n} \binom{d_i}{r}$ stars. Now, we count the stars by their sets of leaves. For each of the $\binom{n}{r}$ subsets of the vertex set of G that are of size r, there may be at most $r-1$ centers, otherwise G would contain a copy of $K_{r,r}$. Therefore,

$$\sum_{i=1}^{n} \binom{d_i}{r} \leq (r-1)\binom{n}{r}. \tag{6.2}$$

As proved in Exercise 3, the function $\binom{x}{r}$ is convex for $x \geq r-1$. (Note that, outside that interval, $\binom{x}{r}$ may be negative.) Since it is possible for the d_i to be less than $r-1$, we introduce the function

$$g(x) = \begin{cases} \binom{x}{r} \text{ if } x \geq r-1, \\ 0 \text{ otherwise.} \end{cases}$$

It is then easy to see (assuming Exercise 3) that g is convex on the set of all real numbers. Because $\binom{x}{r} = g(r)$ for all natural numbers, (6.2) and Theorem 6.14 imply that

$$(r-1)\binom{n}{r} \geq \sum_{i=1}^{n} \binom{d_i}{r} \geq \sum_{i=1}^{n} g(d_i) \geq ng\left(\frac{d_1 + \cdots + d_n}{n}\right) = ng\left(\frac{2e(G)}{n}\right), \tag{6.3}$$

where $e(G)$ is the number of edges of G.

Remember, we are trying to prove an upper bound on $e(G)$, that is, we are trying to prove that $e(G) = O(n^{2-\frac{1}{r}})$. The only time $e(G)$ appears in the previous equation is in $g\left(\frac{2e(G)}{n}\right)$, so we are going to take a closer look at this term. The definition of g is by cases, but one of them can be handled very easily. Indeed, if $2e(G)/n < r-1$, then $e(G) < n(r-1)/2$, so in that case, $e(G) = O(n)$, which is much stronger than the statement we are proving.

So we can assume that $2e(G)/n > r-1$, and therefore $g\left(\frac{2e(G)}{n}\right) = \binom{2e(G)/n}{r}$. Using this, (6.2) and (6.3) imply

$$n\binom{2e(G)/n}{r} \leq (r-1)\binom{n}{r}.$$

We would like to use this inequality to get an upper bound for $e(G)$ in terms of n and r. Fortunately, the upper bound to be proved is not sensitive to constant factors in n, so we can modify the inequality so as to make it easier to handle. In particular, let us replace the right-hand side by $(r-1)n^r/r!$ (this preserves the inequality, since the right-hand side grows), and let us replace the left-hand side by $n \cdot \frac{\left(\frac{2e(G)}{n}-r\right)^r}{r!}$ (this also preserves the inequality, since the left-hand side decreases). This leads to

$$n \cdot \frac{\left(\frac{2e(G)}{n}-r\right)^r}{r!} \leq \frac{(r-1)n^r}{r!},$$

$$\left(\frac{2e(G)}{n}-r\right)^r \leq (r-1)n^{r-1}.$$

Taking rth roots, and rearranging, we get

$$\begin{aligned} e(G) &\leq \frac{n}{2} \cdot ((r-1)^{1/r} \cdot n^{(r-1)/r} + r) \\ &= n^{2-\frac{1}{r}} \cdot \frac{(r-1)^{1/r}}{2} + \frac{rn}{2} \\ &= O(n^{2-\frac{1}{r}}). \end{aligned}$$

◊

Quick Check

1. At most how many edges can a simple graph on n vertices have if it does not contain a path of three vertices (a copy of $K_{1,2}$)?
2. At most how many edges can a simple graph on n vertices have if it does not contain a copy of $K_{1,3}$?
3. At most how many edges can a simple graph on n vertices have if it does not contain a path of four vertices?

6.2 Hypergraphs

In the previous section, we looked at *graphs* that were extremal from some viewpoint. A graph was in fact nothing other than a set of edges, or, in other words, *two-element subsets* of a vertex set.

This concept can be generalized if we allow *larger* subsets. These new collections of subsets are our main topics in the present section.

Definition 6.28 *A* hypergraph *on the set $[n]$ is a collection of subsets of $[n]$. The elements of this collection, which are subsets of $[n]$, are called the* edges *of the hypergraph, while the elements of $[n]$ are called the* vertices *of the hypergraph.*

So graphs are just a special case of hypergraphs, namely, the special case in which each edge consists of two vertices. Hypergraphs are often called *set systems* or *families of subsets* as well. They are often denoted by calligraphic letters, such as $\mathcal{F}$.

In this section, we will be looking at hypergraphs that are optimal from some point of view, for instance, they have the highest number of edges possible under certain conditions. For a hypergraph $\mathcal{F}$, let $|\mathcal{F}|$ denote the number of edges of $\mathcal{F}$.

6.2.1 Hypergraphs with pairwise intersecting edges

Theorem 6.29 *Let $\mathcal{F}$ be a hypergraph on vertex set $[n]$ that does not contain two* disjoint *edges. Then $|\mathcal{F}| \leq 2^{n-1}$.*

Proof: Let us arrange the 2^n subsets of $[n]$ into pairs, matching each subset with its complement. This will result in 2^{n-1} pairs. Each pair consists of *disjoint* edges, so $\mathcal{F}$ can contain at most one edge of each pair. This proves that $\mathcal{F}$ cannot have more elements than the number of these pairs, that is, 2^{n-1}. ◊

The above proof was not very difficult. This might suggest that, with a little extra effort, we can improve the result further. However, that hope is false, as the following theorem shows.

Theorem 6.30 *There exists a hypergraph $\mathcal{F}$ on vertex set $[n]$ that has 2^{n-1} edges so that no two elements of $\mathcal{F}$ are disjoint.*

Proof: Let $\mathcal{F}$ be the hypergraph whose edges are the subsets of $[n]$ that contain the vertex 1. ◊

We would like to point out that the hypergraph $\mathcal{F}$ defined in the above proof is actually even better than we needed it to be. It does not only have the property that any two of its elements intersect, but also the property that *all* its elements intersect!

See Exercise 13 for an alternative proof of Theorem 6.30.

One can now ask how large $\mathcal{F}$ can be if we require that any two edges of $[n]$ which are in $\mathcal{F}$ intersect in at least k elements, as opposed to just at least one element. At this point, the reader might suggest that we define $\mathcal{F}$ to be the hypergraph whose edges are the subsets of $[n]$ containing the set $[k]$. That family has 2^{n-k} elements, and it certainly satisfies the criterion on the size of pairwise intersections. The question is whether we can do better. The following construction shows that, if n is large enough, then we actually *can.*

Example 6.31 *Let k be such that $n+k$ is even, and let $\mathcal{F}$ be the hypergraph on vertex set $[n]$ whose edges are the subsets of $[n]$ that have $(n+k)/2$ elements.*

Then any two edges in $\mathcal{F}$ intersect in at least k elements, and if n is large enough, then $|\mathcal{F}| > 2^{n-k}$.

Solution: Indeed, any two edges together have $n+k$ elements, so they must share at least k vertices. On the other hand, the number of edges of $\mathcal{F}$ is $\binom{n}{(n+k)/2}$, and we will show that if n is large enough, then

$$\binom{n}{(n+k)/2} > 2^{n-k}. \tag{6.4}$$

First, note that, if $n = 3k$, then by Stirling's formula (1.13) the left-hand side is equal to

$$\binom{3k}{k} \sim 6.75^k \frac{\sqrt{3}}{2\sqrt{n\pi}},$$

while the right-hand side is just 4^k. Second, note that, if n grows to $n+2$, then the left-hand side grows by a factor of $\frac{2(n+2)(n+1)}{n+k+2}$, whereas the right-hand side simply grows by a factor of four. For large enough n, such as $n = 3k$, the growth rate of the left-hand side is larger, so the left-hand side of (6.4) stays larger than its right-hand side. ◊

Let us now refine our interest and look for optimal hypergraphs in which each edge has the same size. This class of hypergraphs occurs so often that it has its own name.

Definition 6.32 *If all edges of the hypergraph $\mathcal{F}$ consist of k vertices, then $\mathcal{F}$ is called k-uniform.*

First, let us look for large k-uniform hypergraphs that do not contain disjoint edges. Just as for general hypergraphs, we can make sure that the non-disjointness criterion is satisfied by simply placing a given entry into all edges of $\mathcal{F}$.

Proposition 6.33 *For all positive integers $k \in [n]$, there exists a k-uniform hypergraph of $\binom{n-1}{k-1}$ edges on vertex set $[n]$ that does not contain two edges which are pairwise disjoint.*

Proof: The hypergraph of all k-element subsets that contain the vertex 1 has the required property. ◊

It will probably not surprise the reader that the simple construction that we discussed is the best one yet again. While the result is very similar to Theorem 6.29, its proof is considerably more difficult.

Theorem 6.34 (Erdős–Ko–Rado theorem) *Let $\mathcal{F}$ be a k-uniform hypergraph on $[n]$ that does not contain two disjoint subsets. Then*

(a) $|\mathcal{F}| \leq \binom{n-1}{k-1}$ *for* $2k \leq n$, *and*

(b) $|\mathcal{F}| \leq \binom{n}{k}$ *for* $2k > n$.

The proof we present is due to Gyula O. H. Katona [48], and it is significantly simpler than the original proof [28] of the theorem.

Proof: Part (b) is true since we can set $\mathcal{F}$ to be the set of *all* k-element subsets of $[n]$. Let us prove part (a).

Think of the elements of $[n]$ as n different people who sit down on n chairs around a circular table that are numbered 0 through $n-1$, as shown in Figure 6.10.

Once a seating is specified, and in that seating k people sit on k consecutive chairs, then we will say that they form a *block* in that seating. More precisely,

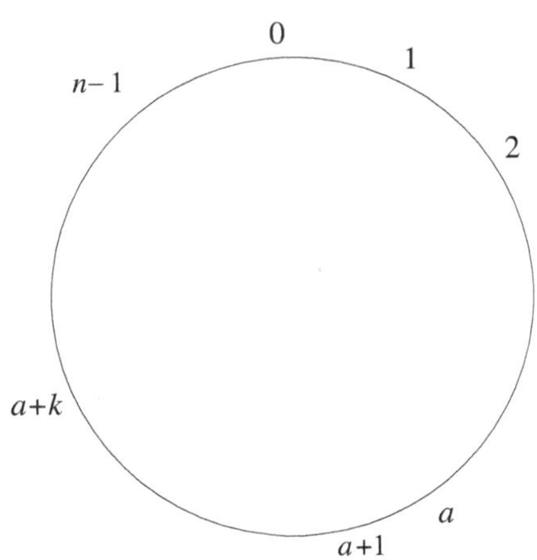

Figure 6.10
Chairs around a table.

if k people sit on chairs $a, a+1, \cdots, a+k-1$ in a given seating, then we will call them the block B_a of that seating. Addition here is meant modulo n, in other words, $n+c=c$.

First, we claim that, for any given seating, $\mathcal{F}$ can contain at most k blocks of that seating. Indeed, let $B_a \in \mathcal{F}$. Then all other blocks in $\mathcal{F}$ have to intersect B_a, so all other blocks in $\mathcal{F}$ have to be of the form B_{a+i} or B_{a-i}, with $i \in [k-1]$. This shows that there are at most $2k-1$ blocks in $\mathcal{F}$, which is not quite what we promised. So we point out that $\mathcal{F}$ cannot even contain all these blocks. In fact, for any $j \in [k-1]$, the blocks B_{a-k+j} and B_{a+j} are disjoint, so $\mathcal{F}$ can contain at most one of them. This shows that, indeed, $\mathcal{F}$ contains at most k blocks.

Now let us count all possible pairs (p, B), where p is a possible seating of our n people around the table and B is a block in that seating that is contained in $\mathcal{F}$. Let P be the number of such pairs.

Let us first count by the seatings. We have just seen that each seating will have at most k blocks which are contained in $\mathcal{F}$. Therefore,

$$P \leq k \cdot n!. \tag{6.5}$$

Let us now count by the blocks. If $B \in \mathcal{F}$, then to seat the people of B in a block of p, first we select the starting chair a of the block in n ways. Then we can seat the k people of B on the chairs $a, a+1, \cdots, a+k-1$ in $k!$ ways, then seat the remaining $n-k$ people on the remaining chairs in $(n-k)!$ ways. Since we can do this for each $B \in \mathcal{F}$, this leads to

$$P = |\mathcal{F}| \cdot n \cdot k! \cdot (n-k)!. \tag{6.6}$$

Comparing (6.5) and (6.6), we get

$$\begin{aligned} |\mathcal{F}| \cdot n \cdot k! \cdot (n-k)! &\leq k \cdot n! \\ |\mathcal{F}| &\leq \frac{k}{n} \cdot \binom{n}{k} \\ &\leq \binom{n-1}{k-1}. \end{aligned}$$

◊

If you think you understand this proof, you can test your understanding by finding the point in the proof at which we used the condition $2k \leq n$.

We would like to point out an important feature of the proof that will be useful many times in the future. The key element of the proof was to count *pairs* (p, B), once by the permutations, once by the blocks. This technique, counting specific pairs, once by their first elements, once by their second elements, and then comparing the results, is a simple but powerful tool.

The Erdős–Ko–Rado theorem has many variations. The following one, in a more general form, was proved by Béla Bollobás [11].

Theorem 6.35 *Let $\mathcal{F}$ be a k-uniform hypergraph on $[n]$ with $2m$ edges so that its edges can all be listed as $X_1, X_2, \cdots, X_m$ and $Y_1, Y_2, \cdots, Y_m$ in a way that $X_i \cap Y_j = \emptyset$ if and only if $i = j$. Then*

$$|\mathcal{F}| \leq \binom{2k}{k}.$$

One could think of this problem as follows: A football coach has n players and wants them to form various teams to play m short games against each other. In each of these games, one team will play offense only, and the other team will play defense only. No player can be on both teams in the same game (hence the $X_i \cap Y_i = \emptyset$ restriction). Furthermore, since this is a combinatorially inclined coach, no offense can be totally disjoint from any defense other than its current opponent, and vice versa, no defense can be totally disjoint from any offense other than its current opponent. (It is not clear that many coaches would worry about violating this last condition, but this one does.)

Note that the upper bound does not even depend on n. No matter how large n is, the size of the best possible $\mathcal{F}$ will not grow. This shows that we are dealing with a much stronger restriction here than in the Erdős–Ko–Rado theorem.

Proof: (of Theorem 6.35) Let us count the number P of pairs (p, i) so that $p = p_1 p_2 \cdots p_n$ is an n-permutation, $i \in [m]$, and all elements of X_i precede all elements of Y_i in p.

First, let us count by the indices $i \in [m]$. Let i be fixed, then X_i and Y_i are also fixed. Let us say that two permutations q and q' are equivalent if the set of entries that belong to $X_i \cup Y_i$ is located in the same positions in the two permutations. Note that we do *not* require that each entry $z \in X_i \cup Y_i$ be located in the same position in q as it is located in q'. What we do require is that the *set* of positions of the entries that belong to $X_i \cup Y_i$ is the same in q as in q'.

Then there are $\binom{n}{2k}$ equivalence classes, and each equivalence class consists of $k! \cdot k!(n-2k)!$ permutations. Indeed, there are $\binom{n}{2k}$ for the set of positions of the entries that belong to $X_i \cup Y_i$. Once that set is chosen, the k elements of X_i have to be in the first half of the positions of that set, in any order, and the k elements of Y_i have to be in the second half, again in any order. Finally, the entries that do not belong to $X_i \cup Y_i$ can be permuted in any order. This argument works for any fixed i, showing that

$$P = m \cdot \binom{n}{2k} \cdot k! \cdot k!(n-2k)! = n! \cdot \frac{m}{\binom{2k}{k}}. \tag{6.7}$$

Now let us count the pairs (p, i) with the required property by the permutation p. Let p be fixed. We claim that there is *at most one* index i so that (p, i) has the required property. Assume not; that is, assume that both i and j are such indices. Assume without loss of generality that the rightmost element of X_i in p is weakly on the right of the rightmost element of X_j in

Figure 6.11
If X_j ends before X_i, then X_j is disjoint from Y_i.

p. That would mean that $X_j \cup Y_i = \emptyset$, which is a contradiction. Indeed, all elements of X_i precede all elements of Y_i in p, so all elements of X_j must precede all elements of Y_i in p, since X_j ends before X_i. See Figure 6.11 for an illustration.

Since there is at most one index i for each permutation p, we certainly have $P \leq n!$. Comparing this inequality with (6.7), we get $n! \cdot \frac{m}{\binom{2k}{k}} \leq n!$, or $m \leq \binom{2k}{k}$ as claimed. ◊

The reader is asked to prove that this upper bound cannot be further improved (Supplementary Exercise 13). And then the reader is asked to generalize the statement of this theorem for nonuniform hypergraphs (Supplementary Exercise 14).

6.2.1.1 Sunflowers

In the constructions that proved to be optimal in the situations discussed in Theorem 6.34 and Theorem 6.30, we saw examples of hypergraphs in which the intersection is nonempty. This concept can be strengthened by requiring that the intersection of any two edges be not only nonempty, but also the *same*. This class of hypergraphs is so important that it has its own name.

Definition 6.36 *A hypergraph $\mathcal{H}$ of nonempty edges is called a* sunflower *if each pair of edges of $\mathcal{H}$ has the same intersection, which is called the* core *of the sunflower. The edges of the sunflower are called the* petals.

Example 6.37 *The hypergraph consisting of edges $\{1, k\}$, where k ranges over the set $\{2, 3, \cdots, n\}$ is a sunflower. See Figure 6.12 for an illustration of this sunflower for $n = 9$, and an explanation for the name "sunflower."*

Note that the core is allowed to be empty. Also note that Definition 6.36 is equivalent to saying that the intersection of any two edges is equal to the intersection of *all edges*.

For instance, if the edges of $\mathcal{F}$ are pairwise disjoint, then $\mathcal{F}$ is a sunflower, even if that is not a particularly exciting example. What is more interesting is that large-enough hypergraphs always contain sunflowers. This is the content of the famous Erdős–Rado lemma. Let us say that $\mathcal{H}$ is a (p, ℓ)-*sunflower* if $\mathcal{H}$ is a sunflower with p petals, none of which consists of more than ℓ vertices.

Figure 6.12
A sunflower with core $\{1\}$.

Theorem 6.38 (Erdős–Rado lemma) *Let $\mathcal{F}$ be a hypergraph having more than $(p-1)^\ell \ell!$ edges and let us assume that each edge of $\mathcal{F}$ contains at most ℓ vertices. Then $\mathcal{F}$ contains a (p, ℓ)-sunflower.*

Proof: We prove the statement by induction on ℓ. For $\ell = 1$, the statement is true, since a set of p distinct singletons is always a $(p, 1)$-sunflower.

Now let us assume that the statement is true for $\ell - 1$, and prove it for ℓ. If $\mathcal{F}$ contains p disjoint edges, then we are done. Otherwise, let $\mathcal{S}$ be a maximal set of disjoint edges in $\mathcal{F}$; then $|\mathcal{S}| \leq p - 1$. Furthermore, each edge of $\mathcal{F}$ intersects at least one edge of S, otherwise $\mathcal{S}$ would not be maximal.

By the pigeonhole principle, this implies that there is at least one edge E in $\mathcal{S}$ that has a nonempty intersection with more than $(p-1)^{\ell-1} l!$ edges of $\mathcal{F}$. Because E has at most ℓ vertices, this in turn implies that there is a vertex $v \in E$ that is contained in more than $(p-1)^{\ell-1}(\ell-1)!$ edges. Now set

$$\mathcal{F}' = \{H - v | H \in \mathcal{F}, v \in H\},$$

where v is the vertex described in the previous sentence. In other words, the edges of $\mathcal{F}'$ are the edges of $\mathcal{F}$ which contained v, with v removed. Then we know that each edge of $\mathcal{F}'$ has at most $\ell - 1$ vertices, and that there are more than $(p-1)^{\ell-1}(\ell-1)!$ edges in $\mathcal{F}'$. Therefore, by the induction hypothesis, $\mathcal{F}'$ contains a $(p, \ell - 1)$-sunflower. Adding v to each petal of that sunflower, we get a (p, ℓ)-sunflower, proving our statement for ℓ. $\diamondsuit$

6.2.2 Hypergraphs with pairwise incomparable edges

Let us try to find a hypergraph $\mathcal{F}$ on vertex set n so that no edge of $\mathcal{F}$ contains another edge of $\mathcal{F}$ as a subset. If two edges have the property that neither contains the other as a subset, then that pair of edges is called *incomparable*.

One idea is to set $\mathcal{F}$ to be the hypergraph of all k-element subsets of $[n]$. This will make sure that none of the edges of $\mathcal{F}$ will contain another one. The size of this hypergraph will be $\binom{n}{k}$. Therefore, if we are curious which k provides the largest such hypergraph, all we have to do is find the value of k for which $\binom{n}{k}$ is maximal. While there are several interesting and far-reaching methods to settle that question, at this point we present the quick, computational solution. We will present much more enlightening proofs in the next chapter.

Proposition 6.39 *Let n be any fixed positive integer. Then the binomial coefficient $\binom{n}{k}$ is maximal when $k = \lfloor n/2 \rfloor$.*

Proof: We prove a stronger statement by showing that the sequence of binomial coefficients $\binom{n}{k}$ increases up to its midpoint, then decreases. In other words, we are going to show that

1. $\binom{n}{k} < \binom{n}{k+1}$ if $k \leq \lfloor n/2 \rfloor - 1$, and
2. $\binom{n}{k} \geq \binom{n}{k+1}$ if $k \geq \lfloor n/2 \rfloor$.

To see this, just consider the ratio

$$\frac{\binom{n}{k+1}}{\binom{n}{k}} = \frac{\frac{n!}{(k+1)!(n-k-1)!}}{\frac{n!}{k!(n-k)!}} = \frac{n-k}{k+1}.$$

This ratio is larger than 1 precisely when $n-k > k+1$, that is, when $n > 2k+1$, proving our statement. $\diamondsuit$

We mention that the fact that a sequence of nonnegative real numbers behaves like the binomial coefficients, that is, it first increases steadily, then it decreases steadily, is often expressed by saying that the sequence is *unimodal.* The formal definition is that the sequence $a_0, a_1, \cdots, a_n$ is called unimodal if there exists an index m so that $a_0 \leq a_1 \leq \cdots \leq a_m$ and $a_m \geq a_{m-1} \geq \cdots \geq a_n$. We will learn more about unimodal sequences in the next chapter.

Proposition 6.39 shows that, as long as we want all edges in $\mathcal{F}$ to have the same size k, we will get the largest hypergraph $\mathcal{F}$ when $k = \lfloor n/2 \rfloor$. What can we say *without* the requirement that $\mathcal{F}$ is k-uniform? The following famous result [69] shows that, even then, the largest hypergraph $\mathcal{F}$ is the hypergraph of all k-element subsets of $[n]$, for $k = \lfloor n/2 \rfloor$.

Theorem 6.40 *[69]* **(Sperner's theorem)** *Let $\mathcal{A}$ be a hypergraph on vertex set $[n]$ so that there are no edges X and Y in $\mathcal{A}$ satisfying $X \subset Y$. Then $|A| \leq \binom{n}{\lfloor n/2 \rfloor}$.*

Proof: Our proof is somewhat similar to that of Theorem 6.34. Let us count the number of ways we can permute the elements of $[n]$ so that our permutation *starts* with the elements of some $X \in \mathcal{A}$ in some order, then continues by the elements of $[n] - X$ in some order. It is *crucial* to note that this can happen for only one X for any given permutation p, because of the condition that no edge of $\mathcal{A}$ contains another one. Let us call this set X the *starting set* of p.

Let $|X| = i$, and let us say that the permutation p starts with the i elements of X in some order. Then there are $i!$ ways to choose that order, and there are $(n-i)!$ ways to choose the rest of p. Therefore, if the number of edges in $\mathcal{F}$ that have i elements is a_i, then the number of permutations whose starting set is of size i is

$$a_i \cdot i! \cdot (n-i)!.$$

Summing over all i, we get that the number of all permutations of length n that have a starting set is

$$\sum_{i=1}^{n} a_i \cdot i! \cdot (n-i)!.$$

Finally, this is certainly not more than the number of all permutations of length n. Therefore,

$$\sum_{i=1}^{n} a_i \cdot i! \cdot (n-i)! \leq n! \tag{6.8}$$

$$\sum_{i=1}^{n} \frac{a_i}{\binom{n}{i}} \leq 1. \tag{6.9}$$

Recall that we are trying to prove that $|\mathcal{A}| \leq \binom{n}{\lfloor n/2 \rfloor}$, and note that $|\mathcal{A}| = \sum_{i=1}^{n} a_i$. Proposition 6.39 shows that $\binom{n}{i} \leq \binom{n}{\lfloor n/2 \rfloor}$ for all i. Therefore, the left-hand side of the second inequality in (6.8) does not increase if we replace all $\binom{n}{i}$ by $\binom{n}{\lfloor n/2 \rfloor}$. This leads to

$$\frac{\sum_{i=1}^{n} a_i}{\binom{n}{\lfloor n/2 \rfloor}} \leq 1$$

$$|\mathcal{F}| = \sum_{i=1}^{n} a_i \leq \binom{n}{\lfloor n/2 \rfloor},$$

completing our proof. ◊

Theorem 6.40 showed that, in the hypergraph $\mathcal{A}$ of that theorem, there existed a maximal (nonextendible) set $\mathcal{H}$ of edges in which no edge contained another edge so that $\mathcal{H}$ consisted of edges *of the same size*. If a hypergraph $\mathcal{A}$ has this property, then we call $\mathcal{A}$ a *Sperner* hypergraph.

Quick Check

1. Let $\mathcal{H}$ be a hypergraph on vertex set $[n]$ in which no two edges are disjoint. Is it true that we can add edges to $\mathcal{H}$ until we reach a total of 2^{n-1} edges so that the obtained hypergraph still does not have two disjoint edges?
2. Let $\mathcal{H}$ be a hypergraph on vertex set $[n]$ so that for any two edges A and B of $\mathcal{H}$, the inequality $|A \cap B| \geq 2$ holds. Prove that $\mathcal{H}$ has at most 2^{n-2} edges.
3. Matchings in graphs are defined in Exercise 34 of Chapter 5. Let G be a simple graph with more than 200 edges. Is it true that if G does not have a vertex of degree at least 11, then G must contain a matching consisting of at least 11 edges?

6.3 Something is more than nothing: Existence proofs

This section is built on the powerful, if not earth-shattering, idea that if the number of certain objects is positive, then it is at least one, so at least one such an object *exists.* Before the reader laughs at us for talking about such trivialities, we mention that we will use this method in situations when the existence of certain constructions is highly nontrivial and therefore needs to be proved.

6.3.1 Property B

As hypergraphs are still fresh in our minds, we start with an example from that area. Property B is a well-studied property of hypergraphs. We explain it by the following example.

Example 6.41 *A football coach wants to try out various offensive formations with his players. The coach has n offensive players, and the day before practice, he designs m offensive formations, each of which will consist of k offensive players. Then, at the beginning of the practice, the coach gives an orange or a blue shirt to each player to wear. Prove that, if $m < 2^{k-1}$, then, no matter how the offensive formations have been designed, the coach can assign the shirts so that both colors will be represented in each offensive formation.*

In general, we say that a hypergraph $\mathcal{F}$ on $[n]$ has *property B* (bi-colorable) if it is possible to color the elements of $[n]$ using two colors so that *no* edge of $\mathcal{F}$ is monochromatic. In this terminology, Example 6.41 says that, if a k-*uniform* hypergraph $\mathcal{F}$ on $[n]$ has less than 2^{k-1} edges, then $\mathcal{F}$ has property B.

The solution of this example will be interesting because of the novelty of the method we use. We will not actually *construct* a good assignment of colors, we will just prove that a good assignment *exists*. We will achieve this by simply showing that the number of all assignments is larger than the number of assignments in which something goes wrong.

Solution: (of Example 6.41) There are 2^n ways to assign shirts to the n players. Let A be one of the offensive formations, and let A_b (for "A is bad") denote the set of shirt distributions to the n players in which all players in A get shirts of the *same color*. This yields

$$|A_b| = 2^{n-k+1},$$

because there is no restriction for the shirt color of the $n-k$ players not in A, while the k players in A can all get orange shirts or all get blue shirts. Recalling that there are $m < 2^{k-1}$ offensive formations, and summing the previous equation over all of them, we get

$$\sum |A_b| = m \cdot |A_b| = m \cdot 2^{n-k+1} < 2^n.$$

Finally, recalling that the size of the union of some sets is never larger than the sum of the sizes of the sets participating in that union,

$$|\cup A_b| \leq \sum |A_b| < 2^n.$$

So the number of shirt distributions in which *something goes wrong* is less than the number of *all* distributions. Therefore, by the subtraction principle, the number of distributions in which nothing goes wrong is positive. In other words, there has to be a distribution in which nothing goes wrong, that is, both colors are represented in all formations. ◊

At this point, the reader could ask why this method is useful if it does not actually construct the assignment but rather just proves that it exists. It is certainly true that a constructive proof, that is, one which constructs a right assignment, is preferable when it is available. However, it is often much easier to provide a nonconstructive proof, especially when the parameters are large. A nonconstructive proof assures us that a good construction exists, so we are not wasting our time if we look for one. Along the same lines, if we have a constructive answer to a question, we can wonder whether we found the best answer, or whether perhaps our answer could be further improved. In this case, we can compare our construction to the nonconstructive answer and sometimes see if there is room for improvement.

6.3.2 Excluding monochromatic arithmetic progressions

Not that we want to insult the reader, but let us recall that an *arithmetic progression* is a sequence of numbers in which the difference of any two consecutive elements is the same number d. For instance, the sequence $b_n = 2n-1$, for $n \geq 1$, is an arithmetic progression since $b_{n+1} - b_n = 2 = d$ for all positive integers n.

In this section, we show that it is possible to color each positive integer smaller than 2017 either red or blue so that no *monochromatic* arithmetic progression of 18 elements is formed. One way to do this would be, of course, by exhibiting a coloring with the required properties. A less ambitious, but often simpler way is by computation. The number of all possible colorings is 2^{2016}. If we can show that the number of *bad* colorings (that is, colorings that contain a monochromatic arithmetic progression of 18 elements) is less than 2^{2016}, then we will have proved that a good coloring exists.

Let A denote an arithmetic progression of length 18. There are two ways the elements of A can be colored, namely, they can be either all red or all blue. There are 2^{1998} ways to color the remaining 1998 elements of [2016]. Therefore, if A_b denotes the set of colorings in which A is "bad" (monochromatic), then

$$|A_b| = 2^{1999}. \tag{6.10}$$

Continuing the line of thinking seen in the proof of Example 6.41, we want to sum the previous equation over all possible candidates for A. In order to do that, we need to know *how many* such candidates there are, that is, how many arithmetic progressions of length 18 exist on the set [2016].

An arithmetic progression $a_1, a_2, \cdots, a_{18}$ is uniquely determined by its first element, a_1, and its difference, $d = a_{i+1} - a_i$. In our case, these two parameters have to be chosen so that $a_{18} = a_1 + 17d \leq 2016$. In other words, for any fixed positive integer value of a_1, there are

$$D = \left\lfloor \frac{2016 - a_1}{17} \right\rfloor \tag{6.11}$$

choices for d. As $2006 = 17 \cdot 118$, this means that the total number of possible (a_1, d) pairs is

$$10 \cdot 118 + 17 \cdot (117 + 116 + \cdots + 1) = 118351,$$

so this is the number of all 18-element arithmetic progressions within the set [2016]. Indeed, there are 10 values of a_1 for which $D = 118$ (the values $1, 2, \cdots, 10$), there are 17 values of a_1 for which $D = 117$ (the next 17 positive integers), and so on, and at the end, there are 17 values of a_1 for which $D = 1$ (namely, $1983, 1984, \cdots, 1999$).

Therefore, summing (6.10) over all 18-element arithmetic progressions A, we see that

$$\sum_A |A_b| = 118531 \cdot 2^{1999} < 2^{2016},$$

since $118531 < 2^{17} = 131072$. Finally,

$$|\cup_A A_b| \leq \sum_A |A_b| < 2^{2016},$$

so the total number of *bad* colorings is less than the total number of *all* colorings, proving our claim.

6.3.3 Codes over finite alphabets

Assume a company wants to assign a unique *codeword* to each of its employees. The codewords can be of arbitrary finite length, but they all use the letters of a finite alphabet (maybe the English alphabet, or the digits from 0 to 9, or the union of these two sets). There is an important requirement, however. No codeword can be a prefix of another one, that is, if 13A2 is a codeword, then 1, 13, and 13A are not allowed as codewords. Let us call the set of all assigned codewords the *code*. Codes satisfying the mentioned criterion (that is, that an initial segment of a codeword is never a codeword) are called *prefix-free*. This is a very reasonable requirement; indeed, otherwise, when the person with codeword 13A2 types in that code, the system may interpret the first few digits of that codeword as another codeword and treat this person as if he were someone else.

The system manager tries to assign codewords to the employees keeping this prefix-free criterion in mind. At the same time, she wants to assign codewords that are as short as possible so that they are easy to remember. What inequality (or inequalities) can describe the constraints she faces?

In some cases, this question is easy to answer. For instance, if the finite alphabet we are using is the binary alphabet (consisting of 0 and 1), and the number of employees is 2^m, then codewords of length m are sufficient, and they are needed as well, since the number of nonempty codewords of length less than m is $2^m - 1$.

In general, the situation can be described by the following theorem.

Theorem 6.42 (Kraft's inequality) *Let N be a finite set of words over a k-element alphabet so that N contains N_j words of length j and no word in N is a prefix of another word in N. Then*

$$\sum_{j\geq 1} \frac{N_j}{k^j} \leq 1. \tag{6.12}$$

Proof: Let j be the smallest positive integer so that $N_j > 0$. That means that, for all $i > j$, the N_j words of length j will prevent $N_j \cdot k^{i-j}$ words of length i from being in N. Let $P(i, j)$ be the set of these words, that is, the words that cannot be in N as they start with one of the N_j words of length j in N. Then the size of $P(i, j)$ is exactly $\frac{N_j \cdot k^{i-j}}{k^i} = \frac{N_j}{k^j}$ times the size of the set of all possible words of length i over our alphabet.

A similar argument works for each other value of j for which $N_j > 0$. The obtained sets $P(i, j)$ will be pairwise disjoint, since for fixed j they consist of words of length i that start with a j-digit word from N. No word could satisfy that criteria for two distinct values of j, since that would mean that there are two words in N so that one is a prefix of another.

Let i be so large that $N_j = 0$ if $j \geq i$. Then the total fraction of all words of length i over our alphabet that are prevented from being in N by one of the shorter words in N is $\sum_{j\geq 1} \frac{N_j}{k^j}$. Indeed, the argument of the previous paragraph shows that the sets $P(i,j)$ are pairwise disjoint since j ranges over all values for which $N_j > 0$ holds. As the fraction of forbidden words cannot be larger than 1, our statement is proved. ◊

This is all very nice, you might say, but why are we discussing it now? Where is the existence proof here? It is easy to turn Kraft's inequality around to answer that question. Assume someone asks us to create a prefix-free code N over a finite alphabet of k letters so that there are N_j letters of length j. Kraft's inequality tells us that this is not possible unless $\sum_{j\geq 1} \frac{N_j}{k^j} \leq 1$. On the other hand, we claim that, if $\sum_{j\geq 1} \frac{N_j}{k^j} \leq 1$, then this *is* possible. Indeed, let $j_1 < j_2 < \cdots < j_m$ be the values of j for which $N_j > 0$. Then we can choose N_{j_1} codewords of length j_1 in some way. Now assume we have already chosen the codewords of length $j_1, j_2, \cdots, j_n$, with $n < m$. Note that

$$\sum_{r\geq 1}^{n} \frac{N_{j_r}}{k^{j_r}} < \sum_{r\geq 1}^{m} \frac{N_{j_r}}{k^{j_r}} \leq 1,$$

so, for $i > j_n$, not all words of length i are prevented from being in N. More precisely,

$$\begin{aligned} \sum_{r\geq 1}^{n} \frac{N_{j_r}}{k^{j_r}} &\leq \left(\sum_{r\geq 1}^{m} \frac{N_{j_r}}{k^{j_r}} \right) - \frac{N_{j_{n+1}}}{k^{j_{n+1}}} \\ &\leq 1 - \frac{N_{j_{n+1}}}{k^{j_{n+1}}}, \end{aligned}$$

so the portion of words of length j_{n+1} that we cannot choose is at most $1 - \frac{N_{j_{n+1}}}{k^{j_{n+1}}}$. Therefore, we can still choose $N_{j_{n+1}}$ codewords of length j_{n+1}.

The above argument, together with Kraft's inequality (Theorem 6.42) leads to the following corollary.

Corollary 6.43 *Let k be a positive integer. Then there exists a prefix-free code over a k-element alphabet in which exactly N_i words are of length i if and only if $\sum_{j\geq 1} \frac{N_j}{k^j} \leq 1$ holds.*

However, we should not read too much into the "if and only if" nature of Corollary 6.43. It is *not true* that, if a code over a finite alphabet satisfies (6.12), then that code is prefix-free. The reader is asked to construct a counterexample in Supplementary Exercise 22.

Besides the scenario described at the beginning of this subsection, prefix-free codes are useful since they are *uniquely decipherable.* That is, when we want to read a message written in the code, there is no danger of ambiguity; every message can be decoded in only one way.

Example 6.44 *If our codewords are* $Y = 111$*,* $E = 001$*,* $S = 010$*, and* $O = 10$*, then the message 111001010 is easy to read. The first codeword is uniquely deciphered as* $Y = 111$*, since nothing else starts with consecutive 1s. Then we are left with the message 001010, the first codeword of which is deciphered as* E*, since nothing else starts with consecutive 0s. This leaves us with 010, which is the letter* S*. So the message 111001010 corresponds to the word YES.*

If we add the codeword $N = 1110010$ *to our previous code, the code loses its prefix-free property. Indeed, now* $Y = 111$ *is a prefix of* $N = 1110010$*, so when we start reading the message 111001010, we do not know whether the first encoded letter is a* $Y = 111$ *(and then the message is YES) or an* $N = 1110010$ *(and then the message is NO), which makes the code less than perfect.*

Note that if a code is prefix-free, like the original one in the above example, then it is stronger than simply uniquely decipherable. That is, we can do better than just unambiguously find out what the message was (after several hours of thought, brilliant ideas, hard work, and lots of coffee). We can *immediately* find out what the codewords are as we read them from left to right. We do not even need to look at the rest of the message when we figure out its next letter. Each codeword is immediately recognized as soon as its last digit is read. Such codes are called *instantaneous.*

Example 6.44 should give a strong hint to the reader on why a code is prefix-free if and only if it is instantaneous. A proof is given in the solution of Exercise 19.

Being uniquely decipherable is a weaker property of codes than being instantaneous, as shown by the following example.

Example 6.45 *The code consisting of codewords* $A = 11$ *and* $B = 110$ *is uniquely decipherable, but not instantaneous.*

Indeed, to decipher any message written in this code, first find all 0s in the message, then mark the two 1s preceding each of those 0s together with the 0s. This will locate all the letters B *in the encoded message, so the rest of the message must consist of letters* A*.*

However, the code is not instantaneous. Indeed, if we receive a message starting with 11, we cannot tell just from these two digits whether the first letter of the encoded message is an A *or a* B*. We will be able to tell that when we read the next digit, but that is too late; in an instantaneous code we must recognize each letter when we finish reading it, and here that is not the case if the first letter is an* A*.*

Based on this example, it would be plausible to think that it is significantly easier to construct uniquely decipherable codes than instantaneous codes, since there are many more techniques allowed to decipher a message in the former. One could furthermore guess that, for that reason, uniquely decipherable codes satisfy some weakened version of Kraft's inequality.

Surprisingly, this turns out to be *false*. Uniquely decipherable codes must satisfy (6.12) just as prefix-free codes do. This is the content of the following theorem due to MacMillan.

Theorem 6.46 (MacMillan's theorem) *Let N be a finite, uniquely decipherable code over a k-element alphabet so that N contains N_j words of length j. Then*

$$\sum_{j\geq 1} \frac{N_j}{k^j} \leq 1.$$

Conversely, if a set of positive integers N_j satisfies the criterion of MacMillan's theorem, then a uniquely decipherable code with N_j words of length j exists. Indeed, we know from Corollary 6.43 that even a *prefix-free* code exists with those parameters.

Quick Check

1. Prove that any simple graph with m edges contains a bipartite subgraph with more than $m/2$ edges.
2. We chose 26 points inside a square of unit side length. Prove that there are two among the chosen points whose distance from each other is less than 0.29.
3. Pattern-avoiding permutations are defined in Exercise 32 of Chapter 4. Prove that for each permutation p of length nine, there exists a pattern q of length five so that p contains at least two copies of q.

6.4 Notes

Readers interested in a more extensive introduction to extremal graph theory, as part of the general theory of graphs, can consult *Graph Theory* [83], by Douglas West. A more specialized source is *Extremal Graph Theory* [10], by Béla Bollobás. Classic books on hypergraphs include *Hypergraphs* [8] by Claude Berge and *Sperner Theory* [27] by Konrad Engel.

A recent source on extremal combinatorics is the identically titled book by Stasys Jukna [43]

Nonconstructive proofs can be made far more powerful if one learns how to use probability theory in them. The method then is called the *probabilistic method.* A basic introduction to that area is Chapter 15 of *A Walk Through Combinatorics* [12] by the present author, and a high-level textbook is *The Probabilistic Method* [3], by Noga Alon and Joel Spencer.

6.5 Chapter review

(A) Extremal graph theory

1. Graphs containing no K_k (Turán's theorem)
   If G is a simple graph on n vertices containing no copies of K_k, then G cannot have more edges than the complete $(k-1)$-partite graph in which the size of color classes is as balanced as possible.
2. Graphs containing no C_k
   (a) If a graph on n vertices contains no triangles, it cannot have more than $a(n-a)$ edges, where $a = \lceil n/2 \rceil$, and this bound is sharp.
   (b) The equality
   $$f_{C_4}(n) = \Theta(n^{3/2})$$
   holds.
   (c) If $k \geq 4$, and k is even, then
   $$f_{C_k}(n) = O(n^{1+\frac{2}{k}}).$$
   (d) If $k \geq 3$, and k is odd, then
   $$f_{C_k}(n) = \Theta(n^2).$$
   (e) If a graph on n vertices contains no odd cycles, it cannot contain more than $a(n-a)$ edges, where $a = \lceil n/2 \rceil$, and this bound is sharp.
3. Graphs containing no $K_{r,r}$
   $$f_r(n) = f_{K_{r,r}}(n) = O(n^{2-\frac{1}{r}}).$$

(B) Extremal hypergraphs

1. The number of edges in a hypergraph $\mathcal{F}$ on $[n]$ in which each pair of edges intersects is
   $$|\mathcal{F}| \leq 2^{n-1},$$
   and this is sharp.
2. Erdős–Ko–Rado Theorem. If $2k \leq n$, then the number of edges in an intersecting k-uniform hypergraph $\mathcal{F}$ is
   $$\mathcal{F} \leq \binom{n-1}{k-1}, \text{ if } k \leq 2n, \text{ and } \binom{n}{k} \text{ otherwise,}$$
   and this is sharp.

3. Sperner's theorem. The number of edges in a hypergraph $\mathcal{F}$ on $[n]$ in which no edge contains another edge is

$$|\mathcal{F}| \leq \binom{n}{\lfloor n/2 \rfloor}.$$

(C) Codes over finite alphabets

1. Kraft's inequality. A prefix-free code over $[k]$ with N_j words of length j exists if and only if

$$\sum_{j \geq 1} \frac{N_j}{k^j} \leq 1. \tag{6.13}$$

2. MacMillan's theorem. A uniquely decipherable code over $[k]$ with N_j words of length j exists if and only if (6.13) holds.

6.6 Exercises

Remember that, unless otherwise stated, all graphs mentioned are *simple graphs.*

1. Let G be a graph on $2n$ vertices and having more than n^2 edges. Prove by induction that G contains a triangle.
2. Let $\delta(A)$ (resp. $e(A)$) denote the minimum degree (resp. number of edges) of the graph A, and let $d_A(x)$ denote the degree of the vertex x in the graph A.
   (a) Prove that, any graph G contains a bipartite subgraph B so that $e(B) \geq e(G)/2$.
   (b) Prove that any graph G contains a bipartite subgraph B so that the vertex sets of B and G are the same and $\delta(B) \geq \delta(G)/2$.
   (c) Prove that any graph G contains a bipartite subgraph B so that the vertex sets of B and G are the same and so that for all vertices $x \in B$, the inequality $d_B(x) \geq d_G(x)/2$ holds.
3. Prove that for all nonnegative integers n, the function $f(x) = \binom{x}{r}$ is convex in the interval $[r-1, \infty)$.
4. Let G be a graph, and let G_1 be the graph obtained by taking two copies of G, then connecting each vertex x of G to the corresponding vertex x' of G_1. See Figure 6.13 for an illustration. Express $\chi(G_1)$ in terms of $\chi(G)$.

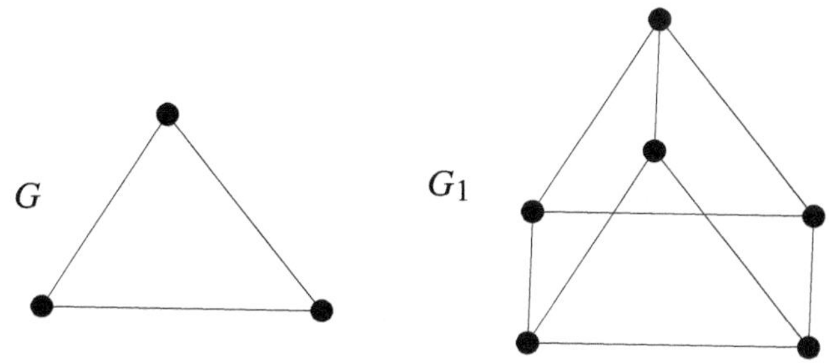

Figure 6.13
An example of the graphs G and G_1.

5. Let G and H be two simple graphs, and define the *weak direct product* $G \times H$ of G and H as follows: The vertex set of $G \times H$ is the set of all ordered pairs (g, h), where g is a vertex of G and h is a vertex of H. There is an edge between (g, h) and (g', h') if there is an edge between g and g' and there is an edge between h and h'. Figure 6.14 shows the weak direct product of K_2 and K_3.

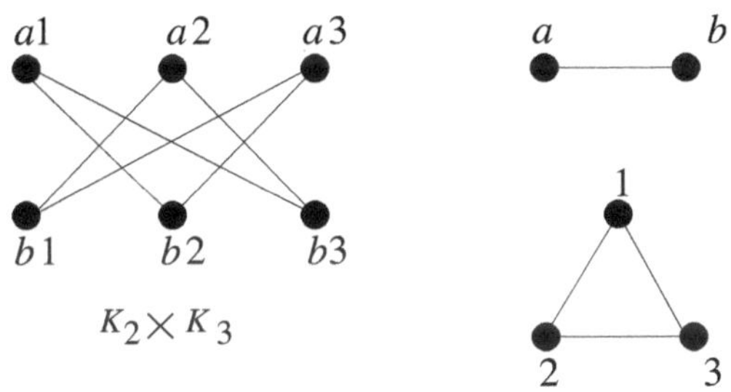

Figure 6.14
The weak direct product of K_2 and K_3.

   (a) Prove that $\chi(G \times H) \leq \min(\chi(G), \chi(H))$.
   (b) Is it true that $\chi(G \times G) = \chi(G)$?

6. A graph G on 10 vertices has degrees 9, 3, 3, 3, and six degrees equal to 2. Prove that $\chi(G) \leq 4$.

7. Prove that, if G is as in Exercise 6, then $\chi(G)$ can be 3 or 4, but not 2.

8. A connected graph is called *k-connected*, or k-vertex-connected, if it stays connected even if we remove any $(k - 1)$ of its vertices and all edges adjacent to them.

At most how many edges can a simple graph on n vertices have if it is not 2-connected?

9. Let G be a graph with maximum degree k.

   (a) Prove that $\chi(G) \leq k+1$.

   (b) Find two infinite classes of graphs for which the bound of part (a) is sharp, that is, $\chi(G) = k+1$.

10. + Let G be a graph with maximum degree k.

    (a) Find *all* graphs G that are 3-connected and for which $\chi(G) = k+1$.

    (b) Find all graphs G for which $\chi(G) = k+1$.

11. The set of workers at a factory altogether have 500 different skills, and for each skill, there are 10 workers who have that skill. Prove that it is possible to schedule the workers in two shifts so that each skill is available in both shifts.

12. Let $\mathcal{F}$ be a hypergraph on $[n]$ so that, for any two edges A and B in $\mathcal{F}$, the inequality $A \cup B \neq [n]$ holds. At most how large can $\mathcal{F}$ be?

13. The proof we gave for Theorem 6.30 showed an example for $\mathcal{F}$ that contained edges of $[n]$ of all sizes from 1 to n. Find an alternative proof that uses edges of a smaller number of different sizes.

14. Let $\mathcal{F}$ be a hypergraph on $[n]$ so that the smallest edge in $\mathcal{F}$ is of size k. Assume that, no matter how we choose $k+1$ edges in $\mathcal{F}$, their intersection is never empty. Can the intersection of *all* edges in $\mathcal{F}$ be empty?

15. + Show an example of a 2-coloring of the set $[2141]$ that does not contain a monochromatic arithmetic progression of length 18.

16. We color each edge of K_6 either red or blue. Prove that the resulting graph contains a triangle with monochromatic edges.

17. + Let k and ℓ be positive integers. Prove that there exists a smallest positive integer $R(k,\ell)$ so that, if we color the edges of $K_{R(k,\ell)}$ red or blue, there will be either a red copy of K_k or a blue copy of K_ℓ.

    Note that $R(k,\ell)$ is called a *Ramsey number.*

18. We color each edge of K_{17} either red or blue or green. Prove that the resulting graph contains a triangle with monochromatic edges.

19. Prove that a code over a finite alphabet is instantaneous if and only if it is prefix-free.

20. Having defined the notion of avoidance for graphs and permutations, let us define avoidance for *matrices whose entries are either 0 or 1* as follows: We say that the $n \times n$ matrix A *avoids* the $k \times l$ matrix B if we cannot find k rows and l columns in A which is

entry-wise at least as large as B. In other words, A does not have a $k \times l$ submatrix that has a 1 in each position where B has a 1.

Let $f(n, B)$ be the highest number of entries equal to 1 that a B-avoiding $n \times n$ matrix having 0s and 1s for entries can have.

(a) Find $f(n, B)$ if $B = (1, 1)$.

(b) + Find $f(n, B)$ if $B = \begin{pmatrix} 1 & 0 \\ 0 & 1 \end{pmatrix}$.

21. The previous exercise might have the reader guess that $f(n, B) = O(n)$ for all matrices B. This is not true, however, as we will see in this exercise.

    Let
    $$B = \begin{pmatrix} 1 & 1 & 0 \\ 1 & 0 & 0 \\ 0 & 0 & 1 \end{pmatrix}.$$

    Let the sequence of matrices A_i be defined by $A_0 = 1$, $A_1 = \begin{pmatrix} 1 & 1 \\ 1 & 0 \end{pmatrix}$, and $A_{n+1} = \begin{pmatrix} I_{2^n} & A_n \\ A_n & 0 \end{pmatrix}$, where I_{2^n} is the identity matrix of size $2^n \times 2^n$. So A_n is of size $2^{n+1} \times 2^{n+1}$.

    (a) Prove that A_n avoids B.

    (b) Determine how many entries equal to 1 the matrix A_n contains, and conclude that $f(n, B) \neq O(n)$.

22. Let B be a *permutation matrix* of size $k \times k$, and assume that n is divisible by k^2. Let A be a B-avoiding $n \times n$ matrix whose entries are equal to 0 or 1, so that A has $f(n, B)$ entries equal to 1. Cut A up into blocks of size $k^2 \times k^2$. Show that at most $f\left(\frac{n}{k^2}, B\right)$ of these blocks can contain nonzero entries.

23. + Continuing the previous exercise, let us call a $k^2 \times k^2$ block of A *tall* if it contains nonzero entries in at least k rows. Similarly, call a block of A *wide* if it contains nonzero entries in at least k columns.

    Keeping in mind that A and B are as defined in the previous exercise, look at any row of blocks of A. Prove that at most $\binom{k^2}{k}(k-1)$ of these n/k^2 blocks can be tall. Formulate and prove the corresponding statement for wide blocks.

24. Use the results of the previous two exercises to prove that, if B is a permutation matrix, then
    $$f(n, B) \leq 2k^4 \binom{k^2}{k} n.$$

    Note in particular that this means that, if B is a permutation matrix, then $f(n, B) = O(n)$.

25. + Recall that $S_n(q)$ is the number of n-permutations avoiding the pattern q. This notion was defined in Exercise 32 of Chapter 4. Use the result of the previous exercise to show that, for all patterns q, there exists a constant c_q so that $S_n(q) < c_q^n$.

26. Let n be a positive integer, and let q be a k-permutation, with $k \leq n$. Prove that there exists an n-permutation containing more than $\binom{n}{k}/k!$ copies of q.

27. A company has n job openings to fill, and n applicants. Each candidate has his own set of qualifications and demands. That is, a candidate may qualify for certain jobs but not for others, and each candidate has a minimum salary expectation for each job he is interested in, meaning that he will not take the job for a salary lower than that expected amount. These amounts vary from candidate to candidate, and from job to job. Nobody can work in more than one job.

    A Human Resources manager wants to fill all n positions so that the total salary cost for the company is as low as possible.

    (a) Express the task of the manager in the language of graphs.
    (b) Assume the manager fills the openings in a greedy way. That is, he first fills the opening for which he finds a qualified candidate at the lowest possible salary, then, from the remaining $n-1$ openings, he first fills the one for which he finds a qualified candidate at the lowest possible salary, and so on. Assuming that the manager succeeds in filling all positions, will he always achieve the lowest possible cost using this strategy?

28. + Let A and B be two cycle-free subgraphs of the same graph K, and assume that A has more edges than B. Prove that there is an edge of A that can be added to B so that B keeps its cycle-free property.

29. + A county wants to build a connected system of roads for its n towns. A case study revealed the costs of building a direct road between each pair of towns in the county. The county commissioner knows that the cheapest connected network will be achieved if the graph G of roads to be built is a tree. Therefore, the commissioner plans to build G in the following greedy way: First, she picks the cheapest road possible between two towns. Then she picks the second cheapest road. In each following step, she picks the cheapest road still not picked that will not create a cycle in the graph of roads already picked. She will stop when it is no longer possible to pick a road without creating a cycle, that is, when the graph G built is a tree.

    Will this greedy strategy always provide the best possible results?

30. + (Basic knowledge of linear algebra required.) Let V be a finite dimensional vector space, and let us associate a positive cost $c(v)$ to each vector $v \in V$. Let us now select a basis for V in the greedy way: First, select the cheapest vector, then select the cheapest vector that is linearly independent of the first chosen vector, and so on. In step i, choose the cheapest vector that is not in the subspace spanned by the $i-1$ vectors that were previously chosen. Stop when $dim\ V$ vectors are selected.

    Will the obtained basis B indeed be a minimum-cost basis?

31. Let A be a 1000×1000 matrix whose entries are equal to 0 or 1 so that there are at least 20000 entries equal to 1. Prove that at least one of the following two statements holds.

    (a) The matrix A has a row that contains at least 100 entries equal to 1, or

    (b) the matrix A has at least 100 rows so that each of them contains at least ten entries equal to 1.

6.7 Solutions to exercises

1. We prove this by induction on n, the initial case of $n = 2$ being easy to verify. Take an edge of G with endpoints A and B. If the sum of the degrees of A and B is at least $2n+1$, then A and B have a common neighbor C, and we are done. Otherwise, remove A, B, and all the edges adjacent to them. The remaining graph has $2n-2$ vertices and at least $n^2+1-(2n-1) = (n-1)^2+1$ edges, so by induction, it contains a triangle.

2. (a) (Part (c) implies this part, but we also provide a different proof.) Take all bipartite subgraphs of G in all possible ways, and count their edges. In other words, count all pairs (e, B) where e is an edge of the bipartite subgraph B of G.

    Let n be the number of vertices of G. On one hand, each edge of G is part of 2^{n-2} bipartite subgraphs, so the number of these pairs is $e(G)2^{n-2}$. Indeed, the number of ways to split the set of $n-2$ vertices disjoint from a given edge into an ordered pair of two subsets is 2^{n-2}. On the other hand, the number of bipartite subgraphs B of G that have at least one edge is $2^{n-1}-1$. Therefore, by the pigeonhole principle, there must be at least one B that has at least $\frac{e(G)2^{n-2}}{2^{n-1}-1} > \frac{e(G)}{2}$ edges.

    (c) (This implies (b).) Partition the vertex set into two blocks, Q and R, and disregard the edges within each block. If the obtained bipartite graph B satisfies the criterion of part (c), then we are done. If not, then there is a vertex x that violates that criterion. That means that more than half of the edges adjacent to x are in the same block as x. Therefore, moving x into the *other block*, the number of edges of B increases. Keep doing this as long as there is a vertex violating the criterion of part (c). The procedure eventually has to stop, since the number of edges of B cannot grow for an infinitely long time, and it does grow in each step. When the procedure stops, there are no vertices x violating the criterion, which proves the statement.

3. It suffices to prove that, on any interval $[a, b]$ within the allowed limits, the function $g(x) = \binom{x}{r} + \binom{a+b-x}{r}$ has a unique minimum at $x = (a+b)/2$. Set $n = a+b$ to simplify the formulae. Then we need to prove that the unique root of $g'(x)$ is at $x = n/2$, and that at that root $g'(x)$ changes from being negative to being positive.

   Apply induction on r, the case of $r = 1$ being true. Note that, for symmetry reasons, we can assume that $x \leq n/2$. Also note that

$$
\begin{aligned}
g'(x) &= \left(\binom{x}{r-1} \cdot \frac{x-r+1}{r} + \binom{n-x}{r-1} \cdot \frac{n-x-r+1}{r}\right)' \\
&= \frac{\binom{x}{r-1} - \binom{n-x}{r-1}}{r} + \binom{x}{r-1}' \cdot \frac{x-r+1}{r} \\
&+ \binom{n-x}{r-1}' \cdot \frac{n-x-r+1}{r}.
\end{aligned}
$$

   It is now routine to verify that $x = n/2$ is indeed a root of $g'(x)$. Note that $\binom{x}{r}$ is increasing if $x > r-1$. This, with the induction hypothesis and the above inequality, imply that $g'(x) > 0$ if $x > n/2$, and $g'(x) < 0$ if $x < n/2$. This completes the proof.

4. We claim that $\chi(G') = \chi(G)$ if G has at least one edge. (Otherwise, $\chi(G) = 1$ and $\chi(G') = 2$.) Indeed, if f is a proper coloring of G, then let $f(x)+1$ be the color of x', where addition is modulo $\chi(G)$. That is, if $f(x) = \chi(G)$, then $f(x') = 1$, and leave the color of x unchanged. This gives a proper coloring of G' with $\chi(G)$ colors.

5. (a) Assume without loss of generality that $k = \chi(G) \leq \chi(H)$. Let f be any k-coloring of G. Let $f(g)$ be the color of all vertices (g, h). This results in a proper k-coloring of $G \times H$ since two vertices of this graph are adjacent only if their G-coordinates are adjacent in G.

   (b) Yes. The induced subgraph of $G \times G$ whose vertices are of the form (g, g) is isomorphic to G, so we do need $\chi(G)$ colors to color that subgraph.

6. Color the vertices with colors from $[4]$ the *greedy way* as follows: Go in nonincreasing order of vertex degrees, and for each vertex, use the

smallest color that keeps the coloring proper. If $d_1 \geq d_2 \geq \cdots \geq d_{10}$ are the degrees, then for vertex i the number of forbidden colors in the above coloring is $\min(d_i, i-1)$. Indeed, vertex i has d_i neighbors, of which at most $i-1$ are colored before i. In the graph at hand, $\max_{i=1}^{10} \min(d_i, i-1) = 3$, and the maximum is taken at the fourth vertex. So there are never more than three forbidden colors, proving that $\chi(G) \leq 4$.

7. In order to see that $\chi(G) = 4$ is possible, take three vertex-disjoint edges and a triangle. This disconnected graph has chromatic number three because of the triangle it contains. Now connect all the nine vertices of this graph to a tenth vertex, creating a graph G with the prescribed degree sequence and chromatic number four.

   To see that $\chi(G) = 3$ is possible, take three vertex-disjoint copies of $K_{2,1}$ and connect all nine of their vertices to a new tenth vertex.

   Finally, if $\chi(G) = 2$ were possible, then there would be a bipartite graph with this degree sequence. That is impossible, since the vertex with degree nine would have to be connected to all other vertices.

8. The graph shown in Figure 6.15 shows that $\binom{n-1}{2} + 1$ edges are possible. On the other hand, more edges are not possible. Indeed, assume G has at least $\binom{n-1}{2} + 2$ edges and a cut-point x. Then $G - x$ has $n-1$ vertices, at least $\binom{n-2}{2} + 1$ edges, and is not connected. This is impossible, as Supplementary Exercise 5 asks you to prove.

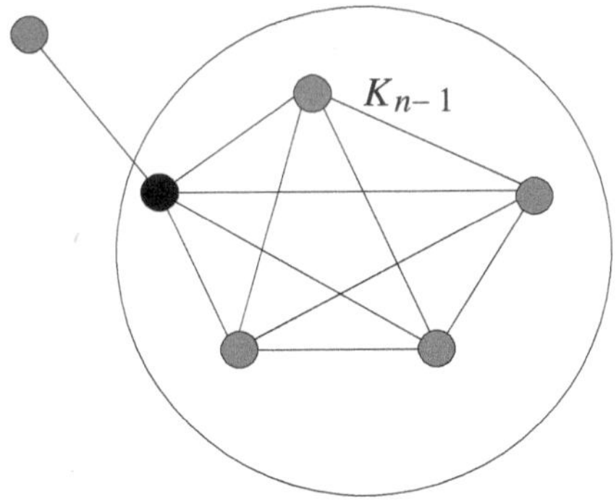

Figure 6.15
A connected, but not 2-connected, graph on n vertices and $\binom{n-1}{2}+1$ edges.

9. (a) Let us use colors $1, 2, \cdots, k+1$, and let us use them in a greedy way as follows: Start coloring the vertices of G one by one, and in each step use the smallest color that can be used without violating existing constraints, that is, the smallest color that does not occur among the colors of the neighbors of our current

vertex. As the maximum degree of G is D, at no step will there be more than D colors forbidden, so there will always be at least one color available for each vertex.

(b) The complete graphs K_n and the odd cycles C_{2m+1} have this property.

10. (a) We claim that, if G is not complete and is not an odd cycle, then $\chi(G) \leq k$.

In order to prove this, note that, if G has a vertex v of degree less than $k = D(G)$, then we are done by induction on the number of vertices. Indeed, remove v to get $G - v$. Then, by the induction hypothesis, $G - v$ has a proper coloring with at most k colors. As v has fewer than k neighbors in G, there is a color that can be used for v. This will provide a proper coloring of G with at most k colors.

It remains to prove the statement for the case when all vertices of G have degree k, in other words, when G is *k-regular*. We can assume that $k \geq 3$, otherwise either the claim is trivially true, or $G = K_2$ or $G = C_{2m+1}$.

In this case, take two vertices b and c in G whose distance is two. Two such vertices must exist, otherwise G would be complete, or not k-regular. Let a be a common neighbor of b and c. Since G is 3-connected, $G - b - c$ is connected. Since $G - b - c$ is connected, it has a subgraph T that is a tree (a *spanning tree* of $G - b - c$, in the language described in the Notes section of Chapter 5). Direct the edges of this tree toward a. Now color the vertices of G by colors from $[k]$ as follows: First, color b, then c, and then color the remaining vertices in any order so that, if the distance of x to a is more than the distance from y to a (along the directed edges of T), then x gets colored before y. So a gets colored last. See Figure 6.16 for an example.

In each step, use the smallest possible color that keeps the coloring proper. Then b and c both get colored 1. The other vertices different from a have at most $k-1$ colored neighbors when they themselves get colored. Finally, a has k colored neighbors when it gets colored, but two of them, b and c, have the same color. Therefore, all vertices can be colored using only colors from $[k]$, as claimed.

(b) Now assume that G is not 3-connected. If G is not 2-connected, then it has a vertex w so that $G - w$ is not connected. Let A be a connected component of $G - w$, and let $B = G - w - A$. Then by induction on the number of vertices, we can assume that both $A \cup w$ and $B \cup w$ have proper k-colorings. Permuting the colors so that w gets the same colors in both graphs yields a proper coloring of G.

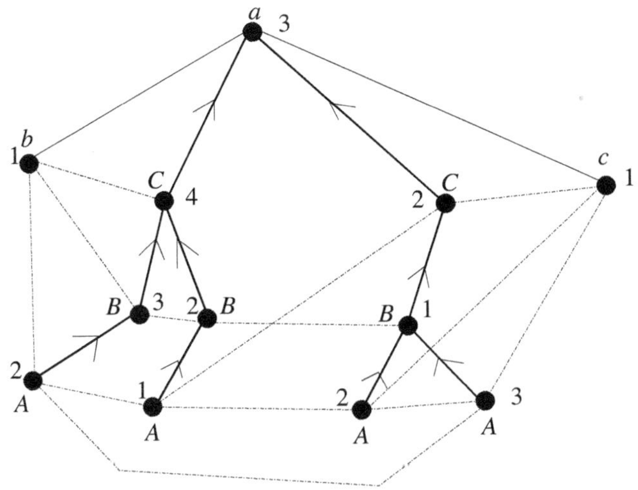

Figure 6.16
In this 4-regular graph G, the edges of T are the thick lines. After coloring b and c, we color the vertices in class A, then class B, then class C, then vertex a. Within each class, any order is allowed, but in this example, we proceeded from left to right.

What is left to consider is the case when G is 2-connected but not 3-connected. That means that G has two vertices, x and y, so that $G - x - y$ is not connected. In other words, there exist two induced subgraphs G_1 and G_2 of G so that $G_1 \cup G_2 = G$ and $G_1 \cap G_2 = \{x, y\}$. Note that both x and y must have at least one neighbor in each G_i, otherwise G would not be 2-connected. Consider the graphs $G_1 \cup (x, y)$ and $G_2 \cup (x, y)$. (We add the edge (x, y) to G_i if it is not there yet.) It follows from the last sentence of the previous paragraph that the degree of each vertex in these graphs is at most k; therefore, these graphs both have proper k-colorings by induction. In each of these two colorings, x and y have different colors since they are adjacent. Therefore, permuting the colors so that the colors of x agree in both colorings and the colors of y agree in both colorings, we get a proper 2-coloring of G.

To summarize, we proved that, if G is not complete and not an odd cycle, then $\chi(G) = k$, where k is the maximum degree in G. This is a famous theorem of Brooks [21].

11. Let A_i be the set of schedules in which skill i is only available in one shift. This yields

$$|A_i| = 2 \cdot 2^{n-10} = 2^{n-9},$$

where n is the number of workers at the company. Indeed, there are two ways to schedule the 10 workers having skill i in a *bad way* (schedule them all in the first shift or all in the second shift), and then there are 2^{n-10} ways to schedule the remaining $n-10$ workers. Therefore,

$$| \cup_i A_i| \leq \sum_i |A_i| = 500|A_i| = 500 \cdot 2^{n-9} < 2^n,$$

since $2^9 = 512$. So the number of schedules in which at least one skill is not represented in both shifts is less than the number 2^n of all schedules. Therefore, there has to be a good schedule.

12. Taking the hypergraph $\mathcal{F}^c$ consisting of the complements of the edges in $\mathcal{F}$, we see that $\mathcal{F}^c$ has to consist of edges no two of which are disjoint. Therefore, Theorems 6.29 and 6.30 imply $|\mathcal{F}^c| = |\mathcal{F}| \leq 2^{n-1}$.

13. If $n = 2m+1$, let $\mathcal{F}$ consist of all edges of $[n]$ having at least $m+1$ vertices. Any two of these edges are too large to be disjoint. If $n = 2m$, let $\mathcal{F}$ consist of all edges of $[n]$ having at least $m+1$ vertices and all subsets of $[n]$ having m vertices and containing the vertex n.

    Note that, in both cases, we pick exactly one vertex from each pair created in the proof of Theorem 6.29.

14. No, it cannot. Let us assume that it can, and let X be a minimum-sized edge in $\mathcal{F}$. Let us assume without loss of generality that $X = [k]$. Then, for each $i \in [k]$, there has to be an edge Y_i in $\mathcal{F}$ so that Y_i does not contain i (otherwise the intersection of all edges in $\mathcal{F}$ contains i). However, this implies that $X \cap Y_1 \cap Y_2 \cdots Y_k = \emptyset$, contradicting our hypothesis.

    Note that this result has a geometric counterpart, called *Helly's theorem.* That theorem says that, if there are $n \geq k+1$ convex sets given in $k+1$-dimensional Euclidean space, and they have the property that every $k+1$ of them intersect, then all of them intersect.

15. Color the integer n red if it is divisible by exactly one of 7 and 17, and color it blue otherwise. We show that this coloring satisfies the requirements.

    Let us assume that our coloring contains a monochromatic arithmetic progression A of length 18. Assume first that our arithmetic progression A has a difference d that is divisible by 17. Then, for A to be monochromatic, either all or no terms of A would have to be divisible by 7. The first is impossible, as that would require d to be at least $7 \cdot 17 = 119$, forcing $a_{18} = 2142$. For the same reason, d

cannot be divisible by 7, which means that exactly one of the first seven terms of A will be divisible by 7. This rules out the possibility that A has no terms divisible by 7, so d must not be divisible by 17.

Therefore, exactly one of the first 17 terms of A will be divisible by 17. If d is divisible by 7, then we are done, since either all or no terms of A are divisible by 7, so there will always be two terms that differ modulo 17, but not modulo 7. If d is not divisible by 7, then there are terms in our sequence not divisible by either 7 or 17, and there are some divisible by only one of them, contradicting our assumption that A is monochromatic.

16. Assume without loss of generality that vertex A has at least three red edges adjacent to it. Let the other endpoints of these edges be B, C, and D. If there is a red edge between any two of these three vertices, then the endpoints of that edge and A form a red triangle. Otherwise BCD is a blue triangle.

17. We prove this by induction on $k + l$. If $k + l = 2$, that is, if $k = l = 1$, then $R(k, l) = 2$, and the statement is true. Assume that the statement is true for $k+l-1$, that is, that $R(k, l-1)$ and $R(k-1, l)$ exist, and prove the statement for $k + l$.

    It suffices to prove that there is *one* integer with the required property. Indeed, in a nonempty set of positive integers, there is always a smallest one.

    We claim that $n = R(k, l-1) + R(k-1, l) - 1$ is such an integer. Indeed, let v be any vertex of K_n, and color each edge of K_n red or blue. Then v has either at least $R(k, l-1)$ blue edges or at least $R(k-1, l)$ red edges adjacent to it. In the first case, let the endpoints of the blue edges adjacent to v form the complete graph X. Then X has $R(k, l-1)$ vertices, so it either contains a red K_k, and we are done, or it contains a blue K_{k-1}, and then we are done again, adding v to this copy of K_{k-1}. The second case can be handled in an analogous way.

18. Let us assume without loss of generality that vertex A has at least six red edges adjacent to it. Look at the other endpoints of these vertices. If there is any red edge between them, we are done. If not, they form a copy of K_6 in which each edge is blue or green. Now apply the result of the previous exercise.

19. If N is prefix-free, then there can be no moment of time during the decoding of a message when we may stop and get a codeword A, or may continue and get a codeword B, since that would mean that A is a prefix of B. Similarly, if N is not prefix-free, and F is a prefix of G, and we try to decode a message starting with the digits of F,

then we do not immediately know whether we should stop and get F, or continue and get G. Therefore, N is not instantaneous.

20. The idea of pattern avoidance for matrices, and the earliest results, come from [35].

    (a) By the pigeonhole principle, if there are $n+1$ entries equal to 1, then two of them must be in the same line. Therefore, $f(n, B) = n$.

    (b) We prove a more general statement. Define $f(n, m, B)$ to be the maximum number of 1s that an $n \times m$ matrix with 0 and 1 entries can contain if it avoids B. We claim that, if B is the 2×2 identity matrix, then $f(n, m, B) = n + m - 1$. This many entries equal to 1 are possible, as can be seen, for instance, by filling up the first row and column with 1s. To see that there cannot be more 1s, use induction on $n+m$. For $n+m=2$, the statement is true. Let us assume that the statement is true for $n+m-1$, and let A be an $n \times m$ matrix avoiding B. Assume A has $n+m$ entries equal to 1. If A has a row or column that contains at most one 1, then we can omit that line, and be done by the induction hypothesis. Otherwise, each row and each column contains at least two 1s. Since the total number of 1s is $n+m$, this means that each row and each column contains *exactly* two 1s. That is, the number of all 1s is $2n = n+m = 2m$, forcing $n = m$.

        Then A is the sum of two distinct permutation matrices. (The reader is invited to prove this directly, possibly using bipartite graphs. A more general statement is proved in Lemma 10.6.) These matrices cannot both correspond to the decreasing permutation, so one must correspond to a permutation that contains a non-inversion. That matrix will then contain B.

21. (a) Proof by induction on n. As A_n avoids B, the only way A_{n+1} could contain B would be if there was a copy of B in A_{n+1} that starts in the upper left quadrant and ends in another quadrant. It is easy to see that this is impossible.

    (b) Let $g(n)$ be the number of 1s in A_n. Then $g(0) = 1$, and $g(n+1) = 2g(n)+n$ for $n \geq 0$. Solving this recurrence, we get $2g(n) = 2^n + n2^{n-2}$. Since A_n is of size $2^n \times 2^n$, setting $m = 2^n$, we see that $f(m, B) = \Omega(m \log m)$.

22. This result, and the results of the next two exercises, were proved in a paper by Adam Marcus and Gábor Tardos [56], and were used to prove a 25-year-old conjecture.

    We claim that the answer is $f(\frac{n}{k^2}, B)$. Call a block of A a *zero* block if all its entries are zeros. Form the matrix A' by replacing all zero blocks of A with a 0 and all nonzero blocks of A with a 1, to get the matrix A'. If A' contains B, then so does A, and the result follows. Note that we do need the fact that B is a permutation matrix, and as such, has only one 1 in each row and column.

23. Let us assume the contrary. There are $\binom{k^2}{k}$ possible k-tuples of rows in which a block can have nonzero elements, so, by the pigeonhole principle, this would mean that there is a k-tuple of rows that contains nonzero elements in k different blocks. However, that implies that A has a $k \times k$ submatrix M *of nonzero blocks.* Therefore, A contains *all* $k \times k$ matrices, including B. Indeed, M essentially simulates a $k \times k$ matrix in which each entry is 1, and the latter certainly contains all $k \times k$ matrices.

    Similarly, in any column of blocks, there can be at most $(k-1)k^2$ wide blocks.

24. First, if n is not divisible by k^2, then we just fill the last rows and columns of A with 1s so that the submatrix A_s of A that has not been filled by 1s has size $k^2 \cdot \lfloor n/k^2 \rfloor \times k^2 \cdot \lfloor n/k^2 \rfloor$. This increases the number of 1s by less than $2k^2 n = O(n)$. We can therefore assume from now on that n is divisible by k^2, that is, $A_s = A$. Indeed, the above argument shows that, if the statement were false, the excess of 1s would have to come from A_s.

    The previous two exercises show that, if A avoids B, and B is a permutation matrix of size $k \times k$, then there can be at most $f(n/k^2)$ nonzero blocks, and among those nonzero blocks, at most $(k-1)\binom{k^2}{k}$ in each row can be tall. Similarly, in each column of blocks, at most $(k-1)\binom{k^2}{k}$ blocks can be wide. If a block is not tall or wide, then the number of 1s in that block is at most $(k-1)^2$, since the block can contain 1s in at most $k-1$ rows and at most $k-1$ columns.

    We can now summarize the number of 1s in our matrix A.

    - There are at most $2\frac{n}{k^2}(k-1)\binom{k^2}{k}$ blocks that are tall or wide, and each of them contains at most k^4 entries equal to 1, for a total of fewer than $2nk^3\binom{k^2}{k}$ entries equal to 1, and
    - there are at most $f(n/k^2, B)$ nonzero blocks that are neither tall nor wide, and each of them contains at most $(k-1)^2$ entries 1, for a total of at most $(k-1)^2 f(n/k^2, B)$ entries 1, and
    - the rest must be zero blocks with no 1s.

    This leads to the recursive formula

    $$f(n, B) \leq 2nk^3 \binom{k^2}{k} + (k-1)^2 f(n/k^2, B),$$

    from which it is routine to prove our statement by induction.

25. This result was first proved by Martin Klazar [49], and then his proof was somewhat simplified by Vincent Vatter and Doron Zeilberger (not formally published). This is the proof we present.

    Let q be a pattern of length k, and let p be an n-permutation that avoids q. Then the permutation matrix of p avoids the permutation matrix B_q of q. Now let $M_n(q)$ be the number of $n \times n$ matrices with 0 and 1 entries that avoid B_q. Then $S_n(q) \leq M_n(q)$. We will now find an upper bound for $M_n(q)$.

    By the result of the previous exercise, we know that there exists a constant b so that $f(n, B_q) \leq bn$. Let A be an $n \times n$ matrix that avoids B_q, and cut A up into 2×2 blocks. (There will be smaller blocks at the end if n is odd.) If a block contains no 1s, then replace it with a 0; if it contains at least one 1, replace it with a 1. This results in a matrix of size $\lceil n/2 \rceil \times \lceil n/2 \rceil$ that is also B_q-avoiding and therefore has at most $f(\lceil n/2 \rceil, B_q) \leq b\lceil n/2 \rceil$ entries equal to 1. Furthermore, each of these ones comes from a nonzero 2×2 block of A. There are $2^4 - 1 = 15$ possibilities for each such block. Therefore, the number $M_n(q)$ of all possible matrices A we could have started with satisfies

$$M_n(q) \leq 15^{b\lceil n/2 \rceil} \cdot M_{b\lceil n/2 \rceil}(q).$$

    Repeat this argument until the right-hand side becomes $M_1(q) = 2$. Then we get $M_n(q) \leq 15^{2bn}$, and therefore $S_n(q) \leq 15^{2bn}$.

26. Count the pairs (p, q'), where p is an n-permutation containing q, and q' is a copy of q in p. There are $\binom{n}{k}$ choices for the position of q', then there are $\binom{n}{k}$ choices for the entries in q', and finally there are $(n-k)!$ choices for the rest of p. So the number of such pairs is $(n-k)! \cdot \binom{n}{k}^2$. The pigeonhole principle shows that at least $1/n!$ of them, that is, at least

$$\frac{(n-k)! \cdot \left(\frac{n!}{k!(n-k)!}\right)^2}{n!} = \frac{\binom{n}{k}}{k!}$$

    pairs, must belong to one permutation. Furthermore, there exist q-avoiding permutations, and no (p, q) pair belongs to them. Therefore, there exists a p with strictly *more* than $\frac{\binom{n}{k}}{k!}$ occurrences of q.

27. (a) Represent the jobs and the candidates by vertices, and connect candidates to jobs for which they are qualified by an edge. The result is a bipartite graph. If aX is an edge of this graph G, then write the number t_{aX} on this edge, where t_{aX} is the lowest salary for which candidate a will accept job X.

    Recall that *perfect matchings* of a graph were defined in Supplementary Exercise 34 of Chapter 5. The task of the manager is then to find the perfect matching of G in which the sum of the numbers written on the edge (the *cost* of the matching) is minimal.

(b) No, this greedy strategy will not always produce the best results. The graph shown in Figure 6.17 is a counterexample. The greedy algorithm provides a matching costing 13, while a matching costing 12 exists.

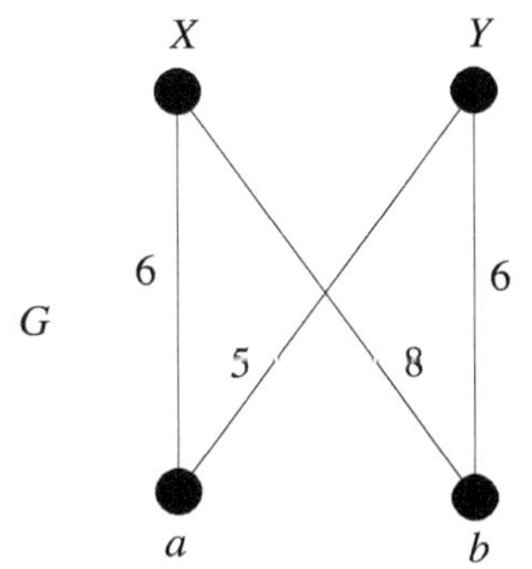

Figure 6.17
The greedy algorithm will not find the minimum-cost perfect matching of G.

28. Both A and B are forests. Because A has more edges, it has fewer components. So A must have an edge between vertices that are in different components of B, and that edge can be added to B without creating a cycle there.

29. Yes, in this case the greedy algorithm will work. Look at all pairs of towns in the county, take a complete graph whose vertices are the towns, and write the cost of construction of each road to the edge corresponding to that road. Using the terminology of the Notes section of Chapter 5, we have to find a *minimum-cost spanning tree* in this graph, that is, a tree on n vertices so that the sum of the numbers written on the edges of the tree is minimal.

    Let G_j be the graph the greedy algorithm builds in j steps. The definitions then imply that G_j is a forest and G_{n-1} is a tree. We claim that no other spanning tree of K_n will have a smaller cost than G. Indeed, assume that $c(T) < c(G)$, where $c(T)$ denotes the cost of T. Let $g_1, g_2, \cdots, g_{n-1}$ and $t_1, t_2, \cdots, t_{n-1}$ be the edges of G and T in nondecreasing order of their costs. As $c(T) < c(G)$, there has to be a smallest index i so that $c(g_i) > c(t_i)$.

    This is a contradiction, however. Indeed, this would imply that $c(g_i) > c(t_j)$ for all $j \leq i$. The previous exercise shows that at least one of the edges t_j, with $j \leq i$, can be added to G_{i-1} without creating a cycle. Therefore, g_i cannot be the edge that the greedy algorithm adds to G_{i-1}, since it is not cost minimal among the eligible edges.

Note that the fact that the underlying graph was a complete graph was not significant. The same greedy algorithm can be used to find a minimum-cost spanning tree in *any* graph.

30. Yes. The key fact is that bases, just as forests, satisfy a "replacement" lemma like the one described in Exercise 28. Structures in which an analogous property holds are called *matroids.*

    Indeed, let X and Y be two sets of linearly independent vectors in V, with $|X| > |Y|$. It is then easy to see that there is a vector in X that can be added to Y without destroying the independence of Y. (Otherwise, X would be contained in the subspace spanned by Y, which is a contradiction of the fact that the dimension of that subspace is $|Y|$.)

    The proof of the exercise is now very similar to that of Exercise 29.

31. Let us assume that neither statement holds. As (b) does not hold, fewer than 100 rows contain ten or more entries equal to 1. As (a) does not hold, none of these fewer than 100 rows can contain 100 or more entries equal to 1. So the total number of 1s in A is not more than

$$99 \cdot 99 + 901 \cdot 9 = 17910,$$

    which contradicts our assumption that A has at least 20000 entries equal to 1.

    This is a special instance of a more general result by Michelangelo Grigni and Michael Sipser [41].

6.8 Supplementary exercises

1. For what graphs H is it true that, if G contains H as a subgraph, then G contains H as an induced subgraph?

2. Prove that every graph G on n vertices contains a subgraph H so that $d(H) \geq e(G)/n$. See Exercise 2 for the relevant notations.

3. In part (a) of Exercise 2, we proved a lower bound on the number of edges that the largest bipartite subgraph H of a graph G can have. Generalize this statement by showing that any simple graph G contains a k-partite subgraph H that has more than $e(G) \cdot \frac{(k-1)}{k}$ edges.

4. Prove that, if the graph G has $2n$ vertices and more than n^2 edges, then G contains at least n triangles.

5. At most how many edges can a simple graph G on n vertices have if G is not connected? Try to give two different proofs.
6. Prove Theorem 6.14 (Jensen's inequality).
7. Let us call a graph G *critically k-chromatic* if $\chi(G) = k$, but $\chi(G') < k$ for all proper subgraphs G' of G. Describe all critically 3-chromatic graphs.
8. Let us define the *strong direct product* $G \cdot H$ of the graphs G and H as follows: The vertex set of $G \cdot H$ is the set of ordered pairs (g, h), where g is a vertex of G and h is a vertex of H. There is an edge between (g, h) and (g', h') when one of the following holds:

   (a) there is an edge between g and g' in G and there is an edge between h and h' in H, or

   (b) $g = g'$ and there is an edge between h and h' in H, or

   (c) $h = h'$ and there is an edge between g and g' in G.

   Prove that $\chi(G \cdot H) \leq \chi(G) \cdot \chi(H)$.
9. Let G be any graph of five vertices and seven edges. Prove that $f_G(n) \geq 7\lfloor n/5 \rfloor$.
10. Generalize the statement of Exercise 6 by finding an upper bound for $\chi(G)$ if the ordered degree sequence of G is $d_1 \geq d_2 \geq \cdots \geq d_n$.
11. Let $\alpha(G)$ be the number of vertices in the largest *empty* (that is, containing no edges) induced subgraph of G. Prove that

$$\frac{|G|}{\alpha(G)} \leq \chi(G).$$

   The number $\alpha(G)$ is called the *independence number* of G.
12. A *tournament* is a directed complete graph. Prove that, in a tournament on n vertices, there are at most $\binom{n+1}{3}/4$ directed 3-cycles.
13. Prove that the upper bound of Theorem 6.35 cannot be improved.
14. Generalize the result of Theorem 6.35 by dropping the condition that $\mathcal{F}$ is k-uniform and showing then that

$$\sum_{i=1}^{m} \frac{1}{\binom{|X_i|+|Y_i|}{|X_i|}} \leq 1.$$

15. Let n be an odd positive integer. Let $\mathcal{F}$ be a hypergraph on $[n]$ that does not contain three edges $A \subset B \subset C$. Find the best possible upper bound for $|\mathcal{F}|$.
16. Let $A_1, A_2, \cdots, A_k$ be subsets of $[n]$. Prove that

$$|A_1 \cap A_2 \cap \cdots \cap A_k| \geq |A_1| + |A_2| + \cdots + |A_k| - (k-1)n.$$

17. The set of workers in a factory has altogether 500 skills, and for each skill there are 30 workers who have that skill. Show that it is possible to schedule the workers into three shifts so that each skill is represented in each shift.

18. Generalize the results of Exercises 16 and 18 to n colors.

19. The Ramsey number $R(k, \ell)$ was defined in Exercise 17. Find the Ramsey numbers $R(3,3)$ and $R(3,4)$.

20. Prove that $R(k, \ell) \leq \binom{k+\ell-2}{k-1}$.

21. Let $\mathcal{F}$ be a hypergraph in which there are no two edges whose intersection consists of exactly one vertex. Is it true that $\mathcal{F}$ has property B?

22. Find an example for a code over a finite alphabet which is not prefix-free, but satisfies (6.12).

23. In a graph, a *Hamiltonian path* is a path that contains each vertex. The same definition applies to *directed* graphs, except that then the path has to be a directed path. Prove that it is possible to direct each edge of K_n so that the obtained tournament has at least $n!/2^{n-1}$ Hamiltonian paths.

24. A Hamiltonian *cycle* in a (directed) graph on n vertices is a (directed) cycle of length n. Formulate and prove the result of the previous exercise for Hamiltonian cycles instead of Hamiltonian paths.

25. Generalize the result of the previous supplementary exercise to complete k-partite graphs.

26. Pattern-avoiding permutations were defined in Exercise 32 of Chapter 4. Let $I_k = 123\cdots k$. Prove that $S_n(I_k) \leq (k-1)^{2n}$. Recall that $S_n(I_k)$ denotes the number of n-permutations that avoid I_k, that is, that do not contain an increasing subsequence of length k.

27. Prove that a permutation of length 25 has to contain at least one of the patterns 1234567 and 54321.

28. Explain how the concept of avoidance for 0–1 matrices is related to the concept of graphs avoiding certain subgraphs.

29. Let $B = \begin{pmatrix} 1 & 1 \\ 1 & 1 \end{pmatrix}$. Find a function $g(n)$ so that $f(n, B) = \Theta(g(n))$.

30. Let $B = \begin{pmatrix} 1 & 1 \\ 1 & 0 \end{pmatrix}$. Prove that $f(n, B) = O(n)$.

31. + Let $\epsilon > 0$ and let $k > (2+\epsilon)\ln n$. Prove that there exists an $n \times n$ matrix whose entries are equal to 0 or 1 that does not contain a $k \times k$ submatrix whose entries are all equal. (By a $k \times k$ submatrix, we mean the k^2 entries in the intersection of k given rows and k given columns. So these entries may be quite far away from each other.)

32. Let us call permutations $p = p_1 p_2 \cdots p_n$ and $q = q_1 q_2 \cdots q_n$ *colliding* if there exists an index i so that $|p_i - q_i| = 1$. Let $f(n)$ denote the size of the largest set S of permutations of length n so that every pair of elements of S is colliding. Prove that $f(n) \leq \binom{n}{\lceil n/2 \rceil}$.

33. Keep the definitions of the previous exercise. Prove that $f(n) \geq f(n-1) + f(n-2)$ if $n \geq 4$.

34. Permutation patterns were defined in Exercise 32 of Chapter 4. A *k-superpattern* is a permutation that contains all permutation patterns of length k. For instance, $p = 25314$ is a 3-superpattern. Let $g(k)$ be the length of the shortest k-superpattern. Prove that, if k is large enough, then $g(n) \geq k^2/e^2$.

35. Prove that the upper bound of Quick Check Exercise 2 of Section 6.3 can be attained.

Part III

An Advanced Method

7

Analytic combinatorics

It often occurs in enumerative combinatorics that obtaining an *exact formula* that answers a question is difficult, or time-consuming, or even impossible, while it is much easier to obtain a formula that approximates the exact solution up to a specified level of precision. In this chapter, we consider examples of this phenomenon, and methods to apply in these situations. Unless otherwise noted, all power series in this chapter are assumed to have *complex* coefficients.

7.1 Exponential growth rates

As we will see, it is often reasonably easy to prove that a sequence grows *roughly* at the speed of the powers of a fixed positive real number. We start our excursion to analytic combinatorics at such sequences.

7.1.1 Rational functions

Rational functions, which are the ratio of two polynomials, are a good starting point to learn our methods.

7.1.1.1 Motivation

What is larger, the number of compositions of 1000 into parts that are equal to 1 or 4, or the number of compositions of 1000 that are equal to 2 or 3?

If a_n denotes the number of the former, and b_n denotes the number of the latter, then it is easy to prove by the methods of Chapter 3 that

$$A(x) = \sum_{n\geq 0} a_n x^n = \frac{1}{1-x-x^4}, \tag{7.1}$$

and

$$B(x) = \sum_{n\geq 0} b_n x^n = \frac{1}{1-x^2-x^3}. \tag{7.2}$$

In theory, nothing prevents us from finding the partial fraction decompositions of $A(x)$ and $B(x)$, deducing exact formulas for a_n and b_n, and then

DOI: 10.1201/9781003304272-7

comparing the values of a_{1000} and b_{1000}. However, in practice, that procedure would be exceedingly cumbersome. The partial fraction decompositions of $A(x)$ and $B(x)$ would contain fractions with complex numbers, and the exact formulas obtained for a_n and b_n would not be particularly pleasant. Comparing the values for $n = 1000$ could pose another problem.

It is therefore a natural question to ask if we can proceed in a simpler way. Is there a way to decide which of the two sequences will grow faster simply by perusing their generating functions, and not computing the actual numbers a_n and b_n?

Fortunately, the answer to that question is in the affirmative. In this section we develop the tools to see why and how.

7.1.1.2 Theoretical background

The following definition will be our crucial tool in describing how fast a given sequence grows.

Definition 7.1 *Let $s_0, s_1, s_2, \ldots$ be an infinite sequence of complex numbers. Let α be a real number. We say that the sequence has* exponential growth rate *α if*

> *the sequence of the numbers s_n does not have an infinite subsequence $r_1, r_2, \cdots$ for which*
>
> $$\lim_{n \geq \infty} \sqrt[n]{|r_n|} > \alpha,$$
>
> *and*
>
> *the sequence of the numbers s_n does have an infinite subsequence $r_1, r_2, \cdots$ for which*
>
> $$\lim_{n \geq \infty} \sqrt[n]{|r_n|} = \alpha.$$

In other words, α is the *largest number* so that there is an infinite subsequence of the sequence $s_1, s_2, \cdots$, (or, in what follows, the sequence s_n) for which the limit of the absolute values of the nth roots of the elements is α. Yet in other words, $\alpha = \limsup_{n \geq 1} \sqrt[n]{|s_n|}$.

Example 7.2 *Let $s_n = n \cdot 2^n$. Then the sequence of the numbers s_n has exponential growth rate 2.*

Solution: This follows from the fact that $\sqrt[n]{s_n} = 2\sqrt[n]{n} \rightarrow 2$. ◊

Example 7.3 *Let $s_n = 2^n/n^{100}$. Then the sequence of the numbers s_n has exponential growth rate 2.*

Solution: This follows from the fact that $\sqrt[n]{s_n} = 2/\sqrt[n]{n^{100}} \rightarrow 2$. ◊

The two examples above show that two sequences can differ by a large polynomial factor and still have the same exponential growth rate. This is because polynomial factors are much smaller than exponential ones. Indeed, as $\lim_{n\to\infty} \sqrt[n]{|p(n)|} = 1$ for any nonzero polynomial $p(n)$, multiplying or dividing a sequence by any nonzero polynomial will not change its growth rate. Note that it follows that, if $f_n = p(n)$ for some nonzero polynomial p, then the exponential order of the sequence f_n is 1.

Example 7.4 *Let $s_n = 3^n + (-3)^n$. Then the sequence of the numbers s_n has exponential growth rate 3.*

Note that, in the last example, every other term is 0, but that does not affect the exponential growth rate of the sequence. It is the infinite subsequence with the highest growth rate that provides the growth rate of the whole sequence. In this case, that is the subsequence $r_n = s_{2n}$.

It goes without saying that not all sequences have a finite growth rate. Recall that the notation $f(x) \sim g(x)$ means that $\lim_{x\to\infty} f(x)/g(x) = 1$. As sequences are functions whose domain is the set of natural numbers, this definition is meaningful for sequences as well.

Example 7.5 *Let $a_n = n!$. Then the sequence a_n has an infinite growth rate.*

Solution: This can be proved by Stirling's formula, which states that $n! \sim (n/e)^n\sqrt{2\pi n}$, so $\sqrt{a_n} \sim n/e$, which clearly diverges to infinity. We first mentioned this formula in (1.13). Alternatively, one can use the fact that $a_{n+1}/a_n = n + 1$, which also diverges to infinity. $\diamondsuit$

This is not to say that if $a_n \geq n!$ holds for the elements of a sequence, then we cannot ask interesting questions about the sequence that involve exponential growth rate. For instance, we can ask *how much faster* the sequence grows than the sequence of factorials, by asking for the exponential growth rate of the sequence $a_n/n!$. This will be a particularly appropriate question in cases when the exponential generating function of the sequence a_n can be computed.

The reader is asked to prove the following propositions in Supplementary Exercise 2.

Proposition 7.6 *Let the sequence f_n have exponential growth rate a, and let the sequence g_n have exponential growth rate b, and let us assume that both sequences consist of complex numbers. Let us assume that $a > b$. Then the sequence $f_n + g_n$ has exponential growth rate a.*

Note that Proposition 7.6 is not true without the condition that $a > b$. A counterexample is $f_n = (-2)^n$ and $g_n = (-2)^{n+1}$. Nevertheless, the following weaker statement always holds.

Proposition 7.7 *Let the sequence f_n have exponential growth rate a, and let the sequence g_n have exponential growth rate b, and let us assume that both sequences consist of complex numbers. Let us assume that $a \geq b$. Then the exponential growth rate of $f_n + g_n$ is at most a.*

Proof: Let us assume the contrary, that is, that the sequence $f_n + g_n$ has an infinite subsequence s_i (where $i \in I$ for some infinite set I of positive integers) that satisfies the inequality $\lim_{i \to \infty} \sqrt[n]{s_i} = a' > a$.

Consider the infinite subsequence $s_i = f_i + g_i$. Let $I_1 \subseteq I$ be the set of indices for which $|f_i| \geq |g_i|$, and let $I_2 \subseteq$ be the set of indices for which $|f_i| < |g_i|$. As I is infinite, so is at least one of I_1 and I_2. Let us assume without loss of generality that I_1 is infinite.

Now notice that, if $i \in I_1$, then

$$|s_i| = |f_i + g_i| \leq |f_i| + |g_i| \leq 2|f_i|,$$

which implies that the sequence f_i (where $i \in I_1$) has exponential order at least a', contradicting our hypothesis. $\diamondsuit$

Now we start putting the growth rate of a sequence in the context of generating functions. Note that, unless otherwise stated, all power series in this section are taken over the field of *complex numbers*, that is, they have complex coefficients. We are only assuming that the reader is familiar with both ways of writing a complex number z, namely, the Cartesian way, which sets $z = x + iy$, where x and y are real numbers, and the polar coordinate way, which sets $z = re^{i\phi}$, where r is a nonnegative real number and ϕ, called the *argument* of z, is an angle measured in radians. The reader is probably familiar with basic complex arithmetics, that is, the reader knows how to add, subtract, multiply, and divide complex numbers, how to take their powers, and how to compute their absolute values. We also assume that the reader is familiar with Euler's formula, stating that

$$e^{i\phi} = \cos\phi + i \sin\phi \tag{7.3}$$

for all real numbers ϕ.

Recall that a rational function is the ratio of two polynomials.

Definition 7.8 *Let $F(x) = P(x)/Q(x)$ be a rational function so that $P(x)$ and $Q(x)$ do not have any roots in common. Then the complex number x_0 is* a singular point *of $F(x)$ if $Q(x_0) = 0$.*

Example 7.9 *The rational function*

$$F(x) = \frac{x^2 + 3x + 2}{(x-1)^2(x^2+1)}$$

has singular points $x_0 = 1$, $x_0 = i$, and $x_0 = -i$.

Singular points of a rational function written in simplest terms are just the roots of the denominator. Still, the name singular point is worth introducing since it will have a meaning for functions that are not rational functions as well.

If r_1 is a root of the polynomial $Q(x)$, then we say that r_1 is a root of *smallest modulus* of $Q(x)$ if there is no root r_2 of $Q(x)$ that satisfies $|r_1| > |r_2|$.

The following theorem describes an easy way to find the exponential growth rate of the sequence of coefficients of a rational function.

Theorem 7.10 *Let*

$$S(x) = \sum_{n\geq 0} s_n x^n = \frac{P(x)}{Q(x)}$$

be a rational function with $Q(0) \neq 0$, let us assume that $P(x)$ and $Q(x)$ do not have any roots in common, and let us also assume that $Q(x)$ has a unique *root r_1 of smallest modulus.*

Then the exponential growth rate of the sequence of the coefficients s_n is equal to $|z_1|$, where $z_1 = 1/r_1$.

Note that the theorem holds even without the assumption that $Q(x)$ has a unique root of smallest modulus, but it would be a bit cumbersome to prove that with our current tools. We will prove that version of the theorem in the next section.

Proof: We can assume without loss of generality that $Q(x)$ has constant term 1. Indeed, dividing a power series by a constant does not change the exponential growth rate of its coefficients.

Now let $Q(x) = (1-z_1x)^{d_1}(1-z_2x)^{d_2}\cdots(1-z_kx)^{d_k}$. In other words, let d_i be the multiplicity of r_i as a root of $Q(x)$. Note that this equation is equivalent to $Q(x) = C(r_1-x)^{d_1}(r_2-x)^{d_2}\cdots(r_k-x)^{d_k}$.

We can assume that the degree of $Q(x)$ is larger than that of $P(x)$. The reader is asked to prove this simple fact in Exercise 1.

Then, just as the reader has seen in calculus, $S(x) = P(x)/Q(x)$ has a partial fraction decomposition of the form

$$S(x) = \frac{C_{1,1}}{1-z_1x} + \frac{C_{1,2}}{(1-z_1x)^2} + \ldots + \frac{C_{1,d_1}}{(1-z_1x)^{d_1}} + \qquad (7.4)$$

$$+ \frac{C_{2,1}}{1-z_2x} + \frac{C_{2,2}}{(1-z_2x)^2} + \ldots + \frac{C_{2,d_2}}{(1-z_2x)^{d_2}} + \ldots \qquad (7.5)$$

$$+ \frac{C_{k,1}}{1-z_kx} + \frac{C_{k,2}}{(1-z_kx)^2} + \ldots + \frac{C_{k,d_k}}{(1-z_kx)^{d_k}}, \qquad (7.6)$$

where the $C_{i,j}$ are constants.

Indeed, multiplying both sides by $Q(x)$, we get an equality involving two polynomials, one of which is obviously the polynomial $P(x)$. Equating coefficients of x^n for all n, we can determine the values of each of the $C_{i,j}$.

So $S(x)$ is a sum of a few power series, each of which is of the form $C_{i,j}/(1-z_ix)^j$. Which of these summands has a coefficient sequence with the highest exponential growth rate? We know from the general version of the binomial theorem (Theorem 3.2) that

$$\frac{1}{(1-z_ix)^j} = (1-z_ix)^{-j} = \sum_{n\geq 0} z_i^n \binom{j+n-1}{j-1} x^n.$$

As $\binom{j+n-1}{j-1}$ is just a polynomial of degree $j-1$ in n, it follows that the exponential growth rate of the power series of this type is $|z_i|$, for *any* j.

Therefore, the sequence s_n (the sequence of coefficients of $S(x)$) is the sum of various sequences whose exponential growth rates $|z_i|$ we know, namely,

$$s_n = \sum_{i=1}^{k}\sum_{j=1}^{d_j} C_{i,j} z_i^n \binom{j+n-1}{j-1}.$$

Recall that r_1 is the only root of $Q(x)$ that is of smallest modulus, so $|z_1| > |z_i|$ for any $i \neq 1$. Therefore, the sum g_n of all these sequences that do not involve z_1 (those for which $i \neq 1$) has an exponential growth rate less than $|z_1|$ by Proposition 7.7. On the other hand, the sum f_n of sequences that do involve z_1 has exponential order $|z_1|$ since it simply equals z_1^n times a nonzero polynomial. Then our claim follows by Proposition 7.6. ◊

It is high time that we returned to the combinatorial example that we mentioned at the beginning of Section 7.1.1.1. That question involved deciding whether the coefficients of $A(x) = 1/(1-x-x^4)$ or the coefficients of $B(x) = 1/(1-x^2-x^3)$ grow faster. Theorem 7.10 tells us that, to find these growth rates, we have to find the roots of smallest modulus of the two denominators. Using our favorite software package, we find that, for $A(x)$, that root is about $r_1 = 0.7245$, whereas for $B(x)$, it is about $r_1 = 0.7549$. Therefore, the exponential growth rate of the sequence a_n is equal to $z_1 = 1/0.7245 = 1.3803$, whereas the exponential growth rate of the sequence b_n is $z_1 = 1/0.7549 = 1.3247$, so the sequence of the numbers a_n grows faster.

Let us consider further examples of finding the exponential growth rate of a combinatorially defined sequence.

Example 7.11 *Let h_n be the number of words of length n over the alphabet $\{A, B, C\}$ that do not contain the subword ABC in consecutive positions. Find the exponential growth rate of the sequence of the numbers h_n.*

Solution: If we take a word w counted by h_n and add one letter to its end, we get a word counted by h_{n+1}, except when w ends in AB and the letter we add to the end of w is C. This leads to the recurrence relation

$$h_{n+1} = 3h_n - h_{n-2}$$

for $n \geq 2$, while $h_0 = 1$, $h_1 = 3$, and $h_2 = 9$. Setting $H(x) = \sum_{n \geq 0} h_n x^n$, the displayed recurrence relation leads to the functional equation

$$H(x) - 9x^2 - 3x - 1 = 3x(H(x) - 3x - 1) - x^3 H(x),$$

$$H(x) = \frac{1}{1 - 3x + x^3}.$$

As any mathematical software package tells us, the root of smallest modulus of the denominator of $H(x)$ is at $x_1 = 0.3473$, so the exponential growth rate of the sequence of the numbers h_n is $1/0.3473 = 2.8794$. ◊

So the exponential growth rate of the sequence h_n is a little bit below 3, which confirms our intuition, since the numbers of *all* words of length n over a 3-element alphabet has exponential growth rate 3.

This is a good time to mention that, in the rest of this chapter, we will often need to find the root of a polynomial $p(x)$ that is closest to 0, for polynomials of degree larger than 2. In such situations, we will simply get that root from a software package.

Example 7.12 *We flip a fair coin n times. Let a_n be the number of possible outcome sequences that do not contain four consecutive heads. Find the exponential order of the sequence of the numbers a_n.*

Solution: We use Theorem 3.25. Let us first consider the acceptable outcome sequences that contain at least one tail. Each such sequence can be broken up into three segments, namely the segment before the last tail, the last tail, and the part after the last tail, which must be all heads, and must be of length at most three.

If an acceptable outcome sequence does not contain a tail, then it is of length at most three, and there is only one such sequence for each such length.

By Theorem 3.25, this leads to the functional equation

$$A(x) = A(x) \cdot x \cdot (1 + x + x^2 + x^3) + 1 + x + x^2 + x^3,$$

$$A(x) = \frac{x^3 + x^2 + x + 1}{1 - x - x^2 - x^3 - x^4}.$$

The root of smallest modulus of the denominator is 0.5188, so the exponential growth rate of the sequence of coefficients of a_n is $1/0.5188 \approx 1.9276$. ◊

We point out that $\lim_{n \to \infty} 1.9276^n / 2^n = 0$, so the probability that a sequence of length n of fair coin flips will contain four consecutive heads converges to 1 as n goes to infinity. The exponential growth rate of the number of sequences without four consecutive heads is less than 2 (the exponential growth rate for unrestricted coin flip sequences), but not by much. The reader is invited to explain what happens if we count sequences that do not contain k

consecutive heads instead of four consecutive heads. How will the exponential growth rate of the sequence change? You are asked to answer this question in Exercise 22.

7.1.2 Singularity analysis

It turns out that the concept of singular points can be extended to a class of functions that is much larger than rational functions, and it is a very useful notion in that large class as well. This is the subject of this section.

7.1.2.1 Motivation

Example 7.13 *A group of n tourists arrives at a restaurant. They sit down around an unspecified number of circular tables. Then the manager counts the tables at which at least one tourist is sitting, finds that there are k of them, and numbers them from 1 through k. Let h_n be the number of different ways all this can happen if two arrangements are considered identical if each person has the same left neighbor in both of them, and each person sits at the same table in both of them. Find the exponential growth rate of the sequence $h_n/n!$.*

The alert reader may notice that the problem only differs from Example 3.31 in that now the tables are numbered. It turns out that this is an important difference. (Recall that without numbering the tables, each arrangement is equivalent to a permutation. If we do number the tables, the number of possibilities clearly increases; the question is by how much, in the exponential sense.)

Let h_n be the number of arrangements we are trying to count. We will use the compositional formula of exponential generating functions (Theorem 3.36). We first partition the set of tourists into an unspecified number of non-empty blocks, then we put a cycle on each block, and then we linearly order the set of blocks. Therefore, $a_k = (k-1)!$ if $k \geq 1$, and $b_k = k!$, so we have

$$A(x) = \sum_{k\geq 1}(k-1)!\frac{x^k}{k!} = \sum_{k\geq 1}\frac{x^k}{k} = \ln\left(\frac{1}{1-x}\right),$$

and

$$B(x) = \sum_{k\geq 0} k!\frac{x^k}{k!} = \sum_{k\geq 0} x^k = \frac{1}{1-x}.$$

Therefore, by the compositional formula, we have

$$H(x) = B(A(x)) = \frac{1}{1-\ln\left(\frac{1}{1-x}\right)}. \tag{7.7}$$

So we have obtained an explicit formula for the generating function $H(x)$, but what good is it? We see that $H(x)$ is not a rational function, so the method of partial fractions will not work. Fortunately, there are stronger tools available.

7.1.2.2 Analytic functions

In what follows, all functions are from **C** to **C**. A *neighborhood* of a point x_0 is any set that contains a circle centered at x_0. The point x is a *boundary point* of the set S if all neighborhoods of x contain points that are in S and also contain points that are not in S. A set is open if it contains none of its boundary points, and closed if it contains all of them. Of course, many sets are neither open nor closed.

A set S in which any two points can be connected by a set of straight lines so that all those lines belong to S is called *connected.* A connected open set is called a *domain*, and a domain, together with a (possibly empty) subset of its boundary points is called a *region.*

The main definition of this section is the following.

Definition 7.14 *A function f that is defined in a region R is called* analytic *at x_0 if it can be written as a convergent power series*

$$f(x) = \sum_{n \geq 0} a_n (x - x_0)^n \tag{7.8}$$

in a neighborhood $U \subseteq R$ of x_0, that is, if at all points $x \in U$, the sum on the right-hand side of (7.8) is convergent and is equal to $f(x)$.

If R is a region, then we say that f is analytic in U if f is analytic at all points of R.

Example 7.15 *The function $f(x) = 1/(1-2x)$ is analytic at $x_0 = 0$.*

Solution: We show that we can choose U to be the open circle centered at 0 whose radius is 1/2. Indeed, let $|x| < 0.5$. Then we have

$$\begin{aligned} f(x) &= \frac{1}{1-2x} \\ &= \sum_{n \geq 0} 2^n x^n, \end{aligned}$$

and the last sum, which is a geometric series, converges since $|2x| < 1$. ◊

Clearly, $f(x) = 1/(1-2x)$ is not analytic (or even defined) at $x = 0.5$. However, it turns out that $x = 0.5$ is the *only* point in the complex plane where this function is not analytic.

Example 7.16 *Let $x_0 \neq 0.5$. Then $f(x) = 1/(1-2x)$ is analytic at x_0.*

Solution: We have

$$\begin{aligned} f(x) &= \frac{1}{1-2x} \\ &= \frac{1}{1-2x_0-(2x-2x_0)} \\ &= \frac{1}{1-2x_0} \cdot \frac{1}{1-\frac{2x-2x_0}{1-2x_0}} \\ &= \frac{1}{1-2x_0} \cdot \sum_{n\geq 0} \left(\frac{2x-2x_0}{1-2x_0}\right)^n \\ &= \sum_{n\geq 0} \frac{2^n}{(1-2x_0)^{n+1}} (x-x_0)^n. \end{aligned}$$

Note that the last form of $f(x)$ is just as it is required to be by (7.8), since the coefficient $a_n = \frac{2^n}{(1-2x_0)^{n+1}}$ does *not* depend on the changing x, just on the fixed x_0 and the index n. The last form of $f(x)$ is $1/(1-2x_0)$ times a geometric series, and as such, it is convergent if the quotient of that series has an absolute value less than 1, that is, when

$$\begin{aligned} \left|\frac{2}{1-2x_0}(x-x_0)\right| &< 1 \\ \left|\frac{x-x_0}{0.5-x_0}\right| &< 1. \end{aligned}$$

In other words, the power series form of $f(x)$ about x_0 is convergent whenever x is *closer to* x_0 *than 0.5 is to* x_0, that is, when x is inside the circle centered at x_0 and of radius $|x_0 - 0.5|$. This shows that f is analytic at x_0, since we have found a region around x_0 where $f(x)$ was representable in the form (7.8). If $x_0 = 0.5$, then no x can be closer to x_0 than 0.5, and so there is no circle centered at x_0 in which $f(x)$ is convergent, hence $f(x)$ is not analytic at that point. ◊

So we have found that the function $f(x) = 1/(1-2x)$ is analytic in every point where it is defined. This is not some exceptional property of that function; this is true for *all rational functions*, as you are asked to prove in Supplementary Exercise 1. For instance, the rational function $g(x) = \frac{1}{1-2x} + \frac{1}{1-3x}$ is analytic everywhere, except at the points $x_0 = 1/2$ and $x_1 = 1/3$.

It is an important property of power series that they are convergent inside a (possibly infinite) circle and divergent outside that circle. This is true for power series about any point, but for the sake of simplicity, we prove it for power series about 0, which is the special case that we will use in applications. Recall that the series $\sum_{n\geq 0} a_n$ is called *absolutely convergent* if the inequality $\sum_{n\geq 0} |a_n| < \infty$ holds. You should spend a minute convincing yourself that, if $\sum_{n\geq 0} a_n$ is absolutely convergent, then it is also convergent.

Theorem 7.17 *Let $f(x) = \sum_{n\geq 0} f_n x^n$ be a power series. If f is not absolutely convergent for all x, then there exists a real number r so that $f(x)$ is absolutely convergent if $|x| < r$, and $f(x)$ is not absolutely convergent for $|x| > r$.*

Proof: Let r be the least upper bound of the absolute values of all numbers for which $f(x)$ is absolutely convergent. Then, by the comparison test, $\sum_{n\geq 0} |f_n||x^n|$ is convergent if $|x| < r$, and not convergent if $|x| > r$, proving our claim. ◊

The number r defined in Theorem 7.17 is called the *convergence radius*, or *radius of convergence* of f, while the circle of radius r centered at the origin is called the *circle of convergence* of f.

Example 7.18 *As we have seen in the discussion following the solution of Example 7.16, the convergence radius of $f(x) = 1/(1-2x)$ is 0.5.*

It is a well-known theorem in complex analysis that f is analytic at x_0 if and only if f is *complex differentiable* at x_0, that is, the limit

$$\lim_{\delta\to 0} \frac{f(x_0+\delta) - f(x_0)}{\delta}$$

exists. Here δ is a complex number, and the limit is independent of the way in which δ goes to 0.

The following lemma will be crucial in determining the exponential growth rate of the sequence of coefficients of power series. Roughly speaking, it shows how closely the properties "convergent" and "analytic" are connected for power series.

Lemma 7.19 *Let $f(x) = \sum_{n\geq 0} f_n x^n$ be a power series that is absolutely convergent at the positive real number m. Then f is analytic in each point of the open circle C given by $|x| < m$.*

Note that the statement of the lemma is not obvious. Let x_0 be a point in C. Then it does follow from the absolute convergence of f at m and Theorem 7.17 that $f(x_0)$ is absolutely convergent *as a power series about 0*. However, the lemma claims that f can be represented by a convergent power series *about* x_0. Also note that most combinatorial generating functions have nonnegative real coefficients, and for such functions, the condition that f is absolutely convergent at m is equivalent to the condition that f is convergent at m.

Proof: Let x_0 be a complex number in the circle C. Then there exists a smaller open circle c with center x_0 and radius $r < m - |x_0|$ that is entirely inside C. Indeed, we can just keep shrinking a circle centered at x_0 until its radius becomes smaller than $m - |x_0|$. See Figure 7.1 for an illustration. We claim that, inside c, the function f can be written as a convergent power series about x_0, which will by definition imply that f is analytic at x_0.

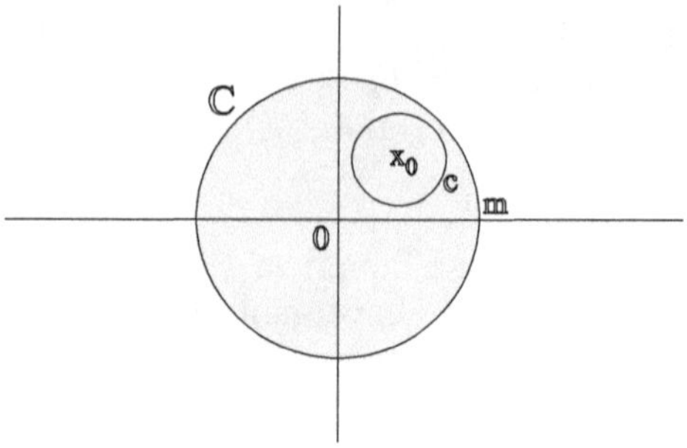

Figure 7.1
The small circle c lies entirely inside the larger circle C.

Let x be a point inside c, and let us notice that $x = x_0+(x-x_0)$. Therefore, by the binomial theorem, we have

$$f(x) = \sum_{n\geq 0} f_n[x_0 + (x - x_0)]^n = \sum_{n\geq 0}\sum_{k=0}^{n} \binom{n}{k} x_0^{n-k}(x - x_0)^k. \tag{7.9}$$

Note that, with the given conditions, the series on the far right-hand side of (7.9) is *absolutely convergent*, that is, the series

$$\sum_{n\geq 0} |f_n| \sum_{k=0}^{n} \binom{n}{k} |x_0^{n-k}|\, |(x - x_0)|^k = \sum_{n\geq 0} |f_n|[|x_0| + |x - x_0|]^n$$

is convergent. This is because $|x_0| + |x - x_0| < |x_0| + m - |x_0| = m$, so $|x_0| + |x - x_0|$ is inside the circle of convergence of f.

Because of the absolute convergence we just mentioned, we can change the order of summation in the power series on the far right-hand side of (7.9), and get the absolutely convergent power series

$$f(x) = \sum_{k\geq 0} \left(\sum_{n\geq 0} f_n \binom{n}{k} x_0^{n-k} \right) (x - x_0)^k.$$

This is a power series about x_0 that is absolutely convergent, hence it is convergent, proving our claim. (Note that, for each k, the coefficient of $(x-x_0)^k$ is a well-defined complex number, since it is a partial sum of an absolutely convergent sum.) ◊

We point out that the "converse" of Lemma 7.19 is not true. That is, if $f(x) = \sum_{n\geq 0} f_n x^n$ is divergent at some x, it can still happen that $f(x)$ is analytic at x. Recall that, in Example 7.16, we showed that $f(x) = 1/(1-2x)$ is analytic in every point, except at $x_0 = 1/2$. On the other hand, the power series form of f about 0, $f(x) = \sum_{n\geq 0} 2^n x^n$, is only convergent when $|x| < 1/2$.

It is a direct consequence of Theorem 7.17 and Lemma 7.19, that if $f(x)$ is a power series that is analytic at $x = 0$, and f has a convergence radius r about 0, then f is analytic in the open circle $|z| < r$. The following fundamental theorem of complex analysis shows that this circle cannot be made any larger without losing the analytic property of f, in the following sense.

Theorem 7.20 *Let $f(x)$ be a power series that is analytic at $x = 0$, and let f have a finite convergence radius r. Then f has a singularity of modulus r.*

We do not prove this theorem here since its proof requires *Cauchy's coefficient formula*, which in turn requires complex integration. The interested reader can consult page 240 of [31] for a proof of this theorem and its following interesting version. If $f(x)$ satisfies the conditions of Theorem 7.20, *and* only nonnegative coefficients, then we can say even more. In that case, *the positive real number r* is actually a singularity of f. Note that power series with nonnegative coefficients are important for us, since all combinatorial generating functions are in their class.

If f and g are both analytic at x_0, then it is easy to see that so are the functions $f+g$, $f-g$, and fg. Furthermore, f/g is analytic at x_0 if $g(x_0) \neq 0$. This is true since the rules that we learned about the derivatives of $f+g$, $f-g$, fg, and f/g in real analysis hold in complex analysis as well.

7.1.2.3 The complex versions of some well-known functions

The reader is certainly familiar with the functions e^x, $\sin x$, $\cos x$, and $\sin x$, $\tan x$, as well as $\ln x$ and $\sqrt{x}$ if x is a *real number*. It turns out that all these functions have an extension to the set of complex numbers. However, just as we have to be somewhat careful in the case of real numbers (we must exclude negative numbers from the domain of $\sqrt{x}$, nonpositive numbers from the domain of $\ln x$, and so on), we must be careful for complex numbers as well. The goal is to extend these functions so that they are defined on a domain D that is as large as possible, the extended functions are analytic in D, and on real numbers the extended functions agree with their usual value.

Let $z \in \mathbf{C}$, and let $z = x + iy$, where x and y are *real* numbers. As a consequence of basic algebraic identities, we have

$$e^z = e^{x+iy} = e^x \cdot e^{iy} = e^x(\cos y + i \sin y).$$

In order to define $\sin z$ and $\cos z$, let us notice that, by Euler's formula (7.3), we have

$$e^{iy} = \cos y + i \sin y,$$

and

$$e^{-iy} = \cos y - i \sin y$$

for all real numbers y. Therefore, adding two equations and dividing by two, we get that $\cos y = (e^{iy} + e^{-iy})/2$, while subtracting the second equation from the first and dividing by $2i$, we get $\sin y = (e^{iy} - e^{-iy})/2i$ for all real numbers y. We use these equations to define sin and cos for complex numbers, by setting

$$\cos z = \frac{e^{iz} + e^{-iz}}{2}$$

and

$$\sin z = \frac{e^{iz} - e^{-iz}}{2i}.$$

We can then define $\tan z = \sin z / \cos z$, $\sec z = 1/\cos z$, and so on.

It is then straightforward to prove that e^z, $\sin z$, and $\cos z$ have the same power series as their real versions. Those power series converge for all real numbers; therefore, by Lemma 7.19, they are *analytic everywhere.* Functions that are analytic in the entire complex plane are called *entire functions.*

We have to be a little bit more careful when we extend the functions $\sqrt{x}$ and $\ln x$ to complex numbers. Clearly, in order to extend $\sqrt{x}$ to the domain of complex numbers, we want to define $\sqrt{z}$ to be a complex number s so that the equality

$$s^2 = z \tag{7.10}$$

holds. The problem is that there are *two* such complex numbers s for each complex number z. Indeed, if $z = re^{i\phi}$, with $r \in \mathbf{R}$, then both $s_1 = \sqrt{r}e^{i\phi/2}$ and $s_2 = \sqrt{r}e^{i(\pi+(\phi/2))}$ are solutions of (7.10). However, we want $z \to \sqrt{z}$ to be a *function*, and therefore we have to choose one of them to be $\sqrt{z}$. Furthermore, we want $z \to \sqrt{z}$ to be an *analytic* function, so we have to make these choices carefully. For instance, as we said, a function that is analytic at z_0 is complex differentiable at z_0, so in particular it is continuous at z_0.

It turns out that there *does not exist* a function $f(z)$ that is defined on all complex numbers, is analytic everywhere, satisfies $f(x) = \sqrt{x}$ if $x \in \mathbf{R}$, and satisfies $f(z)^2 = z$ for all $z \in \mathbf{C}$. We leave the proof of this fact to textbooks in complex analysis, though readers are certainly welcome to think about it on their own. For the purposes of this book, we make the following definition.

Definition 7.21 *Let z be a complex number that is* not a negative real number. *This means that $z = re^{i\phi}$ for some nonnegative real number r and a real number $\phi \in (-\pi, \pi)$. Then we set $\sqrt{z} = \sqrt{r}e^{i\phi/2}$.*

Note that $\sqrt{z}$ defined as above is sometimes called the *principal* square root of z.

Example 7.22 *If $z = 4i = 4e^{i\pi/4}$, then $\sqrt{z} = 2e^{i\pi/8}$.*

It can be proved that the function $f(z) = \sqrt{z}$ defined above is analytic at z_0 if and only if z_0 is not a nonpositive real number. It is clear that $f(z)$ is not analytic at negative real numbers, since it is not even defined there.

Furthermore, and this is important for us, $\sqrt{z}$ is also not analytic at $z_0 = 0$, since there is no neighborhood U of that point so that f is defined in all of U. It follows that, for instance, the function $g(z) = \sqrt{1-4z}$ is not analytic at $z = 1/4$.

Note that the values of $\sqrt{z_1}$ and $\sqrt{z_2}$ are *very* different if $|z_1| = |z_2|$, and z_1 is a little bit above the negative half-axis, while z_2 is a little bit below that half-axis. This gives an intuition about why the square root function does not have an analytic extension to the entire complex plane.

In other words, we managed to extend the square root function to the *slit* complex plane, that is, the complex plane minus a half line. We chose that half line to be the half line of negative real numbers, which was arbitrary. Any other half line starting at the origin would work, but the obtained function would never be analytic at the origin, since it would never be defined in a full neighborhood of the origin.

The situation with logarithmic functions is similar. If we want to define the natural logarithm of the complex number z, we are looking for a complex number $s = x + iy$ so that

$$z = e^s = e^x \cdot e^{iy} = e^x(\cos y + i \sin y). \tag{7.11}$$

However, if we change y by $2k\pi$ for some integer π, then $\cos y + i \sin y$ does not change. So there are infinitely many complex numbers s that satisfy (7.11). Therefore, we again need new restrictions in order to make " complex logarithm" a *function.*

Definition 7.23 *Let $z \in \mathbf{C}$. If $s = x + iy$ so that $x \in \mathbf{R}$, while $y \in (-\pi, \pi]$ and $z = e^{x+iy}$ holds, then we say that s is the logarithm of z, and we write*

$$\operatorname{Log}(z) = s.$$

Note that the image of $\mathbf{C}$ under Log is an infinite band of height 2π that is closed at the top and open at the bottom. Also note that Log(0) is not defined since $e^x \neq 0$ for any real numbers x.

Example 7.24 *As $-e = e^{1+\pi i}$ by Euler's formula (7.3), we have* $\operatorname{Log}(-e) = 1 + \pi i$.

Note that $\operatorname{Log}(z)$ is sometimes called the *principal* branch of the complex logarithm. We write $\operatorname{Log}(z)$ to distinguish the complex logarithm from its real version, $\ln x$. Note in particular that, keeping the notation of Definition 7.23, the equality

$$\operatorname{Log}(z) = \ln(e^x) + y$$

holds, so if $z \in \mathbf{R}$, then $z = e^x$ and $y = 0$, and therefore, $\operatorname{Log}(z) = \ln z$. In particular, formula (7.7) remains true if we replace ln by Log in it.

It is again not difficult to prove that the function $\operatorname{Log}(z)$ is analytic in all *nonzero* points $z = x + iy$ that satisfy $z \in \mathbf{R}$ and $y \in (-\pi, \pi)$.

7.1.2.4 Singularities

If f is not analytic at a point z_0, then we say that z_0 is a *singularity* or *singular point* of f. Note that this definition generalizes the way in which we defined singularities of rational functions. Singularities of power series are crucial tools in finding the growth rate of their coefficients.

There are numerous reasons that may cause f not to be analytic at z_0. If f is given in power series form about a point z_0, then of course f is not analytic at z_0 if it is not convergent there. If f is given in a *closed* form, then of course it can happen that f is not defined in a certain point z_0, because the formula for f involves a division by 0 at z_0. Then f is certainly not analytic at z_0.

Example 7.25 *The function* $f(z) = \frac{1}{(1-z)(1-2z)}$ *has singularities at* $z_1 = 1$ *and* $z_2 = 1/2$.

Example 7.26 *The function* $f(z) = \sqrt{1-4z}$ *has a singularity at* $z = 1/4$.

Indeed, we have just seen that the square root function is not analytic at 0.

Example 7.27 *The function* $f(z) = \text{Log}(1-z)$ *has a singularity at* $z = 1$.

Indeed, we saw in Definition 7.23 that Log(0) is not even defined.

7.1.2.5 The main theorem of exponential asymptotics

We have seen in Theorem 7.10 that if f is a power series that corresponds to a rational function, and f has a *unique* singularity of smallest modulus r_1, then the exponential growth rate of the coefficients of f agrees with the $1/|r_1|$. Crucially, this holds for a much larger class of functions. This is the content of the next theorem, which is sometimes called the main theorem of exponential asymptotics.

Theorem 7.28 *Let* $f(x) = \sum_{n\geq 0} f_n x^n$ *be a power series that is analytic at 0, and let* M *be the modulus of a singularity of* f *that is closest to 0. Then the sequence* f_n *has exponential growth rate* $1/M$.

Proof: Let us first assume that the sequence f_n has exponential growth rate $\alpha < 1/M$. Let α_1 be such that $\alpha < \alpha_1 < 1/M$. Then, for all but a finite number of indices n, the inequality $|f_n| \leq \alpha_1^n$ holds. So, for these indices and for every complex number x of modulus M, the inequality

$$|f_n x^n| \leq \alpha_1^n M^n = (\alpha_1 M)^n$$

holds. Let M_1 be a positive real number that is "just a tiny bit larger" than M. Formally speaking, let $M_1 > M$ so that $M_1\alpha_1 < 1$ still holds. Such a number M_1 always exists, since $M\alpha_1 < 1$.

Then the inequality $\alpha_1 M_1 < 1$ implies that $f(x) = \sum_{n\geq 0} f_n x^n$ is *convergent* at all points x of modulus M_1, so in particular, $f(x)$ is convergent at the positive real number $M_1 > M$. Therefore, by Lemma 7.19, we know that f is analytic at all points z with modulus less than M_1. This is a contradiction, since f has a singularity of modulus M.

Now let us assume that the sequence f_n has exponential growth rate $\beta > 1/M$. Then there is a positive constant c so that there are infinitely many indices n so that $|f_n| > c\beta^n$. Then, for all those indices n, the inequality

$$\left| f_n \cdot \left(\frac{1}{\beta}\right)^n \right| > c$$

holds, which implies that $\sum_{n\geq 0} f_n x^n$ is divergent at the positive real number $1/\beta < M$. That means that $M > 1/\beta > r$, where r is the radius of convergence of f. That is a contradiction, since by Theorem 7.20, we have $M = r$. ◊

Now we are in a position to answer the question that we asked in Example 7.13. In order to find the exponential growth rate of the sequence $h_n/n!$, all we need to do is to find the singularity of $H(x)$ that is of smallest modulus. It is easy to see that $H(x)$ will have a singularity at $x = 1$, and another one when $1 = \ln(1/(1-x)$, that is, when $e = 1/(1-x)$, or $x = 1 - e^{-1}$. Clearly, the latter is the singularity of smallest modulus, showing that the exponential growth rate of the sequence $h_n/n!$ is

$$\frac{1}{1-e^{-1}} = \frac{e}{e-1} \sim 1.582.$$

So numbering the tables increases the exponential growth rate of the number of possibilities by 1.582.

Example 7.29 *Let r_n be the number of lattice paths from $(0,0)$ to (n,n) that use only the steps $(1,0)$, $(0,1)$, and $(1,1)$ that never go above the main diagonal. Find the exponential growth rate of the sequence r_n.*

Solution: We have seen in Exercise 24 of Chapter 3 that

$$R(x) = \sum_{n\geq 0} r_n x^n = \frac{1 - x - \sqrt{1 - 6x + x^2}}{2x}.$$

It is easy to check that $x = 0$ is a root of multiplicity one of both the numerator and the denominator of $R(x)$, so $R(x)$ is analytic at $x = 0$. Therefore, the singularity of smallest modulus of $R(x)$ is the smallest root of the expression under the square root sign, namely, $M = 3 - 2\sqrt{2}$. Therefore, by Theorem 7.28, the sequence r_n (the Schröder numbers) has exponential growth rate $1/M = 3 + 2\sqrt{2} \sim 5.828$. ◊

Example 7.30 *Let h_n be the number of ways an n-person race can end if two-way and three-way ties are allowed. Find the exponential order of the sequence $h_n/n!$.*

Solution: In other words, h_n is the number of ways to partition $[n]$ into any number of blocks of size at most three, then to put an order on the *set* of the blocks. We can use the compositional formula of exponential generating functions to compute $H(x) = \sum_{n\geq 0} h_n \frac{x^n}{n!}$. There is no task to carry out on each block, other than making sure that its size is at most three. This means that $a_k = 1$ if $1 \leq k \leq 3$, and $a_k = 0$ otherwise, so $A(x) = x + \frac{x^2}{2} + \frac{x^3}{3}$. As the blocks are to be linearly ordered, we have $b_m = m!$, so $B(x) = \sum_{m\geq 0} m! \frac{x^m}{m!} = 1/(1-x)$. Therefore, the compositional formula implies

$$H(x) = B(A(x)) = \frac{1}{1 - x - \frac{x^2}{2} - \frac{x^3}{3}}. \tag{7.12}$$

Using a software package, we find that the root of the denominator that is closest to the origin is at $M \sim 0.6725$, so the exponential growth rate of the sequence $h_n/n!$ is $1/M \sim 1.487$. ♢

Finally, let us take a moment to mention the new, more general version of Theorem 7.10, which is now a special case of Theorem 7.28.

Theorem 7.31 *Let*

$$S(x) = \sum_{n\geq 0} s_n x^n = \frac{P(x)}{Q(x)}$$

be a rational function with $Q(0) \neq 0$, and let us assume that $P(x)$ and $Q(x)$ do not have any roots in common. Let r_1 be a root of $Q(x)$ that is of smallest modulus.

Then the exponential growth rate of the sequence of the coefficients s_n is equal to $|z_1|$, where $z_1 = 1/r_1$.

Note that, in contrast to Theorem 7.10, we no longer need the condition that $Q(x)$ has a *unique* root of smallest modulus.

Quick Check

1. Find the exponential growth rate of the sequence $a_n = n^n/n!$.
2. Find the exponential growth rate of the coefficients of the rational function $f(x) = 1/(4x^2 - 5x + 1)$.
3. Find the exponential growth rate of the coefficients in the power series form of
$$f(x) = \frac{1}{1 - \sqrt{2 - 4x}}.$$

7.2 Polynomial precision

Sometimes it is possible to determine the growth rate of a sequence not only with "exponential," but also with "polynomial precision," and sometimes even more precisely. In this section, we will see examples of this phenomenon.

7.2.1 Rational functions again

Let us assume that two functions $f(x) = \sum_{n\geq 0} f_n x^n$ and $g(x) = \sum_{n\geq 0} g_n x^n$ agree in their singularity s of smallest modulus. By Theorem 7.28, that means that their sequence of coefficients has the same exponential growth rate. This does not mean that $f_n \sim g_n$, and there could be two reasons for that.

It can happen that it is the *multiplicity* of s that differs in f and g. An example for this is when $f(x) = 1/(1-3x)$ and $g(x) = 1/(1-3x)^2$. In the first case, we have $f(x) = \sum_{n\geq 0} 3^n x^n$, while in the second case, we have the somewhat different formula $g(x) = \sum_{n\geq 0}(n+1)3^n x^n$. So $g_n/f_n = n+1$ for all n, which implies that $g_n/f_n \to \infty$.

It could also happen that the *type* of s as a singularity is different in f and g, for example, when $f(x) = 1/(1-x)$, $g(x) = 1-\sqrt{1-x}$, or even $h(x) = \ln(1/(1-x))$. All three functions have a singularity of multiplicity 1 at $x = 1$, but the *type* of that singularity is different in each case. In $f(x)$, it is caused by a division by 0 in a rational function; in $g(x)$, it is caused by the fact that the square root function is not analytic at 0; and in $h(x)$, it is caused by the fact that logarithm as a function is not analytic at 0.

Note that $f(x) = \sum_{n\geq 0} x^n$, $g(x) = \sum_{n\geq 1} \frac{(2n-3)!!}{2^n \cdot n!}$ (by the binomial theorem, with exponent $1/2$), and $h(x) = \sum_{n\geq 1} \frac{x^n}{n}$, so the coefficient sequences are quite different in size, even though they all have exponential growth rate 1.

The analysis of how the type of a singularity influences that growth rate of the coefficient sequence of a power series goes beyond the scope of this book. However, for rational functions, we will analyze the effect of the *multiplicities* of singularities of rational functions.

7.2.1.1 Motivation

Let us assume that we want to know the number of ways to pay n dollars using only 1-dollar, 2-dollar, and 5-dollar bills. We have seen in Example 3.17 that, if h_n denotes the number of ways to do that, then

$$H(x) = \sum_{n\geq 0} h_n x^n = \frac{1}{(1-x)(1-x^2)(1-x^5)}. \tag{7.13}$$

So $H(x)$ has a singularity at z whenever the first, second, or fifth power of z is equal to 1. Clearly, all these singularities are at a distance 1 from 0, so, by Theorem 7.28, the exponential order of the numbers h_n is 1.

This is not a very satisfying result. Let us, for instance, consider the following. Let g_n be the number of ways to pay n dollars using only 1-dollar, 2-dollar, 5-dollar, *and* 10-dollar bills. Then clearly,

$$G(x) = \sum_{n\geq 0} g_n x^n = \frac{1}{(1-x)(1-x^2)(1-x^5)(1-x^{10})}. \tag{7.14}$$

The numbers $g(n)$ also have exponential order 1, since all singularities of their generating function are of modulus 1. Still, it is clear that, for large n, the number g_n will be much larger than the number h_n, since pairs of 5-dollar bills can be turned into 10-dollar bills.

What this shows is that two sequences can be quite different even if they have the same exponential growth rate. The exponential growth rate is not subtle enough to express smaller differences between two sequences.

Therefore, we introduce a new, more refined way to measure the growth rate of sequences. We will say that we have described the growth rate of a sequence f_n at *polynomial precision* if we found the exponential growth rate α of f_n, together with the largest real number k so that f_n has an infinite subsequence f_{n_i} for which

$$f_{n_i} \sim C\alpha^n n^k \tag{7.15}$$

holds, for some constant C. In other words,

$$\lim_{n\to\infty} \frac{f_{n_i}}{\alpha^n n^k} = C.$$

Note that not all sequences will fit in this framework. For instance, the sequence $f_n = \ln(n+1)$ will not, since it grows slower than any nonconstant polynomial. However, the growth of many sequences can be determined at polynomial precision.

Example 7.32 *Let* $f_n = (n^3+3)2^n/(2n^2-7)$. *Then the formula*

$$f_n \sim \frac{1}{2} \cdot 2^n \cdot n$$

describes the growth rate of f_n *at polynomial precision.*

Example 7.33 *Let* $f_n = \binom{2n}{n}/(n+1)$, *that is, let* f_n *denote the nth Catalan number. Then the formula*

$$f_n \sim \frac{1}{\sqrt{\pi}} \cdot 4^n \cdot n^{-3/2}$$

describes the growth rate of f_n *at polynomial precision.*

Solution: This follows from Stirling's formula, stating that $n! \sim (n/e)^n\sqrt{2\pi n}$, and the fact that $\binom{2n}{n} = \frac{(2n)!}{n!n!}$. The reader is invited to work out the computational details. ◊

7.2.1.2 Multiple singularities in rational functions

The following theorem helps us describe the growth of the sequence of coefficients of a rational function at polynomial precision.

Theorem 7.34 *Let*

$$f(x) = \sum_{n \geq 0} f_n x^n = \frac{P(x)}{Q(x)}$$

be a rational function so that $P(x)$ and $Q(x)$ do not have roots in common, and let $Q(0) = 1$. Let the roots of $Q(x)$ be $r_1, r_2, \cdots, r_k$, and let the multiplicity of r_i as a root of $Q(x)$ be a_i. Let us further assume that, if $Q(x)$ has more than one root of smallest modulus, then one of those roots has a higher multiplicity than all the other roots of smallest modulus.

Then there exist polynomials $p_1, p_2, \cdots, p_k$ so that, if n is large enough, then

$$f_n = \sum_{i=1}^{k} p_i(n) r_i^{-n}. \tag{7.16}$$

Furthermore, the degree of p_i is $a_i - 1$.

Therefore, the growth rate of f_n at polynomial precision is $C(1/|r_t|)^n n^{a_t - 1}$, where C is a constant, r_t is a root of $Q(x)$ that has smallest modulus, and, among roots of smallest modulus, has maximal multiplicity a_t.

Proof: Set $z_i = 1/r_i$. As f is a rational function, it has a partial fraction decomposition, as we have seen in Section 7.1.1. That is,

$$\begin{aligned} f(x) &= p(x) + \frac{C_{1,1}}{1 - z_1 x} + \frac{C_{1,2}}{(1 - z_1 x)^2} + \ldots + \frac{C_{1,a_1}}{(1 - z_1 x)^{a_1}} && (7.17) \\ &+ \frac{C_{2,1}}{1 - z_1 x} + \frac{C_{2,2}}{(1 - z_1 x)^2} + \ldots + \frac{C_{2,a2}}{(1 - z_1 x)^{a_2}} + \ldots && (7.18) \\ &+ \frac{C_{k,1}}{1 - z_k x} + \frac{C_{k,2}}{(1 - z_k x)^2} + \ldots + \frac{C_{k,a_k}}{(1 - z_k x)^{a_k}}, && (7.19) \end{aligned}$$

where $p(x)$ is a polynomial function which is the zero polynomial if the degree of P is less than the degree of Q, and the $C_{k,i}$ are constants.

If n is large enough, then $p(x)$ does not contain a term of degree n or higher, so $p(x)$ will not contribute to the coefficient f_n of x^n.

Just as we have seen in the proof of Theorem 7.10, we notice that the binomial theorem implies

$$\begin{aligned} \frac{1}{(1 - z_i x)^j} &= (1 - z_i x)^{-j} && (7.20) \\ &= \sum_{n \geq 0} \binom{-j}{n} z_i^n (-1)^n && (7.21) \\ &= \sum_{n \geq 0} \binom{n + j - 1}{j - 1} z_i^n && (7.22) \\ &= \sum_{n \geq 0} \binom{n + j - 1}{j - 1} r_i^{-n}. && (7.23) \end{aligned}$$

Now fix i, and consider the ith row of (7.17). Each summand in that row is of the required form (a polynomial times a power of the root r_i with a negative exponent); therefore, their sum is also of the required form. The exponential growth rate of such a sequence is largest when r_i is smallest, that is, when r_i is a root of smallest modulus. Furthermore, we have just seen that the polynomial that occurs in the sum form of $\frac{1}{(1-z_i x)^j}$ is of degree $j-1$, where j is the multiplicity of r_i as a root of Q. So, on the set of roots of smallest modulus, this degree is maximized when r_i is the unique root of smallest modulus that has maximal multiplicity. ◊

Note that Theorem 7.34 remains true if we drop the condition that there is a root of Q of smallest modulus that has a higher multiplicity than all other roots of smallest modulus, but the proof is a little bit more complicated.

Now we can solve the initial example of this section. In that example, $Q(x) = (1-x)(1-x^2)(1-x^5)$, whereas $P(x) = 1$. So $P(x)$ and $Q(x)$ have no roots in common (P has no roots at all), and the root $r_1 = 1$ of $Q(x)$ has multiplicity three, while the other five roots of $Q(x)$ have multiplicity 1. So Theorem 7.34 applies, and it yields

$$h_n = p_1(n) + \sum_{i=2}^{6} p_i r_i^{-n},$$

where p_1 is a polynomial of degree two, while all the other p_i are constants (polynomials of degree 0). Note that, for all i, we have $|r_i| = 1$, so $|r_i^{-n}| = 1$, and so

$$\left| \sum_{i=2}^{6} p_i r_i^{-n} \right| \leq \sum_{i=2}^{6} \left| p_i r_i^{-n} \right| = \sum_{i=2}^{6} |p_i| = C.$$

That is, the term $\sum_{i=2}^{6} p_i r_i^{-n}$ will never have absolute value more than the constant $\sum_{i=2}^{6} |p_i| = C$, meaning that, for large n, the value of h_n will be very close to the value of the quadratic polynomial $p_1(n)$.

It follows from (7.17) and (7.20) that $p_1(n) = C_1 \sum_{n\geq 0} \binom{n+2}{2}$ for some constant C_1. Given the relatively simple form of $P(x)/Q(x)$, the exact value of C_1 can be determined, even without actually computing the whole partial fraction decomposition of $P(x)/Q(x)$. You will be asked to do so in the Exercises. A technique we will cover in Section 7.3 may be helpful.

Note that, if all the roots of $Q(x)$ are of multiplicity 1, then (7.16) simplifies to

$$f_n = \sum_{i=1}^{k} p_i r_i^{-n},$$

where the p_i are constants.

Example 7.35 *A utility crew is painting a row of n poles proceeding from east to west. In their first shift, they paint each pole red or blue or black. In* *their second shift, they paint each pole gray or orange. In their third and last shift, they paint each pole yellow or green or pink. Let h_n be the number of ways in which the crew can proceed. Find the growth rate of h_n at polynomial precision.*

Solution: Let $H(x) = \sum_{n \geq 0} h_n x^n$. It follows from the product formula (Theorem 3.14) of ordinary generating functions that

$$H(x) = \frac{1}{1-3x} \cdot \frac{1}{1-2x} \cdot \frac{1}{1-3x}.$$

Therefore, $H(x)$ is a rational function and the roots of its denominator are $r_1 = 1/3$ with multiplicity 2, and $r_2 = 1/2$ with multiplicity 1. So Theorem 7.34 shows that

$$h_n = p_1(n)3^n + p_2(n)2^n,$$

where $p_1(n)$ is a polynomial of degree 1, and p_2 is a constant. Therefore, there exists a constant C so that $h_n \sim Cn3^n$. ◊ We point out that in the

previous example, there were no lower bounds on the number of poles that the crew had to paint each day. However, if there were such bounds, that would not have changed the asymptotics of h_n, as only the numerator, and not the denominator, of $H(x)$ would have changed.

Example 7.36 *Let s_n be the number of ways to schedule the next n days of the US congress with the following constraints.*

- *The schedule must contain exactly three sessions of at least one day each, and*
- *there has to be a break of at least one day between any two sessions, and*
- *when Congress is in session, each day must be devoted either to a plenary meeting or to committee work.*

Find the growth rate of the sequence s_n at polynomial precision.

Solution: Just as in the preceding example, we use the product formula (Theorem 3.14). The schedule will start with a break of any (possibly 0) length, and there is only one way to schedule a break of length k, leading to the generating function $A(x) = 1/(1-x)$. After this, there will be a session of length at least one, leading to the generating function $B(x) = 2x/(1-2x)$, since there are 2^k ways to schedule a session of k days. Following this session, there will be a break of length at least 1, leading to the generating function $C(x) = x/(1-x)$. Continuing in this way, and noting that, after the third and last session, there will be a break of any (possibly zero) length, the product formula yields

$$S(x) = \sum_{n \geq 0} s_n x^n = \frac{1}{1-x} \cdot \frac{2x}{1-2x} \cdot \frac{x}{1-x} \cdot \frac{2x}{1-2x} \cdot \frac{x}{1-x} \cdot \frac{2x}{1-2x} \cdot \frac{1}{1-x}.$$

Therefore,

$$S(x) = \frac{8x^5}{(1-x)^4(1-2x)^3}.$$

So $r_1 = 1/2$ is a root of multiplicity 3, and $r_2 = 1$ is a root of multiplicity 4. Hence, by Theorem 1, $s_n = p_1(n)2^n + p_2(n)$, where p_1 is a polynomial of degree 2, and p_2 is a polynomial of degree 3. This implies that there exists a constant C so that $s_n \sim Cn^2 2^n$. ◊

Quick Check

1. Let
$$F(x) = \frac{1}{(1-x)^2(1-2x)}.$$
Find the growth rate of the coefficients of $F(x)$ at polynomial precision.
2. Let
$$G(x) = \frac{1}{(1-4x)^3(1+4x)^3}.$$
Find the growth rate of the coefficients of $G(x)$ at polynomial precision.
3. Let $H(x) = \sum_{n \geq 0} h_n x^n$ so that $h_n \geq 0$ for all n, let us assume that the exponential growth rate of the sequence h_n is at least 1, and let us assume that the sequence h_n is supermultiplicative, that is, the inequality $h_n h_m \leq h_{n+m}$ holds for all n and m. Prove that if H is rational, then it has a *unique* singularity of smallest modulus.

7.3 More precise asymptotics

We can sometimes refine our asymptotic results even further than polynomial precision. We will consider two situations in which this is possible.

7.3.1 Entire functions divided by $(1-x)$

Some power series have a sequence of coefficients that is *convergent* as a sequence, and we can compute the limit of that sequence without computing the coefficients themselves.

Recall that we call a permutation a *derangement* if it does not contain any 1-cycles. Similarly, let us call a permutation a 2-derangement if it does not contain any 1-cycles or 2-cycles. Let $D_{2,n}$ be the number of 2-derangements of length n. What can we say about the ratio $D_{2,n}/n!$? In other words, what is the probability that a randomly selected permutation of length n is a 2-derangement?

Let $D_2(x) = \sum_{n\geq 0} D_{2,n}\frac{x^n}{n!}$ be the exponential generating function of the numbers $D_{2,n}$, where we set $D_{2,0} = 1$. We can obtain $D(x)$ using the exponential formula. Indeed, to get a permutation counted by $D_{2,n}$, we need to partition $[n]$ into blocks of size at least two, then put a k-cycle on each block of size k. There are $(k-1)!$ ways to do that if $k \geq 3$, and 0 ways otherwise. This leads to the generating function

$$\begin{aligned} A(x) &= \sum_{k\geq 3}(k-1)!\frac{x^k}{k!} = \sum_{k\geq 3}\frac{x^k}{k} \\ &= \ln\left(\frac{1}{1-x}\right) - x - \frac{x^2}{2}. \end{aligned}$$

Therefore, by the exponential formula,

$$D(x) = e^{A(x)} = \frac{e^{-x-\frac{x^2}{2}}}{1-x}. \tag{7.24}$$

So, we have an explicit formula for $D(x)$, but how do we use it? Writing up an explicit formula for the numbers $D_{2,n}$ is too complicated, since the numerator of $D(x)$ involves two summation signs, and the division by $1-x$ would mean further summation. It is clear that $D(x)$ has only one singularity, at $x = 1$, but all that means is that the coefficients of $D(x)$, that is, the numbers $D_{2,n}/n!$, have exponential growth rate 1. That is not too surprising, since all sequences that converge to a positive constant have exponential growth rate 1. In order to achieve our goal, that is, to find $\lim_{n\to\infty}\frac{D_{2,n}}{n!}$, we need to learn a new technique. The following theorem is our main tool toward that goal.

Theorem 7.37 *Let f be an entire function, and let*

$$\frac{f(x)}{1-x} = \sum_{n\geq 0} a_n x^n.$$

Then $a_n \sim f(1)$. In other words, $\lim_{n\to\infty} a_n = f(1)$.

Proof: Note that

$$\frac{f(x)}{1-x} = \frac{f(x)-f(1)}{1-x} + \frac{f(1)}{1-x} = \frac{f(x)-f(1)}{1-x} + f(1)\sum_{n\geq 0} x^n. \tag{7.25}$$

Observe that the function $F(x) = \frac{f(x)-f(1)}{1-x}$ is an entire function. Indeed, as $f(x)$ is an entire function, it converges everywhere, so we have $f(x) = \sum_{n\geq 0} f_n x^n$, and, in particular, $f(1) = \sum_n f_n$. Therefore,

$$\begin{aligned} F(x) &= \frac{f(x)-f(1)}{1-x} \\ &= \frac{\sum_{n\geq 0} f_n(x^n-1)}{1-x} \\ &= -\sum_{n\geq 0} f_n\left(1+x+\cdots+x^{n-1}\right) \\ &= \sum_{n\geq 0}\left(-\sum_{k\geq n+1} f_k\right)x^n. \end{aligned}$$

In the last form of $F(x)$, all coefficients are convergent sums, since even the sum $\sum_{k\geq 0} f_k = f(1)$ is finite. For that same reason, the sequence of partial (tail) sums $\sum_{k\geq n+1} f_k$ converges to 0 as n goes to infinity.

In order to prove our claim, let us now compare the coefficient of x^n in the far left and far right terms of (7.25). In the far left, it is, by definition, a_n. The far-right expression is a sum, the second summand of which has coefficient $f(1)$ for each power of x, and the first summand of which has a sequence of coefficients that (as we just proved) converges to 0. ♢

Corollary 7.38 *As n goes to infinity, the probability that a randomly selected permutation of length n is a 2-derangement converges to $e^{-3/2} \sim 0.22313$.*

Proof: Just use Theorem 7.37 with $f(x) = e^{-x-x^2/2}$. ♢

It is also straightforward to recover the result that we proved in Chapter 4, namely, that, for large n, roughly $1/e$ of all permutations of length n are derangements. Indeed, if D_n denotes the number of derangements of length n, then we have seen in (4.11) that

$$\sum_{n\geq 0} D_n \frac{x^n}{n!} = \frac{e^{-x}}{1-x}.$$

So using Theorem 7.37 with $f(x) = e^{-x}$, we get that $\frac{D_n}{n!} \approx e^{-1}$.

Example 7.39 *Let h_n be the number of ways to partition $[n]$ into two blocks A and B so that B can be empty, but A cannot, and then to take a partition of A and a permutation of the elements of B. Find $\lim_{n\to\infty} \frac{h_n}{n!}$.*

Solution: It follows from a routine application of the exponential formula that the exponential generating function for the number of partitions of $[n]$ is $G(x) = e^{e^x-1}$. (Indeed, we do not put any structure on the blocks, except that the blocks are required to be nonempty.) We can now use the products formula of exponential generating functions with $A(x) = e^{e^x-1} - 1$ since A cannot be empty, and $B(x) = \sum_{n\geq 0} n! \frac{x^n}{n!} = 1/(1-x)$.

This leads to

$$H(x) = A(x)B(x) = \frac{e^{e^x-1} - 1}{1-x}.$$

As the numerator $A(x)$ of $H(x)$ is clearly an entire function, Theorem 7.37 applies and so $\lim_{n\to\infty} h_n/n! = A(1) = e^{e-1} - 1 \approx 4.5749$. $\Diamond$

7.3.2 Rational functions one more time

If $h(x) = \sum_{n\geq 0} h_n x^n$ is a rational function, then we can, in theory, compute its partial fraction decomposition and obtain an exact formula for the numbers h_n. Sometimes this is too cumbersome, and a rather precise estimate can be obtained without computing that exact formula.

Recall that, in Section 7.2.1, we studied the power series

$$H(x) = \sum_{h\geq 0} h_n x^n = \frac{1}{(1-x)(1-x^2)(1-x^5)},$$

and proved that there exists a constant C so that

$$h_n \sim C \cdot n^2, \tag{7.26}$$

since all singularities of $H(x)$ are of modulus 1, and the singularity $x = 1$ has multiplicity three.

Now we will discuss a simple method to compute the constant C in (7.26). Our tool is the following theorem, which is an enhanced version of Theorem 7.34.

Theorem 7.40 *Let*

$$f(x) = \sum_{n\geq 0} f_n x^n = \frac{P(x)}{Q(x)}$$

be a rational function so that $P(x)$ and $Q(x)$ do not have roots in common, and $Q(0) \neq 0$. Let us assume that $Q(x)$ has a root r_1 of smallest modulus that is of higher multiplicity a_1 than any other roots of $Q(x)$ that are of highest modulus.

Then we have

$$f_n \sim \frac{K}{(a_1-1)!} \cdot n^{a_1-1} |r_1|^{-n} \tag{7.27}$$

where

$$K = \lim_{x \to r_1} (1 - z_1 x)^{a_1} f(x),$$

with $z_1 = 1/r_1$.

Proof: The proof is somewhat similar to that of Theorem 7.37. The key observation is that the difference between the functions $f(x)$ and $\frac{K}{(1-z_1x)^{a_1}}$ is relatively small in a neighborhood of r_1.

In order to see this, let us introduce the notation $f(x) = g(x)/(1 - z_1x)^{a_1}$. In other words, $g(x)$ is all of $f(x)$ except those factors involving z_1. Consider the rational function

$$T(x) = f(x) - \frac{K}{(1 - z_1x)^{a_1}} = \frac{g(x) - K}{(1 - z_1x)^{a_1}}.$$

Crucially, $T(r_1) = 0$, since, by definition, $K = \lim_{r_1} g(x)$, and $g(x)$ is continuous at r_1. Therefore, the polynomial in the numerator of $T(x)$ is divisible by $1 - z_1x$, and so we can simplify $T(x)$ by at least one factor $(1 - z_1x)$. However, that implies, by Theorem 7.34, that the coefficients of $T(x)$ grow at least a linear factor slower than the coefficients of $f(x)$. Therefore, the coefficients of $f(x)$ and that of $f(x) - T(x) = \frac{K}{(1-z_1x)^{a_1}}$ are asymptotically equal. In symbols,

$$\begin{aligned} f_n &\sim [x^n]\frac{K}{(1 - z_1x)^{a_1}} \\ &\sim \binom{n + a_1 - 1}{a_1 - 1} r_i^{-n}, \end{aligned}$$

proving our claim. The reader may revisit (7.20) for more details on the last computation. ◊

Example 7.41 *Continuing with our example from Section 7.2.1, let* $f(x) = \frac{1}{(1-x)(1-x^2)(1-x^5)}$. *As we have discussed, we have* $r_1 = 1$ *and* $a_1 = 3$. *This yields*

$$K = \lim_{x \to 1} (1 - x)^3 f(x) = \frac{1}{(1 + x)(1 + x + x^2 + x^3 + x^4)} = \frac{1}{10}.$$

Since $a_1 = 3$, *Theorem 7.40 implies that*

$$f_n \sim \frac{n^2}{20}.$$

Quick Check

1. Let H_n denote the number of permutations of length n that contain no 2-cycles and no 3-cycles. Compute $\lim_{n \to \infty}(H_n/n!)$.

2. Let G_n be the number of ways to split $[n]$ into two blocks (which are allowed to be empty), and choose a permutation with no 2-cycles on the second block. Compute $\lim_{n\to\infty}(G_n/n!)$.

3. Let f_n be the number of ways to pay n dollars with using 1-dollar, 2-dollar, 5-dollar, and 10-dollar bills. Find constants C and k satisfying $f_n \sim Cn^k$.

7.4 Notes

Our goal in this chapter was to introduce the reader to some techniques in analytic combinatorics, even though we assumed that the reader has not taken a course in complex analysis. Therefore, we did not cover any topics that needed complex integration. In particular, that meant that we could not use famous theorems such as the residue theorem or Cauchy's coefficient formula. For that reason, one important problem that we did not address in this chapter was to find the growth rate of the coefficients of *entire functions*, that is, functions that have no singularities, such as e^x, or $e^{x+\frac{x^2}{2}}$, the exponential generating function of involutions of length n, or e^{e^x-1}, the exponential generating function of partitions of $[n]$. It goes without saying that a completely new method is needed for such functions, since there are no singularities to analyze.

If the reader wants to learn the relevant material in complex analysis, the reader should consult a textbook on that topic, such as *Complex Analysis* by Serge Lang [51], or *Complex Variables and Applications* by Ruel V. Churchill and James Ward Brown [24].

Once the reader is familiar with the mentioned important theorems in complex analysis, Section 5.4 of the classic book *generatingfunctionology* [84] (where both the single word of the title and its lowercase first letter were coined by the author Herb Wilf) provides a quick introduction to a technique that enables us to compute the growth rate of the coefficients of some entire functions. As its name shows, the book *Analytic Combinatorics* [31] by Philippe Flajolet and Robert Sedgewick is a high-level, comprehensive treatment of the area. Among the many useful skills one can acquire by reading that book, we just mention one. In this chapter, we have seen how to find estimates of polynomial precision for the growth rate of the coefficient sequence of certain power series whose singularities were of the *rational type*, that is, they were caused by the presence of a nonconstant polynomial in the denominator. In [31], one can learn how to obtain similar results for power series whose singularities are of the *square root type*, like $\sqrt{1-2x-3x^2}$ or $1/\sqrt{1-6x+x^2}$, and also for power series whose singularities are of the *logarithmic type*, such as $\ln(1-x)$ or $1/\ln(1-2x)$.

7.5 Chapter review

1. The exponential growth rate of the sequence $a_0, a_1, \cdots$ is equal to $\limsup \sqrt[n]{|a_n|}$.
2. The exponential growth rate of the sequence of coefficients f_n of a power series $F(x) = \sum_{n\geq 0} f_n x^n$ is equal to $1/M$, where M is the minimal distance between 0 and any singularity of F.
3. If $F(x)$ is a rational function, we can get more precise estimates of the growth rate of the sequence f_n from the *multiplicities* of the singularities of F that are of smallest modulus.

7.6 Exercises

1. Let $P(x)$, $Q(x)$, $A(x)$, and $R(x)$ be polynomials so that $P(x) = Q(x)A(x) + R(x)$, where the degree of $R(x)$ is less than the degree of $Q(x)$. In other words, $R(x)$ is the remainder of $P(x)$ when divided by $Q(x)$.

   Show that the coefficients of the rational functions $P(x)/Q(x)$ and $R(x)/Q(x)$ have the same exponential growth rate.
2. Use an argument from analysis (no software packages) to show that the root r_0 of smallest modulus of $1 - x^k - x^\ell$, where k and ℓ are relatively prime positive integers, is a positive real number.
3. Use an argument from analysis (no software packages) to show that the root r_0 of smallest modulus of the polynomial $Q_a(x) = 1-x-x^4$ is smaller than that of the polynomial $Q_b(x) = 1 - x^2 - x^3$.
4. Let h_n be the number of surjections from $[n]$ to $[k]$ where k varies from 1 to n. Set $h_0 = 1$. Find an explicit formula for $H(x) = \sum_{n\geq 0} h_n \frac{x^n}{n!}$, then find the exponential growth rate of the coefficients $h_n/n!$ of $H(x)$.
5. Explain what the result of the previous exercise means by comparing the numbers of surjections defined on $[n]$ and the number of bijections on $[n]$.
6. Decreasing *nonplane* 1-2 trees were defined in Section 5.5.2.2. For easy reference, such a tree on vertex set $[n]$ is a tree in which each vertex has at most two children, and each child has a label that is less than the label of its parent. The trees are not plane trees, so there is no *left child* or *right child*. Let T_n be the number of such trees. Note that $T_0 = T_1 = T_2 = 1$, while $T_3 = 2$ and $T_4 = 5$.

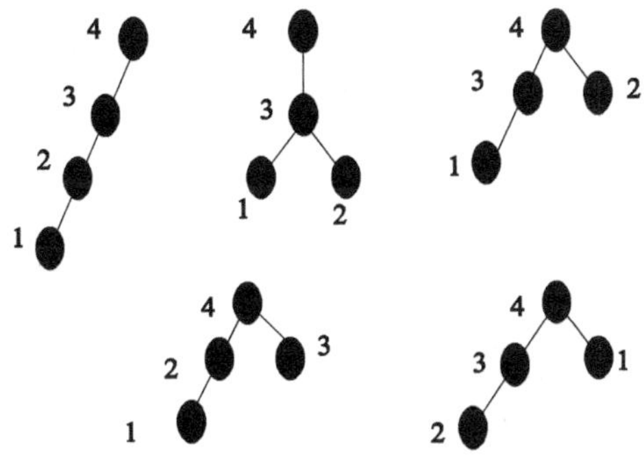

Figure 7.2
The five rooted 1-2 trees on vertex set [4].

See Figure 7.2 for an illustration. Let $T(x) = \sum_{n \geq 0} T_n \frac{x^n}{n!}$ be the exponential generating function of the sequence of the numbers T_n. Find the exponential growth rate of the sequence $T_n/n!$.

7. Recall that the Stirling number of the second kind, $S(n,k)$, denotes the number of partitions of $[n]$ into k blocks. Let $F_k(x) = \sum_{n \geq k} S(n,k)x^k$. In other words, $F_k(x)$ is the generating function of the Stirling numbers $S(n,k)$ with k fixed.

   (a) Prove that, if $k \geq 2$, then
   $$F_k(x) = \frac{x}{1-kx} F_{k-1}(x).$$

   (b) Compute the exponential growth rate of the sequence $\{S(n,k)\}_{n \geq k}$.

8. Let t_n be the number of rooted plane trees on n unlabeled vertices in which each vertex has at most two children, and let $t_0 = 0$. See Figure 7.3 for an illustration. Find the exponential growth rate of

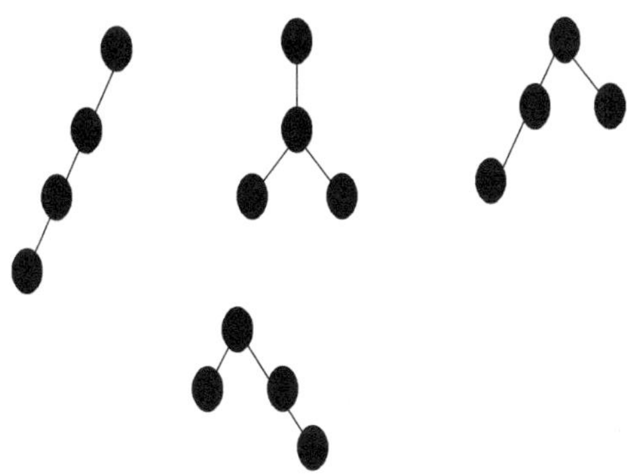

Figure 7.3
The four plane 1-2 trees on four unlabeled vertices.

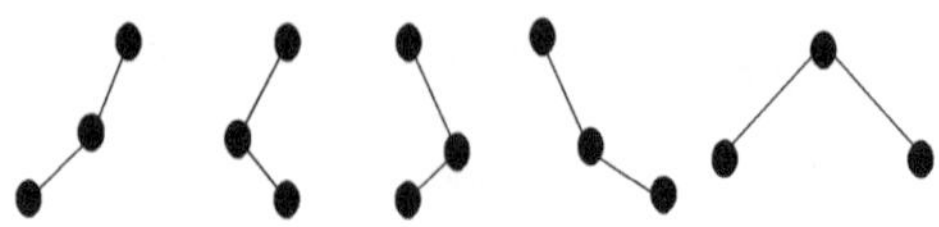

Figure 7.4
The five binary plane trees on three vertices.

the sequence of the numbers t_n. (Note that we have computed the numbers t_n in Exercise 32 of Chapter 5 using the Lagrange Inversion Formula, but do not use that result here; use generating functions instead and singularity analysis instead.)

9. Let T_n be the number of *binary* plane trees on n unlabeled vertices. In such trees, each vertex has at most two children, and a child of a vertex can be a left child or a right child, even if it is an only child. Figure 7.4 shows the five binary plane trees on three vertices. Set $T_0 = 1$. Find the exponential growth rate of the sequence of the numbers T_n. Where have you seen these numbers?

10. Let b_n be the number of rooted plane trees on n unlabeled vertices in which any non-leaf vertex can have any number of children. These trees were formally defined in Definition 5.26. Set $b_0 = 0$. Find the exponential growth rate of the sequence of the numbers b_n.

11. Let m_n be the number of ways to have a group of n people split into nonempty subsets, to have each subset sit down around a circular table, then to arrange the tables in a circle. Find the exponential growth rate of the sequence $m_n/n!$. Two arrangements are considered identical if each person has the same left neighbor in them, and each table has the same left neighbor in them.

12. Let t_n be the number of undirected graphs on vertex set $[n]$ in which each connected component is a cycle of length at least three. Find the exponential generating function for the numbers t_n, then find the exponential growth rate of the sequence $t_n/n!$.

13. Let p_n be the number of undirected graphs on vertex set $[n]$ in which each connected component is a path. Find the exponential growth rate of the sequence $p_n/n!$. (Components that consist of a single vertex are allowed.)

14. A baseball coach asks his n players to split into a few subsets that consist of at least one person each, then he asks that the set of groups stand around a cycle. (There is no structure imposed on each set of players.) Let c_n be the number of ways in which this can happen. Find the exponential order of the sequence $c_n/n!$. Two arrangements are considered the same if the partition of the players is the same in both, and each set of players has the same set as its left neighbor in both.

15. A football coach partitions the set of his n players into blocks of players so that each block consists of an odd number of players. Then he asks each block to form a line. Let L_n be the number of ways in which this can happen. Find the exponential growth rate of the sequence $L_n/n!$. (There is no structure imposed on the set of blocks.)

16. Let us modify the situation described in the previous exercise by imposing a linear order on the blocks of players (so that there is now a first block, a second block, and so on). In other words, we arrange the set of blocks in a line. Let w_n be the number of ways that the combined task can be carried out. Find the exponential growth rate of the sequence $w_n/n!$.

17. A group of football players stand in a line. The coach walks by the line and splits the line into a few segments. (The coach has the option to leave the entire line as one segment.) Then she asks each segment to choose three captains. Let b_n be the number of ways in which this can happen. Find the exponential growth rate of the sequence of the numbers b_n.

18. Let $f(n)$ be the number of words of n letters over the alphabet $\{A, B\}$ that do not contain a subword ABA in consecutive positions. Find the exponential growth rate of the sequence $f(n)$.

19. Consider the undirected cycle C_n on vertex set $[n]$, and let us call a subset S of its vertices a *maximal independent set* if there is no edge between any two vertices in S, but adding any new vertex to S would destroy that property. For $n \geq 3$, let P_n denote the number of maximal independent sets of C_n. Find the exponential growth rate of the sequence P_n.

20. Let A_n be the number of *all* independent sets of any size in the path on vertex set $[n]$. Find the exponential growth rate of the sequence A_n. An independent set is a set of vertices so that there are no edges between vertices of the set.

21. Let $n \geq 3$, and let B_n be the number of all independent sets of any size in the circle C_n on vertex set n. Find the exponential growth rate of the numbers B_n.

22. Generalizing Example 7.12, let $a_{n,k}$ be the number of outcome sequences of n coin flips that do not contain k consecutive heads. Find the closed form of the exponential generating function $A_k(x) = \sum_{n\geq 0} a_{n,k}x^n$, and explain how its singularity of smallest modulus will change if k changes.

23. A *decreasing plane 1-2 tree* on vertex set $[n]$ is a plane tree whose vertices are bijectively labeled by the elements of $[n]$ so that each vertex has a smaller label than its parent. Note that the only difference between these trees and the decreasing binary trees that were

introduced in Definition 5.31 is that, in decreasing plane 1-2 trees, if a child is the only child of its parent, then it does not have a "direction," that is, it is not a left child or a right child. Set $h_0 = 0$, and let h_n be the number of decreasing plane 1-2 trees for $n \geq 1$. Find the exponential growth rate of the sequence $h_n/n!$. You may use a software package to solve any differential equations that you need for your solution.

7.7 Solutions to exercises

1. This is because
$$\frac{P(x)}{Q(x)} = A(x) + \frac{R(x)}{Q(x)},$$
so those two rational functions only differ in a polynomial, that is, their coefficients of x^n only differ for a finite number of exponents n.

2. Let us assume that z is not a positive real number, and $z^k + z^\ell = 1$. Then, by the triangle inequality, we have $|z|^k + |z|^\ell > 1$. Consider the increasing function $f(t) = t^k + t^{ell}$ defined on the positive reals. As $f(|z|) > 1$, and $f(0) = 0$, it follows from the continuity of f that there exists a $t \in (0, |z|)$ for which $f(t) = 1$.

   The triangle inequality is strict, unless z^k and z^ℓ are vectors that point into the same direction, that is, unless $z^{k-\ell}$ is a positive real number. However, as $1 = z^k + z^\ell = (1 + z^{k-\ell})z^\ell$, that would imply that z^ℓ is also a positive real number. That means that both ℓ and $k - \ell$ are multiples of $2\pi/Arg(z)$, a contradiction.

3. It suffices to prove that, for all positive real numbers x, the inequality $x^4 + x > x^3 + x^2$ holds. That inequality holds since it is equivalent to
$$\begin{aligned} x^4 + x - x^3 - x^2 &> 0 \\ (x+1)x(x-1)^2 &> 0, \end{aligned}$$
which obviously holds for all positive real numbers x. Now we know from the previous exercise that the root of smallest modulus of both $Q_a(x)$ and $Q_b(x)$ is a positive real number. Let r_0 be that root of $Q_a(x)$, and let r_1 be that root of $Q_b(x)$. Then $r_0 + r_0^4 = 1$, and it follows from the inequality that we just proved that $r_1^2 + r_1^3 < 1$. As $f(x) = x^2 + x^3$ is obviously an increasing function for positive x, the statement is proved.

4. We use the compositional formula of exponential generating functions, with $A = e^x - 1$ and $B(x) = 1/(1-x)$. Indeed, we split $[n]$ into k blocks, and then there is no task to carry out on these blocks other than making sure that they are nonempty. Then, we order the set of these blocks linearly, and we map the elements of the first block into 1, the elements of the second block into 2, and so on. There are $k!$ ways to linearly order k blocks, showing that $b_k = k!$, and so $B(x) = \sum_{k\geq 0} k! \cdot \frac{x^k}{k!} = 1/(1-x)$.
   Therefore, the compositional formula yields
$$H(x) = B(A(x)) = \frac{1}{1-A(x)} = \frac{1}{1-(e^x-1)} = \frac{1}{2-e^x}.$$
   The function e^x is an entire function; therefore, the only way in which $H(x)$ can have a singularity is by $e^x = 2$, There are infinitely many complex numbers x that satisfy $e^x = 2$, but the one with the smallest modulus is the real number $\ln 2$. Therefore, the exponential order of the numbers h_n is $1/\ln 2 \approx 1.4427$. Note that the numbers h_n are called the *surjection numbers*.
5. As the number of bijections on $[n]$ is $n!$, the ratio $h_n/n!$ is the ratio of all surjections defined on $[n]$ (and mapping to any $[k]$) and the bijections on $[n]$. The result of the previous exercise says that, for large n, there will be about 1.4427^n times as many surjections as bijections.
6. We have seen in Section 5.5.2.2 that $T(x) = \sum_{n\geq 0} T_n \frac{x^n}{n!} = \sec x + \tan x$. As both $\sec x$ and $\tan x$ have their singularities of smallest modulus at distance $\pi/2$ from 0, it follows that the exponential growth rate of the sequence $T_n/n!$ is $2/\pi$.
7. (a) Consider the recurrence relation proved in Theorem 2.10. Multiply both sides by x^n, then sum over all $n \geq k$ to get $F_k(x) = xF_{k-1}(x) + kxF_k(x)$, which is equivalent to our claim.
   (b) Note that $F_1(x) = x/(1-x)$. Then the recurrence we proved in part (a) and induction on k leads to the explicit formula
$$F_k(x) = \frac{x^k}{(1-x)(1-2x)\cdots(1-kx)}.$$
   So the singularity of smallest modulus of $F_k(x)$ is at $x_0 = 1/k$, proving that the exponential growth rate of the sequence of its coefficients is k.
8. Let $t(x)$ be the generating function of the sequence. Removing the root of such a tree, we get either the empty set, or one such tree, or the ordered pair of two such trees. This leads to the functional equation
$$t(x) = x + xt(x) + xt(x)^2.$$

This is a quadratic equation for $t(x)$, whose solutions are

$$t_{1,2}(x) = \frac{1 - x \pm \sqrt{(x-1)^2 - 4x^2}}{2x}.$$

We know that $t(0) = 0$, so we need the solution that satisfies that requirement, that is, the one with the *negative* sign before the square root.

This yields

$$t(x) = \frac{1 - x - \sqrt{1 - 2x - 3x^2}}{2x}.$$

So the singularity of $t(x)$ that is of smallest modulus is at $1/3$, hence the exponential growth rate of the sequence t_n is 3. It may seem at first that $x = 0$ is a singularity of t. However, we have defined $t(0) = 0$, and that definition keeps t analytic at 0. That is the only definition of $t(0)$ that achieves that, since $\lim_{x\to 0} t(x) = 0$. Note that the numbers $t_n = M_{n-1}$, where the numbers M_n are the *Motzkin numbers* that we saw in the solution of Exercise 23 of Chapter 3.

A word on terminology: We say that $x = 0$ is a *removable* singularity of the function $s(x) = \frac{1-x-\sqrt{1-2x-3x^2}}{2x}$ since s becomes analytic at 0 if we define its value at 0 properly. For general s, this is possible if and only if s is analytic in a neighborhood of 0 except at 0 itself, and $\lim_{x\to 0} s(x) = L$ exists and is finite. In that case, setting $s(0) = L$ will make s analytic at 0. For instance, $s(x) = -\ln(1-x)/x$ has a removable singularity at 0, since $\lim_{x\to 0} s(x) = 1$. So we can set $s(0) = 1$.

9. Removing the root of such a tree (which we can do if the tree is not empty), we get an ordered pair of such trees, some or both of which may be empty. This leads to the functional equation

$$T(x) - 1 = xT(x)^2.$$

Solving this quadratic equation for $T(x)$, we get

$$T(x) = \frac{1 \pm \sqrt{1-4x}}{2x}.$$

We know that $T(0) = 1$, so we need the solution that satisfies that requirement, that is, the one with the *negative* sign before the square root. Therefore, $T(x) = \frac{1-\sqrt{1-4x}}{2x}$, and the singularity of the smallest modulus of $T(x)$ is $1/4$, hence the exponential growth rate of the sequence T_n is 4. Note that $x = 0$ is not a singularity, since we have defined $T(0) = 1$, which equals $\lim_{x\to 0} T(x)$. Just as in the previous exercise, the reader should explain why $x = 0$ is not a singular point of T. In other words, 0 is a removable singularity of $\frac{1-\sqrt{1-4x}}{2x}$ (and not a singularity of $T(x)$).

It follows from the closed form of the generating function $T(x)$ that the numbers T_n are the Catalan numbers.

10. We proved in Section 5.5.2.1 that

$$B(x) = \frac{1-\sqrt{1-4x}}{2},$$

so $B(x)$ has only one singularity, at $x = 1/4$. Therefore, the exponential order of the sequence b_n is 4. Note that $b_n = c_{n-1}$, the $(n-1)$st Catalan number.

11. We use the compositional formula of exponential generating functions. The inside function is $\ln(1/(1-x))$ and the outside function is $1 + \ln(1/(1-x))$. Therefore, we get

$$M(x) = 1 + \ln(1 - \ln(1/(1-x))).$$

Clearly, $M(x)$ has a singularity at $x = 1$, and also when $1 = \ln(1 - \ln(1/(1-x))$, which occurs if $1 = \ln(1/(1-x))$, that is, when $1/(1-x) = e$, or $x = 1 - e^{-1}$. This second singularity is closer to 0 than the first, so the exponential growth rate of the sequence of the m_n is $1/(1-e^{-1}) = 1 + \frac{1}{e-1} \approx 1.582$.

12. We will use the exponential formula. There are $(k-1)!$ ways to put a *directed* cycle on k vertices, so there are $(k-1)!/2$ ways to put an undirected cycle on them, for $k \geq 3$. Therefore,

$$\begin{aligned} A(x) &= \sum_{k\geq 3} \frac{(k-1)!}{2} \cdot \frac{x^k}{k!} \\ &= \sum_{k\geq 3} \frac{x^k}{2k} \\ &= \frac{1}{2}\ln\left(\frac{1}{1-x}\right) - \frac{x}{2} - \frac{x^2}{4}. \end{aligned}$$

Therefore, the exponential formula implies that

$$T(x) = \sum_{n\geq 0} t_n \frac{x^n}{n!} = \exp(A(x)) = \frac{e^{-\frac{x}{2}-\frac{x^2}{4}}}{\sqrt{1-x}}.$$

So $T(x)$ has only one singularity, at $x = 1$, and therefore the exponential growth rate of the sequence $t_n/n!$ is 1.

13. This is similar to the previous exercise. For $n \geq 2$, there are $n!/2$ ways to put an undirected path on n vertices, whereas there is one way to put a path on one vertex. This yields

$$A(x) = x + \frac{1}{2}\sum_{n\geq 2} \frac{n!}{2}\frac{x^n}{n!} = x + \frac{1}{2}\frac{1}{1-x} - \frac{x}{2} - \frac{1}{2} = \frac{1}{2(1-x)} + \frac{x-1}{2}.$$

As $A(x)$ has only one singular point, at $x = 1$, so does $P(x) = \sum_{n\geq 0} p_n x^n = \exp(A(x))$, so the sequence of the numbers $p_n/n!$ has exponential growth rate 1.

14. We apply the compositional formula of exponential generating functions. The inside function is $A(x) = \sum_{n\geq 1} \frac{x^n}{n!} = e^x - 1$, since there is no structure on each set of players. As the sets of players are arranged in a cycle, the outside function is $B(x) = 1 + \sum_{n\geq 1} x^n/n = 1 + \ln(1/(1-x))$. Therefore, the compositional formula implies

$$H(x) = \sum_{n\geq 0} h_n \frac{x^n}{n!} = B(A(x)) = 1 + \ln\left(\frac{1}{1-(e^x-1)}\right) = 1 + \ln\left(\frac{1}{2-e^x}\right).$$

So the singularity of smallest modulus is at $r_1 = \ln 2$; therefore, the exponential growth rate of the coefficients $h_n/n!$ is $1/\ln 2 \approx 1.4427$.

15. As there is no structure imposed on the set of blocks, we use the exponential formula. We have $A(x) = \sum_{n\geq 0} \frac{x^{2n+1}}{(2n+1)!}(2n+1)! = x/(1-x^2)$. Therefore, the exponential formula implies

$$L(x) = \sum_{n\geq 0} L_n \frac{x^n}{n!} = \exp(A(x)) = \exp\left(\frac{x}{1-x^2}\right),$$

so $L(x)$ has two singularities, both of modulus 1. Hence the exponential growth rate of the coefficients $L_n/n!$ is 1.

16. Now we have to use the compositional formula for exponential generating functions, with $A(x)$ being the same as in the previous exercise, but $B(x) = \sum_{n\geq 0} n!\frac{x^n}{n!} = 1/(1-x)$. We obtain

$$W(x) = \sum_{n\geq 0} w_n \frac{x^n}{n!} = \frac{1}{1-\frac{x}{1-x^2}} = \frac{1-x^2}{1-x-x^2}.$$

The singularity of smallest modulus is $r_0 = (\sqrt{5}-1)/2$, so we obtain the exponential growth rate of the sequence $w_n/n!$ as $1/r_0 = (\sqrt{5}+1)/2 \approx 1.618$.

17. As the players are standing in a line, we need to use *ordinary* generating functions, namely, Theorem 3.25. The inside function is

$$\begin{aligned} A(x) &= \sum_{k\geq 3} \binom{k}{3} x^k \\ &= x^3 \sum_{k\geq 3} \binom{k}{3} x^{k-3} \\ &= \frac{x^3}{6} \cdot \left(\frac{1}{1-x}\right)''' \\ &= \frac{x^3}{(1-x)^4}. \end{aligned}$$

Hence Theorem 3.25 yields

$$B(x) = \sum_{n\geq 0} b_n x^n = \frac{1}{1-A(x)} = \frac{(1-x)^4}{1-4x+6x^2-5x^3+x^4}.$$

So $B(x)$ is a rational function. Its singularity of smallest modulus is at $r_1 = 0.4503$, so the exponential growth rate of the coefficients b_n is $1/r_1 \approx 2.2207$.

18. We will make repeated use of Theorem 3.25. Let $f(0) = 1$, and let $F(x) = \sum_{n\geq 0} f(n)x^n$. If an allowed word starts with a letter B, then that letter B can be followed by any allowed word. Such words are counted by the generating function $xF(x)$. Otherwise, let us look for the first letter B in our word w. This letter B is immediately preceded by a string of letters A, so it cannot be immediately followed by a letter A. Therefore, after the first letter B, the word w either ends or continues with another letter B, and then with any allowed word. In the first case, w is counted by the generating function $\sum_{n\geq 2} x^n = x^2/(1-x)$ since w is just a nonempty sequence of letters A, and then a single B. In the second case, w is counted by the generating function $\frac{x^2}{1-x} \cdot x \cdot F(x)$. Finally, w can be the empty word, which has generating function 1, or the word that consists of letters A only, which has generating function $x/(1-x)$.

This leads to the functional equation

$$F(x) = xF(x) + \frac{x^2}{1-x} + \frac{x^2}{1-x} \cdot x \cdot F(x) + 1 + \frac{x}{1-x}.$$

Solving this equation, we get

$$F(x) = \frac{1+x^2}{1-2x+x^2-x^3}.$$

The singularity of $F(x)$ that has smallest modulus is $r_1 \approx 0.5698$, so the exponential growth rate of the sequence $f(n)$ is $1/r_1 \approx 1.755$.

19. We claim that $P(n) = P(n-3) + P(n-2)$ if $n \geq 6$, with $P(3) = 3$, $P(4) = 2$, and $P(5) = 5$. In order to prove this claim, let $n \geq 6$. Let i be the smallest vertex in S, and let j be the second smallest vertex in S. Then either $j - i = 2$ or $j - i = 3$. In either case, contract the entire path $i, i+1, \cdots j$ into one vertex, call that vertex i, and decrease the label of all vertices larger than j by $j - i$. Form a new set S' from the new vertex i and the images of the other vertices that were in S. Then S' is a maximal independent set in the obtained C_{n-2} or C_{n-3}. Furthermore, this construction is bijective, since, starting with a maximal independent set T of a C_{n-2} or C_{n-3}, we can find its preimage by finding the smallest i vertex in T, and then replacing it by a path of three or four vertices, the two endpoints of which are in the preimage of T.

    In order to find the generating function $P(x) = \sum_{n\geq 0} P(n)x^n$, we may set $P(2) = 2$, $P(1) = 0$, and $P(0) = 3$ to keep the recurrence true for all values of $n \geq 3$. Set $P(x) = \sum_{n\geq 0} P(n)x^n$. Then the recurrence leads to the functional equation

$$P(x) - 2x^2 - 3 = x^3 P(x) + x^2(P(x) - 3),$$

$$P(x) = \frac{-x^2 + 3}{1 - x^2 - x^3}.$$

    Finding the singularity of smallest modulus of this rational function, we conclude that the exponential growth rate of the numbers $P(n)$ is 1.3247. The numbers $P(n)$ are called the *Perrin numbers*.

20. Let $A(x) = \sum_{n\geq 0} A_n x^n$. Clearly, we have $A_0 = 1$, $A_1 = 2$, (note that the empty set is independent as well), and $A_n = A_{n-1} + A_{n-2}$ for $n \geq 2$, since there are A_{n-1} such sets not containing n, and A_{n-2} such sets containing n. This leads to the functional equation

$$A(x) - 2x - 1 = x(A(x) - 1) + x^2 A(x),$$

$$A(x) = \frac{x+1}{1 - x - x^2},$$

    hence the exponential growth rate of the sequence A_n is $(\sqrt{5}+1)/2$. Note that the number A_n is the $(n+1)$st Fibonacci number.

21. Independent sets S of the path on vertex set $[n]$ will be independent sets in C_n if and only if they do not contain *both* 1 and n. If S does contain both 1 and n, then the rest of S is an independent set in the path from vertex 3 to vertex $n-2$, so in a labeled path of $n-4$ vertices, if $n \geq 5$. So for $n \geq 5$, we have $B_n = A_n - A_{n-4}$. Setting

$B_0 = B_1 = B_2 = 0$, and noting that $B_3 = 4$ and $B_4 = 7$, we have (with $B(x) = \sum_{n\geq 0} B_n x^n$),

$$B(x) = A(x)\left(1 - x^4\right) - 1 - 2x - 3x^2 - x^3,$$

so $B(x)$ has the same set of singularities as $A(x)$, hence the sequence B_n has the same exponential growth rate as the sequence A_n, and we have computed that one in the solution of the previous exercise.

Note that the numbers B_n are called the *Lucas numbers* and they satisfy the recurrence relation $B_n + B_{n+1} = B_{n+2}$ for $n \geq 3$, with the initial conditions mentioned above.

22. Analogously to the solution of Example 7.12, we get the functional equation

$$A_k(x) = A_k(x)x\left(1 + x + \cdots + x^{k-1}\right) + \left(1 + x + \cdots + x^{k-1}\right).$$

This yields the closed formula

$$A_k(x) = \frac{\left(1 + x + \cdots + x^{k-1}\right)}{1 - x - x^2 - \cdots - x^k}.$$

That is, the singularity of smallest modulus of $A_k(x)$ is the root of $f_k(x) = x + \cdots + x^{k-1} - 1$ that is closest to 0. It is easy to prove, and you are asked to do so in Supplementary Exercise 4 in more general circumstances, that $f_k(x)$ has a unique such root, and that root is a positive real number r_k. It follows that $r_{k+1} < r_k$, since f_{k+1} has an additional summand over f_k, namely, x^k. Therefore, the exponential growth rate of the sequence $a_{n,k+1}$ is larger than that of the sequence $a_{n,k}$. This is what we expected, since it is easier to avoid a sequence of $k + 1$ consecutive heads than it is to avoid k consecutive heads.

23. The procedure is quite similar to the one that we applied in Section 5.5.2.2 to find the number of *nonplane* decreasing 1-2 trees on vertex set $[n]$. Let $H(x) = \sum_{n\geq 0} h_n x^n/n!$. If $n \geq 2$, then removing the root of a plane decreasing 1-2 tree in vertex set $[n]$, we either get an *ordered* sequence of two trees with combined vertex set $[n-1]$, or just one tree with vertex set $[n-1]$. If $n = 1$, then we get the empty set. This leads to the differential equation

$$H'(x) = H(x)^2 + H(x) + 1.$$

Using the initial condition $H(0) = 0$, we can solve this differential equation with our favorite software package. After simplification, we get the answer

$$H(x) = -\frac{1}{2} + \frac{\sqrt{3}}{2} \cdot \tan\left(\frac{\pi}{6} + \frac{\sqrt{3}}{2}x\right).$$

Therefore, the candidates for the singularities of smallest modulus of $H(x)$ are the points for which

$$\frac{\pi}{6} + \frac{\sqrt{3}}{2}x = \pm\pi/2.$$

If the right-hand side is $\pi/2$, we get $x = \frac{2\cdot\sqrt{3}}{9}\pi$, whereas if the right-hand side is $-\pi/2$, we get $x = -\frac{4\cdot\sqrt{3}}{9}\pi$. So the former is the singularity of smallest modulus and therefore, the sequence $h_n/n!$ has exponential growth rate

$$\frac{1}{\frac{2\cdot\sqrt{3}}{9}\pi} = \frac{3\sqrt{3}}{2\pi} \approx 0.827.$$

Note that this growth rate is about halfway between that of the corresponding growth rate $2/\pi$ for *decreasing nonplane 1-2 trees* that we computed in Exercise 6 and the corresponding growth rate 1 for decreasing binary trees that follows from the exact enumeration formula we got for such trees in Proposition 5.32.

7.8 Supplementary exercises

1. Prove that all rational functions are analytic in all points where they are defined.
2. Prove Proposition 7.6.
3. Let $Q(x)$ be a polynomial with real coefficients so that $Q(x)$ has a unique root r of smallest modulus. Prove that r is a real number.
4. Let $P(x) = \left(\sum_{i=1}^{n} a_i x^i\right) - m$ be a polynomial such that all a_i are positive real numbers ($1 \leq i \leq n$), and m is also a positive real number. Prove that $P(x)$ has a unique root of smallest modulus, and that root is a *positive real* number.
5. Let f be a power series with nonnegative real coefficients. Let us assume that there exists a complex number z of modulus r so that the sum $f(z)$ is divergent. Prove that then $f(r)$ is divergent.
6. Let $F(x) = \sum_{n\geq 0} f_n x^n = \frac{p(x)}{(1-x)^2}$, where $p(x)$ is a polynomial so that $p(1) \neq 0$. Prove that $\lim_{n\to\infty}(f_n/(n+1)) = p(1)$.
7. Will the claim of the previous exercise remain true if we drop the condition that $p(1) \neq 0$?
8. Let f_n be the number of compositions of the integer n that do not contain parts equal to 2 or 3. Find the exponential growth rate of the sequence f_n.

9. Let $f(n)$ be the number of ways to pay n dollars using only 1-, 2-, 5-, 10-, 20-, 50-, and 100-dollar bills. Find the asymptotics of the sequence $f(n)$ with polynomial precision.

10. Let $f(n)$ be defined as in the previous exercise. Find a positive real number α, a positive constant K, and a positive integer k so that

$$f(n) \sim K \cdot \frac{\alpha^n}{n^k}.$$

11. Let $A(x) = \sum_{n \geq 0} a_n x^n$ be a power series so that, for all $n \geq 0$, the equality

$$\sum_{k=0}^{n} a_k a_{n-k} = 5^n$$

holds. Find the exponential growth rate of the sequence a_n.

12. Let c_n be the number of surjections from $[n]$ to $[k]$ where k varies from 1 to $\lfloor n/2 \rfloor$ in which each element of $[k]$ has a preimage of size at least two. Set $c_0 = 1$. Find an explicit formula for $C(x) = \sum_{n \geq 0} c_n \frac{x^n}{n!}$, then find the exponential growth rate of the coefficients of $C(x)$.

13. Let $k \geq 2$ be a fixed positive integer, and let $c_k(n)$ be the number of permutations of length n in which each cycle length is divisible by k. Find the exponential growth rate of the sequence $c_k(n)$.

14. Let r_n be the number of rooted trees on vertex set $[n]$ in which every non-leaf vertex has exactly two children. Find the exponential order of the numbers $r_n/n!$.

15. Let t_n be the number of rooted trees on vertex set $[n]$ in which every non-leaf vertex has at most two children. Find the exponential order of the numbers $t_n/n!$.

16. Let H_n be the number of permutations of length n that do not contain a 2-cycle. Find an explicit formula for $H(x) = \sum_{n \geq 0} H_n \frac{x^n}{n!}$, then find $\lim_{n \to \infty} H_n/n!$.

17. Let K_n be the number of permutations of length n in which every cycle length is even. Find the exponential growth rate of the sequence $K_n/n!$.

18. Let $k \geq 2$ be a fixed positive integer. Find the exponential growth rate for the numbers $a_k(n)$ of all compositions of n into parts that are congruent to 1 modulo k.

19. A group of n tourists arrives at the starting point of a ride at an amusement park. Each car in the ride can seat either one or two people. Let $f(n)$ be the number of ways in which tourists go on this ride if they are to go in a string of consecutive cars. (The order of two people in the same car is insignificant.) Find the exponential order of the numbers $f(n)/n!$.

20. A group of n tourists arrives at a restaurant. They sit down around circular tables in groups of three or four people, after which the only server at the restaurant will visit all tables in some order. Let $g(n)$ be the number of ways all this can happen. Find the exponential order of the numbers $g(n)/n!$.

21. A group of n tourists arrives at a restaurant. They sit down around circular tables in groups of size of at least one and at most three. The circular tables themselves are arranged in a circle. Let $g(n)$ be the number of ways in which this can happen. Find the exponential order of the numbers $g(n)/n!$.

22. Let $B(n)$ denote the number of all partitions of the n-element set. Prove that, for any fixed real number k, the exponential growth rate of the numbers $B(n)$ (the Bell numbers) is larger than k.

23. Recall from Section 2.3 that the number of partitions of the integer n is denoted by $p(n)$, and that

$$\sum_{n\geq 0} p(n)x^n = \prod_{i\geq 1} \frac{1}{1-x^i}$$

holds, where we set $p(0) = 1$.

   (a) Use the methods learned in this chapter to determine the exponential growth rate of the sequence $\{p(n)\}_{n\geq 0}$.

   (b) Use the methods learned in this chapter to prove that $p(n)$ grows faster than any polynomial P, that is, prove that $\lim_{n\to\infty} P(n)/p(n) = 0$.

24. Let f_n be the number of ways to partition the set $[n]$ into two nonempty intervals $[1, i]$ and $[i+1, n]$, then to partition the first interval into two nonempty blocks, and to partition the second interval into three nonempty blocks. Find the growth rate of the sequence f_n at polynomial precision.

25. Let p_n be the probability that a randomly selected permutation of length n does not have a fixed point. Let f_n be the probability that a randomly selected function from $[n]$ to $[n]$ has exactly one fixed point. Is it true that

$$\lim_{n\to\infty} p_n = \lim_{n\to\infty} f_n?$$

Part IV

Special Topics

8

Symmetric structures

8.1 Designs

In Chapter 6, we considered families of sets from the perspective of extremal combinatorics. Keeping the tradition of that field, we called these families of sets hypergraphs, alluding to the fact that they were generalizations of graphs.

In the present chapter, we will consider families of sets again, but we will be interested in their *symmetries* instead. Following the tradition of this line of work, in this chapter, we will call our families of sets *designs.*

That is, a *block design* B is a finite set of elements V, called the *vertices* of the design, together with a collection E of nonempty subsets of V, called the *blocks.* We will keep a few words from the terminology of the previous chapter. In particular, we call a block design B *uniform* or k-uniform if each block of B contains the same number k of vertices.

In real life, we often need more symmetry than that in a block design. For instance, if a soccer coach wants to try out his v new offensive players in various k-striker formations, he may want to give equal amounts of playing time to each of his players. In that case, he would need a *regular*, or r*-regular* block design, that is, a design in which each vertex appears in the same number r of blocks. It goes without saying that the vertices are the players and the blocks are the formations.

The coach may be even more sensitive to fairness, and may want to make sure that any *pair* of strikers plays together in the same number λ of trials. If that happens, then we say that the design the coach uses is *balanced.*

If a design is balanced, uniform, and regular, then it is called a *block design.* Finally, if $k < v$, so the blocks of a block design do not contain all vertices, then that design is called a *balanced incomplete block design*, or BIBD.

This can certainly all be achieved if the coach tries out all $\binom{v}{k}$ possible formations, but that may not end before the season is over. It is therefore interesting to look for designs with a smaller number of blocks that still satisfy all these criteria.

In what follows, we will take a look at designs with one or more of the symmetric properties described above. Let us set the following notation for the rest of this section.

DOI: 10.1201/9781003304272-8

In a design $\mathcal{F}$,

1. the number of vertices will be denoted by v,
2. the number of blocks will be denoted by b,
3. if $\mathcal{F}$ is uniform, then the number of vertices of any (equivalently, each) of its blocks will be denoted by k,
4. if $\mathcal{F}$ is regular, then the number of blocks in which any (equivalently, each) of its vertices appears will be denoted by r, and
5. if $\mathcal{F}$ is balanced, then the number of blocks in which any (equivalently, each) pair of its vertices appear together will be denoted by λ.

We often mention the parameters of a BIBD by saying that it is a (b, v, r, k, λ)-BIBD.

For the rest of this chapter, we will assume that our designs have more than one block, otherwise a design on $[n]$ could consist of one block that is $[n]$ itself. It turns out that this design would constitute a unique counterexample for many theorems. Recall from the previous chapter that we do *not* allow designs to have repeated blocks.

Example 8.1 *The design $\mathcal{F}$ whose blocks are the k-element subsets of $[n]$ is regular, uniform, and balanced.*

Solution: Each vertex occurs in $r = \binom{n-1}{k-1}$ blocks, so $\mathcal{F}$ is regular. Each block contains k vertices, so $\mathcal{F}$ is uniform. Finally, any two vertices appear together in $\lambda = \binom{n-2}{k-2}$ blocks, so $\mathcal{F}$ is balanced. ◊

It is not surprising that this many symmetries seriously restrict the structure of a design. The following two propositions start exploring the restrictions on the five parameters themselves.

Proposition 8.2 *Let $\mathcal{F}$ be a uniform regular design. Then*

$$bk = vr.$$

Proof: Let us count all pairs (a, B) so that a is a vertex of the block B. If we count by the blocks, then this number is bk, since each of the b blocks contains k vertices. If we count by the vertices, then this number is vr, since each of the v vertices is contained in r blocks. ◊

We can verify Proposition 8.2 for the design $\mathcal{F}$ of Example 8.1. In that example, $b = \binom{n}{k}$, $v = n$, and the other parameters are computed in the proof of that example. Then we get the equation

$$\binom{n}{k} \cdot k = n \cdot \binom{n-1}{k-1},$$

which is indeed correct.

Proposition 8.3 *Let $\mathcal{F}$ be a* balanced *uniform regular design. Then*

$$\lambda(v-1) = r(k-1).$$

Proof: Let a be any fixed vertex of $\mathcal{F}$. Let us count all pairs (B, c), where a and c are two vertices of the block B. If we count by the vertices c, then, for each of the $v-1$ choices for c, there are λ blocks containing both a and c, which explains the left-hand side. If we count by the blocks B, then each of the r blocks containing a contains $k-1$ other vertices that can play the role of c. This explains the right-hand side. $\diamondsuit$

Verifying this proposition again for $\mathcal{F}$ of Example 8.1, we get the equality

$$\binom{n-2}{k-2}(n-1) = \binom{n-1}{k-1}(k-1).$$

It is important to point out that Propositions 8.2 and 8.3 only establish *necessary* conditions on the existence of certain designs, not *sufficient* conditions. For instance, there is no balanced uniform regular design with $v = 16$, $b = 8$, $k = 6$, $r = 3$, and $\lambda = 1$, even if these numbers satisfy the equalities of the mentioned two propositions. Theorem 8.12 will show why there is no such design.

At this point, the reader may hope for a theorem that establishes a sufficient and necessary condition for the existence of a design with parameters v, b, r, k, and λ. Unfortunately, no such theorem is known. In fact, there are quite a few 5-tuples of parameters (v, b, r, k, λ) when it is not known whether a design with those parameters (or, for shortness, a (v, b, r, k, λ)-design) exists. (The simple fact that the parameters r, k, and λ are present assumes that we are considering balanced uniform regular designs.) in a bit more detail in the next section. Until then, we challenge the reader to decide whether a $(7, 7, 3, 3, 1)$-design exists. In the language of our original example, We will discuss these difficult problems we are asking if a soccer coach with seven strikers can try out seven attacking formations so that each formation consists of three strikers, each striker gets three chances to play, and any pair of players gets to play together exactly once.

We have considered three different properties of designs so far, that is, we considered balanced, uniform, and regular designs. These are all strong properties, which can seriously restrict the structure of a design. Exercise 1 shows that, if a design is balanced and uniform, then it is also regular. After learning this fact, the reader might suspect that a design that is balanced and regular is necessarily uniform (since being regular means roughly the same thing for the vertices as being uniform means for the blocks). This conjecture is false, however, as the following example shows.

Example 8.4 *Let $\mathcal{F}$ be the design on $[n]$ whose blocks are the 2-element and 3-element subsets of $[n]$. Then $\mathcal{F}$ is regular (with $r = n-1+\binom{n-1}{2} = \binom{n}{2}$), and balanced (with $\lambda = 1+n-2 = n-1$), but not uniform.*

At this point, the reader may wonder what causes this subtle difference between uniformity and regularity, which seem to be duals of each other. The answer lies in the nature of the *balanced* property. That property ensures that any two vertices appear together in the same number of blocks. However, we have not defined the dual of this property yet, that is, we have not discussed designs in which the intersection of any pair of blocks has the same size. The following definition fills that gap.

Definition 8.5 *A design is called* linked *if any two of its blocks intersect in the same number μ of vertices.*

Example 8.6 *The design $\mathcal{F}$ of Example 8.1 is linked when $k = 1$ (and in that case, $\mu = 0$) and when $k = n-1$ (and in that case, $\mu = n-2$). For other values of k, the mentioned design is not linked.*

We can now state the counterpart of Exercise 1.

Proposition 8.7 *Let $\mathcal{F}$ be a design that is regular and linked. Then $\mathcal{F}$ is uniform.*

Hopefully, the reader is now thinking that maybe we will not need to work out an independent proof of this proposition; maybe we will be able to deduce it from the result of Exercise 1 more or less effortlessly. If that is the case, we have good news for the reader. There exists a general technique that is very useful in translating statements into their *duals.* (We will finally be able to explain precisely what that word means.)

The crucial definition is the following.

Definition 8.8 *Let $\mathcal{F}$ be a design with vertices $a_1, a_2, \cdots, a_v$ and blocks $e_1, e_2, \cdots, e_b$. Then the* incidence matrix *of $\mathcal{F}$ is the $v \times b$ matrix $M = M_{\mathcal{F}}$ for which*

$$M_{i,j} = \begin{cases} 1 \text{ if } a_i \in e_j, \\ 0 \text{ if } a_i \notin e_j. \end{cases}$$

In other words, the *rows* of $M_{\mathcal{F}}$ correspond to the vertices of $\mathcal{F}$, and the columns of $M_{\mathcal{F}}$ correspond to the blocks of $\mathcal{F}$. The intersection of a row and a column is 1 if the vertex corresponding to the row is contained in the block corresponding to the column.

Example 8.9 *If $\mathcal{F}$ is the design whose vertex set is $[4]$ and whose blocks are $e_1 = \{1,2\}$, $e_2 = \{1,3\}$, $e_3 = \{2,4\}$, $e_4 = \{2,3,4\}$, and $e_5 = \{1,4\}$, then*

$$M_{\mathcal{F}} = \begin{pmatrix} 1 & 1 & 0 & 0 & 1 \\ 1 & 0 & 1 & 1 & 0 \\ 0 & 1 & 0 & 1 & 0 \\ 0 & 0 & 1 & 1 & 1 \end{pmatrix}.$$

One great advantage of incidence matrices is that now we can easily define the *dual* of a design.

Definition 8.10 *The* dual $\mathcal{F}^d$ *of a design* $\mathcal{F}$ *is the design whose incidence matrix is the* transpose *of* $M_{\mathcal{F}}$.

If we had stated this definition before Chapter 5, the reader might have asked, What do you mean the dual of $\mathcal{F}$ is *the* design with incidence matrix $M_{\mathcal{F}}^T$? What if there are several designs with the same incidence matrix? By now, however, the reader easily dismisses that concern, noting that all designs with the same incidence matrix are *isomorphic.*

We have not formally defined isomorphism of designs yet, but the definition is very similar to that of graph isomorphism. That is, let $\mathcal{F}$ and $\mathcal{G}$ be two designs. If there exists a bijection f from the vertex set of $\mathcal{F}$ onto the vertex set of $\mathcal{G}$ that takes blocks of $\mathcal{F}$ into blocks of $\mathcal{G}$, then we say that f is an isomorphism and that $\mathcal{F}$ and $\mathcal{G}$ are isomorphic.

Example 8.11 *In order to construct the dual* $\mathcal{F}^d$ *of the design* $\mathcal{F}$ *given in Example 8.9, we first take the transpose of* $M_{\mathcal{F}}$ *and get the matrix*

$$M_{\mathcal{F}}^T = \begin{pmatrix} 1 & 1 & 0 & 0 \\ 1 & 0 & 1 & 0 \\ 0 & 1 & 0 & 1 \\ 0 & 1 & 1 & 1 \\ 1 & 0 & 1 & 1 \end{pmatrix}.$$

Then we read the column j *of this matrix to see which vertices are contained in* e_j. *We get* $e_1 = \{1,2,5\}$, $e_2 = \{1,3,4\}$, $e_3 = \{2,4,5\}$, *and* $e_4 = \{3,4,5\}$.

It is straightforward to translate the discussed properties of designs into the language of incidence matrices. For instance, $\mathcal{F}$ is regular if all rows of $M_{\mathcal{F}}$ contain the same number r of 1s; $\mathcal{F}$ is uniform if all columns of $M_{\mathcal{F}}$ contain the same number k of

1s; and so on. Therefore, $\mathcal{F}$ is regular if and only if $\mathcal{F}^d$ is uniform, while $\mathcal{F}$ is balanced if and only if $\mathcal{F}^d$ is linked, and so on. Proving Proposition 8.7 is now a breeze.

Proof: (of Proposition 8.7) If $\mathcal{F}$ is regular and linked, then $\mathcal{F}^d$ is uniform and balanced, and therefore, by Exercise 1, $\mathcal{F}^d$ is regular. Therefore, the dual of $\mathcal{F}^d$, that is, $\mathcal{F}$, is uniform. $\Diamond$

The concept of incidence matrices is useful in proving much more difficult results than Proposition 8.7, as we will see shortly.

We have proved two results (Propositions 8.2 and 8.3) stating equalities involving products of parameters of designs, and Supplementary Exercise 1 provides another such result. What about inequalities, though? Is it true that some parameters are always larger than others? There are, of course, the trivial inequalities $k \leq v$ in each uniform design, and $r \leq b$ in each regular design.

If $\mathcal{F}$ is balanced, then $\lambda \leq r$, since the total number of blocks containing a vertex x is at least as large as the number of blocks in which x occurs together with another vertex y. In fact, $\lambda = r$ is only possible if all vertices occur in all blocks, a case that is really not very exciting.

The following is a much deeper inequality among parameters of a design.

Theorem 8.12 **(Fisher's inequality)** *Let $\mathcal{F}$ be a balanced uniform design with at least two blocks. Then $v \leq b$.*

Proof: Let M be the incidence matrix of $\mathcal{F}$, and assume that $b < v$. Then M has more rows then columns. Add $v - b$ zero columns to the end of M so that M becomes the $v \times v$ matrix A. Then we see that $MM^T = AA^T$, since the recently added zeros at the end of each row will not change anything. Therefore,

$$\det(MM^T) = \det(AA^T) = \det(A) \cdot \det(A^T) = 0, \tag{8.1}$$

since A has at least one zero column.

On the other hand, the solution of Exercise 3 shows that

$$MM^T = \begin{pmatrix} r & \lambda & \lambda & \cdots & \lambda \\ \lambda & r & \lambda & \cdots & \lambda \\ \lambda & \lambda & r & \cdots & \lambda \\ \cdots & \cdots & \cdots & \cdots & \cdots \\ \lambda & \lambda & \cdots & \cdots & r \end{pmatrix}.$$

Before the reader asks what r is, let us point out that Exercise 1 shows that a balanced uniform design is *regular*, so r is the number of blocks in which each vertex is contained.

Let us now compute $\det(MM^T)$ by transforming this matrix into triangular form, using elementary row and column operations. First, subtract the first column from all other columns. This does not change the determinant of the matrix, yielding that

$$\det(MM^T) = \det \begin{pmatrix} r & \lambda - r & \lambda - r & \cdots & \lambda - r \\ \lambda & r - \lambda & 0 & \cdots & 0 \\ \lambda & 0 & r - \lambda & \cdots & 0 \\ \cdots & \cdots & \cdots & \cdots & \cdots \\ \lambda & 0 & 0 & \cdots & r - \lambda \end{pmatrix}.$$

In other words, we will know $\det MM^T$ if we can compute the determinant above. That is encouraging, since the matrix above is almost lower triangular; indeed, its only nonzero entries above the main diagonal are in the first row.

Let us note that the sum of each column, except for the first column, is zero. Therefore, if we add each row to the first, each element of the first row will become zero except for its first element. As this operation again does not change the determinant of the matrix, we get

$$\begin{aligned}\det(MM^T) &= \det\begin{pmatrix} r+(v-1)\lambda & 0 & 0 & \cdots & 0 \\ \lambda & r-\lambda & 0 & \cdots & 0 \\ \lambda & 0 & r-\lambda & \cdots & 0 \\ \cdots & \cdots & \cdots & \cdots & \cdots \\ \lambda & 0 & 0 & \cdots & r-\lambda \end{pmatrix} \\ &= (r+(v-1)\lambda)(r-\lambda)^{v-1} \\ &\neq 0.\end{aligned}$$

Indeed, the determinant of a triangular matrix is obtained by taking the product of its diagonal entries. In order to see that this product is not zero, note that $r > \lambda$ since $\mathcal{F}$ has at least two blocks. So we get that $\det(MM^T) \neq 0$, contradicting (8.1). That is, the assumption that $b < v$ leads to a contradiction, proving that $v \leq b$. $\diamondsuit$

An easy consequence of Fisher's inequality is the fact that the dual of a BIBD cannot be a BIBD unless $v = b$, and therefore; $r = k$. Such BIBDs are called *symmetric*, and it turns out that the dual of a symmetric BIBD is indeed a BIBD. You are asked to prove this fact in Exercise 5.

Quick Check

1. Let $\mathcal{D}$ be a BIBD with parameters (b, v, r, k, λ). What relation holds between r and k?
2. Let $\mathcal{D}$ be a BIBD in which $v = 12$, $r = 4$, and $\lambda = 1$. Determine the value of b and k.
3. Does there exist a BIBD with parameters $v = 8$, $b = 70$, and $r = 35$?

8.2 Finite projective planes

In this section, we are going to discuss *finite projective planes*. This name suggests that the objects at hand are geometrical. There is some truth to that suggestion; indeed, as we will see, finite projective planes consist of lines and points. However, in contrast to Euclidean geometry, any two lines will intersect. There will be no parallel lines. On the other hand, we will see that finite projective planes are in fact designs with many of the symmetric properties we studied in the previous section.

Definition 8.13 *A finite projective plane is a collection of a finite set $\mathcal{P}$ of* *points and a finite set $\mathcal{L}$ of lines, where the lines themselves are subsets of $\mathcal{P}$, satisfying the following axioms:*

- *Any two points are in exactly one common line.*
- *Any two lines intersect in exactly one point.*
- *The set $\mathcal{P}$ contains four distinct points, no three of which are on a common line.*

The following classic example is called the *Fano plane*. Soon, we will be able to prove that it is actually the *smallest* finite projective plane.

Example 8.14 *Figure 8.1 shows a finite projective plane in which $\mathcal{P} = [7]$ and the lines are $\{1, 2, 4\}$, $\{2, 3, 5\}$, $\{1, 3, 6\}$, $\{1, 5, 7\}$, $\{2, 6, 7\}$ $\{3, 4, 7\}$, and $\{4, 5, 6\}$.*

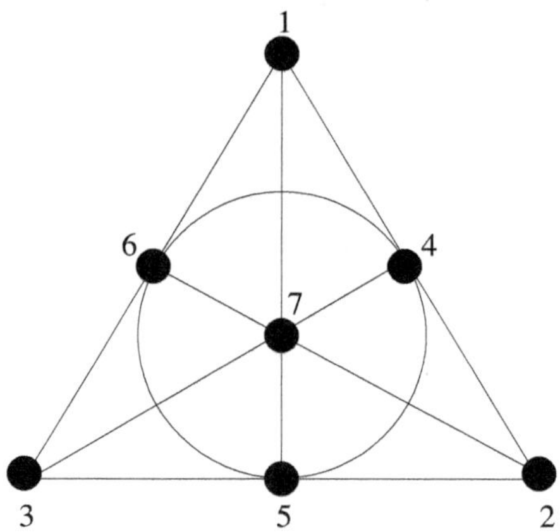

Figure 8.1
The Fano plane.

This example shows that, if we want to represent a finite projective plane by a figure, we can make the task easier if we do not use *straight* lines to represent the lines.

Definition 8.13 shows that a finite projective plane is in fact a *balanced linked design*, in which the points are the vertices, the lines are the blocks, and $\lambda = \mu = 1$. In what follows, we show that this design is uniform and regular as well.

Lemma 8.15 *In a finite projective plane, all points are contained in the same number of lines.*

Proof: Let s and t be two points, and let L be a line that does not contain either one of them. Such an L must exist because of the third axiom in the definition of finite projective planes.

Say L consists of $n+1$ points. Then there is a bijection f between the set of these points and the lines containing s. Indeed, each line S containing s must intersect L in a point a_S. Then the map defined by $f(S) = a_S$ is a bijection with the mentioned properties. Therefore, s is contained in $n+1$ lines. See Figure 8.2 for an illustration.

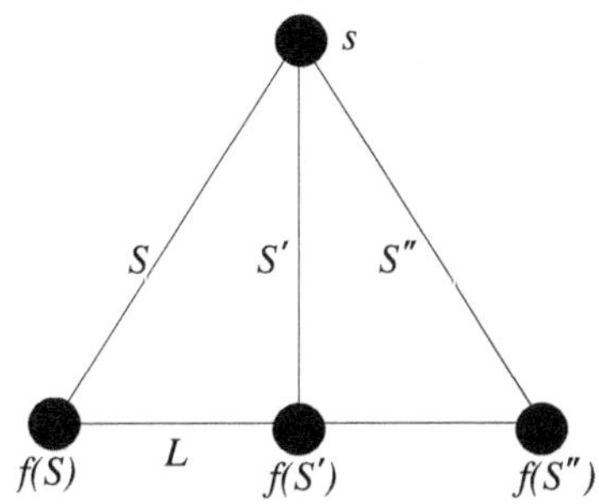

Figure 8.2
The bijection f maps the lines containing s into points of L.

An analogous argument (in which s is replaced by t) shows that t is contained in $n+1$ lines as well. Iterating this argument, we see that any point is contained in the same number $n+1$ of lines as the point s, proving our claim. ◊

Corollary 8.16 *In a finite projective plane, all lines consist of the same number of points. This number $n+1$ is equal to the number of lines containing each point.*

Proof: This follows from the existence of the bijection f just constructed in the proof of Lemma 8.15. ◊

Therefore, each finite projective plane is a regular uniform design, with $k = r = n+1$. By Proposition 8.2, this implies that $v = b$, that is, each finite projective plane has as many points as lines. But *how many*? It is not difficult to answer that question.

Proposition 8.17 *If the lines of a finite projective plane $\mathcal{H}$ consist of $n+1$ points each, then $\mathcal{H}$ has n^2+n+1 points and n^2+n+1 lines.*

Proof: Consider all lines containing point h of $\mathcal{H}$. By Lemma 8.15, there are $n+1$ such lines. By Corollary 8.16, each of these lines contains n points other than h. By the first and second axioms, any point other than h lies in exactly one of these lines. Therefore, these lines together contain $(n+1)n$ points other than h. See Figure 8.3 for an illustration. ◇

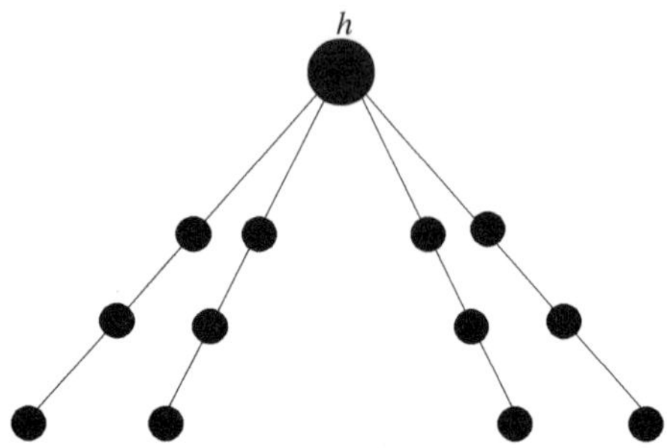

Figure 8.3
Removing h from $\mathcal{H}$, we get $n+1$ disjoint lines with n points in each line.

Consequently, each finite projective plane is an $(n^2+n+1, n^2+n+1, n+1, n+1, 1)$-design for some n. The number n is called the *order* of the projective plane. Then $n = 1$ is not possible, since that would mean that the projective plane has only three points, in violation of the third axiom. So the Fano plane, which corresponds to the special case of $n = 2$, is indeed minimal among all finite projective planes.

It turns out that the opposite is also true, that is, a $(n^2+n+1, n^2+n+1, n+1, n+1, 1)$-design is always a finite projective plane. See Exercise 6 for an explanation.

Let us point out again that it is *not* true that there exists a finite projective plane of order n for all positive integers $n > 1$. In fact, it is one of the most long-standing open problems of combinatorics to characterize the values of n for which a finite projective plane of order n exists. The most general, and best-known, result is that, if $n = p^t$, where p is a prime and t is a positive integer, then a finite projective plane of order n exists.

8.2.1 Finite projective planes of prime power order

If the reader took a course in abstract algebra, the reader probably knows what a *field* is and that all finite fields have $n = p^t$, where p is a prime, and t is a positive integer. It is also known that, if $n = p^t$, then a finite field of n elements exists. We will now build a finite projective plane $\mathcal{H}$ of order $n = p^t$ using these fields.

Let F be a finite field of n elements. Let us look at ordered triples (x_1, x_2, x_3), where $x_i \in F$ for $i \in [3]$ and not all x_i are equal to zero. There are $n^3 - 1$ such triples. Let us say that two such ordered triples are *similar* if one is a constant multiple of the other, that is, (x_1, x_2, x_3) and (y_1, y_2, y_3) are similar if there exists a $\lambda \in F - \{0\}$ so that $(x_1, x_2, x_3) = \lambda(y_1, y_2, y_3)$. The reader is then invited to verify that similarity is an equivalence relation and that there are

$$\frac{n^3 - 1}{n - 1} = n^2 + n + 1$$

equivalence classes under this equivalence relation. The number $n^2 + n + 1$ reminds us of the parameters of projective planes of order n. Therefore, it should now not come as a surprise that these equivalence classes will be the points (and also, the lines) of the projective plane $\mathcal{H}$ to be created.

Now we are ready to define $\mathcal{H}$.

1. Let the set of points of $\mathcal{H}$ be the set of equivalence (or similarity) classes x of triples (x_1, x_2, x_3) just defined, and
2. let the set of lines of $\mathcal{H}$ be the set of equivalence (or similarity) classes v of triples (v_1, v_2, v_3) just defined, and
3. let the point x be contained in the line v if

$$x_1v_1 + x_2v_2 + x_3v_3 = 0.$$

It is not difficult to show that the design $\mathcal{H}$ defined this way indeed satisfies the axioms of Definition 8.13. For the first axiom, let (x_1, x_2, x_3) and (x_1', x_2', x_3') be two triples that belong to different equivalence classes; we can then choose these triples so that $x_3 = x_3' = 1$. We will now show that there is exactly one line in $\mathcal{H}$ that contains both of these points. This is equivalent to saying that the system of equations

$$x_1v_1 + x_2v_2 + v_3 = 0 \tag{8.2}$$

$$x_1'v_1 + x_2'v_2 + v_3 = 0 \tag{8.3}$$

always has a unique solution (v_1, v_2, v_3).

In order to prove this, note that, since $(x_1, x_2, 1)$ and $(x_1', x_2', 1)$ are not equivalent, they must differ in one coordinate, and we can assume that $x_1 \neq x_1'$. Subtracting (8.2) from (8.3) and then rearranging, we get

$$v_1 = -v_2\frac{x_2 - x_2'}{x_1 - x_1'}.$$

Substituting this into (8.2) and rearranging, we get

$$v_3 = x_1v_2 \cdot \frac{x_2 - x_1'}{x_1 - x_1'} - x_2v_2.$$

Because we are within a field, the computed values of v_1, v_2, and v_3 exist, and are uniquely determined, proving our claim.

An analogous argument proves the second axiom. The third axiom follows from the result of Exercise 6, now that we know that $\mathcal{H}$ satisfies the conditions of that exercise.

Interestingly, *no finite projective plane of order n is known* if n is not a power of a prime; it is in fact conjectured that in that case no finite projective planes exist. For small values of n, this means that finite projective planes of order n exist for $n = 2, 3, 4, 5, 7, 8, 9, 11$. So the first question arises when $n = 6$. In this case, we can prove that the answer is negative as follows: Assume there exists a finite projective plane of order six. That plane would have 43 vertices. The Bruck–Ryser theorem, given in Exercise 8, shows that, in that case, the equation

$$x^2 + z^2 = 6y^2 \tag{8.4}$$

would have a solution in which x, y, and z are all integers and not all equal to zero. Let us assume that such a solution exists, and let (x, y, z) be a solution in which $|x| + |y| + |z|$ is minimal. Note that, in any solution, each of x and z has to be divisible by three. Indeed, otherwise their square would be of the form $3k + 1$, and therefore the left-hand side would not be divisible by three, which is a contradiction. However, as both x and z are divisible by three, the left-hand side is divisible by nine, and so y must also be divisible by three. This implies that $(x/3, y/3, z/3)$ is also an integer solution to (8.4) and that $|x/3| + |y/3| + |z/3| < |x| + |y| + |z|$, which is a contradiction.

Therefore, no finite projective plane of order six exists. The next value of n to discuss is $n = 10$. This is a very difficult case, but the answer is still known to be negative. See Supplementary Exercises 9 and 10 for further negative results on the existence of finite projective planes.

Quick Check

1. Prove that the Fano plane cannot be drawn in the plane so that each edge corresponds to a straight line segment.
2. Let $n \geq 3$, and let S be a set of n points in the plane so that there is no straight line that contains all points in S. Is it true that there are two points $x, y \in S$ so that the straight line spanned by x and y does not contain any other points in S?
3. Let G be a complete graph on six vertices. Consider the design whose vertices are the 15 edges of G and whose blocks are the 15 perfect matchings and 20 triangles in G. Prove that this design is balanced and regular.

8.3 Error-correcting codes

8.3.1 Words far apart

We could probably tell apart a bar and a dentist's office, even if both lacked a few of the usual features. These places are so different that it suffices to see parts of them to tell which one is the bar and which one is the dentist's office. This idea turns out to be of crucial importance in Coding Theory.

Let us assume we want to send a message from our cell phone using just the two-letter binary alphabet consisting of the letters 0 and 1. Say the message that we want to send is a YES or NO message. We could agree with the recipient that 1 means yes, and 0 means no. This is simple enough if we are both sure that we will not make any mistakes in typing.

However, if mistakes are possible, then this way of encoding messages will not be efficient. Indeed, one single mistake could totally turn the meaning of the message into its opposite. One way to make sure that our message is not misunderstood is to send it over and over again, in consecutive bits. Say that we will send our message three times. If the message is YES, then we will send the digits 111, and if the message is NO, then we will send the digits 000. These two codewords are not at all similar to each other. Therefore, if we are sure that at most one typing mistake will be made, we can rest assured that our message will be understood properly. Indeed, if we want to send the codeword 111 (resp. 000), and at most one mistake will be made, then the received word will contain at least two 1s (resp. at least two 0s). So as long as at most one bit is erroneous in each codeword, *all errors can be corrected.*

This simple example can be generalized in many different directions. First, it could be that there are more than just two possible messages to send. Second, it could also be that there are more than two digits in our coding alphabet. Third, more than one mistake may be made during typing. Nevertheless, the main idea of our simple example is crucial. This idea is that *if the codewords are sufficiently dissimilar* from each other, then we can tell them apart *even if a few mistakes are made.*

It is time that we made the notions of "sufficiently dissimilar" and "few mistakes" more precise.

Definition 8.18 *Let x and y be words of the same length over the same finite alphabet. Then the* Hamming distance *of x and y, denoted by $d(x, y)$, is the number of positions in which x and y differ.*

Example 8.19 *If $x = 100101$ and $y = 001100$, then $d(x, y) = 3$, since x and y differ in the first, third, and sixth positions.*

Now that we have a notion of distance between two words, we can define the concept of *spheres* and *balls* as well. These will be defined in a way analogous to Euclidean geometry.

Definition 8.20 *Let x be a word of length k over a finite alphabet A. Then the sphere $S_r(x)$ of center x and radius r is the set of k-letter words over A that are of distance r from x.*

Similarly, the ball $B_r(x)$ of center x and radius r is the set of k-letter words over A that are of distance at most *r from x.*

Example 8.21 *Let A be the binary alphabet, let $k = 4$, and let $x = 1010$. Then*

$$S_1(x) = \{0010, 1110, 1000, 1011\},$$

while

$$B_1(x) = \{1010, 0010, 1110, 1000, 1011\}.$$

It is now easy to express our previous observations in a more precise way.

Proposition 8.22 *Let C be a code over a finite alphabet that consists of codewords of the same length. Let us assume that, for every pair of codewords x and y of C, the equality $B_r(x) \cap B_r(y) = \emptyset$ holds. Then, as long as at most r digits are erroneous in each codeword, all errors can be corrected.*

If a code C has the property that all errors can be corrected if at most r mistakes per codeword are made, then we will call that code *r-error-correcting.*

Fine, the reader could say, but how can we quickly check that the balls $B_r(x)$ and $B_r(y)$ are indeed disjoint for every pair (x, y) of codewords? The reader has surely learned the *triangle inequality* in high school. This inequality said that, in Euclidean geometry, the sum of two sides of a triangle is always larger than its third side, essentially because the shortest path between two points is by a straight line. Fortunately, something very similar is true for words over a finite alphabet.

Lemma 8.23 **(Triangle inequality for words)** *Let x, y, and z be words of the same length over a finite alphabet. Then*

$$d(x, y) \leq d(x, z) + d(z, y).$$

A proof is given in Exercise 10.

Corollary 8.24 *Let C be a code consisting of words of the same length over the same finite alphabet. Let us assume that, for every pair (x, y) of codewords in C, the inequality $d(x, y) \geq 2r + 1$ holds. Then C is r-error-correcting.*

Proof: It suffices to prove that, for each pair (x, y) of words in C, the equality $B_r(x) \cap B_r(y) = \emptyset$ holds, and our claim will follow from Proposition 8.22.

Now let us assume that $z \in B_r(x) \cap B_r(y)$. Then the triangle inequality implies that

$$d(x, y) \leq d(x, z) + d(z, y) \leq r + r = 2r,$$

which is a contradiction. So such z does not exist. ◊

Very good, you might say again, but I still have to check that no two codewords are closer than $2r + 1$ to each other. How can I do that quickly? This would not be a problem if space were not a concern. Indeed, it is very easy to construct very long codewords that are far away from each other. For instance, let C be a code over the binary alphabet consisting of n words of length n^2. Let us denote these words by $w_1, w_2, \cdots, w_n$, and let w_i consist of 0s everywhere except between $(i-1)n+1$ and in. It then follows from the definition of the words w_i that $|w_i - w_j| = 2n$ as long as $i \neq j$.

The reader surely feels that the mentioned code C is quite wasteful. It uses words of length n^2 to send just n different messages. It is more challenging to construct better, that is, less wasteful error-correcting codes. This is the content of the next subsection, where we can use the highly symmetric structures that we learned earlier in this chapter.

8.3.2 Codes from designs

The highly regulated structure of some designs allows us to use these designs to construct error-correcting codes.

Example 8.25 *Let M be the incidence matrix of the Fano plane, introduced in Example 8.14. Then the rows of M form a 1-error-correcting code.*

Solution: We have seen that the Fano plane is a balanced uniform regular design with $v = b = 7$, $r = k = 3$, and $\lambda = 1$. Since $\lambda = 1$, any pair of rows agrees in one position in which they both contain a 1. The remaining two 1s in each of these two rows must be in different positions, leaving two positions in which the two rows can agree and have zeros. So the Hamming distance of any two rows is four. $\Diamond$

The ideas used above can be generalized to other designs, as in the following corollary.

Corollary 8.26 *Let M be the incidence matrix of a balanced uniform (and therefore regular) design. Then the rows of M form an $r-\lambda-1$-error-correcting code.*

The proof is analogous to that of Example 8.25.

8.3.3 Perfect codes

The codes discussed in Example 8.25 and Corollary 8.26 are not as wasteful as the code discussed just before them, but hopefully the reader still feels that they are not as efficient as they could be. For one, the Hamming distance between any two codewords of Example 8.25 is four, which is not enough for the code to be 2-error-correcting. One could have different opinions on whether this code is highly efficient or not. On one hand, it is nice and symmetric, and we know that it cannot be extended by adding another vertex and still retain all the symmetries of the Fano plane. On the other hand, the code is 1-error-correcting, but that would have been achieved by a minimum Hamming distance of three between any two codewords. So there is a waste there, and therefore there might be an opportunity to add codewords without losing the 1-error-correcting property.

After this intuitive discussion of efficiency of codes, we may feel the need to define codes that are optimal in some precise sense. This discussion will have to contain quite a few parameters, but it will otherwise not be difficult. So we ask that the reader please bear with us.

Let C be an m-error-correcting code over a q-element alphabet, containing w codewords, each of length n. That means that the balls of radius m centered at each of the w codewords are pairwise disjoint. Therefore, the total number W of words in all of these spheres satisfies

$$W = ws,$$

where s is the number of words in any one of the mentioned balls. Let us now determine s. Let $c \in C$, and let f be a word over our alphabet so that $d(c, f) = d$. Then there are $\binom{n}{d}$ choices for the set of positions in which c and f differ, and in each of these positions, there are $(q-1)$ choices for the letter of f there. Since our balls have radius m, this yields

$$s = \sum_{d=0}^{m} \binom{n}{d} (q-1)^d.$$

Comparing this with the previous displayed equality, we get

$$W = ws = w \sum_{d=0}^{m} \binom{n}{d} (q-1)^d.$$

On the other hand, W is the number of words of length n over our alphabet that belong to the union of our balls, so W cannot be larger than the number of *all* of words of length n over our alphabet, that is, q^n. This yields

$$w \sum_{d=0}^{m} \binom{n}{d} (q-1)^d \leq q^n.$$

If equality holds, then there is no word of length n over our alphabet that belongs to no ball of radius m centered at a codeword. This justifies the following definition.

Definition 8.27 *The code C is called* perfect *if it satisfies*

$$w \sum_{d=0}^{m} \binom{n}{d} (q-1)^d = q^n. \tag{8.5}$$

In other words, no word of the right size is wasted for a perfect code; each word belongs either to the code itself or is used as "padding" between two codewords.

If C is perfect, then the spheres of radius m centered around each codeword are disjoint, but they cover the entire set of words of length n over our alphabet without a gap. Try that in Euclidean space!

We would like to point out that, just because some set of parameters satisfies (8.5), it does not necessarily follow that a perfect code with those parameters exists. A famous example for this was constructed by Golay, who showed that, while $w = 2^{78}$, $m = 2$, $q = 2$, and $n = 90$ satisfy (8.5), there is no perfect code with these parameters. In other words, (8.5) is only a necessary condition for the existence of a perfect code, not a sufficient one.

It is easy to check that the code of Example 8.25 is *not* perfect, since for that code (8.5) becomes $7 \cdot (1 + 7) < 2^7$, or $56 < 128$.

Constructing perfect codes, or sometimes even deciding whether they exist for a given set of parameters, is usually a difficult task, and we will leave it to more specialized books, such as [45].

Let us consider the simplest possible alphabet, the one with $q = 2$ elements. For codes over this alphabet, (8.5) becomes

$$w \sum_{d=0}^{m} \binom{n}{d} = 2^n. \tag{8.6}$$

As all divisors of 2^n are powers of 2, this means that w and $\sum_{d=0}^{m} \binom{n}{d}$ must both be powers of 2. Looking at the even more special case of $m = 1$, that of 1-error-correcting codes, 8.6 becomes

$$w(1 + n) = 2^n.$$

So there must be a nonnegative integer k so that $2^k = n + 1$ (in other words, $n = 2^k - 1$). This leads to $w = 2^{2^k - k - 1}$. Here $k = 0$ and $k = 1$ lead to the uninteresting parameter $w = 1$, which means a "code" consisting of one word.

For $k = 2$, we get $w = 2$ and $n = 3$. The reader is urged to find a perfect code over the two-element alphabet with these parameters.

See the Notes section at the end of this chapter for more on codes over a two-element alphabet.

Now let us consider perfect codes over a three-element alphabet instead of a two-element alphabet. For $q = 3$ and $m = 1$, (8.5) becomes

$$w(1 + 2n) = 3^n.$$

So there has to be a nonnegative integer k so that $3^k = 2n + 1$ (or $n = (3^k - 1)/2$) and therefore $w = 3^{n-k} = 3^{(3^k - 2k - 1)/2}$. Again, $k = 0$ and $k = 1$ lead to $w = 1$. For $k = 2$, we get $w = 9$ and $n = 4$. It takes a little time to construct a perfect code with these parameters. Below is one example.

- 1100, 1010, 1001,
- 2211, 2121, 2112,
- 0022, 0202, 0220.

Again, this example can be generalized for larger values of k, and we will discuss that more in the Notes section.

Quick Check

1. Let $\mathcal{D}$ be a balanced uniform (and therefore regular) design. Is there an analogous version of Corollary 8.26 that uses the *columns* of the incidence matrix M of $\mathcal{D}$ instead of the rows?
2. We have mentioned that the code of Example 8.25 is not perfect. Now let us extend that code by adding nine new codewords, namely the all-0 word, the all-1 word, and the complement of each of the original seven codewords. (The complement of a codeword w is the word in which each letter in w is replaced by the opposite letter.) Is the new code C still 1-error correcting?
3. Is the code C of the previous Quick Check exercise perfect?

8.4 Counting symmetric structures

The reader may remember that, in Chapter 5, when we counted graphs, we typically counted graphs with *labeled* vertices, which prevented symmetries. We mentioned that the enumeration of *unlabeled* graphs is usually harder. This is true for other structures as well. To see why, let us consider the job of painting each of the six sides of a cube with one of k colors. If the sides are all considered different, for instance, because they are labeled 1 through 6, then the number of ways to do this is 6^k. However, if the faces are indistinguishable, the problem is much more difficult. Say we consider two colorings identical if one can be transformed into the other by a series of rotations. We cannot simply count the number of all possible colorings regardless of rotations and then divide by the number of all possible rotations, since the number of possible rotations is not the same for each coloring. (Compare a coloring that uses only one color and a coloring that uses all six.) If we allow reflection through planes in addition to rotations, the situation is even more complex. Again, the problem is that not all equivalence classes will have the same size.

In order to be able to discuss the machinery relevant to problems like the one above, we need to introduce basic notions of *Group Theory*, in particular the *theory of permutation groups*. Readers who have taken a class in abstract algebra before will probably be familiar with these notions.

Definition 8.28 *A* group *is a set G of elements and an operation that we will call* multiplication *on the set of ordered pairs of G so that the following axioms hold.*

(1) There exists an identity *element in G, that is, there exists an element $e \in G$ so that $a \cdot e = e \cdot a = a$ for all $a \in G$.*

(2) The set G is closed under multiplication, that is, if $a \in G$ and $b \in G$, then $a \cdot b \in G$.

(3) Multiplication is associative, that is, $(a \cdot b) \cdot c = a \cdot (b \cdot c)$.

(4) Each element of G has a unique inverse, *that is, for each $a \in G$, there exists a unique element $b \in G$ so that $ab = ba = e$. We then write $b = a^{-1}$.*

Note that the operation "multiplication" can be defined in any way that satisfies the axioms; that is, it does not have to be what we typically call multiplication when dealing with real numbers.

If this is the first occasion the reader has heard about groups, then the reader should take the time and prove (by verifying that all axioms hold) that the set of real numbers with addition as the operation form a group and that the set of nonzero real numbers with traditional multiplication as the operation form a group. After this, the reader should explain why the set of *all real numbers* does not form a group with traditional multiplication as the operation.

While an entire chapter could be filled with examples of interesting groups, we will focus on the group of all *permutations* of length n. First, of course, we have to prove that this set indeed forms a group with the operation that naturally comes to mind, that is, multiplication of permutations as defined in Chapter 4. Recall, in Definition 4.13, we simply said that the product of n-permutations f and g is simply their composition as bijections from $[n]$ to $[n]$.

Proposition 8.29 *The set of all n-permutations, equipped with multiplication of permutations as the operation, forms a group.*

As we mentioned in Chapter 4, this group is called the *symmetric group*, denoted by S_n.

Proof: We will check that all the axioms hold.

1. The permutation $p = (1)(2) \cdots (n)$ is the identity element of this group.
2. The product of two n-permutations is an n-permutation, since the composition of two bijections from $[n]$ to $[n]$ is a bijection from $[n]$ to $[n]$.

3. Let f, g, and h be three permutations, and let $f(i) = j$, $g(j) = k$, and $h(k) = m$. Then

$$((f \cdot g)) \cdot h)(i) = h(g(f(i))) = h(g(j)) = h(k) = m,$$

and

$$(f \cdot (g \cdot h))(i) = (g \cdot h)(f(i))) = h(g(j)) = h(k) = m.$$

4. The inverse of the permutation f is simply the inverse of f as a bijection.

◊

We say that a subset H of elements of the group G forms a *subgroup* of G if H is a group itself with the same operation as G. Subgroups of the symmetric group S_n are called *permutation groups*. This is because their elements are permutations.

The crucial concept is that permutations *move* (permute) objects. In our examples so far in this chapter, as well as those in Chapter 4, these objects were most often simply elements of the set $[n]$. However, permutations can act on other sets as well. Recall the definition of an automorphism of a graph M on vertex set $[n]$ from Chapter 5. Such an automorphism was a bijection f from the vertex set of G onto itself so that $f(a)$ and $f(b)$ were adjacent if and only if a and b were. In other words, an automorphism is a *permutation* of the vertices of M. It is straightforward to show that all automorphisms of M form a group. Call this group $Aut(M)$; then $Aut(M)$ is a *permutation group*; indeed, it is a subgroup of S_n.

Graphs and their automorphism groups provide a simple way to visualize the next notion we are going to discuss, the *orbit* of an object under the action of a permutation group.

The reader is asked to consider the graph G shown in Figure 8.4. The group $Aut(M)$ permutes the vertices of this graph among themselves. However, no matter what element $f \in Aut(M)$ we choose, $f(4) = 4$ will always hold, since 4 is the only vertex of G with degree five, and we know that the degrees of 4 and $f(4)$ must agree since an automorphism must preserve degrees. By an analogous argument, $f(1)$ must be either 1, 2, or 3, since these are the only vertices of degree three. The same holds for $f(2)$ and $f(3)$. Similarly, $f(5)$ must be equal to 5 or 6, and the same holds for $f(6)$, since 5 and 6 are the only vertices of degree one.

Finally, any one of these choices is indeed possible, that is, there exists an $f \in Aut(M)$ so that $f(1) = 3$, there exists an $f \in Aut(M)$ so that $f(1) = 2$, and so on.

The phenomena described above are so important that they deserve a name.

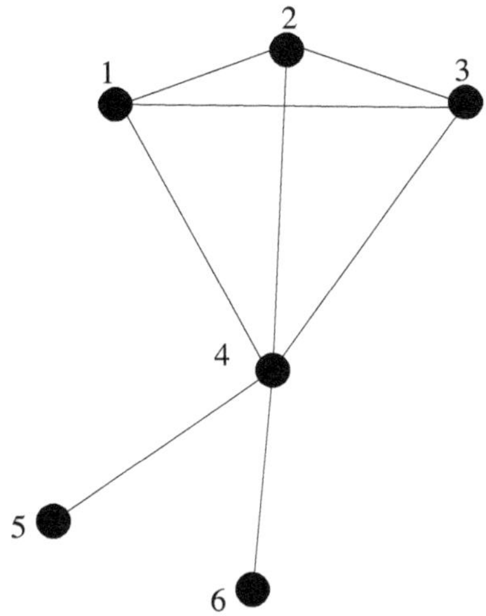

Figure 8.4
The orbits of the vertex set of this graph M are $\{1, 2, 3\}$, $\{4\}$, and $\{5, 6\}$.

Definition 8.30 *Let G be a permutation group acting on a set S. Let $i \in S$. Set*

$$i^G = \{g(i)|g \in G\}.$$

In other words, i^G is the set of all vertices into which i can be mapped by an element of G. Then i^G is called the orbit *of i under G.*

Example 8.31 *If S is the vertex set of the graph M shown in Figure 8.4, and $G = Aut(M)$, then the orbits of the elements of M under $Aut(M)$ are as follows:*

- $1^G = 2^G = 3^G$,
- 4^G, *and*
- $5^G = 6^G$.

The reader might have noticed that the orbits of two elements are either equal or disjoint; it never happens that they have a few common elements and a few different elements. As the following lemma shows, this is always the case.

Lemma 8.32 *Let G be a permutation group acting on a set S. Let i and j be two distinct elements of S. Then either $i^G = j^G$ or $i^G \cap j^G = \emptyset$.*

In other words, the orbits form a *partition* of S if we remove repeated copies of the same orbit.

Proof: (of Lemma 8.32) First assume that $j \in i^G$, that is, there exists an $f \in G$ so that $f(i) = j$. Assume now that $k \in j^G$, since $h(j) = k$ for some $h \in G$. Then we also have $k \in i^G$, since $(f \cdot h)(i) = h(f(i)) = h(j) = k$. Therefore, $j^G \subseteq i^G$. Now note that $f^{-1}(j) = i$, so $i \in j^G$. Therefore, we can switch the roles of i and j in the previous argument and show that $i^G \subseteq j^G$, yielding that $i^G = j^G$.

Now assume that $j \notin i^G$. We claim that then i^G and j^G must be disjoint. Assume not; that is, assume that there exists an element $s \in S$ so that $p(i) = s$ and $q(j) = s$ for some p and q in G. Then $q^{-1}(s) = j$, so $(p \cdot q^{-1})(i) = q^{-1}(p(i)) = q^{-1}(s) = j$, contradicting the assumption that $j \notin i^G$. ◊

That is, the relation "i and j have the same orbit" is an equivalence relation on S. The number of equivalence classes is just the number of distinct orbits. In our quest to enumerate structures that are not equivalent, we will try to reduce our problems to the problem of counting orbits of a permutation group acting on a set.

There is one last notion we need to introduce before our main result. It is a natural counterpart of the notion of orbits. The size of the orbit of an element i told us into how many different elements i could be mapped by a given permutation group G. We could also ask how many elements of G will keep i where it was, or we could ask how many elements of S a given element g will keep fixed. We need both of these useful notions.

Definition 8.33 *Let G be a permutation group acting on a set S, and let $i \in S$. Then the set*

$$G_i = \{g \in G | g(i) = i\}$$

is called the stabilizer *of i.*

In Supplementary Exercise 15, the reader is asked to verify that G_i is always a *subgroup* of G.

It is reasonable to conjecture at this point that, when G and S are fixed, the larger i^G is, the smaller G_i will be. Indeed, the more places i can go to, the fewer elements should fix it. We would also like to point out that Example 8.31 suggests that $|i^G|$ is always a *divisor* of G. The following lemma will show that both of these observations hold in the general case.

Lemma 8.34 *Let G be a finite permutation group acting on a set S, and let $i \in S$. Then*

$$\frac{|G|}{|G_i|} = \left|i^G\right|.$$

Proof: If $g \in G$, let gG_i denote the set $\{h \in G | gx = h \text{ for some } x \in G_i\}$. These sets gG_i are called the *cosets* of G_i in G. We now claim that two cosets of G_i in G are either equal or disjoint. The proof of this claim is very similar to that of Lemma 8.32, and therefore will be left for Exercise 12.

We ask the reader to verify that all cosets gG_i have the same size as G_i. Indeed the elements of gG_i are the products gx, where x ranges over all elements $x \in G_i$. This means there are $|G_i|$ products, and they are all different. If they were not, then $gx = gx'$ would hold, which, after multiplying by g^{-1} from the left, would result in $x = x'$.

As the distinct cosets of G_i partition G, this proves that $|G_i|$ divides $|G|$. Now that we have discussed the *size* of these cosets, let us turn our attention to their *number*. We claim that the number of the cosets of G_i is $|i^G|$, the size of the orbit of i.

We will prove this claim by constructing a bijection α from i^G onto the set of all cosets of G_i in G. Elements of i^G can be written in the form $g(i)$, with $g \in G$ where g is not necessarily unique. Now set $\alpha(g(i)) = gG_i$.

Before we try to prove that this map α is indeed a bijection, we have to prove that it is indeed well-defined, that is, if g and g_1 are such that $g(i) = g_1(i)$, then $\alpha(g(i)) = \alpha(g_1(i))$. This will show that α is indeed a map that is defined on the elements of the orbit of i and which does not depend on anything else, such as the choice of g.

To see this, note that $g(i) = g_1(i)$ is equivalent to $g_1^{-1}g(i) = i$, meaning that $g_1^{-1}g \in G_i$. That leads to $g_1^{-1}gG_i = G_i$, and therefore $gG_i = g_1G_i$; so $\alpha(g(i)) = \alpha(g_1(i))$ as claimed.

Finally, to prove that α is a bijection, let us construct its inverse. Let gG_i be a coset of G_i. Then all elements of this coset are of the form gx, where $x(i) = i$. Therefore, for all elements gx of this coset, $gx(i) = g(i)$. Then $g(i) \in i^G$ is the preimage of gG_i under α. This preimage is unique, since, if $\alpha(g(i)) = \alpha(g_1(i))$, then $gG_i = g_1G_i$, yielding that $g_1^{-1}g \in G_i$. That would lead to $g(i) = g^{-1}(i)$, meaning that $g(i)$ and $g^{-1}(i)$ are in fact identical as elements of i^G.

So the number of cosets of G_i is $|i^G|$, proving our lemma. ◊

Note that the simple fact that two cosets of G_i in G must be either disjoint or equal is true in a more general way. See Exercise 12 for that more general version.

Definition 8.35 *Let G be a permutation group acting on a set S, and let $g \in G$. Let*

$$F_g = \{i \in S | g(i) = i\}.$$

Now we are in a position to announce and prove the main result of this section.

Theorem 8.36 *Let G be a permutation group acting on a set S. Then the number of orbits of S under the action of G is equal to*

$$\frac{1}{|G|} \sum_{g \in G} |F_g|.$$

This theorem is a classic, and it has many names, such as Frobenius' theorem, Cauchy's theorem, Burnside lemma, or a name consisting of a nonempty subset of the previous three names.

The following example provides an opportunity to practice the notions of this theorem before proving it.

Example 8.37 *Let S be the vertex set of the graph M shown in Figure 8.4, and let $G = Aut(M)$. Then $|G| = 6 \cdot 2 = 12$, since vertices 1, 2, and 3 can be permuted among each other in any way, vertices 5 and 6 can be permuted among each other in any way, and there are no other automorphisms. Each of these 12 automorphisms will map 4 to 4. Half of them will map 5 to 5 and 6 to 6, and the other half will map 5 to 6 and 6 to 5. On the set $\{1, 2, 3\}$, two of them will have three fixed points, six of them will have one fixed point, and four of them will have no fixed point.*

So altogether, there is one element of G (the identity) with six fixed points; there are four with four fixed points (one of which fixes 1, 2, 3, and 4, and one of which fixes i, 4, 5, and 6 for some $i \in [3]$); there are two with three fixed points (fixing 4, 5, and 6); there are three with two fixed points (fixing i and 4, for some $i \in [3]$); and there are two with one fixed point (fixing 4).

Therefore, Theorem 8.36 says that the number of orbits of G on S is

$$\frac{1}{12}\sum_{g \in G} |F_g| = \frac{1 \cdot 6 + 4 \cdot 4 + 2 \cdot 3 + 3 \cdot 2 + 2 \cdot 1}{12} = \frac{36}{12} = 3,$$

which is in accordance with what we have seen, that is, that the orbits are $\{1, 2, 3\}$, $\{4\}$, and $\{5, 6\}$.

Proof: (of Theorem 8.36) The number of orbits of G is certainly equal to $\sum_i \frac{1}{|i^G|}$, where i ranges over all elements of S. Indeed, the total contribution of each orbit to this sum will be 1. Now let us transform this sum as follows:

$$\sum_i \frac{1}{|i^G|} = \sum_i \frac{|G_i|}{|G|} = \frac{1}{|G|}\sum_i |G_i|.$$

We applied Lemma 8.34 in the first step. The last displayed equation is promising, since it already contains the factor $\frac{1}{|G|}$, which is multiplied by a sum. The only problem is that this sum is over all values of i, not g. However, taking a closer look at the sum $\sum_i |G_i|$, we see that it in fact counts all pairs (g, i) so that $g \in G$, $i \in S$, and $g(i) = i$. Therefore, this sum does not change if we compute it by summing over values of i first and g second. This shows that

$$\sum_i |G_i| = \sum_{g \in G} |F_g|$$

and proves our theorem. $\Diamond$

In a typical application of Theorem 8.36, we count structures that are nonequivalent according to some definition. It then turns out that this nonequivalence means that they belong to different orbits under some group action, and that the number of orbits is the number of these nonequivalent structures. Then we count the orbits using Theorem 8.36 and conclude that it equals the number of nonequivalent structures.

Let us start with a very simple example.

Example 8.38 *We have a rectangular, fenced backyard of size* 90×100. *We want to color the four sides of the fence using red and blue paint, using only one color on each side. In how many different ways can we do this if two colorings are considered equivalent if they differ only by a 180-degree rotation?*

Solution: In this example, the group R of rotations consists of just two elements, the identity and the 180-degree rotation. This group acts on the set of all 16 possible colorings (if we consider the sides all distinguishable and do not identify colorings that differ by a rotation, then there are $2^4 = 16$ colorings). The orbits of this action are precisely the nonequivalent colorings, since two colorings are identical precisely when they can be rotated into each other. The identity will fix all 16 possible colorings, while the 180-degree rotation will fix those four colorings in which opposite pairs are of the same color. Therefore, by Theorem 8.36, the number of nonequivalent colorings is

$$\frac{1}{2}(16+4) = 10.$$

◊

The following example is somewhat more difficult.

Example 8.39 *We color the sides of a square either red or blue. We consider two colorings equivalent if there is symmetry (rotation or reflection) that takes one coloring into the other. How many nonequivalent colorings are there?*

Solution: It is not hard to see that the square has eight symmetries: four rotations (counting the identity, we call them r, r^2, r^3, and id) and four reflections. Two of these reflections (a and b) are through diagonals, and the other two are through lines bisecting opposite sides (c and d).

These symmetries permute the sides of our square, and while doing that, they permute colorings among each other. So they are the permutations acting on the set of all colorings. Now we are going to find out how many colorings each of them fixes. Each symmetry fixes the colorings that use only one color. Rotations r and r^3 do not fix anything else, since they map sides into adjacent sides (so if they fixed a coloring, that coloring would have to use one color only). Rotation r^2 fixes two 2-colorings, those in which opposite sides are of the same color. Reflections c and d fix two 2-colorings each, those in which sides intersecting each other on the reflection axis are monochromatic. Reflections a and b fix six 2-colorings each (in these colorings, the sides parallel to the axis of reflection are monochromatic). See Figure 8.5 for a list. Finally, *id* fixes all 16 colorings.

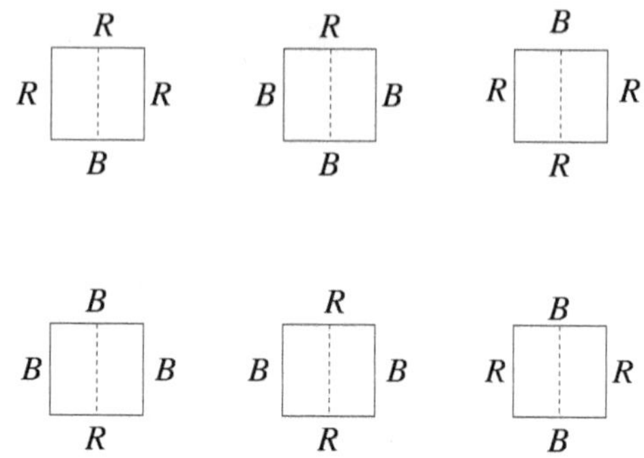

Figure 8.5
The six 2-colorings fixed by reflection a.

Applying Theorem 8.36, and remembering that all symmetries fix the two colorings that use only one color, we get that the number of nonequivalent colorings is

$$\frac{1}{8}(2 \cdot 2 + 3 \cdot 4 + 2 \cdot 8 + 1 \cdot 16) = \frac{48}{8} = 6.$$

◊

The reader should not get the impression that coloring sides of polygons is the only application of Theorem 8.36. In Supplementary Exercise 16, we ask the reader to use Theorem 8.36 to re-prove the fact that the average n-permutation has one fixed point. In Exercise 13, we ask the reader to prove a classic number-theoretical result by Fermat using the same technique. Exercise 14 claims that Theorem 5.7 is just a special case of Lemma 8.34.

Quick Check

1. Let A be a complete graph on four vertices with one edge removed. How many ways are there two color the edges of A using only the colors red, blue, and green if two colorings are considered identical if there is an automorphism that takes one coloring to the other?
2. Find the number of ways to color each side of a regular pentagon red, blue, or green if two colorings are considered identical if there is a rotation or reflection that takes one coloring into the other.
3. Let A be a graph that is a C_5 with a diagonal added. Find the number of ways to color each edge of A red, blue, or green if two

colorings are considered identical if there is a rotation or reflection that takes one coloring into the other.

8.5 Notes

We have seen that, in a balanced design, each pair of vertices occur together in the same number λ of blocks. We could instead require that each t-element subset of vertices occur together in the same number of blocks for a given t. Such a design is called a *t-design.*

One infinite family of perfect error-correcting codes over a two-element alphabet are the *Hamming* codes. These have parameters $n = 2^k - 1$, $w = 2^{n-k} = 2^{2^k-k-1}$, and $m = 1$, for some $k \geq 3$. So the smallest element of this family is a code consisting of 16 binary words of length seven each that is 1-error correcting.

The perfect code over a three-element alphabet that we gave earlier can also be generalized. In fact, it is known that there exists such a code with $3^{(3^k-2k-1)/2}$ words of length $(3^k - 1)/2$ each that is 1-error correcting. This can be generalized even further, from alphabets of size three to alphabets whose size is a power of a prime. The interested reader is invited to consult *A First Course in Coding Theory* [45], by R. Hill, for further details.

8.6 Chapter review

(A) Designs

The following hold for balanced uniform regular designs (block designs):

1. $bk = vr$;
2. $\lambda(v - 1) = r(k - 1)$; and
3. $v \leq b$ (Fisher's inequality).

(B) Finite projective planes

1. They are $(n^2 + n + 1, n^2 + n + 1, n + 1, n + 1, 1)$-designs.
2. They are known to exist for prime power values of n.

(C) Error-correcting codes

1. Triangle inequality:

$$d(x, y) \leq d(x, z) + d(z, y).$$

2. Necessary condition for existence of a perfect m-error correcting code over a q-element alphabet, consisting of w words of length n each:

$$w \sum_{d=0}^{m} \binom{n}{d} (q-1)^d = q^n.$$

(D) Counting symmetric structures

1. Let G be a finite permutation group acting on a set S, and let $i \in S$. Then

$$\frac{|G|}{|G_i|} = \left|i^G\right|.$$

2. If G acts on the set S, then the number of orbits of G is

$$\frac{1}{|G|} \sum_{g \in G} |F_g|,$$

where F_g is the number of fixed points of $g \in G$.

8.7 Exercises

1. Prove that, if a design is balanced and uniform, then it is also regular.
2. Is there a balanced uniform design in which r divides v?
3. + Let M be the incidence matrix of a balanced uniform design $\mathcal{F}$. Express the matrix product MM^T by the parameters of $\mathcal{F}$, the $v \times v$ identity matrix I_v, and the $v \times v$ matrix J_v whose entries are all equal to 1.
4. Let M be the incidence matrix of a BIBD with parameters (b, v, r, k, λ). Prove that the eigenvalues of MM^T are $r - \lambda$ with multiplicity $v - 1$, and $r + \lambda(v - 1) = rk$ with multiplicity 1.
5. + Prove that the dual of a symmetric BIBD is a BIBD.
6. Let $n > 1$. Prove that an $(n^2+n+1, n^2+n+1, n+1, n+1, 1)$-design is a finite projective plane.
7. Prove that the dual design of a finite projective plane is also a finite projective plane.
8. A design is called symmetric if $v = b$. Let $\mathcal{F}$ be a balanced uniform symmetric design, with v even. Prove that $r - \lambda$ has to be a perfect square.

Hint: Look at the determinant $\det(MM^T)$ *of the incidence matrix of this design, which we computed in the proof of Theorem 8.12.*

Note: This is the easy part of the famous Bruck–Ryser theorem. That theorem also says that, if v is odd, then the equation

$$x^2 = (r - \lambda)y^2 + (-1)^{(v-1)/2}\lambda z^2 \tag{8.7}$$

must have an integer solution (x, y, z), where not all of x, y, and z are zero.

9. Uniform designs with $k = 1$ are called *sets*, while uniform designs with $k = 2$ are called *graphs*. The next category, uniform designs with $k = 3$, are often called *triple systems*, and a balanced triple system with $\lambda = 1$ is called a *Steiner triple system*. For instance, the Fano plane is a Steiner triple system.

   (a) Prove that there is no Steiner triple system in which v is even.

   (b) Prove that there is no Steiner triple system in which $v + 1$ is divisible by 6.

10. Prove Lemma 8.23.

11. Will the error-correcting code constructed from a finite projective plane using the method explained in Example 8.25 ever be a perfect code?

12. Prove that, if G is a group, and H is a subgroup of G, then two cosets of H in G are either disjoint or equal. In other words, prove that the cosets of H partition G. (Note that this implies that $|H|$ divides $|G|$.)

13. Use Theorem 8.36 to prove the *small Fermat theorem*, which states that, if x is a positive integer, and p is prime, then $x^p - x$ is divisible by p.

14. Explain why Theorem 5.7 is a special case of Lemma 8.34.

15. Find the number of ways to color each side of a square red or blue or green if two colorings are considered equivalent if they can be moved into each using rotations.

16. Find the number of ways to color the six edges of a regular tetrahedron (that has four sides, each of which is a regular triangle) using k colors if two colorings are considered identical if they can be moved into each other using rotations.

17. + Find the number of ways to color the 12 edges of a cube using k colors if two colorings are considered identical if they can be moved into each using rotations.

8.8 Solutions to exercises

1. The proof is similar to that of Proposition 8.3. Fix a vertex a that is in r_a blocks, and count all pairs (B, c), where a and c are two vertices of the block B. Then, by the same argument as in the mentioned proof, we get that $\lambda(v-1) = r_a(k-1)$, so $r_a = \lambda(v-1)/k$, which is independent from a.

2. No. By the result of the previous exercise, such a design would be regular as well, and $\lambda(v-1) = r(k-1)$ would hold. As r divides v, the number r is relatively prime to $v-1$, so λ would have to be a multiple of r. This is impossible, since $\lambda < r$.

3. The (i, j) entry of the MM^T is the dot product of the ith and jth rows of M. So, if $i \neq j$, this entry is λ, since that is the number of blocks containing both a_i and a_j. If $i = j$, then this entry is simply the number of blocks containing a_i, that is, r. Indeed, the solution of Exercise 1 shows that $\mathcal{F}$ is regular. Therefore, we get

$$MM^T = \lambda J_v + (r-\lambda)I_v. \tag{8.8}$$

4. This follows from (8.8) that we obtained at the end of the solution of the previous exercise. Indeed, J_v has $v-1$ eigenvalues equal to 0, since its rank (number of linearly independent rows) is 1. Its last eigenvalue is λv, since the sum of all the eigenvalues of a matrix equals its trace, which is λv in this case. On the other hand, $(r-\lambda)I_v$ is a constant multiple of the identity matrix, so all its eigenvalues are equal to $r-\lambda$. Finally, if a vector $\mathbf{x}$ is an eigenvector of J_v with eigenvalue a, then $\mathbf{x}$ is also an eigenvector of $\lambda J_v + (r-\lambda)I_v$, with eigenvalue $a+r-\lambda$. Setting $a=0$ ($v-1$ times) and $a = \lambda v$ (once) implies the claim of the exercise.

5. The dual of a regular, uniform design is obviously regular and uniform, even if the design is not symmetric. It remains to prove that, if $\mathcal{F}$ is symmetric, then it is *linked*, because that is the sufficient and necessary condition for $\mathcal{F}^d$ to be balanced. In order to prove that $\mathcal{F}$ is linked, it suffices to prove that $M^TM = MM^T$, where M is the incidence matrix of $\mathcal{F}$. Indeed, all nondiagonal entries in MM^T are identical.

   Note that, in the solution of the previous exercise, we computed the eigenvalues of the incidence matrix of *any* BIBD and found that they were not 0. So, if M is a square matrix, then $\det M \neq 0$, and hence M^{-1} exists. Multiply both sides of (8.8) by M^{-1} from the left, and M by the right, to get

$$M^TM = M^{-1}(\lambda J_v + (r-\lambda)I_v)M = \lambda J_v(r-\lambda)I_v.$$

On the other hand, (8.8) shows that $MM^T = \lambda J_v(r-\lambda)I_v$, proving our claim.

6. Because such a design $\mathcal{F}$ is balanced with $\lambda = 1$, the first axiom is satisfied. If blocks A and B intersected in at least two vertices, say a and b, then a and b would be in at least two common blocks, which is not possible. So $|A \cap B| \leq 1$.

   Now let us assume that A and B are disjoint. Let x be a vertex that is neither in A nor in B. Such a vertex exists, since $2(n+1) < n^2 + n + 1$ since $n > 1$. Because $\mathcal{F}$ is balanced with $\lambda = 1$, there would have to be a block joining x to each of the $2(n+1)$ vertices of $A \cup B$, and all these $2(n+1)$ blocks would be distinct (why?). That is a contradiction, since $\mathcal{F}$ has only $n+1$ lines. So $|A \cap B| = 1$ for any two blocks A and B, and therefore the second axiom is satisfied.

   Finally, the third axiom is satisfied since we can pick two blocks, C and D, and remove their common vertex y. We are left with $2n \geq 4$ vertices, n in each block. Choose one pair of vertices from each block, then the obtained four vertices have the property that no three of them are in the same block (otherwise, there would be a pair of vertices that are contained together in at least two blocks). This shows that the third axiom is satisfied as well.

7. The only nontrivial part is that the third axiom will hold. This is equivalent to the fact that, in the *original* finite projective plane $\mathcal{H}$, there exist four lines, no three of which intersect in one point. We will now prove that this is true.

   Let a and b be two points of $\mathcal{H}$. Let L be the unique line joining them. We know that each of a and b are part of $n+1$ lines, where $n > 1$. Therefore, we can choose two lines A_1 and A_2 distinct from L that contain a (and not b) and two lines B_1 and B_2 distinct from L that contain b (but not a). We claim that these four lines have the required property that no three of them intersect in one point. Assume not; then, without loss of generality, we can assume that A_1, A_2, and B_1 intersect in one point. However, A_1 and A_2 already intersect in a, so this intersection point must be a. That implies $a \in B_1$, which is a contradiction.

8. On one hand,

   $$\det(MM^T) = \det(M)\det(M^T) = (\det(M))^2,$$

   so $\det(MM^T)$ is a perfect square.

   On the other hand, we computed in the proof of Theorem 8.12 that

   $$\det(MM^T) = (r + (v-1)\lambda)(r-\lambda)^{v-1}.$$

   Note that, since our design $\mathcal{F}$ is balanced and uniform, it is regular, so the use of r is justified. We know from Proposition 8.3 that

$(v-1)\lambda = r(k-1)$; therefore, the first factor of the right-hand side of the last displayed equation is equal to $r + r(k-1) = rk = r^2$. Indeed, since $\mathcal{F}$ is symmetric, $r = k$ by Proposition 8.2. So we get

$$(\det(M))^2 = \det(MM^T) = r^2 \cdot (r-\lambda)^{v-1}.$$

This implies that $(r-\lambda)^{v-1}$ is a perfect square. However, $v-1$ is odd, so $r-\lambda$ must be a perfect square.

9. (a) Because a Steiner triple system is balanced and uniform, it is regular, so $\lambda(v-1) = r(k-1)$ holds. However, in our special case, $k = 3$ and $\lambda = 1$, leading to $v = 2r+1$.

   (b) For Steiner triple systems, the identity $bk = vr$ becomes $3b = vr$. Comparing this to the identity $v = 2r+1$ of part (a), we get
   $$r(2r+1) = 3b.$$
   So either r is divisible by 3, and then $2r+1 = v$ is of the form $6t+1$, or $2r+1 = v$ itself is divisible by 3, in which case, being odd, v must be of the form $6t+3$. Note that it was proved by Kirkman in 1847 that, if v is of one of these two forms, then a Steiner triple system on v vertices does exist.

10. We can get from x to y by first changing $d(x,z)$ digits to turn x to z, then by changing $d(z,y)$ digits to turn z to y. During this procedure, at most $d(x,z) + d(z,y)$ digits change, so the number of different digits of x and y is at most that.

11. No. For that code to be perfect, we would need $v(1+v) = 2^v$. That is impossible, since one of v and $1+v$ must be odd (and at least 7), and a power of 2 cannot have a proper odd divisor.

12. Let aH and bH be two cosets of H in G. Let us assume first that $a \in bH$. That means that there exists $h \in H$ so that $a = bh$. Then $aH = bhH = bH$. Indeed, $hH = H$, since $hH \subseteq H$ (as H is closed under multiplication) and $|hH| = |H|$.

    Now let us assume that $a \notin bH$. We claim that then $aH \cap bH = \emptyset$. Assume not; that is, assume rather that $c \in aH \cap bH$. That means $c = ah_1 = bh_2$, with $h_1, h_2 \in H$. However, that implies $a = bh_2h_1^{-1}$, which contradicts $a \notin bH$.

13. Let us color the sides of a regular p-gon with colors from the set $[x]$ so that at least two colors are used. Let us say that two colors are equivalent if they are rotated images of each other. If all sides were distinguishable, the total number of colorings would be $x^p - x$, since we have to use at least two colors.

    There are p possible rotations, $r, r^2, \cdots, r^p = id$. The nonidentity rotations will not fix any allowed coloring. Indeed, if r^i fixed a coloring C, then the first side of the polygon would have to have the same color as its $i+1$st side, and that would have to have the same color as its $2i+1$st side, and so on. As p is prime, p is relatively prime to i, so that would imply that all sides of the polygon are of the same color, which is a contradiction.

Therefore, Theorem 8.36 yields that the number of nonequivalent colorings is

$$\frac{1}{p}(1 \cdot (x^p - x) + (p-1) \cdot 0) = \frac{x^p - x}{p}.$$

As this is the number of certain colorings, it must be an integer, proving our claim.

14. Let G be the *symmetric* group acting on the set $[n]$ of vertices of our graph. Then G acts on the set of all *labelings* as well. In the latter action, $|i^G|$ is just the number of all labelings of the graph, and $|Aut(H)|$ is the number of orbits of the action of G.

15. Of the four rotations r, r^2, r^3, id, two, namely, r and r^3, only preserve the colorings in which all sides are of the same color. The identity preserves all 81 colorings. Finally, r^2 preserves those in which opposite sides are of the same color, which happens in nine cases. By Theorem 8.36, this shows that there are

$$\frac{1}{4}(1 \cdot 81 + 1 \cdot 9 + 2 \cdot 3) = 24.$$

16. The tetrahedron has 12 symmetries that can be achieved by a series of rotations, as we showed in Exercise 36 of Chapter 4. The easiest one is the identity, which will fix all k^6 colorings. Then there are the eight rotations around an axis that is perpendicular to one face F of the tetrahedron and contains the fourth vertex D; these will only fix the colorings in which the three blocks of F are monochromatic and the three blocks adjacent to D are monochromatic. There are k^2 such colorings. Finally, there are three transformations which interchange two pairs of vertices. Without loss of generality, consider the transformation that interchanges both A and B and C and D. This transformation fixes blocks AB and CD, interchanges blocks BC and AD, and interchanges blocks AC and BD. Therefore, this transformation will preserve colorings in which BC and AD have the same color, and AC and BD have the same color. There are k^4 such colorings.

    By Theorem 8.36, this yields that the number of nonequivalent colorings is

$$\frac{1}{12} \cdot (k^6 + 8k^2 + 3k^4).$$

    The reader is invited to find the 12 nonequivalent colorings for $k = 2$.

17. In this case, there are 24 symmetries that can be obtained by rotations. They are as follows:

    - The identity, which fixes all k^{12} colorings.
    - Six rotations, by ± 90 degrees each, around axes joining the centers of opposite faces. These fix the k^3 colorings in which the blocks of each face intersecting the axis of rotation are monochromatic (as are the four blocks parallel to the axis).
    - Three rotations, by 180 degrees, around the same axes. Note that these are fixed point–free involutions on the set of blocks. They fix the k^6 colorings in which pairs of blocks that are interchanged with each other are monochromatic.
    - Eight rotations, by 120 degrees (positive and negative), around the longest diagonals. They fix k^4 colorings, those in which the relevant triples of blocks are monochromatic.
    - Six rotations by 180 degrees, around axes that join midpoints of opposite blocks. These are involutions, but they have two fixed blocks. Therefore, they fix k^7 colorings.

    Therefore, Theorem 8.36 shows that the number of nonequivalent colorings is

$$\frac{1}{24}(k^{12} + 6k^3 + 3k^6 + 8k^4 + 6k^7).$$

8.9 Supplementary exercises

1. Let $\mathcal{F}$ be a linked regular design. Prove that $\mu(b-1) = r(k-1)$.
2. Find an example for a design on vertex set $[n]$ that is uniform and regular, but not balanced. Find the smallest possible n for which such a design exists.
3. Let M be the incidence matrix of the linked regular system $\mathcal{F}$. Express the $b \times b$ matrix $M^T M$ by the parameters of $\mathcal{F}$, the $b \times b$ identity matrix I_b, and the $b \times b$ matrix J_b whose entries are all equal to 1.
4. Prove that, in a balanced uniform design,

$$\lambda \cdot \binom{v}{2} = b \cdot \binom{k}{2}.$$

    Try to give two proofs as follows:

    (a) one by a direct combinatorial argument, and

(b) one by the learned identities on these parameters.

5. Prove that, in a balanced uniform (and therefore, regular) design with at least two blocks, $r \leq k$.

6. Prove that, in a linked regular (and therefore uniform) design, $b \leq v$.

7. Let us assume that a finite projective plane of order n exists. Prove that there then exists a balanced uniform design with $v = n^2$, $b = n^2 + n$, $k = n$, $r = n + 1$, and $\lambda = 1$.

8. Deduce Corollary 8.16 from Lemma 8.15 using only results on designs, not the axioms of projective planes.

9. Prove that no finite projective plane of order 14 exists.

10. (a) + Let n be of the form $4k+1$ or $4k+2$, and let n have a prime divisor p of the form $4k+3$. Assume furthermore that p^2 is not a divisor of n. Prove that no finite projective plane of order n exists. (Note this shows in particular that no finite projective planes of order 6 or 14 exist.)

    (b) Replace the condition that p^2 does not divide n by the condition that, in the prime factorization of n, the exponent of the prime p is odd. Prove that no finite projective plane of order n exists.

In Exercises 11-13, assume that no block is repeated.

11. Let $\mathcal{F}$ be a regular design in which each vertex occurs in r blocks. Let $\mathcal{F}^c$ be the design whose vertex set is the same as that of $\mathcal{F}^c$ and whose blocks are the *complements* of the blocks of $\mathcal{F}$. Then $\mathcal{F}^c$ is called the *complementary* design of $\mathcal{F}^c$.

    (a) Is $\mathcal{F}^c$ regular?

    (b) Now let $\mathcal{F}^c$ be balanced and regular. Is $\mathcal{F}^c$ balanced and regular?

12. Prove that there are as many k-uniform regular designs on v vertices in which each vertex occurs in r blocks as there are k-uniform regular designs on v vertices in which each vertex occurs in $\binom{v-1}{k-1} - r$ blocks.

13. Complete the following sentence: There are as many balanced k-uniform regular designs on v vertices in which any two vertices occur together in λ blocks as there are balanced k-uniform regular designs on v vertices in which any two vertices occur together in $\cdots$ blocks.

14. Is the reverse of the implication proved in Exercise 7 true?

15. Prove that if the permutation group G acts on a set S, and $i \in S$, then G_i is a subgroup of G.

16. Use Theorem 8.36 to prove that the average number of fixed points in a randomly selected n-permutation is 1.

17. Find all six nonequivalent colorings of Example 8.39.

18. Find the number of ways to color each side of a regular hexagon either red or blue if two colorings are considered equivalent if they can be rotated into each other.

19. + Find the number of ways we can color the *faces* of a cube using k colors if two colorings are considered equivalent if they can be rotated into each other.

20. Find the number of all $(7, 7, 3, 3, 1)$-BIBDs on vertex set $[7]$. For this exercise, two BIBDs are considered different if their incidence matrices are different. Isomorphic BIBDs count as different, unless they have the same incidence matrices.

21. Show an example of two BIBDs with the same parameters that are not isomorphic.

22. Let $\mathcal{F}$ be a symmetric BIBD, and let B be a block of $\mathcal{F}$. Prove that the sets $B \cap B_i$, where B_i ranges all other blocks of $\mathcal{F}$, form a BIBD. This is called the *derived design* of $\mathcal{F}$.

23. Let $\mathcal{F}$ be a symmetric BIBD, and let B be a block of $\mathcal{F}$. Prove that the sets $B - B_i$, where B_i ranges all other blocks of $\mathcal{F}$, form a BIBD. This is called the *residual design* of $\mathcal{F}$.

24. Find the number of ways to color the *vertices* of a square using only colors from the set $[r]$ if two colorings are considered equivalent if they can be translated into each other by rotations or reflections.

9

Sequences in combinatorics

In this chapter, we will consider combinatorial properties of entire sequences, as opposed to their individual elements.

9.1 Unimodality

In the previous chapter, we encountered the following problem in a different context. A professor is preparing questions for an oral examination of a large number of students. Being a math professor, he puts together a list of n questions. He wants to be fair and ask each student the same number k of questions. He does not want to ask the same set of questions twice. What should k be in order to maximize the number of students the professor can test with these conditions?

As $[n]$ has $\binom{n}{k}$ subsets of k elements, the professor needs to find the value of k for which $\binom{n}{k}$ is maximal. A look at Figure 9.1 (which you may remember from Chapter 1), in which the entries in the nth row are the values of $\binom{n}{k}$ for fixed n, suggests that the numbers $\binom{n}{0}, \binom{n}{1}, \cdots$ increase steadily until $k = \lfloor n/2 \rfloor$, and then they decrease steadily. This pattern is an important property of sequences that has its own name.

$$\begin{array}{ccccccccccc} & & & & & 1 & & & & & \\ & & & & 1 & & 1 & & & & \\ & & & 1 & & 2 & & 1 & & & \\ & & 1 & & 3 & & 3 & & 1 & & \\ & 1 & & 4 & & 6 & & 4 & & 1 & \\ 1 & & 5 & & 10 & & 10 & & 5 & & 1 \end{array}$$

Figure 9.1
The values of $\binom{n}{k}$. Note that for fixed n, the values $\binom{n}{0}, \binom{n}{1}, \cdots$ form the nth row of the Pascal triangle.

Definition 9.1 *The sequence $a_0, a_1, \cdots, a_n$ of nonnegative real numbers is called* unimodal *if there exists an index m so that $a_0 \leq a_1 \leq \cdots \leq a_m \geq a_{m+1} \geq \cdots \geq a_{n-1} \geq a_n$.*

DOI: 10.1201/9781003304272-9

In other words, a sequence is unimodal if its elements increase steadily, then decrease steadily.

It turns out that our observation about the unimodality of the sequence $\binom{n}{k}_{0 \leq k \leq n}$ always holds true.

Proposition 9.2 *For any fixed positive integer n, the sequence* $\binom{n}{k}_{0 \leq k \leq n}$ *is unimodal.*

There are many proofs of this well-known fact. We start with the simplest one, which is computational.

Proof: Let $k \leq \lfloor (n-1)/2 \rfloor$. Then we claim that $\binom{n}{k} \leq \binom{n}{k+1}$, so our sequence is (weakly) increasing. Indeed, we have

$$\frac{\binom{n}{k+1}}{\binom{n}{k}} = \frac{(n-k)!k!}{(n-k-1)!(k+1)!} = \frac{n-k}{k+1} \geq 1. \tag{9.1}$$

Furthermore, the computation above shows that, if $k > \lfloor (n-1)/2 \rfloor$, then $\binom{n}{k} > \binom{n}{k+1}$, so our sequence is decreasing. This proves our claim. ◊

Thus, the professor can use his questions most efficiently if he sets $k = \lfloor n/2 \rfloor$.

Note that our sequence $\binom{n}{k}_{0 \leq k \leq n}$ is not simply unimodal; it is also symmetric, since $\binom{n}{k} = \binom{n}{n-k}$. This implies that the maximum of the sequence must be in the middle.

The above proof was simple, but not very enlightening. It strongly depended on the fact that we had a formula for the number of k-element subsets of $[n]$, and a simple one at that. We would like to develop other methods that can be used to prove unimodality results in cases when this is not the case. We will first illustrate our method, called the *reflection principle*, by using it to re-prove Proposition 9.2. As a bonus, we will get a *combinatorial* proof of the fact that $\binom{n}{k} \leq \binom{n}{k+1}$ for $k \leq \lfloor (n-1)/2 \rfloor$.

To that end, let us return to Example 1.26 of Chapter 1, where we implicitly defined a simple bijection between the set of all k-element subsets of $[n]$ and all northeastern lattice paths from $(0,0)$ to $(n-k,k)$. For easy reference, the bijection f works as follows: If the elements of a k-element subset $A \subseteq [n]$ are $a_1, a_2, \cdots, a_k$, then $f(A)$ is the northeastern lattice path whose k north steps are the a_1st, a_2nd, ..., a_kth steps of the lattice path.

Example 9.3 *Figure 9.2 shows the images of two 4-element subsets of* $[8]$.

Because of this bijection, we can present the rest of our proof in terms of lattice paths instead of k-element subsets of $[n]$. We propose the following strategy. Let L_k be the set of northeastern lattice paths from $(0,0)$ to $(n-k,k)$, and let $k \leq \lfloor (n-1)/2 \rfloor$. We will show that, in that case, there exists an *injection*, that is, a one-to-one map $g : L_k \rightarrow L_{k+1}$. This will prove that $|L_k| \leq |L_{k+1}$, which then implies the unimodality of our symmetric sequence $\binom{n}{k}_{0 \leq k \leq n}$.

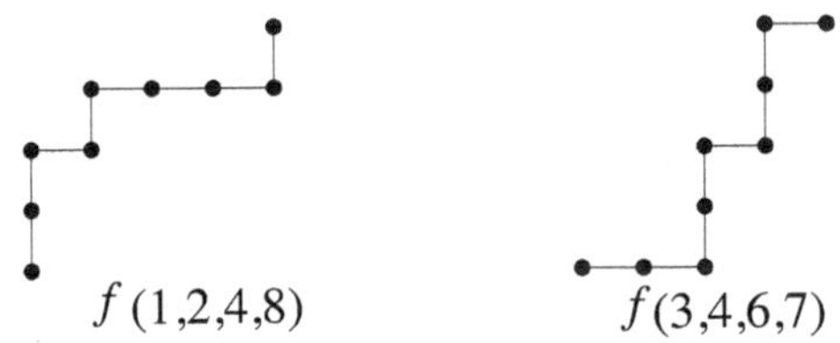

Figure 9.2
The images of $\{1,2,4,8\} \subseteq [8]$ and $\{3,4,6,7\} \subseteq [8]$.

Let $p \in L_k$. Since $k \leq \lfloor n-1/2 \rfloor$, this means that the endpoint $P = (n-k, k)$ of p is below the main diagonal. Let Q be the point $(n-k-1, k+1)$, the point where the paths belonging to L_{k+1} end. Let t denote the bisector of the segment PQ. Then t is the line $y = x + (n-2k) - 1$, and the reader is invited to verify that the two endpoints of p are on *opposite sides* of t. Indeed, the intersection of t and the horizontal axis is the point $(0, n-2k-1)$, which is in the positive half of the horizontal axis thanks to our condition on k.

Therefore, p and t intersect. Let X be the last (most northeastern) of their intersection points. Let us *reflect* the partial path XP through t to get a partial path XQ. Now define $g(p)$ as the part of p that goes from $(0,0)$ to X, followed by the path XQ we just obtained by reflection. This method of proving unimodality statements is called the *reflection principle* and is illustrated with additional examples in Bruce Sagan's survey article [63].

Example 9.4 *Figure 9.3 shows how g maps the lattice path associated with* $\{3,4,8\} \subset [8]$ *into the lattice path associated with* $\{3,4,6,7\}$.

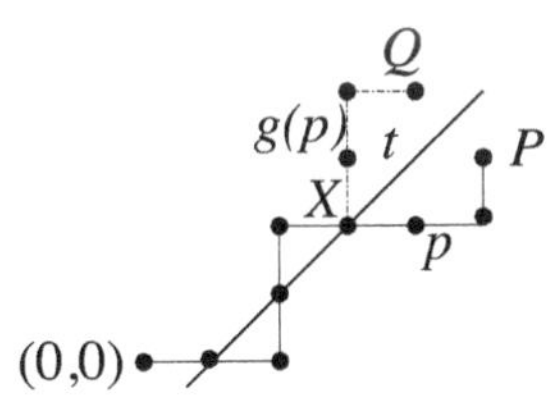

Figure 9.3
The reflection principle at work.

Note that $g(p)$ and t always intersect. The following lemma shows that the reflection principle indeed works.

Lemma 9.5 *For all positive integers $k \leq \lfloor (n-1)/2 \rfloor$, the map $g : L_k \to L_{k+1}$ defined above is an injection.*

Proof: Let $q \in L_{k+1}$, and let P, Q, and t be defined as above. If q and t do not intersect, then q is not in the range of g. Otherwise, let X be the most northeastern intersection point of q and t. Then the unique preimage of q under g is obtained by reflecting the part of q between X and Q through t, and leaving the rest of q unchanged. ◊

The reflection principle can be applied to more complicated structures as well. The interested reader should look at Chapter 8 of [13], where this technique is applied to decreasing binary trees to prove a unimodality result.

Quick Check

1. Let $\{a_k\}_{0\leq k\leq n}$ and $\{b_k\}_{0\leq k\leq n}$ be unimodal sequences. Is it true that $\{a_k + b_k\}_{0\leq k\leq n}$ is a unimodal sequence?
2. Is it true that all subsequences of a unimodal sequence are unimodal?
3. Let n be a fixed positive integer, and let a_k be the number of compositions of n into k parts. Is $\{a_n\}_{0\leq k\leq n}$ a unimodal sequence?

9.2 Log-concavity

Interestingly, there is a property of sequences, called *log-concavity*, that is stronger than unimodality, yet often is easier to prove.

9.2.1 Log-concavity implies unimodality

Let n be a fixed positive integer, and let $a_{n,k}$ denote the number of involutions of length n that have k 2-cycles. It is straightforward to compute (see Example 4.36) that

$$a_{n,k} = \frac{n!}{k!(n-2k)! \cdot 2^k}. \tag{9.2}$$

Consider the sequence $a_{n,0}, a_{n,1}, \cdots, a_{n,\lfloor n/2\rfloor}$ for the first few values of n. See Figure 9.4.

From these data, it seems that the sequences $\{a_{n,k}\}_{0\leq k\leq \lfloor n/2\rfloor}$ are unimodal for each fixed n. However, proving this directly seems difficult, even if there is an exact formula for the numbers $a_{n,k}$, since it is not clear where the *peak* of the sequences is. Indeed, for $n = 4$ and $n = 6$, the peak is not at the end of the sequence, while for the other values of n, it is, though sometimes tying the next-to-last element of the sequence. Not knowing where the peak is caused difficulty, since without the position of the peak we cannot prove unimodality by proving that the sequence increases until it reaches the peak, and decreases afterward.

In similar situations, it is often easier to prove a stronger property of sequences.

n				
1	1			
2	1	1		
3	1	3		
4	1	6	3	
5	1	10	15	
6	1	15	45	15
7	1	21	105	105

Figure 9.4
The sequences $\{a_{n,k}\}_{0\leq k\leq \lfloor n/2\rfloor}$, for $n \leq 7$.

Definition 9.6 *The sequence* $d_0, \cdots, d_m$ *of positive real numbers is called* log-concave *if, for all* $k \in [m-1]$,

$$d_{k-1}d_{k+1} \leq d_k^2. \tag{9.3}$$

The terminology "log-concave" is easy to explain. If a sequence is log-concave, then the sequence of the *logarithms* of its elements (in any fixed base) is *concave*. Indeed, let us take the logarithm of both sides of (9.3) in some fixed base and divide by 2, to get

$$\frac{\log d_{k-1} + \log d_{k+1}}{2} \leq \log d_k,$$

for all $k \in [m-1]$. That is, the graph of the sequence $\log d_k$ is *concave*; in other words, $\log d_k$ is above the line connecting $\log d_{k-1}$ and $\log d_{k+1}$. See Figure 9.5 for an illustration.

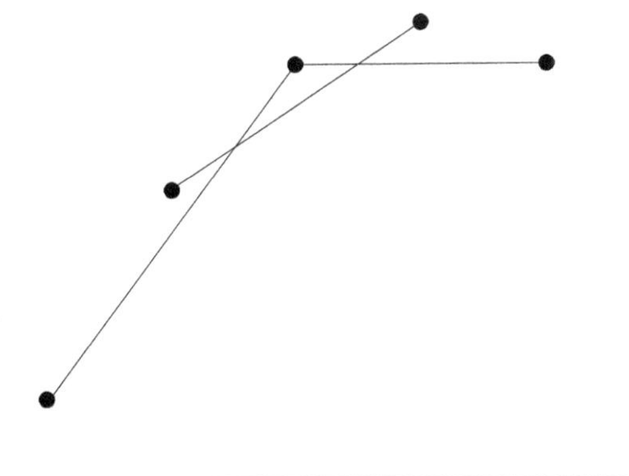

Figure 9.5
A concave sequence. Each element is larger than the average of its two neighbors.

The crucial property of log-concavity is presented in the following proposition.

Proposition 9.7 *If a sequence of positive real numbers is log-concave, then it is unimodal.*

Proof: On one hand, the sequence $d_0, d_1, \cdots, d_m$ is unimodal if and only if it is true that, once the ratio d_{k+1}/d_k dips below 1, it stays there. On the other hand, the sequence $d_0, d_1, \cdots, d_m$ is log-concave if and only if the sequence d_{k+1}/d_k is monotone decreasing. (This can be seen by dividing both sides of (9.3) by $d_k d_{k+1}$.) Because the second condition is stronger, our statement is proved. ◊

It follows from the previous proof that log-concavity is a *strictly* stronger property than unimodality. Indeed, the sequence 1, 1, 2 is unimodal, but not log-concave.

Let us use our fresh knowledge to prove that the sequence $\{a_{n,k}\}_k$ is unimodal.

Proposition 9.8 *Let $a_{n,k}$ be defined as above. Then, for any fixed n, the sequence $a_{n,0}, a_{n,1}, \cdots, a_{n,\lfloor n/2 \rfloor}$ is log-concave and therefore unimodal.*

Proof: Using Formula 9.2, we get

$$b_{k+1} = \frac{a_{n,k+1}}{a_{n,k}} = \frac{\frac{n!}{(k+1)!(n-2k-2)!2^{k+1}}}{\frac{n!}{k!(n-2k)!\cdot 2^k}} = \frac{(n-2k)(n-2k-1)}{2(k+1)},$$

and

$$b_k = \frac{a_{n,k}}{a_{n,k-1}} = \frac{\frac{n!}{k!(n-2k)!2^k}}{\frac{n!}{(k-1)!(n-2k+2)!\cdot 2^{k-1}}} = \frac{(n-2k+1)(n-2k+2)}{2k}.$$

Therefore, comparing the preceding two equations, we get

$$\frac{b_{k+1}}{b_k} = \frac{n-2k-1}{n-2k+1} \cdot \frac{n-2k}{n-2k+2} \cdot \frac{k}{k+1} < 1,$$

since each of the three factors in the middle is less than 1.

Therefore, the ratio $\frac{a_{n,k+1}}{a_{n,k}}$ is monotone decreasing, which is equivalent to the log-concavity of the sequence $a_{n,k}$. ◊

Once again, we would like to point out that, in this proof of log-concavity, we did *not* need to know where the *peak* of the sequence was.

9.2.2 The product property

Another great advantage of log-concavity over unimodality is that it is preserved when products of polynomials are taken.

Theorem 9.9 *If the polynomial $P(x)$ has log-concave coefficients, and the polynomial $Q(x)$ has log-concave coefficients, then so does the polynomial $P(x)Q(x)$.*

Note that the corresponding statement is *false* for unimodal polynomials. Indeed, let $P(x) = 1 + x + 5x^2$, and let $Q(x) = 1 + x + 6x^2$. Then

$$P(x)Q(x) = 1 + 2x + 12x^2 + 11x^3 + 30x^4,$$

which is not a unimodal polynomial.

In order to prove Theorem 9.9, we first need a classic result from linear algebra.

Lemma 9.10 (Cauchy–Binet formula) *Let A and B be two $n \times n$ matrices. For k-element subsets I and J of $[n]$, and the $n \times n$ matrix C, let $C_{I,J}$ denote the $k \times k$ submatrix of C that is obtained from C by taking the intersection of the rows indexed by I and the columns indexed by J.*

Then

$$\det(AB)_{I,J} = \sum_K \det(A_{I,K}) \det(B_{K,J}),$$

where K ranges over all k-element subsets of $[n]$.

The proof of this formula is somewhat tedious, but not particularly interesting. We will leave it for standard textbooks in matrix algebra. Note that, for $k = 1$, the statement is true since it reduces to the definition of matrix multiplication. Starting from this point, it is conceptually not difficult to build up an induction proof, expanding $\det(AB)_{I,J}$ with respect to its first row.

Let us now return to the proof of the product property of log-concave polynomials (for brevity, this is what we will call polynomials with log-concave coefficients).

Proof: (of Theorem 9.9) Let P be of degree p and let Q be of degree q. Let us set $P(x) = \sum_{k=0}^{p+q} p_k x^k$ and $Q(x) = \sum_{k=0}^{p+q} q_k x^k$. In other words, if $k > p$, then $a_k = 0$ and if $k > q$, then $b_k = 0$.

We will now construct two matrices which will multiply together very similarly to the polynomials $P(x)$ and $Q(x)$. Define the $(p+q) \times (p+q)$ upper triangular matrices as follows:

$$P = \begin{pmatrix} p_0 & p_1 & p_2 & \cdots & p_{p+q} \\ 0 & p_0 & p_1 & \cdots & p_{p+q-1} \\ 0 & 0 & p_0 & \cdots & p_{p+q-2} \\ \cdots & \cdots & \cdots & \cdots & \cdots \\ 0 & 0 & \cdots & 0 & p_0 \end{pmatrix},$$

and

$$Q = \begin{pmatrix} q_0 & q_1 & q_2 & \cdots & q_{p+q} \\ 0 & q_0 & q_1 & \cdots & q_{p+q-1} \\ 0 & 0 & q_0 & \cdots & q_{p+q-2} \\ \cdots & \cdots & \cdots & \cdots & \cdots \\ 0 & 0 & \cdots & 0 & q_0 \end{pmatrix}.$$

Then

$$\begin{aligned} P(x)Q(x) &= \begin{pmatrix} p_0q_0 & p_1q_0 + p_0q_1 & p_2q_0 + p_1q_1 + p_0q_2 & \cdots \\ 0 & p_0q_0 & p_1q_0 + p_0q_1 & \cdots \\ \cdots & \cdots & \cdots & \cdots \end{pmatrix} \\ &= \begin{pmatrix} r_0 & r_1 & r_2 & \cdots \\ 0 & r_0 & r_1 & \cdots \\ 0 & 0 & r_0 & \cdots \\ \cdots & \cdots & \cdots & \cdots \end{pmatrix}, \end{aligned}$$

where r_i is the coefficient of x^i in $P(x)Q(x)$. In other words, the rows of P (resp. Q, PQ) are given by the sequences of coefficients of $P(x)$ (resp. $Q(x)$, $P(x)Q(x)$), but they are gradually shifted.

The log-concavity of the polynomials $P(x)$ and $Q(x)$ means that all 2×2 submatrices of P and Q have a nonnegative determinant. By the Cauchy–Binet formula (Lemma 9.10) with $k = 2$, this implies that all 2×2 submatrices of PQ must have a nonnegative determinant. That, in turn, is equivalent to saying that the coefficients of $P(x)Q(x)$ are log-concave. ◊

The following is a classic application of the product property.

Corollary 9.11 *For all positive integers n, the sequence $\{b(n,k)\}_{1\leq k\leq \binom{n}{2}}$ of the numbers counting n-permutations according to their number of inversions is log-concave.*

Proof: We have seen in Theorem 4.54 that

$$\sum_{k=0}^{\binom{n}{2}} b(n,k)x^k = (1+x)(1+x+x^2)\cdots(1+x+\cdots+x^{n-1}).$$

Since each of the $n-1$ factors of the right-hand side is log-concave, our claim follows from Theorem 9.9. ◊

In Supplementary Exercise 14, you are asked to find an injective proof of this result.

9.2.3 Injective proofs

It is challenging to give a combinatorial proof of Corollary 9.11. First of all, how can we prove log-concavity *combinatorially*? Since log-concavity is defined by the inequality $a_{k-1}a_{k+1} \leq a_k^2$, we have to prove that inequality in a combinatorial way. This can be done if we find an injection from a set with $a_{k-1}a_{k+1}$ elements into a set with a_k^2 elements.

It turns out that the following equivalent definition of log-concavity is often helpful.

Proposition 9.12 *The sequence $\{a_k\}_{0\leq k\leq n}$ is log-concave if and only if, for all k and l satisfying $1 \leq k \leq l \leq n-1$,*

$$a_{k-1}a_{l+1} \leq a_k a_l.$$

Proof: Both statements are equivalent to the statement that the sequence a_{k+1}/a_k is decreasing. ◊

As a first example of our next method, let us prove the log-concave property of the binomial coefficients combinatorially.

Let $A_{n,m}$ denote the set of all m-element subsets of $[n]$. We are going to define an injection $f_k : A_{n,k-1} \times A_{n,k+1} \to A_{n,k} \times A_{n,k}$ recursively as follows: Assume that for all positive integers $m < n$, we already know that the sequence $\binom{m}{k}_{0\leq k\leq m}$ is log-concave; therefore, we know that an injection $g_{m,k,l} : A_{m,k-1} \times A_{m,l+1} \to A_{m,k} \times A_{m,l}$ exists for each pair (k,l) satisfying $1 \leq k \leq l \leq m-1$.

Let $(S,T) \in A_{n,k-1} \times A_{n,k+1}$. There are four cases.

1. If $n \notin S$ and $n \in T$, then let $S' = n \cup S$ and let $T' = T - \{n\}$. Then set $f_k(S,T) = (S',T')$. Note that S' contains n, and T' does not.
2. If both S and T contain n, then remove n from both sets and look at the remaining sets S_1 and T_1 as subsets of the set $[n-1]$. Take $g_{n-1,k-2,k}(S_1,T_1)$, then reattach n to the end of both sets to get the image $(S',T') = f(S,T)$. Note that both S' and T' contain n.
3. If neither S nor T contains n, then they are both subsets of $[n-1]$, so we define $f(S,T) = (S',T') = g_{n-1,k-1,k+1}(S,T)$. Note that neither S' nor T' contains n.
4. If $n \in S$ and $n \notin T$, then let $Z = S - \{n\}$. Take $g_{n-1,k-2,k}(Z,T) = (X,Y)$, which is defined since Z is three elements smaller than T. Now $|X| = k-1$, and $|Y| = k$. Attach n to the end of X, then *swap* the role of X and Y by setting $(S',T') = (Y,X)$. Note that S' does not contain n, and T' does.

Example 9.13 *If $n = 2$, and $(S,T) = (\emptyset, \{1,2\})$, then the above rules yield $f_1(S,T) = (\{2\},\{1\})$.*

Example 9.14 *Let $n = 4$ and $k = 2$. Then each of the four examples below shows an instance of a case of the definition of f_k.*

- $f(\{3\},\{2,3,4\}) = (\{3,4\},\{2,3\})$.
- $f(\{4\},\{2,3,4\}) = (\{3,4\},\{2,4\})$.
- $f(\{3\},\{1,2,3\}) = (\{2,3\},\{1,3\})$.
- $f(\{4\},\{1,2,3\}) = (\{1,2\},\{3,4\})$.

In the last example, we assumed that $g_{3,0,3}(\emptyset,\{1,2,3\}) = (3,\{1,2\})$.

Proposition 9.15 *The function* $f_k : A_{k-1} \times A_{k+1} \rightarrow A_k \times A_k$ *defined above is an injection.*

Proof: Let $(U,V) \in A_k \times A_k$. The containment relations between n and U and n and V uniquely determine which one of the four rules must be applied to (S,T) so that $f(S,T) = (U,V)$ could hold. Since each of the four rules is injective, the statement is proved. ♢

We are now ready to use our recursive method to give a nongenerating function proof of Corollary 9.11. Let n be fixed, and let $I_{n,k}$ be the set of all n-permutations with k inversions. We will achieve our goal by proving the following theorem.

Theorem 9.16 *For all integers k and l satisfying* $0 \leq k \leq n-2 \leq \binom{n}{2} - 2$, *there exists an injection* $f_{n,k,l}$ *from* $I_{n,k} \times I_{n,l}$ *to* $I_{n,k+1} \times I_{n,l-1}$.

The smallest value of n for which the domains of the maps $f_{n,k,l}$ are not all empty is $n = 3$. In this case, $f_{n,k,l}$ is defined for the (k,l)-pairs (0,2), (0,3), and (1,3). In those cases, we define

$$f_{3,0,2}(123,231) = (213,132), \quad f_{3,0,2}(123,312) = (213,213),$$

$$f_{3,0,3}(123,321) = (213,231), \quad f_{3,1,3}(132,321) = (231,231),$$

and

$$f_{3,1,3}(213,321) = (312,231).$$

These definitions may look arbitrary now, but they will seem quite logical once the general structure of f is discussed.

Now let $n \geq 4$, and assume we have defined $f_{n,k,l}$ for $n-1$ and for all allowed values of k and l.

Let $(p,q) \in I_{n,k} \times I_{n,k+2}$, with $p = p_1 p_2 \cdots p_n$ and $q = q_1 q_2 \cdots q_n$. Proceed as follows:

(Rule 1) If $p_1 < n$ and $q_1 > 1$, increase p_1 by one, and decrease the entry of p that was one larger than p_1 by one. Let the obtained permutation be p'. Similarly, decrease q_1 by one, and increase the entry of q that was one larger than q_1 by one. Let the obtained permutation be q'. Set $f_{n,k,k+2}(p,q) = (p',q')$.

Note that p' starts with an entry larger than 1, and q' starts with an entry less than n.

Example 9.17 *If* $p = 2134$ *and* $q = 3142$*, then* $f_{4,1,3}(p,q) = (3124, 2143)$.

(Rule 2) If $p_1 = n$ or $q_1 = 1$, then remove these entries to get the permutations $p*$ and $q*$. (After natural relabeling, these are both permutations of length $n-1$.) Because of the extreme values of at least one of the omitted elements, we have $i(q*) - i(p*) \geq i(q) - i(p) \geq 2$. Therefore, there exist positive integers r and s, with $r \leq s-2$, so that $(p*, q*)$ is in the domain of $f_{n-1,r,s}$.

Take $f_{n-1,r,s}(p*, q*) = (\bar{p}, \bar{q}) \in I_{n-1,r+1} \times I_{n-1,s-1}$. Now, prepend $\bar{p}$ by p_1, and prepend $\bar{q}$ by q_1. In both cases, entries larger than or equal to the prepended entry have to be increased by one. Call this new pair of n-permutations $(p_1\bar{p}, q_1\bar{q})$. Finally, set $f_{n,k,k+2}(p,q) = (q_1\bar{q}, p_1\bar{p})$. We point out that we *swapped* p and q.

Note that either $q_1\bar{q}$ starts in 1 or $p_1\bar{p}$ starts in n.

Example 9.18 *If* $p = 1324$ *and* $q = 1432$*, then we have* $(p*, q*) = (213, 321)$*; therefore, recalling that we have already defined* $f_{3,1,3}$ *for 3-permutations,* $f_{3,1,3}(p*, q*) = (\bar{p}, \bar{q}) = (312, 231)$. *Reinserting the removed first entries, we get* $(p_1\bar{p}, q_1\bar{q}) = (1423, 1342)$. *Finally, swapping the two permutations, we get* $f_{4,1,3}(p,q) = (1342, 1423)$.

Lemma 9.19 *The map* $f_{k,k+2}$ *is an injection from* $I_{n,k} \times I_{n,k+2}$ *into* $I_{n,k+1} \times I_{n,k+1}$.

Proof: First, $f_{n,k,k+2}$ maps into $I_{n,k+1} \times I_{n,k+1}$ since both rules increase the number of inversions of the first permutation by one and decrease the number of inversions of the second permutation by one.

Now we prove that $f_{n,k,k+2}$ is one-to-one. We achieve this by induction on n, the statement holding in the initial case of $n = 3$. Let us assume that the statement is true for $n-1$.

Let $(t,u) \in I_{n,k+1} \times I_{n,k+1}$, with $t = t_1t_2\cdots t_n$ and $u = u_1u_2\cdots u_n$. We show that (t,u) can have at most one preimage under $f_{n,k,k+2}$. There are two cases.

1. If $t_1 > 1$ and $u_n < n$, then (t,u) could only be obtained as a result of applying $f_{n,k,k+2}$ if Rule 1 was used. In that case, we have $f^{-1}(t,u) = ((t_1-1)t_2\cdots t_n, (u_1+1)u_2\cdots u_n)$.
2. If $t_1 = 1$ or $u_1 = n$, then (t,u) could only be obtained as a result of applying $f_{n,k,k+2}$ if Rule 2 was used. In that case, to get the preimage of (t,u), we need to remove the first entry of t and the

first entry of u, swap the permutations, and find the preimage of the resulting pair $(\bar{u}, \bar{t})$.

However, the preimage of $(\bar{u}, \bar{t})$ is unique by the induction hypothesis; therefore, so is $f^{-1}(t, u)$.

◊

Consequently, the sequence of the numbers $|I_{n,k}| = b(n,k)$, where $k = 0, 1, \cdots, \binom{n}{2}$ is log-concave, and thus the injections $f_{n,k,l}$ exist for permutations of length n and for all values k and l satisfying $0 \leq k \leq l \leq \binom{n}{2} - 2$.

9.2.3.1 Lattice paths again

Lattice paths are at least as useful in proving log-concavity results as they are in proving unimodality results. The classic example is again the log-concavity of the binomial coefficients $\binom{n}{k}$, for any fixed n.

Let us return to our well-tested lattice path representation of the subsets of $[n]$ by northeastern lattice paths of n steps. In the very unlikely case that the reader does not remember this representation, it was first defined in Example 1.26, then used again in Example 9.3. Using the notations introduced after Example 9.3, we want to find an injection from the set $L_{k-1} \times L_{k+1}$ into the set $L_k \times L_k$. The existence of such an injection will show that $|L_{k-1}||L_{k+1}| \leq |L_k|^2$, or in other words, $\binom{n}{k-1}\binom{n}{k+1} \leq \binom{n}{k}^2$.

In order to find such an injection f, we are going to modify the techniques introduced after Example 9.3 in a minor, but crucial way. Keeping the notation introduced there, let $p \in L_{k-1}$ and let $q \in L_{k+1}$. Now *translate* q by the vector $(1, -1)$ so now it starts at $(1, -1)$. Then p is still a path from $(0, 0)$ to $(n-k+1, k-1)$, but q is a path from $(1, -1)$ to $(n-k, k)$. Therefore, p and q now *intersect*. Let P be their first (most southwestern) intersection point, and let us say that P cuts p into segments p' and p'', and P cuts q into segments q' and q''. Then we define $f(p, q) = (p'q'', q'p'')$. In other words, we *swap* the ending segments of p and q. The obtained pair $f(p, q)$ is in $L_k \times L_k$ (after translation), since both of its paths have exactly k north steps. Swapping their ending segments balanced out the difference in their number of north steps. The map f is an injection, since, given $f(p, q)$, we can find the first intersection point of the two paths of that pair and then recover the unique preimage of that pair by swapping the ending segments back.

Example 9.20 *If $n = 6$, $S = \{1, 5\}$, and $T = \{1, 3, 4, 5\}$, then our injection is shown in Figure 9.6.*

This method of "swapping" segments of lattice paths after their intersection points is a powerful tool for proving log-concavity results. What is often needed is an enhanced version of the northeastern lattice path representation that can represent elements of a sequence. Finding this enhanced lattice path model can be the most difficult part of the proof. The interested reader should

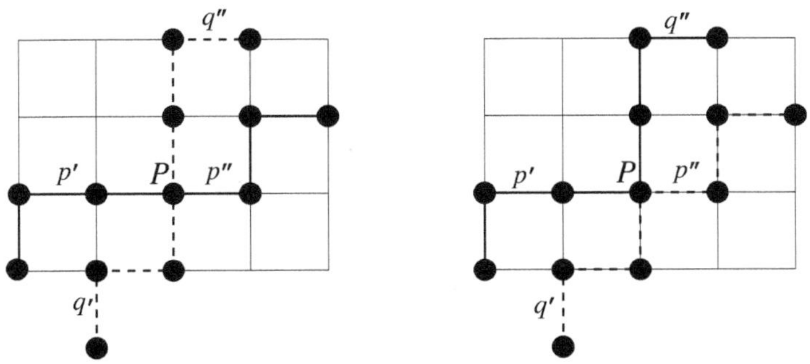

Figure 9.6
The injection $f : L_{k-1} \times L_{k+1} \to L_k \times L_k$.

consult Chapter 1 of *Combinatorics of Permutations* [13], where this method is used to prove that the sequence $A(n,1), A(n,2), \cdots, A(n,n)$ is log-concave for any fixed n.

For now, we show a simple example of enhanced lattice paths to illustrate the concept. Assume we want to decorate our backyard with n plants, which will form a line. Some of these can be flowers, others can be bushes. We have two kinds of flowers and three kinds of bushes of which we can select one for each spot.

Let a_k be the number of ways we can do this if we will have k flowers in our backyard. It is then straightforward to see that $a_k = \binom{n}{k} 2^k \cdot 3^{n-k}$. From this, it is easy to prove computationally that the sequence $a_0, a_1, \cdots, a_n$ is log-concave. However, we will now present a combinatorial proof.

Let us represent each line of plants with a *labeled* northeastern lattice path as follows: Let a flower correspond to a north step and let a bush correspond to an east step. Furthermore, label each edge with the type of the plant that is put in the corresponding slot, that is, by 1 or 2 for north steps, and 1, 2, or 3 for an east step. See Figure 9.7 for an illustration.

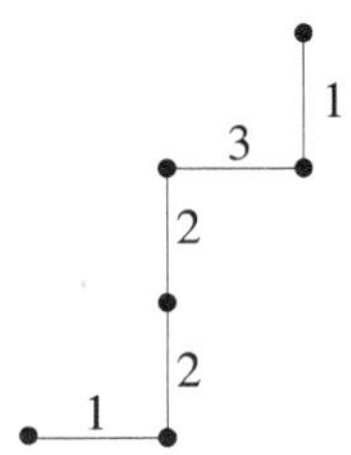

Figure 9.7
The labeled lattice path corresponding to the five-plant line $B_1F_2F_2B_3F_1$.

The reader can easily verify that this representation is bijective. That is, the set of n-plant lines containing k flowers has as many elements as the set of northeastern lattice paths consisting of n steps, k of which are north steps in which north steps are labeled with elements of [2], and east steps are labeled with elements of [3]. Therefore, we can now continue our argument in the language of these lattice paths, instead of plant lines.

The rest of the proof is analogous to the above proof of log-concavity of the binomial coefficients. Define L_k just as in that proof, except that the lattice paths are now labeled. Then define the map $f : L_{k-1} \times L_{k+1} \to L_k \times L_k$ as in that proof as well, that is, by swapping the parts of paths that come after their first intersection point. The fact that the paths are labeled does not change anything, since the validity of a label only depends on the direction of the step to which the label is assigned. So f is an injection by the same argument, and log-concavity is proved.

Quick Check

1. Let $f_{n,m}(x) = (1+x)^n \cdot I_m(x)$. Prove that $f_{n,m}(x)$ is a log-concave polynomial.
2. Recall that $A(n,k)$ denotes an Eulerian number. Prove that the sequence $\{A(n,2)\}_{n\geq 2}$ is log-concave.
3. Let $D(n)$ denote the number of derangements of length n. Prove that the sequence $\{D_n\}_{n\geq 2}$ is log-convex, that is, the inequality $D(n+1)D(n-1) \geq D_n^2$ holds, for all $n \geq 3$.

9.3 The real zeros property

There is a related property of sequences that is even stronger than log-concavity. We say that the sequence $a_0, a_1, \cdots, a_n$ of positive real numbers *has real zeros only* if the polynomial $\sum_{k=0}^{n} a_k x^k$ has real roots only. Note that, in this case, these roots cannot be positive, since for $x > 0$, we have $\sum_{k=0}^{n} a_k x^k > 0$.

Example 9.21 *For any fixed positive integer n, the sequence of binomial coefficients $\binom{n}{0}, \binom{n}{1}, \cdots, \binom{n}{n}$ has real zeros only.*

The reader will be asked to prove this simple statement in Supplementary Exercise 5.

The following theorem is the reason that we discuss the real zeros property in this chapter, but it is not the only interesting consequence of the real zeros property. Other interesting consequences will be mentioned in the Notes section.

Theorem 9.22 *If a sequence of positive real numbers has real zeros only, then it is log-concave.*

Note that the converse is not true. The sequence $1, 1, 1$ is a counterexample.

In order to prove Theorem 9.22, we will need the following classic result from real analysis.

Theorem 9.23 (Rolle's theorem) *Let $f : \mathbf{R} \rightarrow \mathbf{R}$ be a function that is continuous on the interval $[a, b]$ and differentiable on the interval (a, b). Assume furthermore that $f(a) = f(b) = y$. Then there exists a point $t \in [a, b]$ so that $f'(t) = 0$.*

Proof: As f' is continuous on the closed finite interval $[a, b]$, it follows that f has a maximum and a minimum value on this interval $[a, b]$. If both these values are equal to y, then f is constant on $[a, b]$, and $f'(x) = 0$ for all $x \in (a, b)$.

If f has a minimum or maximum value c on $[a, b]$ that is not equal to y, then let us assume without loss of generality that c is a maximum value, and that $f(t) = c$, where $t \in (a, b)$. Then $f'(t) = 0$ because f has a maximum in t. $\diamondsuit$

The following special case is the one that is most often used in combinatorial applications.

Corollary 9.24 *If a polynomial $r(x)$ with real coefficients has real zeros only, then so does its derivative $r'(x)$.*

Proof: First assume for simplicity that all n roots of r are *distinct*. Then, by Rolle's theorem, applied with $y = 0$, the polynomial r' must have a root between any two consecutive roots of r. This provides $n - 1$ real roots, so all $n - 1$ roots of $r - 1$ are real. See Figure 9.8 for an illustration.

Now let us assume r has some repeated roots as well. Let $a_1, a_2, \cdots, a_k$ be the multiplicities of the k distinct roots of r, with $\sum_{i=1}^{k} a_i = n$. It follows from the derivation rule of products that, if r_i is a root of r with multiplicity a_i, then r_i is a root of r' with multiplicity $a_i - 1$. Therefore, the roots r_i of r will provide a total of $\sum_{i=1}^{k}(a_i - 1) = n - k$ real roots of r'. Then, we can apply Rolle's theorem for each pair of consecutive distinct roots of r, and we get $k - 1$ additional real roots. This completes the proof. $\diamondsuit$

We can now prove Theorem 9.22.

Proof: (of Theorem 9.22) Let $p(x) = \sum_{i=0}^{n} a_i x^i$ be a polynomial with real coefficients that has real zeros only. Consider the two-variable polynomial

$$q(x, y) = \sum_{i=0}^{n} a_i x^i y^{n-i} = p(x/y)y^n.$$

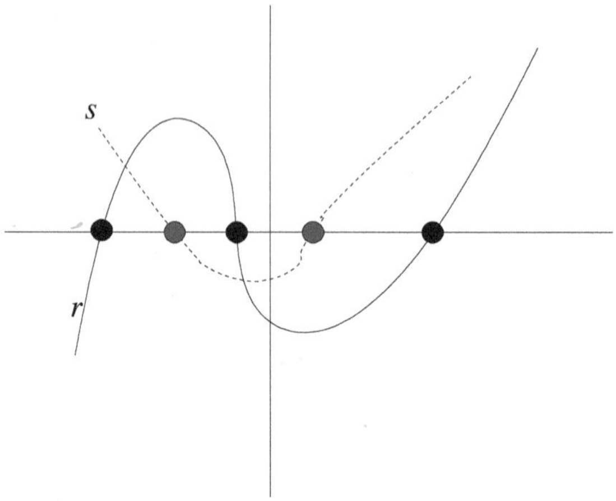

Figure 9.8
The roots of r and $s = r'$ are real and interlacing.

Then q may have roots (x, y) which contain complex numbers, but then, even in these roots, the ratio x/y must be real. Indeed, the above decomposition shows that, if $q(x, y) = 0$ and x or y are not real, then $p(x/y)$ must be 0, forcing the ratio of x and y to be real.

Now, for any fixed real x, we can look at the function $q(x, y)$ as a function of y. This function has real zeros. Therefore, if we differentiate this function with respect to y, the obtained function will have real zeros only, by Corollary 9.24. This will then imply that, for any roots (x, y) of $\partial q/\partial y$, the ratio x/y has to be real. A similar argument can be applied to x instead of y. Iterating this argument shows that the partial derivatives $\partial^{a+b} q/\partial x^a \partial y^b$ also have real zeros only, as long as they are not identically zero, that is, as long as $a + b \leq n - 1$.

Since we want to prove that $a_j^2 \geq a_{j-1}a_{j+1}$, that is, we want to prove an inequality involving three parameters, it is plausible to look for quadratic polynomials deduced from $q(x, y)$. Such polynomials can be obtained by differentiating q a total of $n-2$ times. So let $a = j-1$, let $b = (n-2)-(j-1) = n-j-1$, and let us consider

$$T(x, y) = \partial^{a+b} q/\partial x^a \partial y^b = \partial^{n-2} q/\partial x^{j-1} \partial y^{n-j-1}.$$

Fine, the reader might say at this point, but what can we tell about such complicated partial derivatives? Fortunately, quite a lot. Note that $\partial x^m/\partial x^{j-1} = 0$ unless $m \geq j-1$, and $\partial y^s/\partial y^{n-j-1} = 0$ unless $s \geq n-j-1$. So the only terms of $T(x, y)$ that do not vanish come from the terms of $q(x, y)$ in which $i \in [j-1, j+1]$. This leads to

$$\begin{aligned} T(x, y) &= \frac{(j-1)!}{2} a_{j-1}(n-j+1)! y^2 + a_j j!(n-j)! xy \\ &+ \frac{(j+1)!}{2} a_{j+1}(n-j-1)! x^2. \end{aligned}$$

We have seen that $T(x,y)$, as a partial derivative of $q(x,y)$, has real zeros only, in the sense that, in any root (x,y) of this equation, the ratio x/y has to be real. On the other hand, $T(x,y)/y^2$ is also a quadratic polynomial in x/y; therefore, the fact that it has real zeros only is equivalent to the fact that its discriminant is nonnegative, that is,

$$[a_j j!(n-j)!]^2 - a_{j-1}a_{j+1}(n-j+1)!(n-j-1)!(j-1)!(j+1)! \geq 0,$$

$$a_j^2 \geq a_{j-1}a_{j+1} \cdot \frac{n-j+1}{n-j}\frac{j+1}{j}, \tag{9.4}$$

which is even stronger than the inequality $a_j^2 \geq a_{j-1}a_{j+1}$, which was to be proved. ◊

The following example shows the power of Theorem 9.22.

Example 9.25 *For all positive integers n, the sequence $\{c(n,k)\}_{1\leq k\leq n}$ of signless Stirling numbers of the first kind has real zeroes only and is therefore log-concave.*

Solution: We have seen in Theorem 4.21 that

$$\sum_{k=1}^{n} c(n,k)x^k = (x+n-1)(x+n-2)\cdots(x+1)x.$$

Therefore, the zeroes of the sequence are $0, -1, \cdots, -n+1$. ◊

So, using Theorem 9.22, we could prove the log-concavity of the sequence $\{c(n,k)\}_{1\leq k\leq n}$ quite effortlessly. Without Theorem 9.22, it is considerably more difficult to prove that log-concavity result. Nevertheless, it is always interesting to find a purely combinatorial proof of a combinatorial fact. Such a proof, that is, an injective proof of the log-concavity of the sequence $\{c(n,k)\}_{1\leq k\leq n}$, was given by Bruce Sagan in [64]. In that same paper, Sagan also provided a combinatorial proof of the fact that the sequence $\{S(n,k)\}_{1\leq k\leq n}$ is log-concave. This sequence is also known to have real zeros, but the proof of that fact is not simple.

While many naturally defined combinatorial sequences, such as sequences of binomial coefficients, Stirling numbers, and Eulerian numbers, are known to have real zeros, there are a few intriguing conjectures in this area. The interested reader should consult [20] for these. The reader is also encouraged to read the upcoming Notes for further applications of the real zeros property.

Quick Check

1. Let n be a fixed positive integer, and let $a_k = \binom{n}{k}2^k$. Prove that the sequence $\{a_k\}_{0\leq k\leq n}$ is log-concave.

2. Let n be a fixed positive integer, and let $b_k = k\binom{n}{k}3^{k-1}$. Prove that the polynomial $B_n(x) = \sum_{k=1}^{n} b_k x^k$ has real roots only.
3. Let $f_n(x) = \sum_{k=1}^{n} c(n,k)2^k x^k$. Prove that $f_n(x)$ has real roots only. This implies that the coefficients of $f_n(x)$ are log-concave. Prove that fact without resorting to the real roots property.

9.4 Notes

A classic survey on this field is [70], which was later updated in [20]. A recent, high-level survey written by Petter Brändén can be found in Chapter 7 of [18].

Of the sequences we have previously encountered in this book, the sequence $\{A(n,k)\}_{1\leq k\leq n}$ of Eulerian numbers is known to have real zeros. Its log-concavity has several combinatorial proofs. See [13] for a survey of these results and other related results on permutations.

A celebrated, and difficult, classic result is that the Gaussian polynomials $\begin{bmatrix} \mathbf{n} \\ \mathbf{k} \end{bmatrix}$ have unimodal coefficients for all values of n and k. This result has various proofs which are either short, but use high-brow techniques, or elementary, but longer.

Log-concavity is not the only important implication of the real zeros property. Darroch's theorem [13] shows that, if a polynomial $A(x)$ has real zeros only, satisfies $A(1) > 0$, and if a_m is maximal among the coefficients of A, then

$$\left| \frac{A'(1)}{A(1)} - m \right| < 1.$$

The reader is invited to check Supplemental Exercise 12 for a nice application of Darroch's theorem.

Another interesting property of polynomials with real zeros is their connection to probability theory. If a polynomial with positive coefficients has *rational* real zeros, then its kth coefficient can be obtained (after some simple transformations) as the probability of k successes in the series of certain n independent trials. The interested reader can consult Jim Pitman's survey paper [60] for a detailed overview of this subject or Chapter 3 of [13] for a few simple examples.

9.5 Chapter review

(A) Let $\{a_k\}$ be a finite sequence of positive real numbers.

1. If $\{a_k\}$ has real zeroes only, then it is log-concave.
2. If $\{a_k\}$ is log-concave, then it is unimodal.

(B) Let $P(x)$ and $Q(x)$ be two polynomials with positive coefficients. If the sequence of coefficients of $P(x)$ is log-concave, and the sequence of coefficients of $Q(x)$ is log-concave, then the sequence of coefficients of $P(x)Q(x)$ is also log-concave.

9.6 Exercises

1. Prove that, for any fixed positive integer n and any $q > 1$, the sequence $\begin{bmatrix} \mathbf{n} \\ \mathbf{0} \end{bmatrix}, \begin{bmatrix} \mathbf{n} \\ \mathbf{1} \end{bmatrix}, \cdots, \begin{bmatrix} \mathbf{n} \\ \mathbf{n} \end{bmatrix}$ is log-concave. (The Gaussian coefficients $\begin{bmatrix} \mathbf{n} \\ \mathbf{k} \end{bmatrix}$ are defined in Exercise 29 of Chapter 4.)

2. (a) Recall that in (5.6) we defined the Narayana numbers $N(n,k)$ by setting $N(n,k) = \frac{1}{n}\binom{n}{k}\binom{n}{k+1}$. Prove that for any fixed positive integer n, the sequence $N(n,0), N(n,1), \cdots N(n,n-1)$ of Narayana numbers is log-concave.

   (b) Prove that, for any fixed positive integer k, the infinite sequence $N(k+1,k), N(k+2,k), \cdots$ is log-concave.

3. We call a sequence $a_0, a_1, \cdots$ of positive real numbers *log-convex*, if for all $k \geq 1$, we have
$$a_{k-1}a_{k+1} \geq a_k^2.$$
Prove by a combinatorial argument that the sequence $c_0, c_1, c_2, \cdots$ of Catalan numbers is log-convex.

4. Let us label the vertices of a square grid as shown in Figure 9.9, and let F be a Ferrers shape in the northwestern corner of this grid.

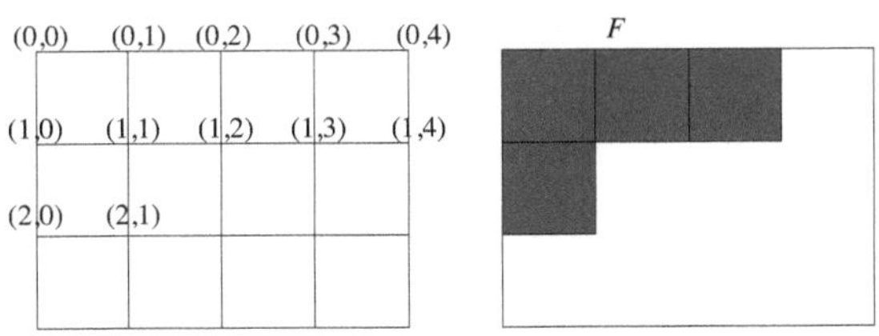

Figure 9.9
How to label the nodes of the grid and how to place the Ferrers shape F.

Let $M_F(m,n) = M(m,n)$ be the number of northeastern lattice paths from $(m,0)$ to $(0,n)$ that *do not go inside* F.

(a) Prove that
$$M(m,n+1)M(m+1,n) \leq M(m,n)M(m+1,n+1). \quad (9.5)$$

(b) + Prove that

$$M(m-1,n+1)M(m+1,n+1) \le M(m,n+1)^2. \quad (9.6)$$

(c) Conclude that

$$M(m-1,n+1)M(m+1,n) \le M(m,n)M(m,n+1). \quad (9.7)$$

5. Keep the notation of the previous exercise.
   (a) Prove that $M(m+1,n-1)M(m,n+1) \le M(m,n)M(m+1,n)$.
   (b) Conclude that $M(m+1,n-1)M(m-1,n+1) \le M(m,n)^2$, that is, that the sequence $\{M(i,n+m-i)\}_{0\le i\le m+n}$ is log-concave.
   (c) Explain how this result generalizes the fact that the binomial coefficients $\binom{r}{k}$ form a log-concave sequence for any fixed r.
   (d) Must the sequence $\{M(i,n+m-i)\}_{0\le i\le m+n}$ have real zeroes only?
6. + Find a more direct combinatorial proof of part (b) of the previous exercise, that is, of the log-concavity of the sequence

$$\{M(i,n+m-i)\}_{0\le i\le m+n}.$$

7. We call the sequence $a_0, a_1, \cdots, a_n$ of positive real numbers *strongly log-concave*, if for $1 \le i \le n-1$, we have

$$(n-(i-1))a_{i-1}\cdot(i+1)a_{i+1} \le (n-i)a_i \cdot ia_i.$$

   Prove that, if the sequence $a_0, a_1, \cdots, a_n$ of positive real numbers has real zeros only, then it is strongly log-concave.
8. Is it true that if a finite sequence of positive real numbers is strongly log-concave, then it has real zeros only?
9. Let G be a bipartite graph with color classes A and B. For any subset $X \subseteq A$, let $N(X)$ denote the set of vertices in B that have a neighbor in X.

   Let us say that A *has a perfect matching into* B if there are $|A|$ vertex-disjoint edges in G, or in other words, if each vertex of A can be matched to an adjacent vertex of B.

   A classic theorem of graph theory, *Philip Hall's* theorem, says that A has a perfect matching into B if and only if $|X| \le |N(X)|$ for all $X \subseteq A$. Use this theorem to prove that the sequence $\{\binom{n}{k}\}_{0\le k\le n}$ is unimodal.
10. + Let n be a fixed positive integer. Find a combinatorial (that is, injective) proof for the unimodality of the sequence of Narayana numbers $N(n,0), N(n,1), \cdots, N(n,n-1)$. *Hint: Try to use the reflection principle.*

11. Find a combinatorial proof of the log-concavity of the infinite sequence $S(2,2), S(3,2), S(4,2), \cdots$. Recall that $S(n,k)$ denotes a Stirling number of the second kind, that is, it is the number of partitions of $[n]$ into k blocks.

9.7 Solutions to exercises

1. Let us compute the ratio $\frac{\begin{bmatrix} n \\ k \end{bmatrix}}{\begin{bmatrix} n \\ k-1 \end{bmatrix}}$. We get the fraction

$$\frac{q^{n-k}-1}{(q^k-1)q^{k-1}},$$

which decreases as k increases. Indeed, the numerator decreases and the denominator increases.

2. (a) Again, look at the ratio

$$\frac{N(n,k)}{N(n,k-1)} = \frac{\binom{n}{k+1}}{\binom{n}{k-1}} = \frac{(k-1)!(n-k+1)!}{(k+1)!(n-k-1)!}$$
$$= \frac{(n-k+1)(n-k)}{k(k+1)},$$

which again decreases as k increases.

(b) We need to prove that $N(n+1,k)^2 \geq N(n,k)N(n+2,k)$, that is,

$$\frac{\binom{n+1}{k}^2\binom{n+1}{k+1}^2}{(n+1)^2} \geq \frac{\binom{n}{k}\binom{n}{k+1}\binom{n+2}{k}\binom{n+2}{k+1}}{n(n+2)}.$$

Multiplying both sides by $(k!(k+1)!)^2$, we get

$$((n)_k(n+1)_k)^2 \geq (n-1)_k(n)_k(n+1)_k(n+2)_k,$$

$$n(n+2-k) \geq (n+2)(n-k).$$

The last inequality holds, either by routine simplification to $0 \geq -2k$, or by noting that the sequence $1, 2, 3, \cdots$ is log-concave.

3. Let $\mathcal{C}_n$ be the set of northeastern lattice paths from $(0,0)$ to (n,n) that do not go above the main diagonal. Then $\mathcal{C}_n = c_n$. An injection $f_k : \mathcal{C}_n \times \mathcal{C}_n \to \mathcal{C}_{n-1} \times \mathcal{C}_{n+1}$ can be defined as follows: If $(p,q) \in \mathcal{C}_n \times \mathcal{C}_n$, translate q by $(-1,-1)$ so it becomes a path q' from $(-1,-1)$ to $(n-1,n-1)$. From here on, the proof is similar to the lattice path proof of the log-concavity of the binomial coefficients that we saw in the text. That is, let X be the first intersection of p and q',

and swap the parts of p and q' that come after X. See Figure 9.10 for an example. This results in a path from $(0,0)$ to $(n-1,n-1)$ and a path from $(-1,-1)$ to (n,n). Since this map is an injection, our claim is proved.

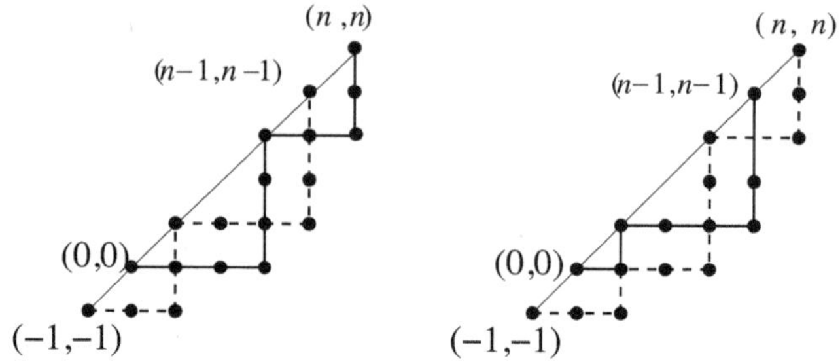

Figure 9.10
Our injection $f_k : \mathcal{C}_n \times \mathcal{C}_n \rightarrow \mathcal{C}_{n-1} \times \mathcal{C}_{n+1}$.

4. The results of this exercise and the following one were all found during research efforts by the author and Bruce Sagan, published in [17], while finding a combinatorial proof of Rodica Simion's conjecture claiming that the sequence $\{M(i, n+m-i)\}_{0\leq i \leq m+n}$ is unimodal. That conjecture, in the stronger form that the sequence is in fact log-concave, now has four proofs, though the other two are not combinatorial.

   (a) This can be done by a simple lattice path argument. Let p be a path counted by $N(m, n+1)$, and let q be a path counted by $N(m+1, n)$. Then these two paths must intersect, and swapping them at their first intersection point, as seen in Example 9.20, we get a pair of paths counted by $M(m,n) \cdot M(m+1, n+1)$. This map is an injection. See Figure 9.11 for an illustration.

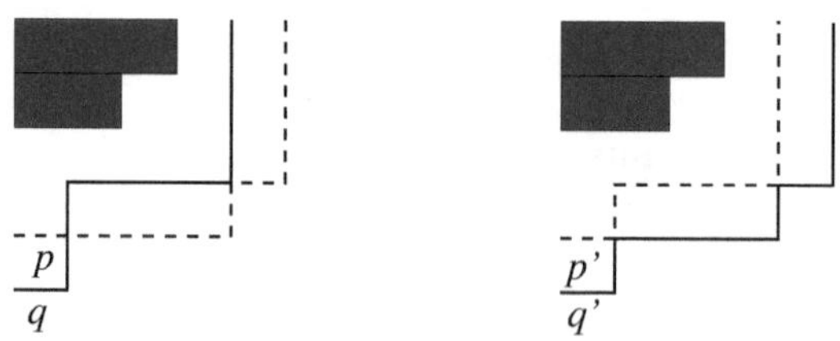

Figure 9.11
Swapping lattice paths at their intersection points.

(b) This part is trickier, since a path counted by $M(m-1,n)$ and a path counted by $M(m+1,n)$ do not necessarily intersect before their endpoint.
Let p be a path counted by $M(m-1,n)$, and let q be a path counted by $M(m+1,n)$. Find the first point P on p that is exactly one unit below a point Q of q. Such a point has to exist since the vertical distance of p and q changes from 2 to 0. Then move the part of p preceding P one unit to the south and attach it to the second part of Q, while moving the part of q preceding Q one unit to the north and attaching it to the second part of p. See Figure 9.12 for an illustration. This constitutes

Figure 9.12
Swapping lattice paths when their vertical distance is 1.

"swapping" the initial segments of p and q and provides the needed injection. The reader is invited to verify the details, that is, that this map is indeed an injection and that this is a valid map (that is, that the obtained paths will never go inside F).

(c) Multiply the inequalities proved in parts (a) and (b) together and simplify.

5. (a) Let F^c be the *conjugate* of the Ferrers shape F. Because the result of part (c) of the previous exercise holds for all Ferrers shapes, it holds for F^c as well. On the other hand, we have $M(a,b) = M_{F^c}(b,a)$. Therefore, if we take (9.7) for F^c instead of F, and then apply the mentioned symmetry, we get

$$M(n+1,m-1)M(n,m+1) \leq M(n,m)M(n+1,m).$$

Now switch m and n.

(b) Take the product of (9.7) and the inequality proved in part (a) and simplify.

(c) If F is the empty Ferrers shape, then the sequence that we prove to be log-concave is the sequence $\{\binom{n+m}{i}\}_{0\leq i\leq m+n}$.

(d) No, it will not, even in very simple cases. Let F consist of one box, and let $m+n=4$. This will lead to the polynomial $3x^3+5x^2+3x$, which has two complex roots.

6. Take a path p that is counted by $M(m-1, n+1)$ and a path q that is counted by $M(m+1, n-1)$. We will map this pair injectively onto a pair (p', q') in which both paths belong to the set enumerated by $M(m, n)$.

   In order to do this, we will cut our paths p and q into *three* parts, instead of two, and we will swap their *middle* parts.

   Similarly to what we have seen in the solution of part (b) of Exercise 4, let P and Q be the first points on p and q so that P is one unit above q. Furthermore, let $P*$ and $Q*$ be the last points on p and q so that $P*$ is one unit to the east of $Q*$. Then the defined four points split p into segments p_1, p_2, p_3 and q into q_1, q_2, q_3. Then we map (p, q) into the pair $(p_1q_2p_3, q_1p_2q_3)$, where p_1 is moved one unit south, p_3 is moved one unit west, q_1 is moved one unit north, and q_3 is moved one unit east. See Figure 9.13 for an example. The reader

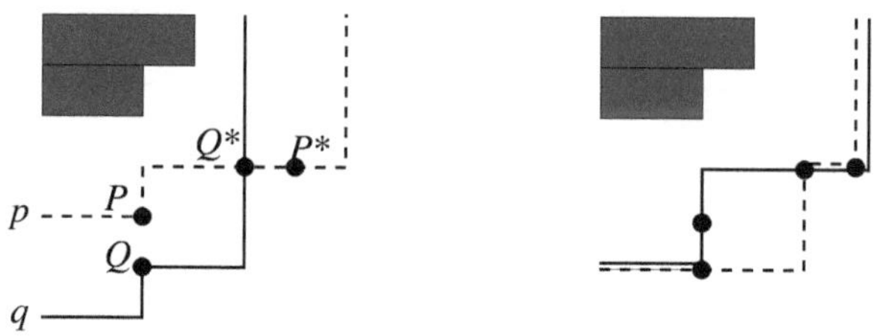

Figure 9.13
We swap the middle parts of p and q.

   is asked to verify that this map is injective and that the obtained paths do not go inside F.

7. This follows from (9.4) by routine rearrangements.

8. No, that is false. A counterexample is the sequence 1, 3, 3.

9. Let $k \leq \lfloor (n-1/2) \rfloor$. Define the graph G to be the bipartite graph whose vertices correspond to the k-element and $(k+1)$-element subsets of $[n]$. Let vertices x and y be adjacent if $x \subset y$, where *strict* containment is required. Then G is bipartite, and its color classes A and B have $\binom{n}{k}$ and $\binom{n}{k+1}$ vertices, respectively, since they correspond to the collections of k-element and $(k+1)$-element subsets of $[n]$. Our claim that $\binom{n}{k} \leq \binom{n}{k+1}$ will be proved if we can show that A has a perfect matching into B.

   It suffices to show that the conditions of Philip Hall's theorem hold. To that end, note that G is a *regular* graph. Indeed, each vertex of A has degree $n-k$, and each vertex of B has degree $k+1$. Because we know that $k \leq \lfloor (n-1/2) \rfloor$, we see that $n-k \geq k+1$, so the vertices in A have a degree at least as high as the vertices in B.

Now look at the induced subgraph H of G whose vertices are the vertices of S and $N(S)$, for some $S \subseteq A$. As vertices of S do not lose any neighbors when we restrict our attention to this subgraph, their degrees are still at least as high as any degree in $N(S)$. Since the sum of all degrees of vertices in S and $N(S)$ must agree, the inequality $|S| \leq |N(S)|$ follows. Therefore, A has a perfect matching into B, and so $|A| \leq |B|$. The rest of unimodality follows from the symmetry of our sequence.

10. We have seen in Exercise 23 that the Narayana number $N(n,k)$ counts unlabeled binary trees on n vertices with k right edges. Let $\mathcal{N}(n,k)$ be the set of such trees, and let $k \leq (n-1)/2$. We will construct an injection $f : \mathcal{N}(n,k) \rightarrow \mathcal{N}(n,k+1)$.

    Let $T \in \mathcal{N}(n,k)$. Let us define a total ordering on the set of vertices of T as follows: Order the vertices first according to their distance from the root (furthest vertices first), and then left to right. Let T_i be the induced subgraph of the first i vertices of T; then T_i is a forest. Furthermore, T_i has either the same number of right edges as T_{i-1} or one more. Since T_0 has the same number of left and right edges, and T_n has at least one more right edge than left edge, the previous sentence implies that there has to be a (smallest) index j so that T_j contains *exactly* one more right edge than left edge.

    Now reflect each component of T_j through a vertical axis that contains its root. Let the obtained tree be $f(T)$. Then $f(T)$ has one more left edge than T, and $f : \mathcal{N}(n,k) \rightarrow \mathcal{N}(n,k+1)$. Indeed, to find the preimage of a tree $T' \in \mathcal{N}(n,k+1)$, just find the smallest index j so that T'_j has one more left edge than right edge (if such an index exists), and reflect each component of T'_j through a vertical axis that contains its root. Figure 9.14 shows an example.

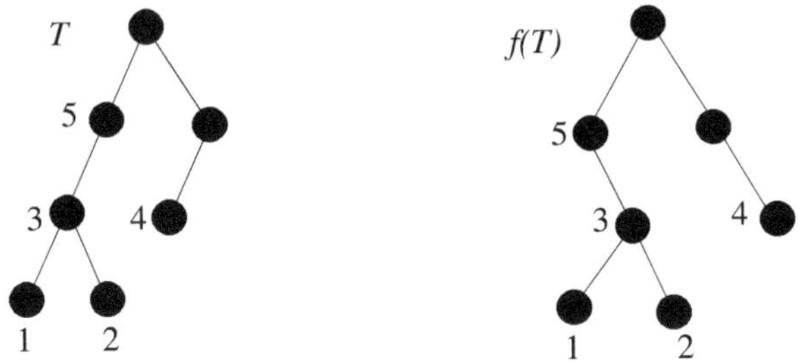

Figure 9.14
In this example, $j = 5$, since T_j has two left edges and one right edge.

11. A partition of $[n]$ into an *ordered pair* of two blocks can be described by a binary vector of length n whose ith coordinate is 1 if i is part

of the first block, and 0 otherwise. Let A_n be the set of such vectors in which all entries are not equal. Let $(x, y) \in A_{n-1} \times A_{n+1}$. Now define $f(x, y) = (u, v) \in A_n \times A_n$ as follows: Let

$$u_i = \begin{cases} x_i \text{ if } i \leq n-1, \\ \\ y_{n+1} \text{ if } i = n. \end{cases}$$

Similarly, let

$$v_i = y_i \text{ for } 1 \leq i \leq n.$$

Note that, since all entries of x were not equal, all entries of u are not equal. However, if all entries of y are equal *except for the last one*, then we have a problem, since all entries of v are then equal. Therefore, we have to define f separately in these cases. Let $x = (x_1, x_2, \cdots, x_{n-1})$, and let $y = (1, 1, 1, \cdots, 1, 0)$. Then set $u = (1, 1, \cdots, 1, 0)$, $v = (x_1, x_2, \cdots, x_{n-1}, 1)$, and $f(x, y) = (u, v)$. Define f in an analogous way if $y = (0, 0, \cdots, 0, 1)$.

The reader is invited to prove that f is an injection. Since $|A_n| = 2S(n, 2)$, our statement is proved.

9.8 Supplementary exercises

1. Find an injective proof of the fact that the sequence $a_n = n!$ is log-convex.
2. Find an injective proof of the fact that the sequence $a_n = \binom{2n}{n}$ is log-convex.
3. Let k be a fixed positive integer. Find a combinatorial proof for the fact that the infinite sequence $\{\binom{n}{k}\}_{n \geq k}$ is log-concave.
4. Let $1 < k < n-1$. Prove that the polynomial $\left[\mathbf{n} \atop \mathbf{k} \right]$ cannot have real zeros only.
5. Prove that, for any fixed n, the sequence $\{\binom{n}{k}\}_{0 \leq k \leq n}$ has real zeros only.
6. Prove, preferably combinatorially, that, for all positive integers n, the sequence $\{\binom{n}{k}\}_{0 \leq k \leq n}$ is strongly log-concave.
7. Let $\{a_0\}_{0 \leq k \leq n}$ have real zeros only. At most how many maximal values can this sequence have?
8. Let n be a fixed integer. Let a_m be the number of n-permutations that have m cycles, each of which is either a 1-cycle or a 2-cycle. Prove that the sequence $a_1, a_2, \cdots, a_n$ is log-concave.

9. Generalize the statement of the previous exercise for permutations whose cycles are all of length k or $k+1$ for some fixed k. Be careful.

10. Is it true that, for any fixed k, the infinite sequence of Stirling numbers $c(k,k), c(k+1,k), c(k+2,k), \cdots$ is log-concave?

11. Let $a_{n,k}$ be the number of acyclic functions on $[n]$ that have k fixed points. Prove that the sequence $a_1, a_2, \cdots, a_n$ is unimodal.

12. How many cycles will a randomly selected n-permutation most likely have?

13. Standard Young Tableaux were defined immediately following Lemma 1.32. Let $a_{n,k}$ denote the number of Standard Young Tableaux on n boxes that consist of at most two rows in which the *second* row is of length k. Prove that, for any fixed n, the sequence $a_{n,0}, a_{n,1}, \cdots, a_{n,\lfloor n/2 \rfloor}$ is log-concave. A combinatorial proof is preferred.

14. + Find an injective proof for the result of Corollary 9.11.

15. Recall that the Fibonacci numbers are defined by $F_0 = F_1 = 1$, and $F_n = F_{n-1} + F_{n-2}$ for $n \geq 2$. Prove that the sequence $F_0, F_2, F_4, \cdots$ of *even-indexed* Fibonacci numbers is log-convex.

16. Prove that the sequence $F_1, F_3, F_5, \cdots$ of *odd-indexed* Fibonacci numbers is log-concave.

17. Is there a positive integer n so that the infinite sequence $F_n, F_{n+1}, \cdots$ of Fibonacci numbers is log-concave?

10

Counting magic squares and magic cubes

In earlier chapters, we considered various distribution problems. Now we are returning to that topic, but our distribution problems will now be, in a sense, multidimensional.

10.1 A distribution problem

Let us assume that we have to distribute 60 building blocks among three children. Let us say that 20 of the blocks are red, 20 are blue, and 20 are green. To preserve world peace, it is our intention to give 20 blocks to each child. In how many different ways can this be done if blocks of the same color are identical? (Note that the problem is not too interesting if even blocks of the same color are different; indeed, in that case all blocks are different, and the answer is $\binom{60}{20} \cdot \binom{40}{20}$.)

This distribution problem is more difficult than the ones we have seen in earlier chapters. In order to understand the problem better, we will represent a possible distribution by a square grid, as shown in Figure 10.1.

	Red	Blue	Green
Miki	7	5	8
Benny	6	8	6
Vinnie	7	7	6

Figure 10.1
A possible distribution.

Note that the sum of the elements in each row is equal to 20 (as each child gets 20 blocks), and the sum of the elements in each column is equal to 20, since there are 20 blocks of each color. This observation motivates the following definition.

DOI: 10.1201/9781003304272-10

Definition 10.1 *A* magic square *is a square matrix with nonnegative integer entries in which all row sums and column sums are equal.*

Other definitions of magic squares exist. For instance, we could require that diagonals also have the same sum as rows and columns, or we could require that no rows or columns contain the same number twice. In this book, however, we will only use Definition 10.1.

Note that, in a magic square, it cannot happen that all row sums are equal to a, and all column sums are equal to b, while $a \neq b$. Indeed, that would mean that the sum of all elements in the magic square is equal to na *when counted by rows* and to nb *when counted by columns*, which would be a contradiction.

This problem can be generalized in at least two directions. Let us call rows and columns by the same word *lines*, and let $H_n(r)$ be the number of magic squares of size $n \times n$ with line sum r. Our example asked what the value of $H_3(20)$ was. Instead, we could ask what the value of $H_3(r)$ is for any given line sum r, or we could ask what the value of $H_n(60)$ is for any side length n. In other words, we could keep the size of the magic square fixed and study how $H_n(r)$ changes in function of r, or we could keep the line sum r fixed and see how $H_n(r)$ changes in function of n.

Quick Check

1. Compute $H_3(1)$ and $H_3(2)$.
2. The solution of the previous Quick Check exercise shows that

$$H_3(2) = \frac{(H_3(1)+1) \cdot H_3(1)}{2}. \tag{10.1}$$

   Explain what causes this equality to hold.
3. If we increase the size of our magic squares from 3 to 4, then the equality analogous to (10.1) will not hold. Explain why.

10.2 Magic squares of fixed size

Let us start with small fixed values of n. If $n = 1$, then there is only one way to construct a magic square of size $n \times n$ having line sum r, namely, by setting its only entry to r. Therefore, $H_n(r) = 1$.

This was not terribly difficult. The case of $n = 2$ is not very complicated either. Indeed, if we know the top left element of a 2×2 magic square with line sum r, we can compute all its elements, as shown in Figure 10.2.

Because x and $r - x$ must be nonnegative integers, we must have $0 \leq x \leq r$, which leaves $r + 1$ possibilities for x. Since x completely determines the magic square, $H_2(r) = r + 1$ follows.

x	$r-x$
$r-x$	x

Figure 10.2
A possible distribution.

10.2.1 The case of $n = 3$

This is where the easy cases end. When $n = 3$, which was the case in our introductory example, the determination of $H_n(r)$ is significantly more difficult. The problem was first solved by Percy MacMahon [55] in 1916. He proved the following theorem.

Theorem 10.2 *For all nonnegative integers* r,

$$H_3(r) = \binom{r+4}{4} + \binom{r+3}{4} + \binom{r+2}{4}. \tag{10.2}$$

MacMahon obtained his answer in a slightly different form. Another proof was given by Anand, Dumir, and Gupta [4] in 1966. The proof we present was found by this author in 1997 [14] in a slightly different form. We do not claim that this is the shortest possible proof. Nevertheless, we decided to present it because it shows why the answer is the sum of three binomial coefficients. Another proof will be given in Example 10.10. That proof will be shorter, but will require some preparation.

Proof: (of Theorem 10.2) If we try to copy the proof for $n = 2$, the first difference we note is that it is no longer sufficient to know what the top left element of a magic square is in order to know all the other elements. It suffices, however, to know the *four* elements in the top left corner, to know the four elements in the lower left corner, or to know four other elements no three of which are in the same line. We propose to choose the four elements a, b, c, and d that are located as shown in Figure 10.3. Then one can compute the remaining elements as shown in Figure 10.3.

The question is, therefore, in how many ways can we choose a, b, c, and d so that we do get a magic square, that is, so that the numbers which will be written in the remaining five boxes are all nonnegative.

In order to answer this question, let us summarize all the conditions that the nonnegative integers a, b, c, and d have to satisfy. Going through the five remaining boxes, we see that we get a magic square if and only if the following hold:

$$a + d \leq r, \tag{10.3}$$

a	d	$r-a-d$
$r+c-(a+b+d)$	b	$a+d-c$
$b+d-c$	$r-b-d$	c

Figure 10.3
These four elements determine the magic square.

$$b+d \leq r, \tag{10.4}$$

$$c \leq a+d, \tag{10.5}$$

$$c \leq b+d, \tag{10.6}$$

$$a+b+d-c \leq r. \tag{10.7}$$

This system of inequalities looks a little bit frightening. The five inequalities seem to be pretty much alike, which does not offer an obvious starting point for attacking them.

Let us not forget that we want to prove that the number of integer solutions to this system of inequalities is *the sum of three binomial coefficients.* Therefore, we would like to *split* the set of all 3×3 magic squares of line sum r into three subsets, and then prove that these subsets have $\binom{r+2}{4}$, $\binom{r+3}{4}$, and $\binom{r+4}{4}$ elements. Since these numbers are close to each other but are not equal, we suspect that the three subsets will be defined by similar, but not identical conditions.

After some experimenting, we arrive at the following split of our set of all 3×3 magic squares of line sum r. Let the first subset consist of those magic squares in which $a \leq b$ and $a \leq c$, let the second subset consist of those magic squares in which $b < a$ and $a \leq c$, and finally, let the third subset consist of those magic squares in which $c < a$ and $c < b$. So, roughly speaking, we classify our magic squares according to the position of their smallest diagonal entry.

We will now enumerate the magic squares in each class. For each class, we will combine all our conditions (10.3)–(10.7) into one chain of inequalities in order to improve our chances of counting their solutions.

(a) First assume that $a \leq b$ and $a \leq c$. Note that, in this case, inequalities (10.3), (10.6), and (10.7) are redundant since they are implied by (10.4) and (10.5). These two inequalities, together with our permanent conditions (10.3)–(10.7), imply

$$a \leq 2a+d-c \leq a+b+d-c \leq b+d \leq r. \tag{10.8}$$

Indeed, the first inequality of the chain is equivalent to (10.5), the second is equivalent to $a \leq b$, the third is equivalent to $a \leq c$, and the last is just (10.4).

In fact, we have shown an even stronger statement. We have shown that (10.8) holds *if and only if* (10.3)–(10.7) hold *and* $a \leq b$ and $a \leq c$.

In other words, if we assume $a \leq b$ and $a \leq c$, we get a magic square if and only if (10.8) holds. So our enumeration problem is reduced to finding the number of 4-tuples of nonnegative integers (a, b, c, d) for which (10.8) holds. These 4-tuples are in bijection with the 4-tuples $(a, 2a+d-c, a+b+d-c, b+d)$ satisfying (10.8). (The reader should prove this for herself, then check the solution of Exercise 1 to see if her argument is correct.)

Finally, the number of 4-tuples $(a, 2a + d - c, a + b + d - c, b + d)$ satisfying (10.8) is $\binom{r+4}{4}$ since we have to choose four elements of the $(r + 1)$-element set $\{0, 1, \cdots, r\}$, with repetitions allowed, and that can be done in $\binom{r+1+4-1}{4} = \binom{r+4}{4}$ ways.

(b) Now assume that $b < a$ and $b \leq c$. Note that, in this case, inequalities (10.5), (10.4), and (10.7) become redundant since they are implied by (10.3) and (10.6). These two inequalities, together with $a > b$ and $a \leq c$, are equivalent to the chain of inequalities

$$b \leq 2b + d - c \leq a + b + d - c - 1 \leq a + d - 1 \leq r - 1. \qquad (10.9)$$

This can be proved as in the previous case and also by using the fact that, for integers x and y, the inequality $x < y$ and the inequality $x \leq y - 1$ are equivalent. Therefore, the number of magic squares that belong to this case equals the number of 4-tuples of nonnegative integers (a, b, c, d) for which (10.9) holds. Just as above, the number of these 4-tuples is equal to that of 4-tuples $(b, 2b+d-c, a+b+d-c-1, a+d-1 \leq r-1)$. Finally, the number of these is $\binom{r+4-1}{4} = \binom{r+3}{4}$ since we have to choose four elements of the set $\{0, 1, \cdots, r - 1\}$, with repetitions allowed.

(c) Finally, assume that $a > c$ and $b > c$. Then inequalities (10.3)–(10.6) are all redundant. (Check this! Verify your solution in Exercise 2.) Condition (10.7) and our assumptions can be collected into the following chain:

$$c \leq b - 1 \leq b + d - 1 \leq a + b + d - c - 2 \leq r - 2. \qquad (10.10)$$

Here the first inequality is equivalent to our assumption $c < b$, the second one says that d is nonnegative, the third one is equivalent to our assumption $c < a$, and the last one is equivalent to (10.7). The four terms of (10.10) determine a, b, c, and d, and they can be chosen in $\binom{r+2}{4}$ ways.

Since we have counted each magic square exactly once, the proof is complete by the addition principle. ◊

10.2.2 The function $H_n(r)$ for fixed n

What can we say about $H_n(r)$ if the fixed number n is larger than 3? Finding an exact formula with methods similar to the method we have just seen becomes extremely tedious as n grows. Nevertheless, an exact formula *can* be found for each n, provided that we have a strong enough computer at hand and that we remember basic facts about polynomials. In order to see this, we will prove two results.

Theorem 10.3 *For any fixed positive integer n, the function $H_n(r)$ is a polynomial function of r.*

Theorem 10.4 *For any fixed positive integer n, the degree of the polynomial $H_n(r)$ is $(n-1)^2$.*

Once we prove these theorems, a formula for $H_n(r)$ can be obtained easily, at least by people like the readers who remember some calculus. Such people know that a polynomial of degree m has m roots (some of which may be complex). Therefore, if polynomials p and q both have degree m and satisfy $p(i) = q(i)$ in $m+1$ different points i, then they must be identical. Indeed, otherwise the polynomial $p-q$ would have $m+1$ roots, which would be a contradiction since its degree is at most m.

Having recalled this, and making good use of the two theorems that we have just presented, we can get a formula for $H_n(r)$ as follows: We compute the values of $H_n(r)$ for $r = 0, 1, \cdots, (n-1)^2$, then take the *unique* polynomial $g(n)$ of degree $(n-1)^2$ that takes these values at $0, 1, \cdots, (n-1)^2$. A strong enough computer can perform both of these tasks easily. Since g and $H_n(r)$ are both polynomials of degree $(n-1)^2$ and agree on $(n-1)^2+1$ values, they must be identical. If you do not remember how to obtain the unique polynomial p of degree m whose values $p(0), p(1), \cdots, p(m)$ are given, see Exercise 3.

Therefore, the mathematical interest lies in proving Theorems 10.3 and 10.4. We start by proving Theorem 10.4 (assuming that Theorem 10.3 is true) because that is the easier of the two.

Proof: (of Theorem 10.4, assuming that Theorem 10.3 is true)

1. First we show that $\deg H_n(r) \leq (n-1)^2$. To show this, it suffices to show that there is a polynomial of degree $(n-1)^2$ whose values are always larger than the corresponding values of $H_n(r)$. An $n \times n$ magic square is completely determined by the $(n-1) \times (n-1)$ elements it has in its top left corner. Each of these elements can range from 0 to r, so there are $r+1$ choices for each of these elements.

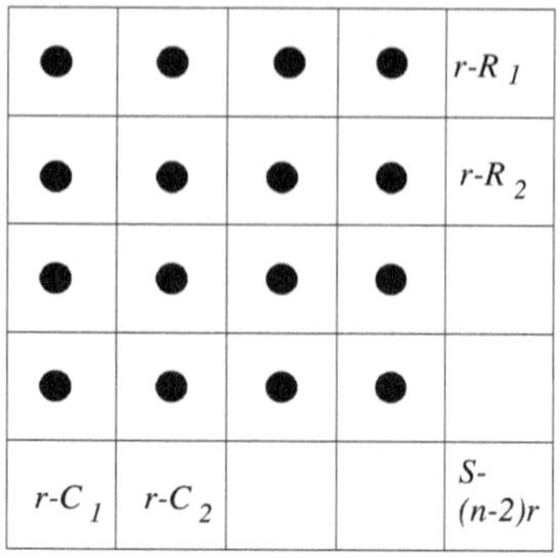

Figure 10.4
Computing the elements in the last row and column.

Therefore, $H_n(r) \leq (r+1)^{(n-1)^2}$ for all r, proving that $\deg H_n(r) \leq (n-1)^2$.

2. Now we show that $\deg H_n(r) \geq (n-1)^2$. To show this, it suffices to show that there is a polynomial of degree $(n-1)^2$ whose values are always smaller than the corresponding values of $H_n(r)$ and whose leading coefficient is positive. How could we find such a polynomial? Let us try to find many different ways to fill up the top left $(n-1) \times (n-1)$ subsquare of our square so that each case leads to a valid magic square. Consider Figure 10.4.

   In this figure, we denoted by R_i (resp. C_i) the sum of the first $(n-1)$ entries in row i (resp. column i) of our magic square, and by S the sum of all these $(n-1)^2$ entries. As the figure shows, we get a magic square if and only if

   i. for all $i \in [1, n-1]$, we have $R_i \leq r$,
   ii. for all $i \in [1, n-1]$, we have $C_i \leq r$, and
   iii. we also have $(n-2)r \leq S$.

   Note that we do not have to require that $S-(n-2)r \leq r$ since this follows by summing $R_i \leq r$ over all $i \in [1, n-1]$.

   To fulfill (i) and (ii), it would suffice to choose *small* numbers for the top left corner. Indeed, if each of these $(n-1)^2$ entries is at most as large as $r/(n-1)$, then (i) and (ii) are satisfied. However, we also have to make sure that the sum of these $(n-1)^2$ entries is *at least* $(n-2)r$ in order to satisfy (iii). This can be achieved by choosing each of them to be at least as large as $(n-2)r/(n-1)^2$.

   Therefore, we get a magic square each time we choose each entry of the top left corner to be at least as large as $(n-2)r/(n-1)^2$, and at most as large as $r/(n-1)$. The number of such choices for each entry is at least

$$\frac{r}{n-1} - \frac{r(n-2)}{(n-1)^2} - 1 = \frac{r}{(n-1)^2} - 1.$$

Therefore, $H_n(r) \geq (\frac{r}{(n-1)^2} - 1)^{(n-1)^2}$, and our claim follows.

Therefore, the degree of $H_n(r)$ is equal to $(n-1)^2$ as claimed. ◊

Note that this general result agrees with the particular results we found so far. For $n = 1$, we found $H_n(r) = 1$, a polynomial of degree zero. For $n = 2$, we found $H_n(r) = r + 1$, a polynomial of degree one. Finally, for $n = 3$, we found $H_3(r) = \binom{r+4}{4} + \binom{r+3}{4} + \binom{r+2}{4}$, a polynomial of degree four. For $n = 4$, the answer would be a polynomial of degree 9.

The proof of Theorem 10.3 is significantly more difficult. It is, in fact, probably the most complicated proof of the chapter. We first show why Theorem 10.3 is relatively easy to prove in the case of $n = 3$ (even if we do not assume the results of the previous section). Then we will show why the methods of that special case are not sufficient to treat the general case. Finally, we will show how to extend those methods so that they apply in the general case as well.

Magic squares of line sum 1 will be very important in the following discussion. Therefore, we take a closer look at them.

Proposition 10.5 *There is a bijection between $n \times n$ magic squares of line sum 1 and permutations of length n. In particular, $H_n(1) = n!$ for all non-negative integers n.*

Proof: Magic squares of line sum 1 must contain exactly one 1 in each row and each column. In other words, they are the *permutation matrices* we discussed in Chapter 4. See Definition 4.46 to recall that concept. Since there is a natural bijection from the set of all n-permutations to that of all permutation matrices of size $n \times n$, our claim is proved. ◊

We now return to the proof of Theorem 10.3. The following is an example of a decomposition of a magic square into the sum of permutation matrices.

$$\begin{bmatrix} 2 & 1 & 0 \\ 1 & 1 & 1 \\ 0 & 1 & 2 \end{bmatrix} = \begin{bmatrix} 1 & 0 & 0 \\ 0 & 1 & 0 \\ 0 & 0 & 1 \end{bmatrix} + \begin{bmatrix} 1 & 0 & 0 \\ 0 & 0 & 1 \\ 0 & 1 & 0 \end{bmatrix} + \begin{bmatrix} 0 & 1 & 0 \\ 1 & 0 & 0 \\ 0 & 0 & 1 \end{bmatrix}. \quad (10.11)$$

If you try to decompose other magic squares into the sum of permutation matrices, you find that you always succeed. The following lemma, which is sometimes called the *Birkhoff–von Neumann theorem*, shows that such decomposition indeed always exists.

Lemma 10.6 *Any magic square of line sum r can be decomposed into the sum of r permutation matrices.*

The proof of Lemma 10.6 is remarkable because we will use a theorem from *graph theory* to obtain it! This is very surprising. Indeed, it would seem that our lemma, or magic squares in general, have nothing to do with graph theory at all. Shortly, we will see that this is a false appearance, and magic squares are in fact deeply connected to graphs.

At this point, the reader may want to review the definition and concept of *bipartite graphs*, which were discussed at the beginning of Section 6.1.1.

If G is a bipartite graph with color classes A and B, then we say that A *has a perfect matching into* B if there exists a set M of $|A|$ pairwise disjoint edges in G. (This concept was mentioned in Exercise 9 of Chapter 8.) Note that, because of the bipartite property of G, this means that each vertex of A is adjacent to exactly one edge of M. See Figure 10.5 for an illustration.

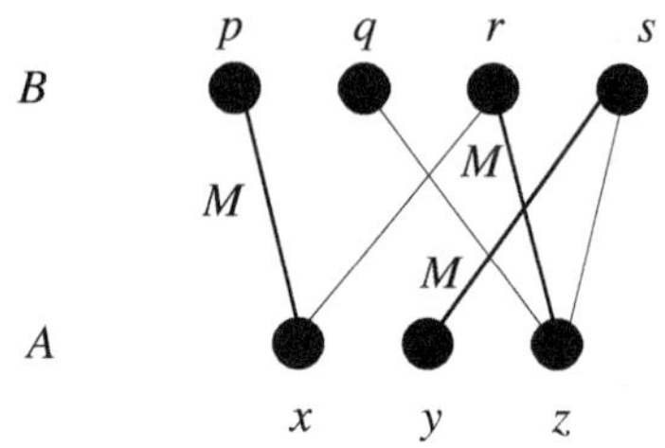

Figure 10.5
A bipartite graph with a perfect matching of A into B.

The following famous theorem of Philip Hall (sometimes called the marriage theorem) characterizes bipartite graphs in which one color class has a perfect matching into the other. The proof of this theorem can be found in most introductory texts on graph theory, such as [12].

Theorem 10.7 *Let G be a bipartite graph with color classes A and B, and for all $X \subseteq A$, let $N(X)$ be the set of all vertices in B that have a neighbor in X.*

Then A has a perfect matching into B if and only if the inequality $|X| \leq |N(X)|$ holds for each subset of vertices $X \subseteq A$.

Note that the "only if" part is straightforward. What is interesting here is the fact that the necessary condition that each X needs at least $|X|$ neighbors is sufficient as well.

You probably cannot wait to hear to which graph we will apply Philip Hall's theorem. We will define that graph at the beginning of the Induction Step in our upcoming induction proof.

Proof: (of Lemma 10.6) We prove our statement by induction on r. The initial case of $r = 1$ is trivial since a magic square of line sum 1 is already decomposed into the sum of one magic square of line sum 1. Now assume that we know the statement for magic squares of line sum $r-1$ and prove it for magic squares of line sum r.

Let H be any magic square of line sum r. All we need to show is that we can always *subtract* a permutation matrix from H without getting negative numbers anywhere. If we can do that, then we are left with a magic square of line sum $r-1$, and for that magic square the induction hypothesis will apply. However, it is not an easy thing to prove that we can always subtract a permutation matrix from H. There is no obvious way to see that our magic square does not have zeros in all the "wrong" places, forbidding the subtraction of any positive entries in those positions.

Now comes the punchline. Define $G_H(A, B)$ to be the bipartite graph in which the vertices of A correspond to the rows of H; the vertices of B correspond to the columns of H; and for $i \in A$ and $j \in B$, the vertices i and j are connected by an edge if and only if the (i, j)-element of H is not zero.

Example 10.8 *If H is the magic square in (10.11), then $G_H(A, B)$ is the graph shown in Figure 10.6.*

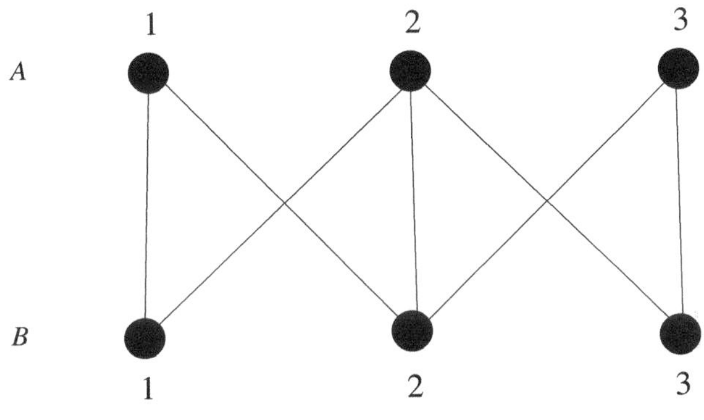

Figure 10.6
The graph $G_H(A, B)$ of the magic square in (10.11).

We claim that the graph $G_H(A, B)$ always has a perfect matching no matter what H is. Indeed, let us assume that the contrary holds. Because of the definition of $G_H(A, B)$, this means that there are s rows in H that contain nonzero elements in only t columns, where $t < s$. Since the sum of elements of any line is equal to the sum of nonzero elements of that line, we get that the sum of the elements in the $s \times t$ submatrix given by these rows and columns is sr when counted by the rows, and at most tr when counted by the columns. This contradicts $t < s$, proving by Philip Hall's theorem that $G_H(A, B)$ has a perfect matching.

Why should we care about $G_H(A, B)$ having a perfect matching? Because a perfect matching of $G_H(A, B)$ corresponds to a set S of squares so that each row and column of H contains one position from S and so that all positions in S contain nonzero entries. However, then we can subtract a permutation matrix from H, namely, the permutation matrix that consists of ones at each position in S and of zeros everywhere else.

We have seen that H can be decomposed into the sum of two magic squares H_1 and H_2, where H_1 is of line sum $r-1$ and H_2 is of line sum 1. Then, by our induction hypothesis, the H_1 can be decomposed into the sum of $r-1$ permutation matrices, and the theorem is proved. ◊

Note that Theorem 10.6 does not say that the decomposition of a magic square into the sum of permutation matrices is unique. In fact, it is not necessarily unique if n, the size of the magic square, is at least three.

Example 10.9 *The following is an example of a matrix that has two distinct decompositions into the sum of permutation matrices. Indeed,*

$$\begin{bmatrix} 1 & 1 & 1 \\ 1 & 1 & 1 \\ 1 & 1 & 1 \end{bmatrix} = \begin{bmatrix} 1 & 0 & 0 \\ 0 & 1 & 0 \\ 0 & 0 & 1 \end{bmatrix} + \begin{bmatrix} 0 & 1 & 0 \\ 0 & 0 & 1 \\ 0 & 0 & 1 \end{bmatrix} + \begin{bmatrix} 0 & 0 & 1 \\ 1 & 0 & 0 \\ 0 & 1 & 0 \end{bmatrix},$$

and also,

$$\begin{bmatrix} 1 & 1 & 1 \\ 1 & 1 & 1 \\ 1 & 1 & 1 \end{bmatrix} = \begin{bmatrix} 0 & 0 & 1 \\ 0 & 1 & 0 \\ 1 & 0 & 0 \end{bmatrix} + \begin{bmatrix} 1 & 0 & 0 \\ 0 & 0 & 1 \\ 0 & 1 & 0 \end{bmatrix} + \begin{bmatrix} 0 & 1 & 0 \\ 1 & 0 & 0 \\ 0 & 0 & 1 \end{bmatrix}.$$

This lack of a unique decomposition is what makes it difficult to prove that $H_n(r)$ is a polynomial. If the decomposition were unique, we could argue as follows: "The number of $n \times n$ permutation matrices is $n!$. We have to choose r of them, *not necessarily distinct ones*, and then add them. This can be done in $\binom{n!+r-1}{n!-1}$ ways, and that is a polynomial of r having degree $n!-1$." The problem with this argument is, of course, that, if $n \geq 3$, then there are different sets of $n \times n$ magic permutation matrices that sum to the *same* magic square.

To correct this mistake caused by overcounting, we will have to see that the *overcount itself* is a polynomial, and therefore the actual function $H_n(r)$ is a difference of two polynomials ("all decompositions" minus "surplus decompositions"), which is again a polynomial.

Let us further explain our plan by working out the case of $n = 3$.

Example 10.10 *For $n = 3$, the above strategy of counting leads to the correct formula*

$$H_3(r) = \binom{r+5}{5} - \binom{r+2}{5}. \tag{10.12}$$

Solution: Let $n = 3$. Then the number of 3×3 permutation matrices is 6, and they are shown in Example 10.9. Therefore, the number of ways to choose r of these magic squares (with repetitions allowed) is $\binom{r+5}{5}$. So this is the number of all decompositions.

Let h_1, h_2, and h_3 denote the permutation matrices in the first equality of Example 10.9, and let h_4, h_5, and h_6 denote those in the second equality. It is routine to check that any five of our six permutation matrices $h_1, h_2, \cdots, h_6$ are linearly independent. Therefore, a magic square H can have more than one decomposition only if one can subtract each of the six h_i from it and still get a magic square; in other words, when H does not contain a 0 entry. Conversely, if H consists of positive entries only, then H has more than one decomposition. Indeed, we can subtract the matrix J_3, (the 3×3 matrix whose entries are all 1) from H, to get $H = J_3 + H'$. We have seen in Example 10.9 that J_3 itself has two decompositions ($J_3 = h_1 + h_2 + h_3 = h_4 + h_5 + h_6$), so H certainly has at least two decompositions.

It remains to count the "surplus" decompositions, that is, the sum of the numbers of decompositions each magic square has in excess of one. Let us first specify what we consider a "surplus" decomposition. Take a decomposition of H in which each of h_4, h_5, and h_6 occurs at least once. Using the identity $h_1 + h_2 + h_3 = h_4 + h_5 + h_6$, we can replace a subset $\{h_4, h_5, h_6\}$ in this decomposition by a subset $\{h_1, h_2, h_3\}$ and get a new decomposition. We can iterate this procedure as long as each of h_4, h_5, and h_6 occurs in our decomposition at least once, after which we have to stop. Therefore, we will say that a decomposition is surplus if and only if the coefficients of h_4, h_5, and h_6 are all positive in that decomposition. (This is an arbitrary choice; we could just as well say that a decomposition is surplus if and only if the coefficients of h_1, h_2, and h_3 are all positive.) The number of such decompositions is $\binom{r-3+5}{5} = \binom{r+2}{5}$, and our claim is proved. ◇

Thus we have obtained a second formula for $H_3(r)$. It is a good way to brush up your binomial coefficient skills to see that they are indeed equivalent, that is, $\binom{r+2}{4} + \binom{r+3}{4} + \binom{r+4}{4} = \binom{r+5}{5} - \binom{r+2}{5}$.

If $n > 3$, then evaluating the surplus decompositions gets more complicated. Remember, however, that all we need to show is that their number is given by a polynomial function. We do not need to find that polynomial function itself.

Let us summarize what we did in the proof of the previous example. We noticed that $h_1 + h_2 + h_3 = h_4 + h_5 + h_6$ held. Then we "eliminated" decompositions that contained each of h_4, h_5, and h_6, by replacing $h_4 + h_5 + h_6$ with $h_1 + h_2 + h_3$ and repeating this procedure as long as we could. We would like to do something along the same lines for larger magic squares as well. This will be possible, but it will be more complicated.

The main reason why the case of $n = 3$ was simple was that there was only one *minimal dependence relation* among permutation matrices of side length 3. As the notion of minimal dependence relations is crucial to our upcoming argument, we are going to make it more precise.

In the following definitions, we will write $m = n!$, for brevity.

Definition 10.11 *Let $h_1, h_2, \cdots, h_m$ be the magic squares of line sum 1 and side length n. If there are positive integers $c_1, c_2, \cdots, c_m$ so that*

$$\sum_{i \in I} c_i h_i = \sum_{j \in J} c_j h_j = U, \tag{10.13}$$

where I and J are disjoint subsets of $[m]$, then we say that the vector $(c_1, \cdots, c_m)$ describes a dependence relation. *In this vector, $c_k = 0$ if $k \notin I \cup J$. A dependence relation $(c_1, \cdots, c_m)$ is called* minimal *if there is no other dependence relation $(c'_1, \cdots, c'_m)$ satisfying $c'_i \leq c_i$ for all i.*

Definition 10.12 *The expressions $\sum_{i \in I} c_i h_i$ and $\sum_{J \in J} c_j h_j$ in (10.13) are called the* sides *of U.*

The bad news is that, for $n > 3$, we will have to deal with more than one minimal dependence relation. The good news is that their number will nevertheless always be finite. See Exercise 21 for a statement implying this.

There is one more complication for matrices larger than 3×3. It could happen that several minimal dependence relations lead to the same magic square U as defined in formula (10.13). In other words, a magic square can have more than two decompositions into sides of minimal dependence relations. The reader should construct an example for this for $n = 4$, then check our construction in the solution of Exercise 10.

We point out that sides of all our minimal dependence relations can be split into equivalence classes, two sides being in the same class if they sum to the same magic square U. Note that it can happen that a class has more than two elements since the magic square U can occur as the sum in more than one minimal dependence relation. (This did not happen when n was 3.)

In order to make the following proof somewhat easier to read, we will say that a linear combination $c_1 h_1 + \cdots + c_m h_m$ *contains* another linear combination $g_1 h_1 + \cdots + g_m h_m$ if $g_i \leq c_i$ for each i.

Example 10.13 *The linear combination $3h_1 + 4h_3 + 7h_5$ contains the linear combination $h_1 + 2h_3 + 3h_5$, but does not contain the linear combination $h_1 + h_2 + h_3$, since the coefficients of h_2 violate the requirements.*

Also, whenever we say *linear combination*, we mean *linear combination with nonnegative integer coefficients.* After all these precautions, we can finally prove Theorem 10.3.

Proof: (of Theorem 10.3) The total number of ways to choose r magic squares out of our r permutation matrices, with repetitions allowed, is $\binom{n!+r-1}{n!-1}$, which is a polynomial function of r. Therefore, our claim will be proved if we can show that the number of surplus decompositions is also a polynomial function of r.

Fix n, and let $D_1, D_2, \cdots, D_t$ be the equivalence classes of sides defined above. Choose one representative d_i from each D_i and call all other sides *bad sides.*

Example 10.14 *In the special case of $n = 3$, there was only one equivalence class of sides; it contained the two sides $h_1 + h_2 + h_3$ and $h_4 + h_5 + h_6$, the latter being the bad side.*

Now let H be any $n \times n$ magic square with line sum r, and let $c_1h_1 + \cdots c_mh_m$ be a decomposition of H. If H contains a bad side from the class D_i, then replace the elements of that bad side by the elements of the representative d_i. (This is what we did in the case of $n = 3$ when we replaced $h_4 + h_5 + h_6$ with $h_1 + h_2 + h_3$.) Repeat this procedure as many times as possible. When we stop, we will have a decomposition of H which does not contain any bad sides.

Just as in the case of $n = 3$, the nonsurplus decompositions are those that do not contain any bad sides. Denote $f_1, f_2, \cdots, f_s$ the bad sides, and let $|f_i|$ be the number of permutation matrices in f_i, with multiplicities counted. Let $B_i(r)$ be the set of linear combinations $c_1h_1+\cdots+c_mh_m$ in which $\sum_{i=1}^{m} c_i = r$ that do contain f_i. Then

$$|B_i(r)| = \binom{r - |f_i| + m - 1}{m - 1}, \tag{10.14}$$

which is a polynomial of r of degree $m - 1$.

Example 10.15 *If $n = 3$, then (10.14) simplifies to $|B_1(r)| = \binom{r-3+6-1}{6-1} = \binom{r+2}{3}$, and we are done since there are no more bad sides.*

In the general case, however, there are more bad sides, and so we have to resort to the sieve formula to find $|B_{i_1}(r) \cup \cdots \cup B_{i_z}(r)|$, the number of linear combinations containing at least one bad side.

For instance, let us compute $|B_i(r) \cap B_j(r)|$, that is, the number of linear combinations $c_1h_1 + \cdots + c_mh_m$, again with $\sum_{i=1}^{m} c_i = r$ that contain both f_i and f_j. Denote by $|f_{i,j}|$ the minimal number of permutation matrices we have to sum to get such a linear combination. For instance, if $|f_i| = 3$ and $|f_j| = 3$, and there is no permutation matrix contained in both f_i and f_j, then $f_{i,j} = 6$. Once we make sure that f_i and f_j are contained in our linear combination, we can choose the remaining $r - f_{i,j}$ magic squares freely, with repetitions allowed, showing that

$$|B_i(r) \cap B_j(r)| = \binom{r - |f_{i,j}| + m - 1}{m - 1},$$

which is again a polynomial of r.

Now we attack the general case, that is, we compute the number of linear combinations that contain all of z bad sides, where z is any positive integer as opposed to 2. Let $K = \{i_1, i_2, \cdots i_z\}$. We want to evaluate $|\cap_{i_j \in K} B_{i_j}(r)| = |B_{i_1}(r) \cap \cdots \cap B_{i_z}(r)|$. Therefore, we define $|f_K|$ to be the minimal number of permutation matrices that one has to sum to obtain a linear combination containing each bad side f_{i_j}, for $1 \leq j \leq z$. So $f_K = 3$ in our running example of $n = 3$, and $K = \{1\}$. Then, by the same argument as above for $|B_i(r) \cup B_j(r)|$, we have

$$|B_{i_1}(r) \cap \cdots \cap B_{i_z}(r)| = \binom{r - |f_K| + m - 1}{m - 1},$$

which is again a polynomial of r. Therefore, by the inclusion-exclusion principle, we obtain

$$|B_1(r) \cup B_2(r) \cup \cdots \cup B_s(r)| = (-1)^{k-1} \sum_{\substack{|K|=k \\ K \subseteq [s]}} |\cap_{i_j \in K} B_{i_j}(r)|.$$

Therefore, $|B_1(r) \cup B_2(r) \cup \cdots \cup B_s(r)|$ is a sum of several polynomials of r, so it is a polynomial r. Thus the number of all surplus decompositions is a polynomial function of r, and since the number $\binom{n!+r-1}{n!-1}$ of all decompositions is also a polynomial function of r, our theorem is proved by the subtraction principle. ◊

Quick Check

1. Find the number of 3×3 magic squares with line sum r that are symmetric to both diagonals.
2. Find the number of 3×3 magic squares with line sum r that have a vertical line of symmetry.
3. Find the number of 3×3 magic squares with line sum r that have both a vertical and a horizontal line of symmetry.

10.3 Magic squares of fixed line sum

Let us now assume that our magic squares have a fixed line sum r and that their side length n is changing. In other words, we are looking at $H_n(r)$ as a function of n, with r fixed. If you prefer the distribution example, more and more children will play, and more and more toy types will be available, but the number of toys each child gets will stay the same, and so will the number of toys of each type.

What can be said for small values of r? If $r = 0$, then $H_n(0) = 1$ for all n, since all entries of the magic square have to equal zero in this case. If $r = 1$, then $H_n(1) = n!$, as we have seen in Proposition 10.5. So, even in this simple case, $H_n(r)$ is *not* a polynomial function of n, but a much faster growing function.

The task of finding a formula for $H_n(2)$ is significantly more difficult. The key element of our proof is the following lemma, which is due to Békéssy.

Lemma 10.16 *Let $T_n(2)$ be the number of $n \times n$ magic squares with line sum 2 which do not contain an entry equal to 2. Then, for all $n \geq 2$,*

$$T_n(2) = \frac{\sum_{k=0}^{n}(-1)^k \binom{n}{k} n!(2n-2k-1)!!}{2^n}. \tag{10.15}$$

Proof: Recall the distribution problem at the very beginning of this chapter. In the language of that problem, our problem can be expressed as follows: We have $2n$ blocks, two of them red, two of them blue, and so on, two of each of n colors. Blocks of the same color are identical. In how many ways can we distribute these $2n$ blocks to n children so that each child gets two blocks *and no child gets two blocks of the same color?*

First, consider the slightly different problem in which even blocks of the same color are different (for example, they are numbered 1 and 2). If the number of all distributions is $T'_n(2)$, then we have $T'_n(2) = 2^n T_n(2)$. This is because, if the two blocks of color i are different, we can swap them and get a new distribution. (Note that here we use the fact that no child has two blocks of the same color.) So we might as well try to find $T'_n(2)$ instead of $T_n(2)$.

We will find $T'_n(2)$ by an inclusion-exclusion argument. Let A be the set of distributions of the $2n$ blocks to the n children so that each child gets two blocks which may or may not be the same color. Then

$$|A| = (2n)!/2^n = (2n-1)!!n!. \tag{10.16}$$

Indeed, we can just line up the $2n$ blocks, then give the first two to the first child, the third and fourth to the second child, and so on. There are $(2n)!$ ways to do this. Each distribution will be counted 2^n times, however, since swapping blocks in positions $(2i-1)$ and $2i$ of the line does not result in a new distribution.

Now let us count how many of these $|A|$ distributions are "bad," that is, give two blocks of the same color to some children. Let A_i be the set of distributions in which child i gets two blocks of the same color. Then

$$|A_{i_1} \cap A_{i_2} \cap \ldots A_{i_k}| = (n)_k (2n-2k)!/2^{n-k} = n!(2n-2k-1)!!.$$

Since we first can choose the k colors of which children $i_1, i_2, \cdots, i_k$ will get their blocks in $(n)_k$ ways, then we can distribute the remaining $2(n-k)$ blocks arbitrarily in $(2n-2k-1)!!(n-k)!$ ways, as explained in the proof of (10.16). Note we use the fact that $(n)_k(n-k)! = n!$.

Using the inclusion-exclusion principle, we get that

$$\begin{aligned}
T'_n(2) &= |A| - |A_1 \cup A_2 \cup \cdots A_n| \\
&= (2n-1)!!n! - \sum_{k=1}^{n} \binom{n}{k} (-1)^{k-1} n!(2n-2k-1)!! \\
&= \sum_{k=0}^{n} \binom{n}{k} (-1)^k n!(2n-2k-1)!!.
\end{aligned}$$

Applying the identity $T_n'(2) = 2^n T_n(2)$, our statement is proved. ♢

The following corollary probably looks a little surprising at first sight.

Corollary 10.17 *Let* $T(x) = \sum_{n=0}^{\infty} T_n(2)\frac{x^n}{n!^2}$, *where we set* $T_0(2) = 1$. *Then*

$$T(x) = \frac{e^{-x/2}}{\sqrt{1-x}}. \tag{10.17}$$

Recall from Section 3.5 that, for a sequence of real numbers a_n, the power series $\sum_{n\geq 0} a_n \frac{x^n}{n!^2}$ is called the *doubly exponential generating function* of the sequence a_n. So $T(x)$ is the doubly exponential generating function of the sequence $T_n(2)$, and Corollary 10.17 shows that it has a fairly compact form.

Proof: (of Corollary 10.17) We will show that the two power series on the two sides of (10.17) are equal by showing that, for all n, the coefficients of $x^n/(n!)^2$ are the same in both power series.

On the left-hand side, this coefficient is equal to

$$\frac{\sum_{k=0}^{n}(-1)^k\binom{n}{k}n!(2n-2k-1)!!}{2^n}$$

by Lemma 10.16. The right-hand side is the product of two power series, $e^{-x/2}$ and $(1-x)^{-1/2}$. The expansion of $e^{-x/2}$ is straightforward, namely,

$$e^{-x/2} = \sum_{k\geq 0}(-1)^k \frac{x^k}{2^k k!}.$$

The expansion of $(1-x)^{-1/2}$ can be obtained by the binomial theorem. Note that

$$\binom{-1/2}{i} = \frac{(-1/2)(-3/2)\cdots((-2i+1)/2)}{i!} = \frac{(-1)^i(2i-1)!!}{2^i i!}.$$

Therefore, by the binomial theorem,

$$(1-x)^{-1/2} = \sum_{i\geq 0}\binom{-1/2}{i}(-x)^i = \sum_{i\geq 0}\frac{(2i-1)!!}{2^i i!}x^i. \tag{10.18}$$

Multiplying the expansions of $e^{-x/2}$ and $(1-x)^{-1/2}$ together, we get that the right-hand side of (10.17) is equal to

$$\sum_{n\geq 0} x^n \sum_{k\geq 0}^{n} \frac{(-1)^k}{2^k k!}\cdot\frac{(2n-2k-1)!!}{(n-k)!2^{n-k}} = \sum_{n\geq 0} x^n \sum_{k=0}^{n}\frac{(-1)^k(2n-2k-1)!!}{k!(n-k)!\cdot 2^n}.$$

The coefficient of x^n on the right-hand side of (10.17) is

$$\sum_{k=0}^{n}\frac{(-1)^k(2n-2k-1)!!}{k!(n-k)!\cdot 2^n},$$

implying that the coefficient of $x^n/(n!)^2$ is

$$(n!)^2 \sum_{k=0}^{n} \frac{(-1)^k(2n-2k-1)!!}{k!(n-k)!2^n} = \frac{n!}{2^n} \sum_{k=0}^{n} (-1)^k \binom{n}{k} (2n-2k-1)!!,$$

which indeed agrees with the corresponding coefficient on the left-hand side. This proves our corollary. ◊

That's fine, you might say, but how did you know in advance that $\frac{e^{-x/2}}{\sqrt{1-x}}$ was the right expression for $T(x)$? This is a very appropriate question. That is, the "problem" with the above proof is that the result was *verified* rather than *deduced.* A proof where the answer is actually *deduced* can be found by solving Exercises 11 and 12.

The following lemma shows the very close connection between the numbers $T_n(2)$ and $H_n(2)$.

Lemma 10.18 *Let* $H(x) = \sum_{n\geq 0} H_n(2)\frac{x^n}{n!^2}$ *be the doubly exponential generating function of the sequence* $H_n(2)$. *Then*

$$H(x) = \frac{e^{x/2}}{\sqrt{1-x}}. \tag{10.19}$$

Proof: Note that by Corollary 10.17, our lemma is equivalent to

$$H(x) = e^x T(x).$$

We will prove the latter by showing that the coefficients of $x^n/(n!)^2$ agree on both sides, for all n.

On the left-hand side, this coefficient is $H_n(2)$. Let us expand the right-hand side as

$$\begin{aligned} e^x T(x) &= \left(\sum_{k\geq 0} \frac{x^k}{k!}\right) \cdot \left(\sum_{i\geq 0} T_i(2)\frac{x^i}{i!^2}\right) \\ &= \sum_{n\geq 0}\sum_{k=0}^{n} \frac{1}{k!} \cdot \frac{T_{n-k}(2)}{(n-k)!^2} x^n. \end{aligned}$$

Therefore, the coefficient of x^n on the right-hand side is $\sum_{k=0}^{n} \frac{1}{k!} \cdot \frac{T_{n-k}(2)}{(n-k)!^2}$, and so that of $x^n/(n!)^2$ is

$$n!^2 \sum_{k=0}^{n} \frac{1}{k!} \cdot \frac{T_{n-k}(2)}{(n-k)!^2} = \sum_{k=0}^{n} \binom{n}{k}^2 k! \cdot T_{n-k}(2). \tag{10.20}$$

Consequently, our lemma will be proved if we can show that the right-hand side of this last equation (10.20) is equal to $H_n(2)$. This can be done as follows:

Let us assume that H is a magic square counted by $H_n(2)$ that contains exactly k entries that are equal to 2. There are $\binom{n}{k}^2 k!$ ways to choose the location of these k entries. Then we must fill out the remaining $(n-k) \times (n-k)$ grid so that each row and column contains two entries equal to 1. This can be done in $T_{n-k}(2)$ ways. This proves that (10.20) holds, completing the proof of the lemma. $\diamondsuit$

Finally, we are in a position to state and prove our formula for the numbers $H_n(2)$.

Theorem 10.19 *For all positive integers $n \geq 1$,*

$$H_n(2) = \frac{n!}{2^n} \sum_{k=0}^{n} \binom{n}{k} n!(2n-2k-1)!!. \tag{10.21}$$

Proof: By Lemma 10.18, $H_n(2)$ is the coefficient of $x^n/(n!)^2$ in $H(x) = e^{x/2}/\sqrt{1-x}$. On one hand, we have

$$e^{x/2} = \sum_{k \geq 0} \frac{x^k}{2^k k!}.$$

On the other hand, we have recently computed $(1-x)^{-1/2}$ in (10.18). Using that formula, we get that

$$H(x) = \sum_{n \geq 0} x^n \sum_{k=0}^{n} \frac{(2n-2k-1)!!}{k!(n-k)! \cdot 2^n}.$$

Note that the computation is the same as in the proof of Corollary 10.17, but with the $(-1)^k$ omitted. This shows that the coefficient $H_n(2)$ of $x^n/(n!)^2$ in $H(x)$ is $\frac{\sum_{k=0}^{n} \binom{n}{k} n!(2n-2k-1)!!}{2^n}$ as claimed. $\diamondsuit$

A couple of questions are in order. Is there a reasonably simple proof of this formula that does not use generating functions? If yes, does that combinatorial proof explain the 2^n term in the denominator? Note that it is certainly not obvious why $n! \sum_{k=0}^{n} \binom{n}{k} n!(2n-2k-1)!!$ should be divisible by 2^n.

The answer to both of these questions is in the affirmative, as shown by a nice argument by W. Griffiths [40]. The reader is invited to consult Exercises 4 and 5, which walk the reader through that interesting proof.

Quick Check

1. How many $n \times n$ magic squares are there with line sum 2 that contain $n-2$ digits 2?

2. How many $n \times n$ magic squares are there with line sum 2 that contain $n - 3$ digits 2?
3. How many $n \times n$ magic squares are there with line sum 3 that contain $n - 3$ digits 3?

10.4 Why magic cubes are different

Now consider the following enhanced version of our original distribution problem. Again, we want to distribute 60 building blocks to three children so that each child gets 20 blocks. 20 blocks are red, 20 are blue, and 20 are green, and blocks of the same color are identical. This is where the original problem ended. Assume, however, that the children come back to play tomorrow, and the day after tomorrow. On each of these three days, we distribute the blocks following the above rule (each child gets 20), however, we are also looking for a long-term balance. Therefore, we require, that at the end of the three-day period, the number of red blocks each child had throughout the entire period be 20, the number of blue blocks each child had throughout this period be 20, and the number of green blocks each child had throughout this period be 20. In how many ways can all this be done?

We know from Section 10.1 that an acceptable distribution of blocks can be represented by a magic square of side length 3 and line sum 20, and therefore the number of acceptable distributions for any given day is $H_3(20)$. However, we cannot just take three magic squares, put them on top of each other, and say that the obtained three-dimensional array of size $3 \times 3 \times 3$ represents a solution to the entire problem. This is because of the extra requirement on the balance of colors. That is, we are looking for distributions so that, if we put the corresponding three magic squares on top of each other, the sum of the entries in each *vertical line* is also 20. One possible distribution is shown in Figure 10.7. Let A, B, and C denote the three children.

	A	B	C
red	5	6	9
blue	6	7	7
green	9	7	4

today

	A	B	C
red	8	6	6
blue	7	5	8
green	5	9	6

tomorrow

	A	B	C
red	7	8	5
blue	7	8	5
green	6	4	10

day after tomorrow

Figure 10.7
A possible three-day distribution.

Such a construction—an $n \times n \times n$ array of nonnegative integers in which the sum of each line, that is, (row, column, and vertical line), is the same—is called a *magic cube*. So the array in Figure 10.7 is an example of a magic cube of side length 3 and line sum 20.

Let $C_n(r)$ be the number of magic cubes of side length n and line sum r. The reader should check that $C_1(r) = 1$ and $C_2(r) = r + 1$. After this, the reader might be thinking that the enumeration of magic cubes is in fact very similar to that of magic squares. This is not true, however, as the smallest nontrivial case, that of $n = 3$, shows. A $2\times2\times2$ subcube completely determines a $3 \times 3 \times 3$ magic cube, so $C_3(r) \leq r^8$. Therefore, if $C_3(r)$ is a polynomial, its degree can be at most eight. It is not difficult (see Exercise 20) to write a computer program that computes the first 10 values of $C_3(r)$. These values are shown in Figure 10.8.

r	$C_3(r)$
0	1
1	12
2	132
3	847
4	3921
5	14286
6	43687
7	116757
8	280656
9	619219

Figure 10.8
The values of $C_3(r)$ for $0 \leq r \leq 9$.

We will now explain how to deduce from those 10 values that $C_3(r)$ is *not* a polynomial.

Let $f : \mathbf{N} \to \mathbf{R}$ be any function, and let $\Delta(f)(n) = f(n+1) - f(n)$. Similarly, $\Delta^2(f)(n) = \Delta(\Delta(f)(n)) = (f(n+2) - f(n+1)) - (f(n+1) - f(n))$, and one can obtain Δ^d along the same lines. The following proposition characterizes polynomials of degree at most d in terms of Δ.

Proposition 10.20 *Let $p : \mathbf{N} \to \mathbf{R}$ be a polynomial. Then the degree of p is at most d if and only if $\Delta^d(p)$ is a constant function.*

Proof: We prove the statement by induction on d. For $d = 0$, the claim trivially holds. Now assume the claim holds for d and prove it for $d + 1$.

(a) (the "only if" part) Let p be a polynomial of degree $d + 1$. Then $p(n) = a_{d+1}n^{d+1} + a_d n^d + \cdots + a_1 n + a_0$. Therefore, $p(n+1) = a_{d+1}(n+1)^{d+1} + a_d(n+1)^d + \cdots + a_1(n+1) + a_0$. From this, we see that n^{d+1} cancels out in $p(n+1) - p(n) = \Delta(p)(n)$, so $\Delta(p)$ is a polynomial of degree at most d. Therefore, by the induction hypothesis, $\Delta^d(\Delta(p))$ is constant, in other words, $\Delta^{d+1}(p)$ is constant.

(b) (the "if" part) Let p be a function so that $\Delta^{d+1}(p)$ is constant. Then $\Delta^d(\Delta(p))$ is constant, so, by the induction hypothesis, $\Delta(p)$ must be a polynomial of degree at most d. This implies that p is of degree at most $d+1$. Indeed, if we have $p(n) = \sum_{i=0}^{h} a_i n^i$, then we also have $\Delta(p)(n) = \sum_{i=0}^{h} a_i(n+1)^i - \sum_{i=0}^{h} a_i n^i$, which contains a term of degree $h-1$.

◊

We point out that Proposition 10.20 remains true if we change its wording a little bit. Instead of requiring that p is a polynomial, we could simply require that p be a function that satisfies the requirement that $\Delta^d(p)$ is constant. Then the above argument would show that p is a polynomial of degree at most d. However, we will not need this stronger form.

Using our numerical data, we can easily compute that $\Delta^8(C_3(r))(0) \neq \Delta^8(C_3(r))(1)$; therefore, $C_3(r)$ cannot be a polynomial of degree at most 8, so $C_3(r)$ cannot be a polynomial at all.

Hopefully, you are wondering now what could possibly cause this unexpected behavior of the function $C_3(r)$. You might also be wondering if $C_3(r)$ is just one exception or if maybe $C_n(r)$ is never a polynomial if $n \geq 3$.

Let us try to copy the proof of Theorem 10.3, the theorem showing that $H_n(r)$ is always a polynomial. The point where we get stuck is Lemma 10.6. That lemma has no three-dimensional analog. In fact, that lemma does not hold for magic cubes, even if we restrict our attention to magic cubes of side length 3 and line sum 2. Figure 10.9 shows a counterexample.

$$\begin{bmatrix} 2 & 0 & 0 \\ 0 & 1 & 1 \\ 0 & 1 & 1 \end{bmatrix} \quad \begin{bmatrix} 0 & 1 & 1 \\ 1 & 0 & 1 \\ 1 & 1 & 0 \end{bmatrix} \quad \begin{bmatrix} 0 & 1 & 1 \\ 1 & 1 & 0 \\ 1 & 0 & 1 \end{bmatrix}$$

Figure 10.9
An irreducible magic cube of line sum 2.

Indeed, the first level can be decomposed in only one way, and its bottom right corner makes any decompositions of the other two levels impossible to fit in a magic cube.

Let us call a magic cube *irreducible* if it cannot be decomposed into the sum of magic cubes of smaller line sums. The above example then shows that not all irreducible magic cubes have line sum 1. It can be shown, however, that all irreducible magic cubes of side length 3 have line sum either 1 or 2. Then it can be proved, in a manner similar to the proof of Theorem 10.3, that both $C_3(2r)$ and $C_3(2r+1)$ are polynomials. Now that we know that there are irreducible magic cubes of side length 3 with line sum 2, this result is not that surprising. Indeed, if we want to build a magic cube of line sum $2r+1$, we have to use at least one of the three-dimensional permutation matrices; if we want to build a magic cube with line sum $2r$, then we can do that by using just the irreducibles of line sum 2.

At this point, we remind the reader that, in the solution of Exercises 19 and 20 of Chapter 2 we saw functions that were similar to $C_n(r)$ in that they were *polynomials on residue classes*, sometimes abbreviated as *porcs*. These functions are also known as *quasi-polynomials*.

If $n > 3$, then it can still be proved that the number of irreducible magic cubes of side length n is always finite. Then one can show that the function $C_n(r)$ is a quasi-polynomial, that is, there exists a k so that $C_n(rk)$ is a polynomial, $C_n(rk+1)$ is a polynomial, and so on; finally, $C_n(rk+k-1)$ is a polynomial. In other words, $C_n(r)$ is a polynomial on any residue class modulo k, so $C_n(r)$ is a quasi-polynomial. These proofs are beyond the scope of this book, but in the Notes section, we will mention sources that the reader can use for further indulgence in these interesting topics. To conclude, we mention that open problems abound in this area. For instance, we do not even have a formula to compute $C_n(1)$ for general n. See Exercise 16 for an alternative interpretation of this number.

Quick Check

1. Find the number of magic bricks of size $2 \times 2 \times 3$ in which every line sum is r.
2. Find the number of $3 \times 3 \times 3$ magic cubes with line sum 2 that contain exactly one entry 2.
3. Find the number of $3 \times 3 \times 3$ magic cubes with line sum 2 that contain exactly three entries 2.

10.5 Notes

As we mentioned while introducing this chapter, there are many different ways to define a magic square or just a highly symmetric way of filling out a square matrix with nonnegative integers or some other symbols. The best-studied of them may well be *Latin squares*, which we will define in Exercise 16. The interested reader can read an introduction to the theory of these in Eric Gossett's book [39] and a more advanced review in the book by van Lint and Wilson [85].

Let us now return to our magic squares. For a higher-level exposition on magic squares, see Section 4.5 of [71] or [75]. This second book uses machinery from commutative algebra to prove some interesting results. For instance, having proved that $H_n(r)$ is a polynomial for any fixed n, one can *define* $H_n(x)$ for any real number x by simply substituting x into the polynomial $H_n(r)$. It is then possible to prove that $H_n(-1) = H_n(-2) = \cdots = H_n(1-r) = 0$. The reader is invited to verify this for $n = 2$ and $n = 3$. The surprising identity $H_n(-n-r) = (-1)^{n-1}H_n(r)$ also holds. These identities are useful for numerical purposes. Indeed, when we want to determine the polynomial $H_n(r)$, we need $(n-1)^2+1$ values of this polynomial. The mentioned two identities decrease this number significantly if n is reasonably small.

For a short overview of algebraic methods on magic cubes, see [15].

Another direction of research is to study *how large* the number $H_n(r)$ is. For fixed n, we know that $H_n(r)$ is a polynomial function, so the question essentially reduces to how large the *leading coefficient* a of this polynomial is. No exact answer has been given to that question, but a is surprisingly large, as shown in [16].

We have seen formulae for the numbers $H_n(1)$ and $H_n(2)$. The next question, naturally, is what can we say about $H_n(3)$? First, it was proved in [42] that a polynomial recursion must exist for these numbers. That is, there must exist an integer k and nonzero polynomials $p_0, p_1, \cdots, p_k$ so that

$$\sum_{i=0}^{k} p_i(r)H_{r+i}(3) = 0. \tag{10.22}$$

Then an exact formula was found by W. Griffiths [40], who proved that

$$H_n(3) = \frac{1}{6^{2n}} \sum_{k=0}^{n} \binom{n}{k} \frac{n!}{(n-k)!} 30^k \left[(3n-3k)! + \sum_{j=0}^{n-k} (2^j - 1)P_j(n-k) \right],$$

where

$$P_j(n) = \sum_{m=j}^{n} (-1)^{m-j} \binom{m}{j} \binom{n}{m}^2 18^m m! \left[\sum_{l=0}^{m} (-1)^l \binom{n}{l} (3n-2m-l)! \right].$$

The proof is not immediately obvious.

Richard Stanley [77] conjectures that an integer k and nonzero polynomials $p_0, p_1, \cdots, p_k$ satisfying (10.22) exist, even if we replace $H_n(3)$ by $H_n(r)$ for any fixed positive integer r (for $H_n(3)$, Griffiths' result proves Stanley's conjecture, as can be shown with some effort).

Finally, a far-reaching generalization of the concept of magic squares is the following: An $n \times n$ symmetric magic square H is equivalent to a complete bipartite graph $K_{n,n}$ with color classes A and B, with the rows corresponding to vertices in A and the columns corresponding to vertices in B. Let us write the number x on the edge between A_i and B_j if the intersection of row i and column j of the magic square H is the entry x. We then get a *magic labeling* of $K_{n,n} = (A, B)$, that is, a labeling of the edges of $K_{n,n} = (A, B)$ so that the sum of the labels adjacent to each vertex is the same. All theorems we proved for magic squares will then hold for magic labelings of this graph. This raises a few natural questions. What if we take a graph other than $K_{n,n}$? It is not difficult to do away with the condition of H being symmetric. Just take two edges between any two points, directed opposite to each other. See [76] for some results in this direction.

10.6 Chapter review

(A) Magic squares of fixed size ($n \times n$) and line sum r

1. They are always decomposable into the sum of r magic squares of line sum 1.
2. Their number, $H_n(r)$, is a polynomial function of r, of degree $(n-1)^2$.
3. We have $H_1(r) = 1$, $H_2(r) = r + 1$, and

$$\begin{aligned} H_3(r) &= \binom{r+2}{4} + \binom{r+3}{4} + \binom{r+4}{4} \\ &= \binom{r+5}{5} - \binom{r+2}{5}. \end{aligned}$$

(B) Magic squares of fixed line sum (r) and size $n \times n$

1. If $r = 1$, then $H_n(1) = n!$.
2. If $r = 2$, then

$$H_n(2) = \frac{n!}{2^n} \sum_{k=0}^{n} \binom{n}{k} n!(2n-2k-1)!!.$$

(C) Magic cubes of fixed size ($n \times n \times n$) and line sum r

1. They are in general *not* decomposable into the sum of r magic cubes of line sum 1.
2. Their number $C_n(r)$ is in general *not* a polynomial on r.

(D) Magic cubes of line sum (r) and size $n \times n \times n$

1. For $r = 1$, their number equals the number of $n \times n$ Latin squares.

10.7 Exercises

1. Prove our claim made in part (a) of the proof of Theorem 10.2 stating that, the 4-tuples of nonnegative integers (a, b, c, d) satisfying (10.8) and the 4-tuples $(a, 2a+d-c, a+b+d-c, b+d)$ satisfying (10.8) are in bijection.

2. Prove our claim made in part (c) of the proof of Theorem 10.2 stating that if $a > c$ and $b > c$, then inequalities (10.3)–(10.6) are all redundant.

3. Assume we know that p is a polynomial of degree m, and also assume we know the values of $p(0), p(1), \cdots, p(m)$. How can we find p? (Define a general strategy that shows that p can be found. Do not discard a method simply because it may require a lot of computation.)

4. Let A be a $2n \times 2n$ matrix with nonnegative integer entries. For any $i \in [n]$, let us say that rows $2i-1$ and $2i$ form a *row pair*, and columns $2i-1$ and $2i$ form a *column pair*. We say that A is a *double magic square* of order n if it satisfies the following criteria:

   (a) The sum of entries of each row pair and each column pair is 2,

   (b) if a row pair contains a 2, then that 2 must be in the *top* row of the row pair, and

   (c) there is at most one positive integer in any row or column.

   See Figure 10.10 for an example. Prove that the number $DM(n)$ of double magic squares of order n is

$$\sum_{k=0}^{n} \binom{n}{k} \frac{n!}{(n-k)!} 2^k (2n-2k)!.$$

0	0	2	0
0	0	0	0
0	1	0	0
1	0	0	0

Figure 10.10
A double magic square of order 2.

5. (a) Prove that there exists a 2^{2n}-to-one map f from the set S of all double magic squares of order n onto the set T of all magic squares of size $n \times n$ having line sum 2.
   (b) Deduce Theorem 10.19.
6. Let $P_3(r)$ be the number of 3×3 magic squares that are symmetric to their main diagonal and have line sum r. Is $P_3(r)$ a polynomial?
7. Prove that $P_{n+1}(1) = P_n(1) + nP_{n-1}(1)$.
8. Find the exponential generating function of the numbers $P_n(1)$.
9. Let P be a magic square that is symmetric to its main diagonal. Is it true that P is the sum of some *symmetric* permutation matrices?
10. Show an example of three disjoint sets of permutation matrices having the same sum.
11. Prove (preferably with a direct argument) the recurrence relation
$$T_{n+1}(2) = n^2(n+1)T_{n-1}(2)/2 + n(n+1)T_n(2).$$
12. Use the result of the previous exercise to find the doubly exponential generating function $T(x) = \sum_{n\geq 0} \frac{T_n(2)}{(n!)^2} x^n$.
13. Prove that
$$\sum_{k=0}^{n} \binom{n}{k}^2 H_k(2)T_{n-k}(2) = n!^2.$$
14. Prove (by any method) that the recurrence relation
$$H_n(2) = n^2 H_{n-1}(2) - \binom{n}{2}(n-1)H_{n-2}(2)$$
holds for $n \geq 2$.
15. Compute $C_3(1)$.
16. A *Latin square* is an $n \times n$ square grid that has been filled out by n letters so that each letter occurs in each row and each column exactly once. Figure 10.11 shows a 3×3 Latin square.

A	B	C
B	C	A
C	A	B

Figure 10.11
A 3×3 Latin square.

Construct a bijection from the set of $n \times n$ Latin squares onto that of magic cubes enumerated by $C_n(1)$.

17. Prove that the only irreducible $3 \times 3 \times 3$ magic cubes of line sum 2 are those that can be obtained from our example shown in Figure 10.9 by permuting lines.
18. Compute $C_3(2)$.
19. Prove that $C_n(1)$ is always divisible by $n!$.
20. Write a computer program in any language that computes the value of $C_3(r)$.
21. Let us call a set $A \subset \mathbf{N}^m$ an *antichain* in $\mathbf{N}^m$ if there are no two points $(a_1, a_2, \cdots, a_m)$ and $(b_1, b_2, \cdots, b_m)$ in A so that $a_i \leq b_i$ holds for all $i \in [m]$.

    Prove that all antichains in $\mathbf{N}^m$ are finite.

10.8 Solutions to exercises

1. Given (a, b, c, d), we can compute $\phi(a, b, c, d) = (a, 2a+d-c, a+b+d-c, b+d)$. On the other hand, given $(e, f, g, h) = (a, 2a+d-c, a+b+d-c, b+d)$, we can compute (a, b, c, d) by $a = e$, $b = e - f + g$, $c = e - g + h$, and $d = (-e + f - g + h)$. Thus ϕ has an inverse, and therefore ϕ is a bijection.
2. Inequalities (10.5) and (10.6) are both redundant since d is nonnegative and c is smaller than a and b. Inequality (10.3) is redundant since it follows from (10.7). Indeed,

    $$r - a - d \geq r - a - d - (b - c) \geq 0$$

    because $b > c$. Finally, (10.4) is also redundant since it also follows from (10.7). Indeed,

    $$r - b - d \geq r - b - d - (a - c) \geq 0$$

    because $a > c$.
3. Let $p(x) = a_m x^m + a_{m-1} x^{m-1} + \cdots a_1 x + a_0$. Substituting the values $0, 1, \cdots, m$ for x, we get a system of $m+1$ linear equations with the $m+1$ unknowns $a_m, a_{m-1}, \cdots, a_0$. We can then use linear algebra to solve this system.
4. The term indexed by k is the number of double magic squares of order n that contain exactly k entries 2. In these double magic squares, there are $\binom{n}{k}$ ways to choose the row pairs in which those

2s will be put, then there are $\frac{n!}{(n-k)!}$ ways to choose the column pairs. As the 2s can be in either column but must be in the top row of each pair, there are 2^k ways to find their exact positions once the pairs are selected. Deleting the row pairs and column pairs containing a 2, we get one of $(2n-2k)!$ permutation matrices.

5. (a) The definition of f is simplicity itself. Given $A \in S$, let $f(A)$ be the magic square whose (i,j)-element is the sum of the four elements in the intersection of the ith row pair and the jth column pair of A. So if A is the double magic square shown in Figure 10.10, then $f(A)$ is the magic square shown in Figure 10.12.

0	2
2	0

Figure 10.12
The magic square $f(A)$.

It then follows from the first criterion in the definition of double magic squares that $f(A) \in T$. It remains to be proved that, for each $B \in T$, there are exactly 2^{2n} elements $A \in S$ satisfying $f(A) = B$.

Let $B \in T$. We will find all the preimages of B under f as follows: First, double the columns of B. That is, turn B into an $n \times 2n$ matrix B' in which the old column j becomes the new column pair $(2j-1, 2j)$, and then the old (i,j)-entry will give us instruction on what the new entries in positions $(i, 2j-1)$ and $(i, 2j)$ are. Namely, if column j had two entries 1, in positions (i,j) and (i',j), then the corresponding column pair gets an entry 1 in one of its columns, in one of the two positions $(i, 2j-1)$ and $(i, 2j'-1)$, as well as in the *opposite* position of $(i, 2j'-1)$ and $(i, 2j')$. So each column without a 2 in it will let us choose one of two possibilities.

If the (i,j)-entry of B was 2, then again, we have two possibilities. We can set either the $(i, 2j-1)$-entry or the $(i, 2j)$-entry of B' to 2. See Figure 10.13 for an example.

Note that each column of B' has at most one nonzero entry, and that the sum of each column pair is 2.

Now let us double the rows of B' according to slightly different rules. Again, if row i of B contained two entries 1, then the corresponding row pair will get two entries 1 in diagonal positions, letting us choose one of two possibilities. In other words, say

$$B = \begin{bmatrix} 2 & 0 & 0 \\ 0 & 1 & 1 \\ 0 & 1 & 1 \end{bmatrix}$$

$$\begin{bmatrix} 2 & 0 & 0 & 0 & 0 & 0 \\ 0 & 0 & 0 & 1 & 0 & 1 \\ 0 & 0 & 1 & 0 & 1 & 0 \end{bmatrix} \quad \begin{bmatrix} 2 & 0 & 0 & 0 & 0 & 0 \\ 0 & 0 & 1 & 0 & 1 & 0 \\ 0 & 0 & 0 & 1 & 0 & 1 \end{bmatrix}$$

$$\begin{bmatrix} 0 & 2 & 0 & 0 & 0 & 0 \\ 0 & 0 & 1 & 0 & 1 & 0 \\ 0 & 0 & 0 & 1 & 0 & 1 \end{bmatrix} \quad \begin{bmatrix} 0 & 2 & 0 & 0 & 0 & 0 \\ 0 & 0 & 0 & 1 & 0 & 1 \\ 0 & 0 & 1 & 0 & 1 & 0 \end{bmatrix}$$

Figure 10.13
The first step in finding the preimages of B.

row i contains a 1 in position j and another in position j'. Then the new matrix B'' will contain a 1 corresponding to the 1 in position (i, j) of B' either in $(2i-1, j)$ or in $(2i, j)$. Whichever choice we make here, we have to make the *other choice* for the 1 in position (i, j') of B'.

However, an entry 2 can either stay in the top row of the row pair, or it can be broken up into two entries 1 located in a diagonal of the 2×2 block corresponding to the original entry 2. See Figure 10.14 for an example.

2	0	0	0	0	0
0	0	0	0	0	0
0	0	0	1	0	0
0	0	0	0	0	1
0	0	1	0	0	0
0	0	0	0	1	0

0	2	0	0	0	0
0	0	0	0	0	0
0	0	0	1	0	0
0	0	0	0	0	1
0	0	1	0	0	0
0	0	0	0	0	0

1	0	0	0	0	0
0	1	0	0	0	0
0	0	0	1	0	0
0	0	0	0	0	1
0	0	1	0	0	0
0	0	0	0	1	0

0	1	0	0	0	0
1	0	0	0	0	0
0	0	0	1	0	0
0	0	0	0	0	1
0	0	1	0	0	0
0	0	0	0	1	0

Figure 10.14
Some possibilities for B'' if B' is the matrix at the top of Figure 10.13.

Note that no matter what sequences of choices we made, B'' is a double magic square in S. Furthermore, it follows from our definitions that $f(B'') = B$, since the sum of the entries in the intersection of row pair i and column pair j of B'' is the (i, j)-entry of B. Again by our definitions, all preimages of B under f can be obtained as B'' by this procedure.

Because during this procedure we had 2^{2n} choices to make (2^n connected to the n columns, then 2^n connected to the n rows), the statement is proved.

(b) Apply the division principle and the result of part (a) to the result of the previous exercise.

6. No, it is not. Entries a, b, and c, as shown in Figure 10.15, completely determine such a magic square. So if $P_3(r)$ were a poly-

a	b	
	c	

Figure 10.15
The entries a, b, and c.

nomial, its degree could not be larger than three. Computing the first five values of $P_3(r)$, we get $P_3(0) = 1$, $P_3(1) = 4$, $P_3(2) = 11$, $P_3(3) = 23$, and $P_3(4) = 42$. It is then straightforward to check that $\Delta^3(P_3)(0) \neq \Delta^3(P_3)(1)$; therefore, $P_3(n)$ cannot be a polynomial.

7. Let P be a magic square enumerated by $P_{n+1}(1)$. Then the first row of P contains one 1. This 1 can be either in the first position, and then we have $P_n(1)$ possibilities for the remaining $n \times n$ symmetric magic square, or it is somewhere else. In that case, it can be at n different positions, each of them outside the main diagonal. Whatever position it is in, however, the mirror image of that position must also contain a 1, and then we have $P_{n-1}(1)$ possibilities for the remaining $(n-1) \times (n-1)$ symmetric magic square.

8. Let $P(x) = \sum_{n\geq 0} \frac{P_n(1)}{n!} x^n$. Then multiplying both sides of the recurrence relation of the previous exercise by x^n/n and summing for all $n \geq 0$ leads to the functional equation

$$P'(x) = (1 + x)P(x),$$

implying that

$$\frac{P'(x)}{P(x)} = 1 + x,$$

$$\ln P(x) = x + \frac{x^2}{2}.$$

Therefore, $P(x) = e^{x+\frac{x^2}{2}}$. Note that this is equal to the exponential generating function of all involutions of length n. This is not surprising. Indeed, a permutation matrix is symmetric to its main diagonal if and only if it is the permutation matrix of an involution.

9. No, it is not. A counterexample for $n = 3$ is shown in Figure 10.16.

0	1	1
1	0	1
1	1	0

Figure 10.16
A counterexample.

Indeed, this magic square cannot be decomposed into the sum of two symmetric permutation matrices. Since all diagonal entries are zero, a symmetric permutation matrix Q that is part of the decomposition would have to have an even number of 1s in it. That is impossible since Q would have to contain two or four 1s, not three as needed to have line sum 1.

10. We have seen that no such example exists when $n = 3$. An example for $n = 4$ is shown in Figure 10.17.

11. Let b_n denote the number of *almost magic squares.* We say that an $n \times n$ matrix with 0–1 entries is an almost magic square if it contains two entries equal to 1 in each row and column except for the first row and the first column, and it contains one entry equal to 1 in the first row, and one entry equal to 1 in the first column.

    We then claim that

$$T_n(2) = \binom{n}{2}(n-1)b_{n-1}. \tag{10.23}$$

Indeed, take a matrix that is enumerated by $T_n(2)$. There are $\binom{n}{2}$ pairs of positions in which the first column of this matrix can contain

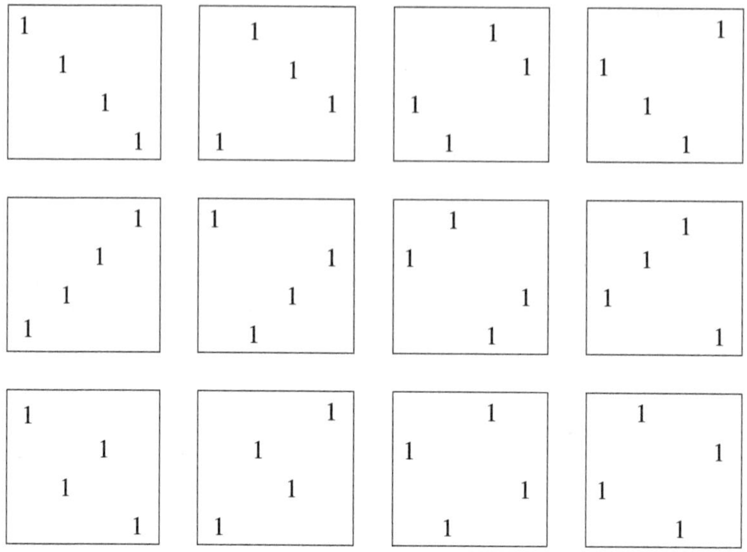

Figure 10.17
All three sets sum to J_4.

its two entries equal to 1. In row R of the first of these two entries, there are $(n-1)$ possible positions for the second entry equal to 1. Finally, if we delete the first column and row R of our matrix, we get an $(n-1)\times(n-1)$ matrix that is an almost magic square, up to a permutation of rows and columns.

On the other hand, we claim that the numbers b_n can also be obtained from the numbers $T_n(2)$. Indeed, we show that

$$b_n = T_{n-1}(2) + (n-1)^2 b_{n-1}. \tag{10.24}$$

Indeed, the two entries equal to 1 in the first row and column of an almost magic square either coincide, in which case the remaining matrix of size $(n-1)\times(n-1)$ is a magic square, or do not coincide. In this case, they can be in $(n-1)^2$ different positions, and, omitting the row and column containing these entries (which are not the first row and not the first column), we get an almost magic square of size $(n-1)\times(n-1)$.

Let us express $T_{n+1}(2)$ by (10.23), then, in the expression thus obtained, express b_n by (10.24) to get

$$T_{n+1}(2) = \binom{n+1}{2} n \cdot b_n = \frac{1}{2}(n+1)n^2 T_{n-1}(2) +$$

$$\frac{1}{2}(n+1)n^2(n-1)^2 b_{n-1}.$$

Finally, note that the last term will become simpler if we replace $n(n-1)^2 b_{n-1}/2$ by $T_n(2)$, as is allowed by (10.23). This yields

$$T_{n+1}(2) = \binom{n+1}{2} n \cdot T_{n-1}(2) + T_n(2)n(n+1),$$

which was to be proved.

12. Multiply both sides of the recursive relation by $x^n/(n!)^2(n+1)$, then sum for all nonnegative integers n to get

$$\sum_{n\geq 0} \frac{T_{n+1}(2)}{(n!)^2(n+1)} x^n = \sum_{n\geq 0} \frac{\frac{1}{2}\cdot n^2 \cdot T_{n-1}(2)}{(n!)^2} x^n +$$

$$\sum_{n\geq 0} \frac{T_n(2)n}{(n!)^2} x^n.$$

Now note that the previous equation can be written in the shorter form

$$T'(x) = xT(x)/2 + xT'(x).$$

From here, we get

$$\frac{T'(x)}{T(x)} = \frac{x}{2(1-x)}.$$

Let us integrate both sides. On the left-hand side, we currently have $(\ln T(x))'$, so integration leads to $\ln T(x)$. The right-hand side is easy to integrate if we notice that $\frac{x}{2(1-x)} = \frac{1}{2}\sum_{n\geq 1} x^n$. This leads to

$$\ln T(x) = \frac{1}{2}\sum_{n\geq 1} \frac{x^{n+1}}{n+1} = -\frac{x}{2} - \frac{1}{2}\cdot \ln(1-x),$$

and $T(x) = \frac{e^{-x/2}}{\sqrt{1-x}}$ follows by taking exponentials.

13. Taking the product of generating functions $H(x) = \sum_{n\geq 0} H_n(2)\frac{x^n}{n!^2}$ and $T(x) = \sum_{n\geq 0} T_n(2)\frac{x^n}{n!^2}$, and remembering Corollary 10.17 and Lemma 10.18, we get

$$H(x)T(x) = \frac{e^{x/2}}{\sqrt{1-x}} \cdot \frac{e^{-x/2}}{\sqrt{1-x}} = \frac{1}{1-x}. \tag{10.25}$$

On the other hand, expanding the product $T(x)H(x)$, we get

$$H(x)T(x) = \sum_{n\geq 0} x^n \sum_{k=0}^{n} \frac{H_k(2)}{k!^2} \cdot \frac{T_{n-k}(2)}{(n-k)!^2}. \tag{10.26}$$

As the left-hand sides of the last two equations are identical, so too must be their right-hand sides, and in particular, so too must be the coefficients of $x^n/(n!)^2$ in the two right-hand sides. On the right-hand side of (10.25), this coefficient is $n!^2$ since $1/(1-x) = \sum_{n\geq 0} x^n$. On the right-hand side of (10.26), this coefficient is

$$\sum_{k=0}^{n} \frac{n!^2 H_2(k) T_{n-k}(2)}{k!^2 \cdot (n-k)!^2} = \sum_{k=0}^{n} \binom{n}{k}^2 H_2(k) T_{n-k}(2),$$

proving our claim.

14. Note that, defining $a_{-1} = 0$, the recursive formula to be proved also holds for $n = 1$. Because the formula holds for all $n \geq 1$, we can take it for all such n, multiply it by $\frac{nx^{n-1}}{n!^2}$, and sum the obtained equations. We get

$$\sum_{n\geq 1} n \cdot \frac{H_n(2)}{n!^2} x^{n-1} = \sum_{n\geq 1} n \cdot \frac{H_{n-1}(2)}{(n-1)!^2} x^{n-1} - \frac{1}{2} \sum_{n\geq 1} \frac{H_{n-2}(2)}{(n-2)!} x^{n-1}.$$

Note that, by term-by-term comparison, this is equivalent to

$$H'(x) = xH'(x) + H(x) - \frac{1}{2} x H(x),$$

which leads to

$$\frac{H'(x)}{H(x)} = \frac{1}{2} - \frac{1}{2} \cdot \frac{1}{1-x},$$

and the solution of this last equation is indeed $H(x)$.

A proof that does not use the explicit formula for $H_n(2)$ can be found in [4].

15. Choose any of the six permutation matrices of size 3×3 for the bottom level. If you choose the matrix π, then there are two matrices that are eligible to be put to the middle and top levels. These are the two matrices that are obtained by multiplying π by the matrices of one of the two fixed point–free permutations of length 3. We can put either of them to the middle level; then we must put the other one to the top level. Therefore, $C_3(1) = 6 \cdot 2 \cdot 1 = 12$.

16. Instead of filling the Latin square with letters, we fill it with numbers, from 1 through n. Let L be such a Latin square. Now we turn all the entries of L into zeros, except for the n entries equal to i, which we turn into ones. This gives us the permutation matrix L_i. Let $f(L)$ be the magic cube whose level i is L_i. It is straightforward to verify that this is a bijection.

17. This can be done by an intelligent proof by cases. If there is a level (or a rotated copy of a level) in a cube that contains three entries

equal to 2, then the cube is not irreducible. There *cannot* be a level with exactly two entries equal to 2. We still have to consider the cases when the highest number m of entries equal to 2 on a level (or rotated copy of it) is 0 or 1. If $m = 0$, then the cube is the difference of the cube full of ones and a cube C of line sum 1, and as such is reducible. (In fact, it is the sum of two circular translates of C.) Finally, if $m = 1$, then we arrive at the example shown in Figure 10.9.

18. Exercise 15 tells us that there are 12 magic cubes of side length 3 and line sum 1. Adding any two of them we get a magic cube enumerated by $C_3(2)$. Moreover, all these two-member sums will be different (why?). This yields $\binom{12+2-1}{2} = \binom{13}{2} = 78$ reducible magic cubes. To count the irreducible ones, consider Figure 10.9. The irreducible magic cube shown there can be transformed into 54 irreducible cubes. Indeed, the sole entry 2 can be placed in 27 different positions, and then the two levels not containing that entry can be permuted in two different ways. The previous exercise then shows that there are no more irreducible cubes either. Therefore, $C_3(2) = 78 + 54 = 132$.

19. **First solution.** Permute the n levels of the magic cube. Each of these $n!$ permutations yields a different magic cube since all n levels are different. Thus, the magic cubes enumerated by $C_n(1)$ can be arranged into subsets, each of which has $n!$ cubes in it.

    Second solution. A magic cube of line sum 1 is just n permutation matrices placed on top of each other. Denote $p_1, p_2, \cdots, p_n$ the permutations corresponding to these matrices. The crucial property of these permutations is that there is no $i \in [n]$ so that $p_j(i) = p_k(i)$ for some $j \neq k$. We claim that this property is preserved if all p_j are multiplied by the same permutation q.

    Indeed, assume that $q(p_j(i)) = q(p_k(i))$ for some $j \neq k$. That is impossible, since the bijection q cannot map the distinct entries $p_j(i)$ and $p_k(i)$ to the same entry. Therefore, for any n-permutation q, the matrices of permutations $qp_1, qp_2, \cdots, qp_n$ also form a magic square. We have thus again arranged our magic cubes into subsets of size $n!$ each.

20. The following is a Maple program computing $C_3(8)$. Changing the value of r in the first line will give other values of $C_3(r)$.

```
r:=8;
                          i:=0;
                          for a from 0 to r do
                          for b from 0 to r do
                          for c from 0 to r do
                          for d from 0 to r do
```

```
                         for e from 0 to r do
                         for f from 0 to r do
                         for g from 0 to r do
                         for h from 0 to r do

if a+b < r+1 and c+d < r+1 and a+c < r+1 and b+d < r+1
and a+b+c+d > r-1 and  e+f < r+1 and g+h < r+1 and
e+g < r+1 and f+h < r+1 and e+f+g+h > r-1 and
   a+e < r+1 and b+f < r+1 and c+g < r+1 and d+h < r+1
and a+b+e+f > r-1 and c+d+g+h > r-1 and a+c+e+g > r-1
and b+d+f+h >  r-1 and  a+b+c+d+e+f+g+h < r+r+r+1 then
                                i := i+1;
                           fi;
                        od;
                        od;
                        od;
                        od;
                        od;
                        od;
                        od;
                        od;
                        i;
                        r;
```

21. We prove the statement by induction on m. For $m = 1$, the statement is true, since between two natural numbers there is always one that is at least as large as the other. Now let us assume we know the statement for $m - 1$, and prove it for m.

    Let us assume that $A = \{d_1, d_2, \cdots\}$ is an *infinite* antichain in $\mathbf{N}^m$. For $\mathbf{a} = (a_1, a_2, \cdots, a_m) \in A$, define $f(a) = (a_1, a_2, \cdots, a_{m-1})$. Let $f(A) = \{f(\mathbf{a}) | \mathbf{a} \in A\}$. Then $f(A) \in \mathbf{N}^{m-1}$, so $f(A)$ cannot contain an infinite antichain. The elements of $f(A)$ are pairwise disjoint, in other words, f is an injection (why?).

    Because $f(A)$ is an infinite set with no infinite antichain in it, $f(A)$ must contain an infinite sequence of elements $\mathbf{x_1}, \mathbf{x_2}, \cdots$ so that $\mathbf{x_1} \leq \mathbf{x_2} \leq \cdots$. (This is not obvious; we ask the reader to explain why.) Now let us look at the sequence $S = f^{-1}(\mathbf{x_1}), f^{-1}(\mathbf{x_2}), \cdots$, whose elements are all from A. The only way S would not violate the antichain condition would be if the last coordinates of the elements of S form a strictly decreasing sequence. However, that is clearly impossible, since these coordinates are nonnegative integers.

10.9 Supplementary exercises

1. Compute $H_4(3)$.
2. Verify that $H_1(r)$ and $H_2(r)$ are polynomially recursive. That is, show that there exist an integer k and polynomials $p_0, p_1, \cdots, p_k$ so that (10.22) holds if $H_3(r)$ is replaced by $H_1(r)$ or $H_2(r)$.
3. Recall that $P_n(r)$ is the number of magic squares of line sum r that are symmetric to their main diagonal. Is it true that, for all $n > 1$, we have $H_n(r) > P_n(r)$?
4. There is a bijection from the set of all magic squares counted by $P_n(1)$ and a certain kind of n-permutations. What are these permutations?
5. (a) Find a polynomial $p(r)$ so that $P_n(r) \leq p(r)$ for all positive integers r.
   (b) Find a polynomial $q(r)$ so that $P_n(r) \geq q(r)$ for all positive integers r. Choose q so that its degree is the same as that of p in part (a).
6. For what fixed n will the function $P_n(r)$ be a polynomial function of r?
7. Let $Z(n)$ be the number of permutations $p = p_1p_2 \cdots p_n$ that are equal to their reverse complements, that is, for which $p_i = n+1-p_{n+1-i}$ for all $i \in [n]$.
   (a) Find a formula for $Z(n)$.
   (b) What does $Z(n)$ have to do with magic squares?
8. Let $V_n(r)$ be the number of magic squares of size $n \times n$ of line sum r that are symmetric to both of their diagonals. Find a formula for $V_n(1)$.
9. Let $F(r)$ be the number of 2×3 matrices with nonnegative integer entries in which each column has sum $2r$ and each row has sum $3r$. Find a formula for $F(r)$.
10. Let $G(r)$ be the number of 2×4 matrices with nonnegative integer entries in which each column has sum $2r$ and each row has sum $4r$. Find a formula for $G(r)$.
11. Let m and n be positive integers that are larger than 1. Let $A_{m,n}(r)$ be the number of $m \times n$ matrices with nonnegative integer entries in which each column has sum mr and each row has sum nr.
    (a) Find a polynomial $p(r)$ so that we have $p(r) \geq A_{m,n}(r)$ for all positive integers n.

(b) Find a polynomial $q(r)$ so that we have $q(r) \leq A_{m,n}(r)$ for all positive integers n. Choose q so that the degree of q is the same as that of p in part (a).

12. Use a combinatorial argument to characterize the nonnegative integers r for which $H_n(r)$ is odd for all nonnegative integers n.

13. Let $M_n(r)$ denote the number of $n \times n$ magic squares with line sum r in which the sum of entries in each of the two main diagonals is also r. Find an explicit formula for $M_3(r)$.

14. Let n be a fixed positive integer, and let us place n nonattacking white rooks on an $n \times n$ chessboard, then s n nonattacking black rooks on the same chessboard. Let A be the set of $2n$ squares where we placed a rook. Now let $M_A(r)$ be the number of magic squares of size $n \times n$ in which all entries are 0, except possibly for the entries in squares that belong to A. Describe the function $M_A(r)$.

15. How is Supplementary Exercise 24 of Chapter 8 connected to magic squares?

16. Recall that $P_n(2)$ denotes the number of $n \times n$ magic squares of line sum 2 that are symmetric to their main diagonal. Find the exponential generating function of the sequence $P_n(2)$, where $n = 1, 2 \cdots$.

Appendix

The method of mathematical induction

This appendix is meant to provide first aid for students who have not seen proofs with the method of mathematical induction before. While this method is based on a simple idea, it takes practice to be able to successfully apply it. Therefore, after reading our quick review of the method, the reader should consult any textbook on the transition to higher mathematics, such as [78] for a more detailed treatment of the method or [12] for a chapter on combinatorial applications of the method.

A.1 Weak induction

If a statement is true today, and from the fact that it is true on one day follows the fact that it will be true the following day, then we can conclude that the statement will be true every day from now on. An example of this is the statement "the 2004 football season is over."

This idea is the centerpiece of a very powerful, and very often used, tool in proving mathematical statements, that of *mathematical induction*. A straightforward translation of the above idea to a more mathematical context is as follows: Let us assume that a statement involving the variable n is true for $n = 1$, and from the fact that it is true for $n = k$ follows the fact that it is also true for $n = k + 1$. Then the statement is true for all positive integers n.

In practice, it is usually easy to prove that the statement is true for $n = 1$ (the initial step), but it is somewhat harder to prove that the fact that the statement is true for $n = k$ implies the fact that it is also true for $n = k + 1$. Once we succeed in proving these two "local" statements, the method of induction will imply that the original "global" statement is true for all positive integers n.

Example A.1 *For all positive integers n,*

$$\sum_{i=1}^{n} i(i-1) = \frac{(n+1)n(n-1)}{3}. \tag{A.1}$$

Proof: We prove our statement by induction on n.

DOI: 10.1201/9781003304272-A

1. The initial step: If $n = 1$, then the statement says that $0 = 0$, so the statement is true.
2. The induction step: Let us assume that the statement is true for $n = k$, that is,

$$\sum_{i=1}^{k} i(i-1) = \frac{(k+1)k(k-1)}{3} \tag{A.2}$$

holds. We need to prove that then the statement also holds for $n = k + 1$, that is,

$$\sum_{i=1}^{k+1} i(i-1) = \frac{(k+2)(k+1)k}{3}. \tag{A.3}$$

In other words, we need to show that (A.2) implies (A.3). In order to achieve that goal, it clearly suffices to show that the *difference* of (A.3) and (A.2) holds, since then we can add that difference to both sides of (A.2) and get (A.3). On the other hand, this difference is the equation

$$(k+1)k = (k+1)k \cdot \frac{(k+2)-(k-1)}{3},$$

which is clearly an identity. This completes the induction step.

Therefore, by induction, (A.1) holds for all positive integers n. ◊

A.2 Strong induction

Sometimes knowing that a statement is true today is not sufficient to conclude that the statement will be true tomorrow; sometimes we need to know that the statement has been true *every day up to now* in order to allow such a conclusion. Similarly, sometimes we need to know that a statement is true for all values $k \leq n$ in order to conclude that it is true for $n + 1$ as well. If that is the case, and the statement is true for $n = 1$, then it must be true for all values of n. Indeed, from the fact that it is true for $n = 1$, it follows that it is true for $n = 2$, since 1 is the only positive integer smaller than 2. Then, as the statement is true for $n = 1$ and $n = 2$, it is true for $n = 3$, and then, as it is true for $n = 1$ and $n = 2$ and $n = 3$, it is true for $n = 4$, and so on.

This way of proving statements for all n using a *stronger* induction hypothesis is called the method of *strong induction*.

Example A.2 *Let $a_0 = 1$, and let*

$$a_n = \sum_{i=0}^{n-1} 2a_i \tag{A.4}$$

if $n \geq 1$. Prove that then $a_n = 2 \cdot 3^{n-1}$ for all positive integers n.

Proof:

1. The initial step: For $n = 1$, formula (A.4) yields $a_1 = 2 \cdot a_0 = 2$, agreeing with the formula to be proved.
2. The induction step: Let us assume that the statement holds for all $k \leq n$. Then (A.4) yields

$$\begin{aligned} a_{n+1} &= \sum_{k=0}^{n} 2a_k \\ &= 2(1 + 2 \cdot \sum_{k=1}^{n} 3^{k-1}) \\ &= 2 + 4 \cdot \sum_{k=1}^{n} 3^{k-1} \\ &= 2 + 4 \cdot \frac{3^n - 1}{2} \\ &= 2 \cdot 3^n. \end{aligned}$$

So our statement holds for $k = n + 1$ as well.

Therefore, by strong induction, the statement holds for all positive integers n. ◊

Induction has many variations. The smallest allowed value of n can be different from 1, in which case the initial step needs to be changed accordingly, or there could be many variables. The reader is encouraged to consult the references mentioned at the beginning of this appendix and practice inductive proofs if the reader has not done so before.

Bibliography

[1] L. Ahmadi, R. G. Aíza, M. D. Ward, "A unified treatment of families of partition functions." *Preprint*, available at `https://arxiv.org/pdf/2303.02240`.

[2] M. Aigner and G. Ziegler, *Proofs from THE BOOK*, sixth edition. Springer, New York, 2018.

[3] N. Alon and J. Spencer, *The Probabilistic Method*, third edition. Wiley-Interscience Series in Discrete Mathematics and Optimization. A Wiley-Interscience Publication. John Wiley & Sons, Hoboken, NJ, 2008.

[4] H. Anand, V. C. Dumir, and H. Gupta, "A Combinatorial Distribution Problem." *Duke Math. Journal* **33** (1966): 757–769.

[5] D. André, "Calcul des probabilités. Solution directe d'un problème résolu par M. J. Bertrand." *Comptes Rendus de l'Académie des Sciences* **105** (1887): 436.

[6] G. Andrews, *The Theory of Partitions.* Cambridge Mathematical Library, Cambridge, UK, 1998.

[7] G. Andrews, K. Eriksson, *Integer partitions*, Cambridge University Press, Cambridge, UK, 2014.

[8] C. Berge, *Hypergraphs.* North-Holland, New York, 1989.

[9] J. Bertrand, "Calcul des probabilités. Solution d'un problème." *Comptes Rendus de l'Académie des Sciences* **105** (1887): 369.

[10] B. Bollobás, *Extremal Graph Theory.* Dover, New York, 2004.

[11] B. Bollobás, "On Generalized Graphs." *Acta Math. Acad. Sci. Hungar.* **16** (1965): 447–452.

[12] M. Bóna, *A Walk Through Combinatorics*, third edition. World Scientific, Rive Edge, NJ, 2011.

[13] M. Bóna, *Combinatorics of Permutations.* CRC Press – Chapman Hall, third edition, Boca Raton, FL, 2022.

[14] M. Bóna, "A New Proof of the Formula for the Number of the 3×3 Magic Squares." *Mathematics Magazine* **70** (1997): 201–203.

[15] M. Bóna, "Sur l'énumeration des cubes magiques." *Comptes Rendus de l'Academie des Sciences* **316** (1993): 636–639.

[16] M. Bóna, "There Are a Lot of Magic Squares." *Studies in Applied Mathematics* **94** (1995): 415–421.

[17] M. Bóna and B. Sagan, "Two Injective Proofs of a Conjecture of Simion." *J. Combin. Theory Ser. A* **102**, no. 1 (2003): 212–216.

[18] M. Bóna (editor), *Handbook of Enumerative Combinatorics*, CRC Press, Boca Raton, FL, 2015.

[19] A. Bondy and M. Simonovits, "Cycles of Even Length in Graphs" *J. Combinatorial Theory* **16** B (1974): 97–105.

[20] F. Brenti, "Log-Concave and Unimodal Sequences in Algebra, Combinatorics, and Geometry: An Update." *Jerusalem Combinatorics. '93, Contemp. Math.* **178** (1994), Amer. Math. Soc., 71–89.

[21] R. L. Brooks, "On Colouring the Nodes of a Network." *Proc. Cambridge Philos. Soc.* **37** (1941): 194–197.

[22] W. G. Brown, "On Graphs That Do Not Contain a Thomsen Graph." *Cana. Math. Bull.* **9** (1966): 281–285.

[23] A. Cayley, "A Theorem on Trees." *Quart. J. Pure Appl. Math.* **23** (1889): 376–378; *Collected Mathematical Papers*. Vol. 13. Cambridge University Press, 1897, 26–28.

[24] R. Churchill and J. W. Brown, *Complex Variables and Applications*, ninth edition, McGraw-Hill, New York, 2013.

[25] J. Désarmenien and M. Wachs, "Descent Classes on Permutations." *J. Combin. Theory A*, **64** no. 2 (1993): 311–328.

[26] R. Diestel, *Graph Theory*. 4th ed. Springer-Verlag, New York, 2010.

[27] K. Engel, *Sperner Theory*. Cambridge University Press, 1997.

[28] P. Erdős, C. Ko, and R. Rado, "Intersection Theorems for Systems of Finite Sets." *Quarterly Journal of Math. Oxford 2* **12** (1961): 313–320.

[29] P. Erdős, M. Rényi, and V. T. Sós, "On a Problem of Graph Theory." *Studia Sci. Math. Hungar.* **1** (1966): 215–235.

[30] P. Erdős, and M. Simonovits, "A Limit Theorem in Graph Theory." *Studia Sci. Math. Hungar.* **1** (1966): 51–57.

[31] P. Flajolet and R. Sedgewick, *Analytic Combinatorics*, Cambridge University Press, Cambridge, UK, 2009.

[32] D. Foata and J. Riordan, "Mappings of Acyclic and Parking Functions." *Aeuqationes Mathematicae* **10** (1974): 10–22.

[33] D. Foata and M. P. Schützenberger, "Major Index and Inversion Number of Permutations." *Math. Nachr.* **83** (1978): 143–159.

[34] W. Fulton, *Young Tableaux.* London Mathematical Society Student Texts, 35. Cambridge University Press, Cambridge, 1997.

[35] Z. Füredi and P. Hajnal, "Davenport-Schinzel Theory of Matrices." *Discrete Math.* **103** (1992): 233–251.

[36] A. Garsia and S. C. Milne, "A Rogers-Ramanujan Bijection." *J. Combin. Theory Ser. A* **31** (1981), no. 3, 289–339.

[37] R. J. Faudree, and M. Simonovits, "On a Class of Degenerate Extremal Graph Problems." *Combinatorica* **3** (1) 1983: 83–93. *J. Combin. Theory A*, **31** (1981): 289–339.

[38] F. Gilbey and L. Kalikow, *Parking Functions, Valet Functions, and Priority Queues*, Discrete Math {**197/198**} (1999) 351–373.

[39] E. Gossett, *Discrete Mathematics with Proof,* second edition. Wiley, 2009.

[40] W. G. Griffiths, "On Integer Solutions to Linear Equations." *Annals of Combinatorics* **12**, no. 1, **2008**, 53–70.

[41] M. Grigni and M. Sipser, "Monotone Separation of Logarithmic Space From Logarithmic Depth." *J. Comput. System Sci.* **50** (1995), no. 3, part 1, 433–437.

[42] I. P. Goulden, D. M. Jackson, and J. W. Reilly, "The Hammond Series of a Symmetric Function, and its Application to P-recursiveness." *SIAM Journal on Algebraic and Discrete Methods* **54** no. 2 (1983): 179–193.

[43] S. Jukna, *Extremal Combinatorics*, second edition. Springer-Verlag, 2011.

[44] F. Harary and E. M. Palmer, *Graphical Enumeration.* Academic Press, 1973.

[45] R. Hill, *A First Course in Coding Theory.* Oxford University Press, 1990.

[46] J. L. W. V. Jensen, "Sur les Fonctions Convexes et les inégalités entre les Valeurs Moyennes." *Acta Math.* **30** (1906): 175–193.

[47] A. Joyal, "Une Théorie Combinatoire des séries Formelles." *Adv. Math* **42** (1981): 1–82.

[48] G. O. H. Katona, "A Simple Proof of the Erdős-Ko-Rado Theorem." *J. Combin. Theory (B)* **13** (1972): 183–184.

[49] M. Klazar, "The Füredi-Hajnal Conjecture Implies the Stanley-Wilf Conjecture." *Formal Power Series and Algebraic Combinatorics*, Springer-Verlag, 2000, 250–255.

[50] T. Kővári, V. T. Sós, and P. Turán, "On a Problem of K. Zarankiewicz." *Coll. Math.* **3** (1954): 50–57.

[51] S. Lang, *Complex Analysis*, fourth edition. Springer, New York, 2003.

[52] D. Lay, *Linear Algebra and Its Applications*, fifth edition Addison-Wesley, New York, 2014.

[53] R. P. Lewis, The number of spanning trees of a complete multipartite graph, *Disc. Math.* **197–198** (1999), 537–541.

[54] L. Lovász, *Combinatorial Problems and Exercises*, second edition. AMS Chelsea Publishing, Providence, RI, 2007.

[55] P. A. MacMahon, *Combinatorial Analysis.* 2 vols. Cambridge, 1916. (Reprinted by Chelsea, 1960.)

[56] A. Marcus and G. Tardos, "Excluded Permutation Matrices and the Stanley-Wilf Conjecture." *J. Combin. Theory Ser. A* **107** no. 1 (2004): 153–160.

[57] R. K. Nagle, E. B. Saff, and A. D. Snider, *Fundamentals of Differential Equations and Boundary Value Problems*, seventh edition Addison-Wesley, Boston, 2017.

[58] K. O'Hara, "Unimodality of Gaussian Coefficients: A Constructive Proof." *J. Combin. Theory A* **53** no. 1 (1990): 29–52.

[59] J. Pitman, "Coalescent Random Forests." *J. Combinatorial Theory, Series A* **85** (1985): 165–193.

[60] J. Pitman, "Probabilistic Bounds on the Coefficients of Polynomials with Only Real Zeros." *J. Combin. Theory Ser. A* **77** no. 2 (1997): 279–303.

[61] H. Prüfer, "Neuer Beweis eines Satzes über Permutationen." *Archiv der Mat. und Physik* (3) **27** (1918): 142–144.

[62] I. Reiman, "Über ein Problem von K. Zarankiewicz." *Acta Math. Acad. Sci. Hungar.* **9** (1958): 269–273.

[63] B. E. Sagan, "Unimodality and the Reflection Principle." *Ars Combin.* **48** (1998): 65–72.

[64] B. E. Sagan, "Inductive and Injective Proofs of Log Concavity Results." *Discrete Math.* **68**, no. 2–3 (1998): 281–292.

[65] B. E. Sagan, *The Symmetric Group*, second edition. Graduate Texts in Mathematics, 203. Springer-Verlag, New York, 2001.

[66] H. I. Scoins, "The Number of Trees with Nodes of Alternate Parity." *Proc. Cambr. Phil. Soc.* **58** (1962): 112–116.

[67] M. Simonovits, "A Method for Solving Extremal Problems in Graph Theory, Stability Problems, 1968." In *Theory of Graphs (Proc. Colloq., Tihany, 1966)*, Academic Press, New York, 279–319.

[68] M. Simonovits, "Extremal Graph Problems with Symmetrical Extremal Graphs. Additional Chromatic Conditions." *Discrete Math.* **7** (1974): 349–376.

[69] E. Sperner, "Ein Satz über Untermengen einer endlichen Menge." *Math. Zeitschrift* **27** (1928): 544–548.

[70] R. P. Stanley, "Log-Concave and Unimodal Sequences in Algebra, Combinatorics, and Geometry." *Graph Theory and Its Applications: East and West. Ann. NY Acad. Sci.* **576** (1989): 500–535.

[71] R. P. Stanley, *Enumerative Combinatorics*, second edition. Volume 1. Cambridge University Press, Cambridge, UK, 2012.

[72] R. P. Stanley, *Enumerative Combinatorics*, second edition. Volume 2. Cambridge University Press, Cambridge, UK, 2023.

[73] R. P. Stanley, "The Chromatic Polynomial with Negative Arguments." *Discrete Math.* **5** (1973): 171–178.

[74] R. P. Stanley, *personal communication*, 2003.

[75] R. P. Stanley, *Combinatorics and Commutative Algebra*, second edition. Progress in Mathematics 41. Birkhäuser, Boston, 2004.

[76] R. Stanley, "Linear Homogenous Equations and Magic Labellings of Graphs." *Duke Math. J.* 40 (1973): 607–632.

[77] R. Stanley, "Differentiably Finite Power Series." *European J. Combin.* **1** (1980): 175–188.

[78] T. Sundstrom, *Mathematical Reasoning: Writing and Proof*, second edition. Pearson, Boston 2006.

[79] L. Takács, "A Generalization of the Ballot Problem and Its Application in the Theory of Queues." *J. Amer. Statist. Assoc.* **57** (1962): 327–337.

[80] L. Takács, "On the Method of Inclusion and Exclusion." *J. Amer. Statist. Assoc.* **62** (1967): 102–113.

[81] T. Thanatipanonda, "Inversions and Major Index for Permutations." *Math. Magazine* **77**, no. 2 (2004): 136–140.

[82] P. Turán, "On an Extremal Problem in Graph Theory." *Mat. Fiz. Lapok* **48** (1941): 436–452.

[83] D. West, *Introduction to Graph Theory*, third edition. Prentice Hall, Hoboken, NJ, 2007.

[84] H. Wilf, *generatingfunctionology*, third edition. Academic Press, Boston, 2005.

[85] J. van Lint, and R. Wilson, *A Course in Combinatorics*, second edition. Cambridge University Press, Cambridge, UK, 2001.

[86] D. Zeilberger, "Garsia and Milne's Bijective Proof of the Inclusion-Exclusion Principle." *Discrete Math.* **51** (1984): 109–110.

Index

Kolibri 360 GmbH
Ferdinand-Jühlke-Straße 7
99095 Erfurt, Deutschland
produktsicherheit@kolibri360.de